Atomic weights are based on carbon-12. Atomic weights in parentheses indicate the most stable or best-known isotope. Slight disagreement exists as to the exact electronic configuration of several of the high-atomic-number elements.

Noble gases

	IB	IIB	IIIA	IVA	VA	VIA	VIIA	Noble gases
								2 Helium **He** 4.00260
			5 Boron **B** 10.81	6 Carbon **C** 12.011	7 Nitrogen **N** 14.0067	8 Oxygen **O** 15.9994	9 Fluorine **F** 18.99840	10 Neon **Ne** 20.179
			13 Aluminum **Al** 26.98154	14 Silicon **Si** 28.086	15 Phosphorus **P** 30.97376	16 Sulfur **S** 32.06	17 Chlorine **Cl** 35.453	18 Argon **Ar** 39.948
ckel **i** .71	29 Copper **Cu** 63.546	30 Zinc **Zn** 65.38	31 Gallium **Ga** 69.72	32 Germanium **Ge** 72.59	33 Arsenic **As** 74.9216	34 Selenium **Se** 78.96	35 Bromine **Br** 79.904	36 Krypton **Kr** 83.80
ladium **d** 6.4	47 Silver **Ag** 107.868	48 Cadmium **Cd** 112.40	49 Indium **In** 114.82	50 Tin **Sn** 118.69	51 Antimony **Sb** 121.75	52 Tellurium **Te** 127.60	53 Iodine **I** 126.9045	54 Xenon **Xe** 131.30
tinum **t** 5.09	79 Gold **Au** 196.9665	80 Mercury **Hg** 200.59	81 Thallium **Tl** 204.37	82 Lead **Pb** 207.2	83 Bismuth **Bi** 208.9804	84 Polonium **Po** (210)[a]	85 Astatine **At** (210)[a]	86 Radon **Rn** (222)[a]

Inner transition elements →

opium **u** 1.96	64 Gadolinium **Gd** 157.25	65 Terbium **Tb** 158.9254	66 Dysprosium **Dy** 162.50	67 Holmium **Ho** 164.9304	68 Erbium **Er** 167.26	69 Thulium **Tm** 168.9342	70 Ytterbium **Yb** 173.04	71 Lutetium **Lu** 174.97
nericium **m** 43)[a]	96 Curium **Cm** (247)[a]	97 Berkelium **Bk** (249)[a]	98 Californium **Cf** (251)[a]	99 Einsteinium **Es** (254)[a]	100 Fermium **Fm** (253)[a]	101 Mendelevium **Md** (256)[a]	102 Nobelium **No** (254)[a]	103 Lawrencium **Lr** (257)[a]

Foundations of College Chemistry

3rd Alternate Edition

Foundations of College Chemistry
3rd Alternate Edition

Morris Hein
MOUNT SAN ANTONIO COLLEGE

Brooks/Cole Publishing Company
PACIFIC GROVE, CALIFORNIA

Chemistry Editor: Sue Ewing
Production Editor: S. M. Bailey
Manuscript Editor: Patricia Cain
Text Design: Wendy Calmenson
Cover Design: Sharon L. Kinghan
Photo Research: Roberta Spieckerman
Illustrations: Cyndie Clarke-Huegel
Typesetting: Polyglot
Cover photograph: Batista Moon Studio

Color Insert Photos: **1, 2, 32, 33,** courtesy of SCM Corporation; **3,** courtesy of Hewlett-Packard Corporation; **4, 19,** © Ray Ellis, Photo Researchers, Inc.; **5, 13,** © John Blaustein, Woodfin Camp & Associates; **6, 7, 8, 12,** courtesy of American Petroleum Institute; **9,** courtesy of Exxon Corporation; **10,** © Mickey Pfleger, 1983; **11,** courtesy of Allied Corporation; **14,** © George Hall, Woodfin Camp & Associates; **15,** © Christopher Springmann, After-Image, Inc.; **16, 17, 18, 20,** courtesy of Chesapeake Corporation; **21,** © Mark Godfrey, courtesy of Pfizer, Inc.; **22, 24,** courtesy of Squibb Corporation; **23,** © Dick Luria, Photo Researchers, Inc.; **25,** © Ian Berry, Magnum Photos, Inc.; **26,** © Guy Gillette, Photo Researchers, Inc.; **27,** © Charles Harbutt, courtesy of Warner Lambert Corporation; **28,** courtesy of Hayes Microcomputer Products, Inc.; **29,** © Martin M. Rotker, Taurus Photos, Inc.; **30,** © Erich Hartmann, Magnum Photos, Inc.; **31,** courtesy of Glidden Coatings and Resins—Division of SCM Corporation.

Printed in the United States of America

10 9 8 7 6 5 4 3 2 1

Library of Congress Cataloging in Publication Data

Hein, Morris.
Foundations of college chemistry.

Includes index.
1. Chemistry. I. Title.
QD33.H45 1986 540 85-29936

ISBN 0-534-082327

To stop learning is to wither on the vine.

To those who are near and dear
Edna, Leslye, Steve, Sam, Darren, Jory, and Shanon

Contents

Preface

The success of *Foundations of College Chemistry* through five editions has been gratifying, for it tells me that the book has been continuously useful and effective for both students and teachers. I sincerely hope that the sixth edition and the third alternate edition continue to be valuable to students beginning their study of chemistry and associated sciences.

This alternate edition contains the material in the first seventeen chapters of the sixth edition of *Foundations of College Chemistry*. The main reason for this edition is to provide a shorter, less expensive text for those courses that do not intend to include the last three chapters of *Foundations of College Chemistry*. For those courses that include Radioactivity and Nuclear Chemistry, and an Introduction to Organic and Biochemistry, the sixth edition of *Foundations of College Chemistry* is available.

The primary purpose of this edition, like its predecessors, is to instruct students in the basic concepts of chemistry so they will be qualified to enter courses in general college chemistry. The text remains designed for a course in beginning or preparatory chemistry. Although a number of changes have been made, the overall sequence of topics and the level of material is approximately the same as in the previous edition. The entire text was carefully reviewed for further improvement in clarity of expression and to provide students with greater assistance in solving problems.

Learning Aids

The following features are intended to serve as learning aids and to make the text convenient to use—a book that students can read, understand, and study by themselves.

- A list of achievement goals is given at the beginning of each chapter as a guide to students.
- Important new terms are set in boldface type where they are defined in the text, and they are also printed in color in the margin.

Chemistry in Industry—And in Our Lives

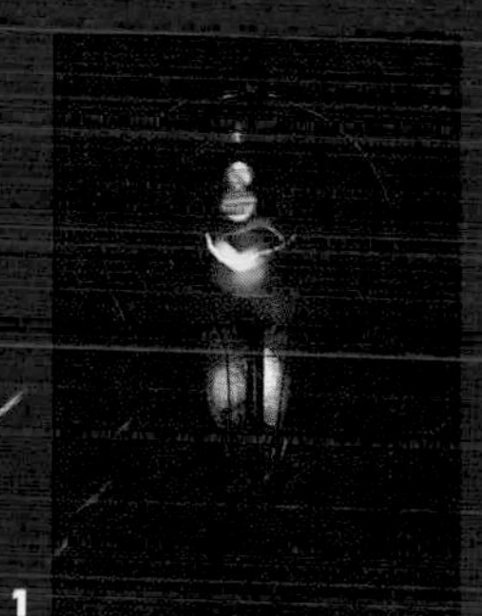

1

Think of the word chemistry. What comes to mind? Lengthy equations? Test tubes? Now think of the light bulb you're probably using to read this page. Why is it still burning? How might it be made to last even longer?

We rarely stop to think about the ways chemisty affects us. Most of the things we use or come in contact with during the course of our daily lives are manufactured by different industries dependent on some aspect of chemistry to make their products.

In the metallurgy industry, for example, protective coatings or chemical processes are often used to prevent metal from rusting, much in the same way we apply wax to our cars to preserve the paint. Because nature often works against our human interests, returning metals to their natural state, modern metallurgists increasingly use chemistry to find new treatments of metal that prevent the natural chemical reactions—or corrosion—from occurring.

The special incandescent light bulb above (1), for instance, uses copper that has been chemically strengthened. Without its copper coating, the filament would burn up instantaneously. Through the use of chemistry, new ways are being discovered to make products like this light bulb last longer.

The thermal-spraying of nickel-based powder onto metal (2) enhances metal's resistance to wear. In the manufacturing of steel (3), a hydrochloric acid "pickling" solution removes rust from new steel before a protective coating of zinc or tin is applied to keep the steel from oxidizing, or corroding. The steel end-products (4) often undergo a final chemical painting process that also prevents corrosion.

2

3

4

5

Now consider this roll of plastic (5). Where does it come from, and how is it made? Plastic, like many other products in our daily lives, is made from chemicals produced by the petroleum industry, another industry dependent on chemistry to make its products. But before a material like plastic can be manufactured, many complex processes must take place.

Petroleum is a thick, flammable, dark-colored liquid fuel composed of mixtures of complex hydrocarbon compounds. Usually collected through drilling deep into the earth (6), it is sometimes also found in surface springs and pools. After the petroleum is obtained, it is transported (7) to the refinery, because it must be purified before being put to further use. The transportation process itself requires petroleum, in the form of bunker oil, to move petroleum-carrying tankers.

At the refinery (9), the mixtures of complex hydrocarbon compounds are separated into different components. During this process they are chemically treated to remove impurities. A refinery, such as the one pictured, can convert some 15,000 barrels of heavy fuel into lighter petroleum products in one day.

The refined petroleum is then put to many different uses. Perhaps the most familiar one is the use of petroleum to make gasoline (8)—the petroleum industry's major product. Like the fuel that heats our homes, the use of gasoline involves combustion (burning the fuel), a simple chemical process.

In addition to fulfilling our fuel needs, petroleum is used in many other ways. Petrochemicals, for example, which are derived from petroleum, go into the making of many materials we encounter every day. Acrylic resins made from petrochemicals can be employed to create plastic objects (10). Polyester fiber (11), also made from petrochemicals, is used in many familiar products such as dacron for clothing, tire cords, and sutures in medicine. Paint, artificial organs, and synthetic pharmaceuticals are additional examples of products in which petrochemicals are used. In fact, if we stop to consider all the products that use petroleum in some way, we see that it would be difficult indeed (if not impossible) to go through even one day without being touched by some aspect of the petroleum industry.

New research is being conducted to discover further applications of petroleum. Many researchers, like this chemist studying synthetic resins to find better ways of using them in the manufacture of adhesives (12), devote their entire careers to the never-ending search for ways to improve our lives.

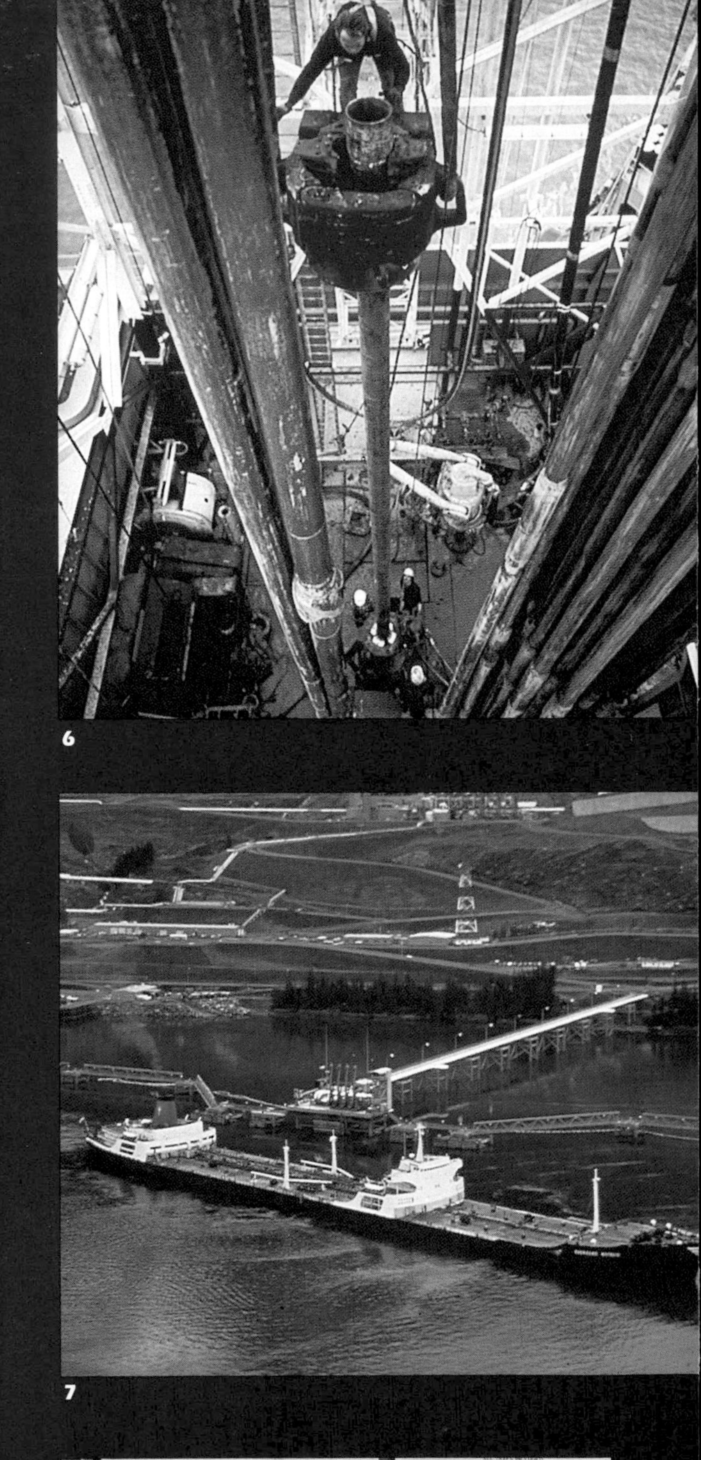

6

7

8

9
10
11
12

13

Have you ever wondered how paper is made—for instance, the paper in this textbook? The paper industry is yet another in which chemistry plays a major role.

The paper-making process begins with the planting of trees (13). Millions of trees must be grown to supply the great demand for paper. It takes about 75,000 pine trees, for example, to publish one Sunday edition of "The New York Times." Once the trees are ready for harvest, they are cut down and transported to the logging mill (14). Even at this early stage chemistry comes into play, for timber is vulnerable to pests. To preserve the logs from possible damage, chemicals in the form of pest-deterring sprays are often used before transporting them to the mill.

At the mill, the logs are cut into small chips (15) which are then mashed into pulp. The pulp is treated with chemicals in preparation for entering the wet end of the paper-making machine (16). The process of chemical pulping dissolves the noncellulose components of the wood, especially the lignins, leaving residual cellulose (the same fiber as in cotton). Because the natural color of pulp is pale brown or yellow, chemical bleaching also occurs at this stage to make the paper white. You may have noticed that the paper in certain books yellows more quickly than in others. This yellowing is caused by residual lignin and acid in the pulp. More and more librarians, concerned about their deteriorating collections, are pressing publishers to print their books on acid-free paper so the books will last.

Eventually the water is squeezed out of the pulp, and the finished paper emerges from the dry end of the machine (17). Huge overhead cranes (18) are used to grasp the rolls of paper for transport to a high-speed coater machine, where papers are sometimes chemically coated to prevent oxidation, impart a glossy appearance to the stock, or further whiten them through dusting with white pigment.

Chemicals are also used in paper so that it does not cause the deterioration of certain inks, or, more often, chemicals are used in inks so they do not cause the paper to disintegrate (19).

Although the chemical processes used in paper manufacturing are relatively simple, the entire operation is very complex. To accomplish everything at a high speed requires a great deal of control. The use of computers (20) has facilitated the paper-making process enormously, enabling manufacturers to handle the process quickly and efficiently.

14

15

16

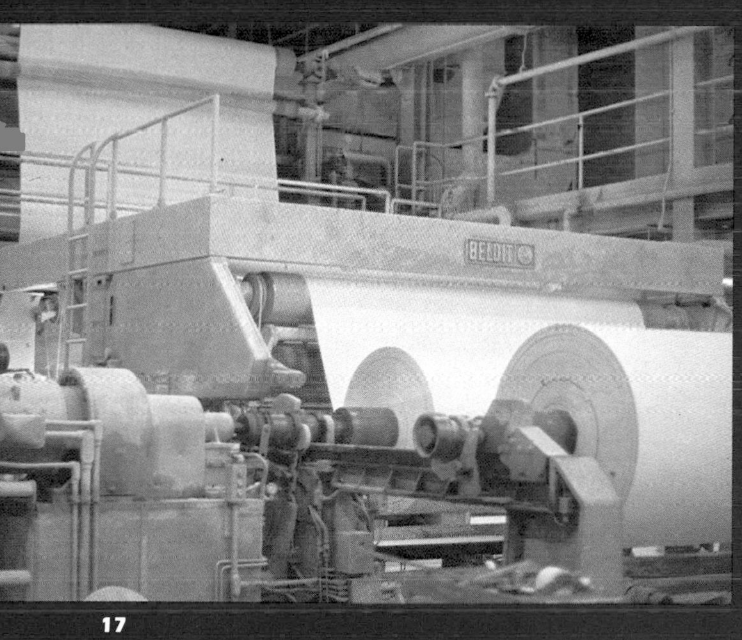

17

18

19

20

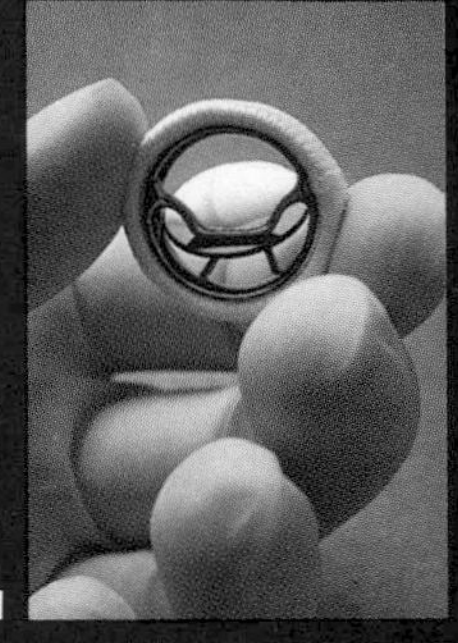

21

What would you do if your heart began to function abnormally? It wasn't so long ago that there wasn't anything you could do. Heart valves such as this one (21) didn't even exist twenty years ago, and this is only one example of the amazing recent advances in medicine and the pharmaceutical industry.

It's easy to see that chemistry must play a major role in pharmaceuticals since most drugs, after all, are made of chemicals. Yet there are many less obvious ways in which chemistry is important to the pharmaceutical industry.

Producing pharmaceuticals demands great amounts of human labor—from the testing of new drugs to their production and storage. Not all pharmaceutical processes are chemical, many involve the isolation of natural drugs. But because the key to whether or not a drug gets produced is whether it can be made inexpensively, and because synthesized drugs are often much less expensive than natural ones, much effort is expended in learning how to chemically synthesize drugs.

Often the first step toward synthesizing a new or existing drug is to determine its chemical structure. A model of the drug (22), using styrofoam balls to represent parts of a molecule, is constructed to visualize, study, and modify the structure of the drug.

In the testing and manufacturing of pharmaceuticals (23), chemistry is used constantly. Researchers perform many complicated tests, like this one being done with an automated spectrometer (24). Because even the humidity in air is enough to affect certain moisture-sensitive chemicals, sometimes a dry box (25) must be used to handle chemicals in an environment in which there is no trace of moisture.

Even at the nonchemical level, the manufacturing of pharmaceuticals is a complex process. Samples must be checked (27) and production monitored (26) to make certain no errors are made. A mistake could literally be a matter of life and death.

Thousands of pharmaceuticals are now produced (28), each one a chemical or mixture of chemicals. More are discovered through new procedures every day. Sometimes when a pharmaceutical can't be made chemically, it can be produced by bacteria. More and more examples of pharmaceuticals interlinking with the body's chemistry, as in an intravenous transfusion (29), are arising in the development of new solutions to human physical and emotional ailments.

22

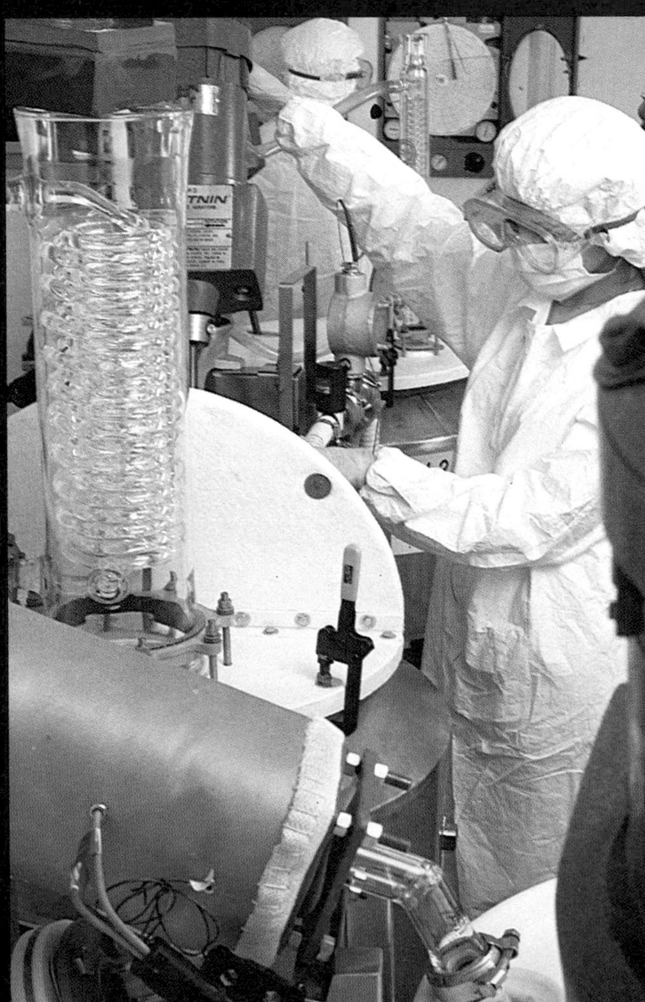

23

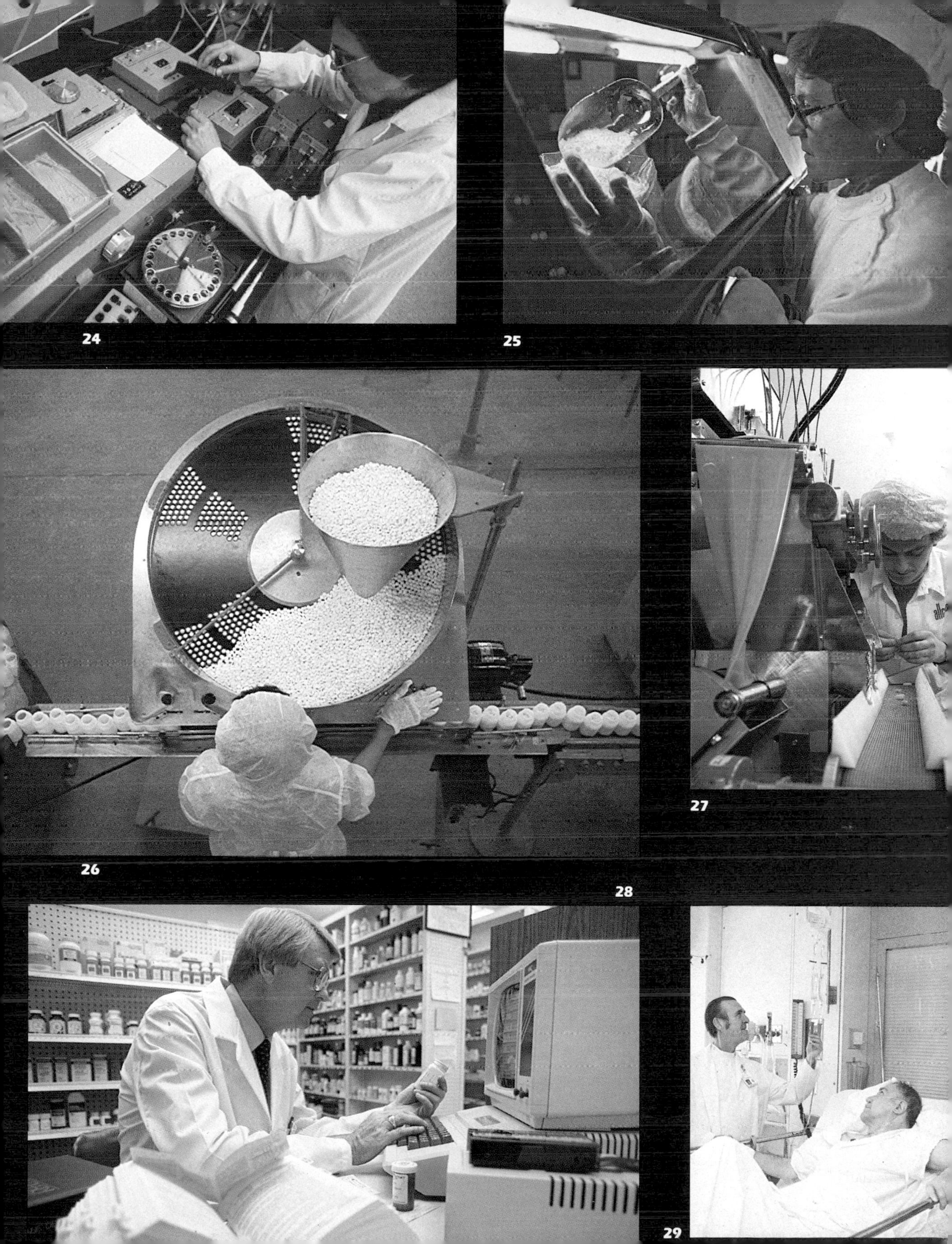
24
25
26
27
28
29

30

Have you ever been fascinated by the hundred different kinds and colors of paint? Just what is it that makes them differ from each other? Why does one pour easily (30), whereas another seems to ooze out of the can?

The paint industry is yet another that depends on chemistry. Most paints include several basic ingredients: a binder, a solvent, a drier, and one or more pigments (colors). Paint is made through experimentation, by mixing different quantities of ingredients to produce the desired results (31). Most of the chemistry that goes into making paints occurs with the binder, which holds the pigment in place. In water-base paints, for example, latex rubber is added as a binder. Other chemical additives make paints dry more rapidly. Oil-base and latex paints, for instance, do not dry by evaporation—a chemical reaction occurs that causes them to dry. Some pigments in paints are metal-containing compounds.

Ingredients that go into paint are also used with a wide range of other materials and products. Plastic can be impregnated with pigments and then tested for color dispersion (32). The pigment titanium dioxide possesses about the greatest hiding power of all inorganic white pigments. It is used in enamels, inks, shoe whitener, ceramics, plastics, and rubber (33).

Paint, pharmaceuticals, paper, petroleum, metallurgy—these are but a few of the many industries in which chemistry plays a major role, directly or indirectly affecting each of us. Far from being confined to test tubes, element tables, or complex equations, the study of chemistry is a fascinating exploration of why things are the way they are—and how we can alter them or develop new ways to improve our lives.

31

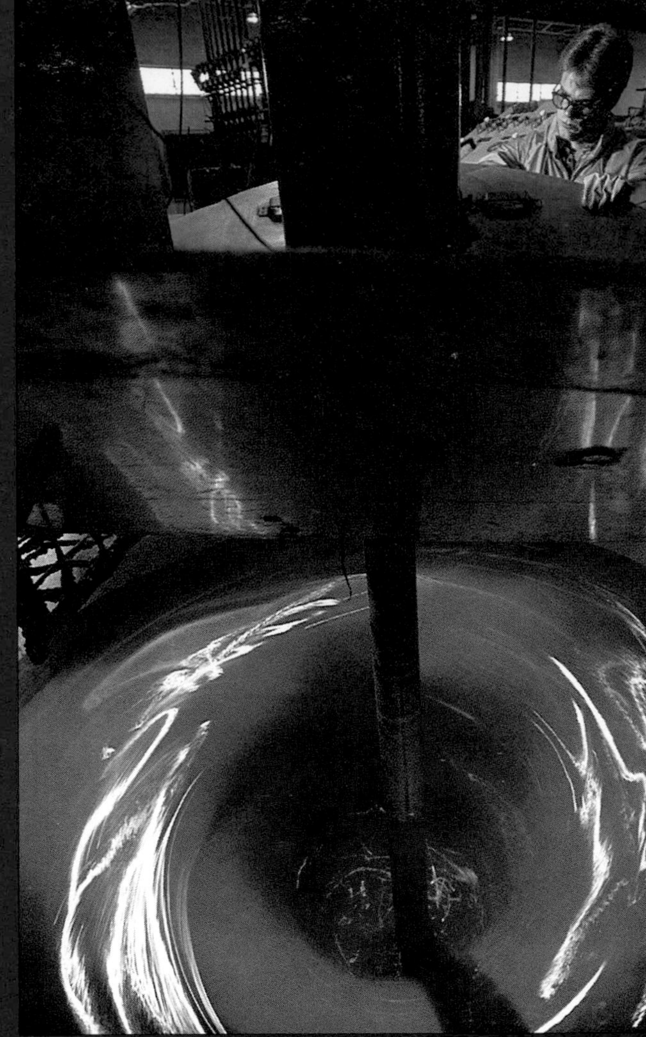

32

33

- A detailed discussion of significant figures, rounding off numbers, scientific notation of numbers, and the use of dimensional analysis in calculations is given in Chapter 2.
- Numerous questions and problems are given in separate groups at the end of each chapter. These questions are generally arranged in the order in which the topics are discussed.
- Answers to all mathematical problems are given in Appendix VI. Complete solutions to all questions and problems are given in the *Student Solutions Manual.*
- The many sample problems are carefully solved in a step-by-step fashion using the conversion-factor dimensional analysis method.
- Six sets of review exercises for student self-evaluation are included.
- A two-color format is used to emphasize noteworthy aspects of the contents and to highlight illustrations and tables.
- A comprehensive glossary is given in Appendix V.

Major Changes and Additions

- A photo essay at the beginning of the text introduces students to some industries that depend on chemistry.
- A simplified method of rounding off numbers is given in Chapter 2.
- Both the joule and calorie are used as energy units, with emphasis on the joule.
- Most of the end-of-chapter problems have been modified or are new.
- The use of K, L, M, . . . for atomic orbitals has been deleted.
- The concepts of the mole and Avogadro's number have been moved from Chapter 5 to Chapter 9 where they complement the material on quantitative composition of compounds.
- The periodic table most recently recommended by the American Chemical Society is shown in Appendix VII.
- Ionic bond is used in place of electrochemical bond.
- The sections on oxidation numbers and oxidation number tables have been moved from the end of Chapter 7 (Chemical Bonds) to the beginning of Chapter 8 (Nomenclature).
- A different approach to limiting reactant problems is given, which includes calculating the amount of unreacted material.
- Several pressure units are introduced in discussing the properties of gases, but emphasis is given to atmospheres and torrs in problems.
- A different method of balancing oxidation-reduction equations in alkaline solutions is given.
- A section on hydrolysis has been added to Chapter 16.

Supplementary Materials

Materials that may be helpful to students and to their teachers have been developed to accompany the text. A short description of them follows.

- *Study Guide* by Peter C. Scott of Linn-Benton Community College includes a self-evaluation section for students to check their understanding of each chapter's objectives, a recap section, and answers to the self-evaluation section.

- *Student Solutions Manual* includes answers and solutions to all end-of-chapter questions and problems.
- *Instructor's Manual with Solutions* includes the answers to the questions and problems in the text, answers to chapter review exercises, test questions, and answers to the test questions.
- *Foundations of Chemistry in the Laboratory, 6th edition,* by Morris Hein, Leo R. Best, and Robert L. Miner, includes twenty-eight experiments for a laboratory program that may accompany the lecture course. Also included are study aids and exercises.
- *Instructor's Manual* to accompany the lab manual includes information on the management of the lab, evaluation of experiments, notes for individual experiments, a list of reagents needed, and answer keys to each experiment's report form.

Acknowledgements

Many people have contributed to this revision of the *Foundations of College Chemistry* teaching package. I am deeply grateful for the critical professional reviews by the following individuals: Caroline L. Ayers, East Carolina University; Paula Ballard, Jefferson State Jr. College; Nordulf W. G. Debye, Towson State University; Robert D. Farina, Western Kentucky University; John A. Grove, South Dakota State University; Robert C. Kowerski, College of San Mateo; Don Pon, Foothill Community College; George H. Potter, Schenectady Community College; and Alexander Vavoulis, Fresno State University.

It is my pleasure to thank Lynn Pendleton of Mt. San Antonio College, who typed, edited, and corrected the *Student Solutions Manual,* and Dr. Peter C. Scott of Linn-Benton Community College who prepared the *Student Study Guide.* Special appreciation goes to my long-standing colleagues and coauthors, Leo R. Best and Robert L. Miner of Mt. San Antonio College, who did most of the revision work for the sixth edition of *Foundations of College Chemistry in the Laboratory.*

No textbook can be completed without the untiring effort of many professionals in publishing. Special thanks are due to the wonderful staff at Brooks/Cole Publishing Company, especially Sue Ewing, Chemistry Editor, Patricia Cain, manuscript editor, and Wendy Calmenson, designer of the text and coordinator of the photo essay.

Last, but not least, I am again grateful to my dear wife, Edna, for her patience and for enduring all the deadlines over the past twenty years.

Morris Hein

CHAPTER 1 Introduction

1.1 The Nature of Chemistry

What is chemistry? A popular dictionary gives this definition: Chemistry is the science of the composition, structure, properties, and reactions of matter, especially of atomic and molecular systems. Another, somewhat simpler dictionary definition is: Chemistry is the science dealing with the composition of matter and the changes in composition that matter undergoes. Neither of these definitions is entirely adequate. Chemistry, along with the closely related science of physics, is a fundamental branch of knowledge. Chemistry is also closely related to biology, not only because living organisms are made of material substances but also because life itself is essentially a complicated system of interrelated chemical processes.

The scope of chemistry is extremely broad: It includes the whole universe and everything, animate and inanimate, in the universe. Chemistry is concerned not only with the composition and changes of composition of matter, but also with the energy and energy changes associated with matter. Through chemistry we seek to learn and to understand the general principles that govern the behavior of all matter.

The chemist, like other scientists, observes nature and attempts to understand its secrets: What makes a rose red? Why is sugar sweet? What is occurring when iron rusts? Why is carbon monoxide poisonous? Why do people wither with age? Problems such as these—some of which have been solved, some of which are still to be solved—are part of what we call chemistry.

A chemist may interpret natural phenomena, devise experiments that will reveal the composition and structure of complex substances, study methods for improving natural processes, or, sometimes, synthesize substances unknown in nature. Ultimately, the efforts of successful chemists advance the frontiers of knowledge and at the same time contribute to the well-being of humanity. Chemistry can help us to understand nature; however, one need not be a professional chemist or scientist to enjoy natural phenomena. Nature and its beauty, its simplicity within complexity, are for all to appreciate.

The body of chemical knowledge is so vast that no one can hope to master it all, even in a lifetime of study. However, many of the basic concepts can be learned in a relatively short period of time. These basic concepts have become part of the education required for many professionals, including agriculturists, biologists, dental hygienists, dentists, medical technologists, microbiologists, nurses, nutritionists, pharmacists, physicians, and veterinarians, to name a few.

1.2 History of Chemistry

People have practiced empirical chemistry from the earliest times. Ancient civilizations were practicing the art of chemistry in such processes as wine-making, glass-making, pottery-making, dyeing, and elementary metallurgy. The early Egyptians, for example, had considerable knowledge of certain chemical processes. Excavations into ancient tombs dated about 3000 B.C. have uncovered workings of gold, silver, copper, and iron, pottery from clay, glass beads, and beautiful dyes and paints, as well as bodies of Egyptian kings in remarkably well-preserved states. Many other cultures made significant developments in chemistry. However, all these developments were empirical; that is, they were achieved by trial and error and did not rest on any valid theory of matter.

Philosophical ideas relating to the properties of matter (chemistry) did not develop as early as those relating to astronomy and mathematics. The Greek philosophers made great strides in philosophical speculation concerning materialistic ideas about chemistry. They led the way to placing chemistry on an intellectual, scientific basis. They introduced the concepts of elements, atoms, shapes of atoms, and chemical combination. They believed that all matter was derived from four elements: earth, air, fire, and water. The Greek philosophers had keen minds and perhaps came very close to establishing chemistry on a sound basis similar to the one that was to develop about 2000 years later. The main shortcoming of the Greek approach to scientific work was a failure to carry out systematic experimentation.

Greek civilization was succeeded by the Roman civilization. The Romans were outstanding in military, political, and economic affairs. They practiced empirical chemical arts such as metallurgy, enameling, glass-making, and pottery-making, but they did very little to advance new and theoretical knowledge. Eventually the Roman civilization was succeeded in Europe by the Dark Ages. During this period European civilization and learning were at a very low ebb.

In the Middle East and in North Africa, knowledge did not decline during the Dark Ages as it did in Western Europe. At this time Arabic cultures made contributions that were of great value to the development of modern chemistry. In particular, the Arabic number system, including the use of zero, gained acceptance; the branch of mathematics known as *algebra* was developed; and alchemy, a sort of pseudochemistry, was practiced extensively.

One of the more interesting periods in the history of chemistry was that of the alchemists (500–1600 A.D.). People have long had a lust for gold, and in those days gold was considered the ultimate, most perfect metal formed in nature. The principal goals of the alchemists were to find a method of prolonging human life indefinitely and to change the base metals, such as iron, zinc, and copper, into gold. They searched for a universal solvent to transmute base metals into gold and for the "philosopher's stone" to rid the body of all diseases and to renew life. In the course of their labors they learned a great deal of chemistry. Unfortunately, much of their work was done secretly because of the mysticism that shrouded their activity, and very few records remain.

Although the alchemists were not guided by sound theoretical reasoning and were clearly not in the intellectual class of the Greek philosophers, they did something that the philosophers had not considered worthwhile. They subjected various materials to prescribed treatments under what might be loosely described as laboratory methods. These manipulations, carried out in alchemical laboratories, not only uncovered many facts of nature but paved the way for the systematic experimentation that is characteristic of modern science.

Alchemy began to decline in the 16th century when Paracelsus (1493–1541), a Swiss physician and outspoken revolutionary leader in chemistry, strongly advocated that the objectives of chemistry be directed toward the needs of medicine and the curing of human ailments. He openly condemned the mercenary efforts of alchemists to convert cheaper metals to gold.

But the real beginning of modern science can be traced to astronomy during the Renaissance. Nicolaus Copernicus (1473–1543), a Polish astronomer, began the downfall of the generally accepted belief in a geocentric universe. Although not all the Greek philosophers had believed that the sun and the stars revolved about the earth, the geocentric concept had come to be generally accepted. The heliocentric (sun-centered) universe concept of Copernicus was based on direct astronomical observation and represented a radical departure from the concepts handed down from Greek and Roman times. The ideas of Copernicus and the invention of the telescope stimulated additional work in astronomy. This work, especially that of Galileo Galilei (1564–1642) and Johannes Kepler (1571–1630), led directly to a rational explanation by Sir Isaac Newton (1642–1727) of the general laws of motion, which he formulated between about 1665 and 1685.

Modern chemistry was slower to develop than astronomy and physics; it began in the 17th and 18th centuries when Joseph Priestley (1733–1804), who discovered oxygen in 1774, and Robert Boyle (1627–1691) began to record and publish the results of their experiments and to discuss their theories openly. Boyle, who has been called the founder of modern chemistry, was one of the first to practice chemistry as a true science. He believed in the experimental method. In his most important book, *The Sceptical Chymist*, he clearly distinguished between an element and a compound or mixture. Boyle is best known today for the gas law that bears his name. A French chemist, Antoine Lavoisier (1743–1794), placed the science on a firm foundation with experiments in which he used a chemical balance to make quantitative measurements of the weights of substances involved in chemical reactions.

The use of the chemical balance by Lavoisier and others later in the 18th century was almost as revolutionary in chemistry as the use of the telescope had been in astronomy. Thereafter, chemistry became a quantitative experimental science. Lavoisier also contributed greatly to the organization of chemical data, to chemical nomenclature, and to the establishment of the Law of Conservation of Mass in chemical changes. During the period from 1803 to 1810, John Dalton (1766–1844), an English schoolteacher, advanced his atomic theory. This theory (see Section 5.2) placed the atomistic concept of matter on a valid rational basis. It remains today as a tremendously important general concept of modern science.

Since the time of Dalton, knowledge of chemistry has advanced in great strides, with the most rapid advancement occurring at the end of the 19th century and during the 20th century. Especially outstanding achievements have been made in determining the structure of the atom, understanding the biochemical fundamentals of life, developing chemical technology, and the mass production of chemicals and related products.

1.3 The Branches of Chemistry

Chemistry may be broadly classified into two main branches: *organic* chemistry and *inorganic* chemistry. Organic chemistry is concerned with compounds containing the element carbon. The term *organic* was originally derived from the chemistry of living organisms: plants and animals. Inorganic chemistry deals with all the other elements as well as with some carbon compounds. Substances classified as inorganic are derived mainly from mineral sources rather than from animal or vegetable sources.

Other subdivisions of chemistry, such as analytical chemistry, physical chemistry, biochemistry, electrochemistry, geochemistry, and radiochemistry, may be considered specialized fields of, or auxiliary fields to, the two main branches.

Chemical engineering is the branch of engineering that deals with the development, design, and operation of chemical processes. A chemical engineer usually begins with a chemist's laboratory-scale process and develops it into an industrial-scale operation.

1.4 Relationship of Chemistry to Other Sciences and Industry

Besides being a science in its own right, chemistry is the servant of other sciences and industry. Chemical principles contribute to the study of physics, biology, agriculture, engineering, medicine, space research, oceanography, and many other sciences. Chemistry and physics are overlapping sciences, since both are based on the properties and behavior of matter. Biological processes are chemical in nature. The metabolism of food to provide energy to living organisms is a chemical process. Knowledge of molecular structure of proteins, hormones, enzymes, and the nucleic acids is assisting biologists in their investigations of the composition, development, and reproduction of living cells.

Chemistry is playing an important role in alleviating the growing shortage of food in the world. Agricultural production has been increased with the use of chemical fertilizers, pesticides, and improved varieties of seeds. Chemical refrigerants make possible the frozen food industry, which preserves large amounts of food that might otherwise spoil. Chemistry is also producing synthetic nutrients, but much remains to be done as the world population increases relative to the land available for cultivation. Expanding energy needs have brought about difficult environmental problems in the form of air and water pollution. Chemists and other scientists are working diligently to alleviate these problems.

Advances in medicine and chemotherapy, through the development of new drugs, have contributed to prolonged life and the relief of human suffering. More than 90% of the drugs and pharmaceuticals being used in the United States today have been developed commercially within the past 45 years. The plastics and polymer industry, unknown 60 years ago, has revolutionized the packaging and textile industries and is producing durable and useful construction materials. Energy derived from chemical processes is used for heating, lighting, and transportation. Virtually every industry is dependent on chemicals—for example, the petroleum, steel, rubber, pharmaceutical, electronic, transportation, cosmetic, garment, aircraft, and television industries. (The list could go on.) Figure 1.1 illustrates the conversion of natural resources by the chemical industry into useful products for commerce, industry, and human needs.

1.5 Scientific Method

Chemistry, as a science or field of knowledge, is concerned with ideas and concepts relating to the behavior of matter. Although these concepts are abstract, their application has had a concrete impact on human culture. This impact is due to modern technology, which may be said to have begun about 200 years ago and which has grown at an accelerating rate ever since.

An important difference between science and technology is that science represents an abstract body of knowledge, and technology represents the physical application of this knowledge to the world in which we live.

Why has the science of chemistry and its associated technology flourished in the last two centuries? Is it because we are growing more intelligent? No, we have absolutely no reason to believe that the general level of human intelligence is any higher today than it was in the Dark Ages. The use of the scientific method is usually credited with being the most important single factor in the amazing development of chemistry and technology. Although complete agreement is

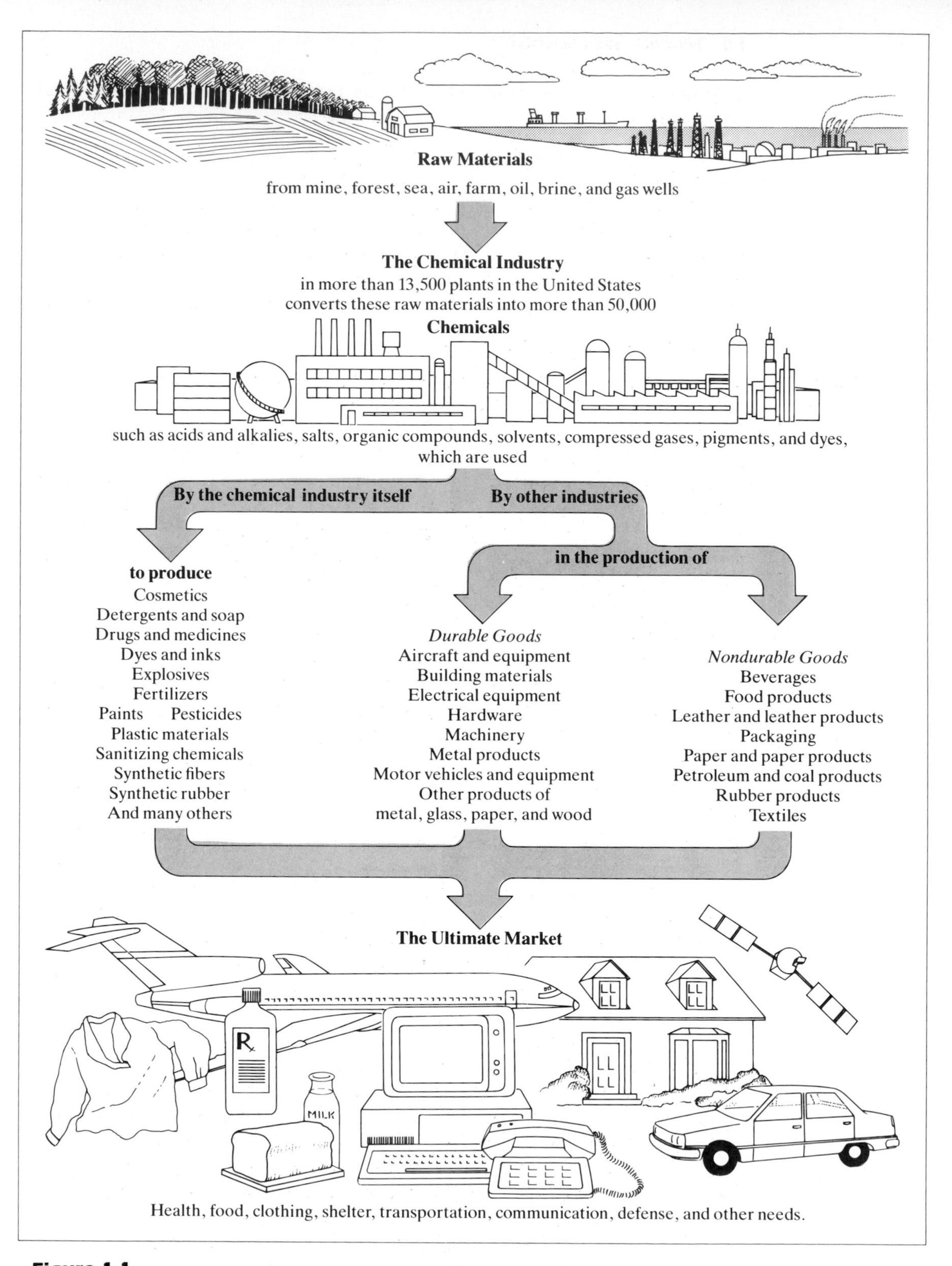

Figure 1.1

Broad scope of the chemical industry today. (Courtesy Chemical Manufacturers Association.)

lacking on exactly what is meant by "using the scientific method," the general approach is as follows:

1 Collect facts or data that are relevant to the problem or question at hand, which is usually done by planned experimentation.
2 Analyze the data to find trends (regularities) that are pertinent to the problem. Formulate a hypothesis that will account for the data that have been accumulated and that can be tested by further experimentation.
3 Plan and do additional experiments to test the hypothesis. Such experiments extend beyond the range that is covered in Step 1.
4 Modify the hypothesis as necessary so that it is compatible with all the pertinent experimental data.

Confusion sometimes arises regarding the exact meanings of the words *hypothesis*, *theory*, and *law*. A hypothesis is a tentative explanation of certain facts that provides a basis for further experimentation. A well-established hypothesis is often called a theory. Thus a theory is an explanation of the general principles of certain phenomena with considerable evidence or facts to support it. Hypotheses and theories explain natural phenomena, whereas scientific laws are simple statements of natural phenomena to which no exceptions are known under the given conditions.

Although the four steps listed in the preceding paragraph are a broad outline of the general procedure that is followed in much scientific work, they are not a recipe for doing chemistry or any other science. But chemistry is an experimental science, and much of its progress has been due to application of the scientific method through systematic research. Occasionally a great discovery is made by accident, but the majority of scientific achievements are accomplished by well-planned experiments.

Many theories and laws are studied in chemistry. They make the study of chemistry or any science easier, because they summarize particular aspects of the science. The student will note that some of the theories advanced by great scientists in the past have since been substantially altered and modified. Such changes do not mean that the discoveries of the past are less significant than those of today. Modification of existing theories in the light of new experimental evidence is essential to the growth and evolution of scientific knowledge.

1.6 How to Study Chemistry

How do you as a student approach a subject such as chemistry with its unfamiliar terminology, symbols, formulas, theories, and laws? All the generally accepted habits of good study are applicable to the study of chemistry. Budget your study time and spend it wisely. In particular, you can spend your time more profitably in regular, relatively short periods of study rather than in one prolonged cram session.

Chemistry has its own language, and learning this language is of prime importance to the successful study of chemistry. Chemistry is a subject of many facts. At first you will simply have to memorize some of them. However, you will also learn these facts by referring to them frequently in your studies and by repetitive use. For example, you must learn the symbols of 30 or 40 common elements in order to be able to write chemical formulas and equations. As with the alphabet, repetitive use of these symbols will soon make them part of your vocabulary.

The need for careful reading of assigned material cannot be overemphasized. You should read each chapter at least twice. The first time, read the chapter rapidly, noting especially topic headings, diagrams, and other outstanding features. Then read more thoroughly and deliberately for better understanding. It may be profitable to underline and abstract material during the second reading. Isolated reading may be sufficient for some subjects, but it is not sufficient for learning chemistry. During the lectures, become an active mental participant and try to think along with your instructor, do not just occupy a seat. Lecture and laboratory sessions will be much more meaningful if you have already read the assigned material.

Your studies must include a good deal of written chemistry. Chemical symbolism, equations, problem solving, and so on, require much written practice for proficiency. One does not become an accomplished pianist by merely reading or listening to music—it takes practice. One does not become a good baseball player by reading the rules and watching baseball games—it takes practice. So it is with chemistry. One does not become proficient in chemistry by only reading about it—it takes practice.

You will encounter many mathematical problems as you progress through this text. To solve a numerical problem, you should read the problem carefully to determine what is being asked. Then develop a plan for solving the problem. It is a good idea to start by writing down the pertinent material—a formula, a diagram, an equation, the data given in the problem. This information will give you something to work with, to think about, to modify, and finally to expand into an answer. When you have arrived at an answer, consider it carefully to make sure that it is a reasonable one. The solutions to problems should be recorded in a neat, orderly, stepwise fashion. Fewer errors and saved time are the rewards of a neat and orderly approach to problem solving. If you need to read and study still further for complete understanding, do it!

QUESTIONS

Review questions

1. Was the concept of a geocentric universe necessarily based on incorrect astronomical observations? Explain.
2. What were the principal goals of the alchemists?
3. What instrument, when first used by chemists, can be considered to be analogous to the use of the telescope by early astronomers? Explain the analogy.
4. Classify the following statements as observation, law, hypothesis, or theory:

(a) When the pressure remains constant, the volume of a gas is directly proportional to the absolute temperature.
(b) The water in a closed test tube boiled at 83°C.
(c) Iron gets heavier when it rusts because it attracts particles of rust from the air.
(d) All matter is composed of tiny particles called atoms.
(e) As it approaches its melting point, glass turns a flame yellow.
(f) Molecules in a gas are always moving.
(g) When wood burns it decomposes to its elements, which all escape as gas.

5. Name at least three products of chemical research you use regularly that probably did not exist 35 years ago.

6. Which of the following statements are correct?
(a) Chemistry is the science that deals with the composition of substances and the transformations they undergo.
(b) Robert Boyle, in the 17th century, clearly distinguished between an element and a compound or mixture.
(c) Both the knowledge and intellectual capacity of Western European people decreased markedly during the Dark Ages.
(d) From 1803 to 1810, John Dalton advanced his atomic theory.
(e) Most of the drugs and pharmaceuticals used in the United States today have been available for at least a century.
(f) A key feature of the scientific method is to plan and do additional experiments to test a hypothesis.
(g) Scientific laws are simple statements of natural phenomena to which no exceptions are known.
(h) Antoine Lavoisier was one of the first chemists to make quantitative measurements using a chemical balance.
(i) The two main branches of chemistry are organic and biochemistry.

CHAPTER 2

Standards for Measurement

After studying Chapter 2, you should be able to

1. Understand the terms listed in Question A at the end of the chapter.
2. Differentiate clearly between mass and weight.
3. Know the basic metric units of mass, length, and volume.
4. Give the numerical equivalents of the metric prefixes deci, centi, milli, micro, deka, hecto, kilo, and mega.
5. Express any number in exponential-notation form.
6. Express the results of arithmetic operations to the proper number of significant figures.
7. Set up and solve problems by the dimensional analysis, or factor-label, method.
8. Convert any measurement of mass, length, or volume in American units to metric units and vice versa.
9. Make conversions between Fahrenheit, Celsius, and Kelvin temperatures.
10. Differentiate clearly between temperature and heat.
11. Make calculations using the equation

$$\text{joules} = (\text{grams of substance}) \times (\text{specific heat of substance}) \times (\Delta t)$$

12. Calculate density, mass, or volume of an object or substance from appropriate data.
13. Calculate the specific gravity when given the density of a substance and vice versa.
14. Recognize the common laboratory measuring instruments illustrated in this chapter.

2.1 Mass and Weight

Chemistry is an experimental science. The results of experiments are usually determined by making measurements. In elementary experiments the quantities that are commonly measured are mass (weight), length, volume, pressure, temperature, and time. Measurements of electrical and optical quantities may also be needed in more sophisticated experimental work.

mass

Although mass and weight are often used interchangeably, the two words have quite different meanings. The **mass** of a body is defined as the amount of matter in that body. The mass of an object is a fixed and unvarying quantity that is independent of the object's location.

weight

The **weight** of a body is the measure of the earth's gravitational attraction for that body. Unlike mass, weight varies in relation to (1) the position of an object on or its distance from the earth and (2) whether the rate of motion of the object is changing with respect to the motion of the earth. Consider an astronaut of mass 70.0 kilograms (154 pounds) who is being shot into orbit. At the instant before blast-off the weight of the astronaut is also 70.0 kilograms. As the distance from the earth increases and the rocket turns into an orbiting course, the gravitational pull on the astronaut's body decreases until a state of weightlessness (zero weight) is attained. However, the mass of the astronaut's body has remained constant at 70.0 kilograms during the entire event.

The mass of an object may be measured on a chemical balance by comparing it with other known masses. Two objects of equal mass will have equal weights if they are measured in the same place. Thus, under these conditions the terms *mass* and *weight* are used interchangeably. Although the chemical balance is used to determine mass, it is said to *weigh* objects, and we often speak of the *weight* of an object when we really mean its mass.

2.2 Measurement and Significant Figures

To understand certain aspects of chemistry it is necessary to set up and solve problems. Problem solving requires an understanding of the elementary mathematical operations used to manipulate numbers. Numerical values or data are obtained from measurements made in an experiment. A chemist may use these data to calculate the extent of the physical and chemical changes occurring in the substances that are being studied. By appropriate calculations the results of an experiment may be compared with those of other experiments and summarized in ways that are meaningful.

The result of a measurement is expressed by a numerical value together with a unit of that measurement. For example,

numerical value

70.0 kilograms = 154 pounds

unit

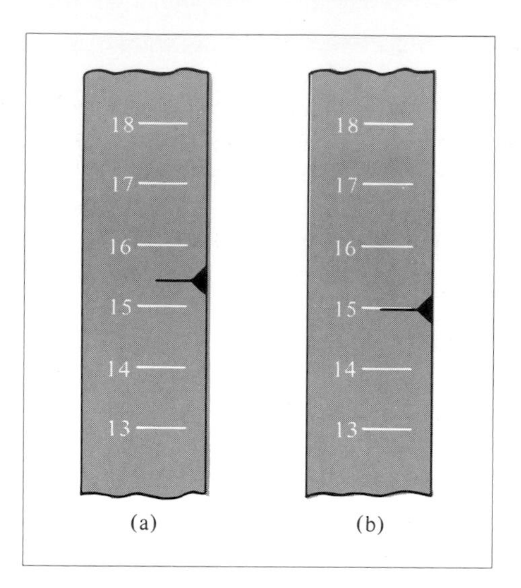

Figure 2.1

(a) Scale on a balance indicating a weight of 15.4 grams. (b) Scale on a balance indicating a weight of 15.0 grams.

Numbers obtained from a measurement are never exact values. They always have some degree of uncertainty due to the limitations of the measuring instrument and the skill of the individual making the measurement. The numerical value recorded for a measurement should give some indication of its reliability (precision). To express maximum precision this number should contain all the digits that are known plus one digit that is estimated. This last estimated digit introduces some uncertainty. Because of this uncertainty every number that expresses a measurement can have only a limited number of digits. These digits, used to express a measured quantity, are known as **significant figures**, or **significant digits**.

significant figures

Suppose we weigh an object on a balance that is calibrated in grams, and we observe that the balance scale stops between 15 and 16 (see Figure 2.1). We then know that the weight of the object is at least 15 grams but less than 16 grams. To express the weight with greater precision, we estimate that the indicator on the balance is about four-tenths the distance between 15 and 16. The weight of the object is, therefore, recorded as 15.4 grams. The last digit (4) has some uncertainty, because it is an estimated value. The recorded weight, 15.4 grams, is said to have three significant figures. Now suppose the balance scale shown in Figure 2.1 stops right on 15. The weight of the object should be recorded as 15.0 grams. It is necessary to use the zero to indicate that the object was weighed to a precision of one-tenth gram. The weight of the object is 15.0 grams, not 14.9 grams or 15.1 grams or 15 grams.

Some numbers are exact and have an infinite number of significant figures. Exact numbers occur in simple counting operations; when you count 25 dollars, you have exactly 25 dollars. Defined numbers, such as 12 inches in 1 foot, 60 minutes in 1 hour, and 100 centimeters in 1 meter, are also considered to be exact numbers. Exact numbers have no uncertainty.

Evaluating Zeros

In any measurement all nonzero numbers are significant. However, zeros may or may not be significant, depending on their position in the number. Rules for determining when zero is significant in a measurement follow.

1 Zeros between nonzero digits are significant:

205 has three significant figures.
2.05 has three significant figures.
61.09 has four significant figures.

2 Zeros that precede the first nonzero digit are not significant. These zeros are used to locate a decimal point:

0.0025 has two significant figures (2, 5).
0.0108 has three significant figures (1, 0, 8).

3 Zeros at the end of a number that include a decimal point are significant:

0.500 has three significant figures (5, 0, 0).
25.160 has five significant figures.
3.00 has three significant figures.
20. has two significant figures.

4 Zeros at the end of a number without a decimal point are ambiguous and may or may not be significant:

1000 (The zeros may or may not be significant.)
590 (The zero may or may not be significant.)

One way of indicating whether these zeros are significant is to write the number using a decimal point and a power of 10. Thus, if the value 1000 has been determined to four significant figures, it is written as 1.000×10^3. If 590 has only two significant figures, it is written as 5.9×10^2 (in this case the zero is not significant).

PROBLEM 2.1 How many significant figures are in each of these numbers?

(a) 4.5 inches (e) 25.0 grams
(b) 3.025 feet (f) 12.20 liters
(c) 125.0 meters (g) 100,000 people
(d) 0.001 mile (h) 205 birds

Answers: (a) 2 (b) 4 (c) 4 (d) 1 (e) 3 (f) 4 (g) unknown but probably 2 (h) 3

2.3 Rounding Off Numbers

rounding off numbers

In calculations we often obtain answers that have more digits than we are justified in using. It is necessary, therefore, to drop the nonsignificant digits in order to express the answer with the proper number of significant figures. When digits are dropped from a number, the value of the last digit retained is determined by a process known as **rounding off numbers**. Two rules will be used in this book for rounding off numbers:

Rule 1 When the first digit after those you want to retain is 4 or less, that digit and all others to its right are dropped. The last digit retained is not changed.

Examples rounded off to four digits:

74.693 = 74.69
This digit is dropped.

1.00629 = 1.006
These two digits are dropped.

Rule 2 When the first digit after those you want to retain is 5 or greater, that digit and all others to the right of it are dropped and the last digit retained is increased by one.

Examples rounded off to four digits:

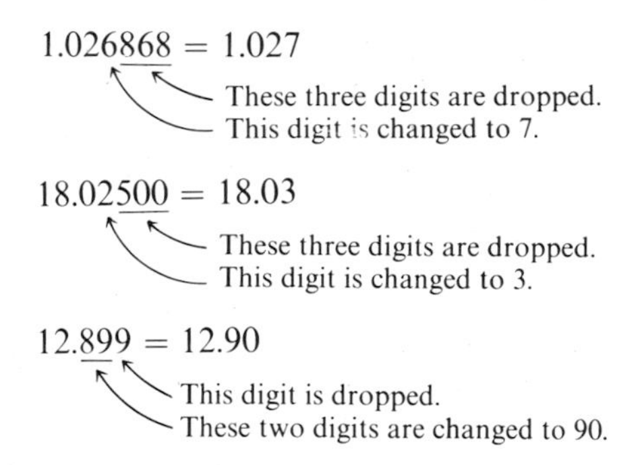
1.026868 = 1.027
These three digits are dropped.
This digit is changed to 7.
18.02500 = 18.03
These three digits are dropped.
This digit is changed to 3.
12.899 = 12.90
This digit is dropped.
These two digits are changed to 90.

PROBLEM 2.2

Round off these numbers to the number of significant digits indicated:

(a) 42.246 (four digits)
(b) 88.015 (three digits)
(c) 0.08965 (three digits)
(d) 0.08965 (two digits)
(e) 225.3 (three digits)
(f) 14.150 (three digits)

Answers: (a) 42.25 (Rule 2) (b) 88.0 (Rule 1) (c) 0.0897 (Rule 2) (d) 0.090 (Rule 2) (e) 225 (Rule 1) (f) 14.2 (Rule 2)

2.4 Scientific Notation of Numbers

The age of the earth has been estimated as about 4,500,000,000 (4.5 billion) years. Because this is an estimated value, let us say to the nearest 0.1 billion years, we are justified in using only two significant figures to express it. To express this number with two significant figures we write it using a power of 10 as 4.5×10^9 years.

Very large and very small numbers are often used in chemistry. These numbers can be simplified and conveniently written using a power of 10. Writing a number as a power of 10 is called scientific notation.

To write a number in scientific notation, move the decimal point in the original number so that it is located after the first nonzero digit. This new number is multiplied by 10 raised to the proper power (exponent). The power of 10 is equal to the number of places that the decimal point has been moved. If the decimal was moved to the left, the power of 10 will be a positive number. If the decimal was moved to the right, the power of 10 will be a negative number.

scientific notation

The **scientific notation** of a number is the number written as a factor between 1 and 10 multiplied by 10 raised to a power. For example,

$$2468 = 2.468 \times 10^3$$

number — scientific notation of the number

Study the examples that follow.

EXAMPLE 2.1 Write 5283 in scientific notation.

5283. (3 2 1) Place the decimal between the 5 and the 2. Since the decimal was moved three places to the left, the power of 10 will be 3, and the number 5.283 is multiplied by 10^3.

5.283×10^3 (Correct scientific notation)

EXAMPLE 2.2 Write 4,500,000,000 in scientific notation (two significant figures).

4 500 000 000. (9 … 1) Place the decimal between the 4 and the 5. Since the decimal was moved nine places to the left, the power of 10 will be 9, and the number 4.5 is multiplied by 10^9.

4.5×10^9 (Correct scientific notation)

EXAMPLE 2.3 Write 0.000123 in scientific notation.

0.0001 23 (4) Place the decimal between the 1 and the 2. Since the decimal was moved four places to the right, the power of 10 will be -4, and the number 1.23 is multiplied by 10^{-4}.

1.23×10^{-4} (Correct scientific notation)

PROBLEM 2.3 Write the following numbers in scientific notation:

(a) 1200 (four digits) (c) 0.0468

(b) 6,600,000 (two digits) (d) 0.00003

Answers: (a) $\underbrace{1200}_{3} = 1.200 \times 10^3$ (b) $\underbrace{6600\,000}_{6} = 6.6 \times 10^6$

(c) $0.\underbrace{04}_{2}68 = 4.68 \times 10^{-2}$ (d) $0.\underbrace{00003}_{5} = 3 \times 10^{-5}$

2.5 Significant Figures in Calculations

The results of a calculation based on measurements cannot be more precise than the least precise measurement.

Multiplication or Division In calculations involving multiplication or division, the answer must contain the same number of significant figures as in the measurement that has the least number of significant figures. Consider the following calculations:

(a) $134 \text{ in.} \times 25 \text{ in.} = 3350 \text{ in.}^2$

The answer should contain only two significant figures because the factor 25 has two significant figures. The answer, therefore, is rounded off and expressed in scientific notation:

$$134 \text{ in.} \times 25 \text{ in.} = 3.4 \times 10^3 \text{ in.}^2 \quad \text{(Answer)}$$

(b) $\dfrac{213 \text{ miles}}{4.20 \text{ hours}} = 50.714285 \text{ miles/hour}$

A hand calculator was used to obtain the eight-digit answer given. Since each factor in the problem has three digits, only three digits can be used in the answer:

$$\frac{213 \text{ miles}}{4.20 \text{ hours}} = 50.7 \text{ miles/hour} \quad \text{(Answer)}$$

(c) $\dfrac{2.2 \times 273}{760} = 0.79026315 = 0.79$ (Answer)

The answer should be rounded off to two significant figures (2.2 is the limiting factor). The correct, rounded-off answer is 0.79.

Addition or Subtraction The results of an addition or a subtraction must contain the same number of digits to the right of the decimal point as in the term with the least number of digits to the right of the decimal point. Consider the following calculations:

(a) Add 125.17, 129.2, and 52.24.

125.17
129.2
52.24
306.61

306.6 (Answer)

The number with the greatest uncertainty is 129.2. Therefore the answer is rounded off to a value having one decimal place.

(b) Subtract 14.1 from 132.56.

132.56
− 14.1
118.46

118.5 (Answer)

14.1 contains the least number of digits to the right of the decimal point. Therefore the answer is rounded off to a value having one decimal place.

PROBLEM 2.4 How many significant figures should the answer contain in each of these calculations?

(a) $14.0 \times 5.2 =$
(b) $0.1682 \times 8.2 =$
(c) $\frac{160 \times 33}{4} =$
(d) $8.2 + 0.125 =$
(e) $119.1 - 3.44 =$
(f) $\frac{94.5}{1.2} =$

Answers: (a) 2 (b) 2 (c) 1 (d) 2 (one decimal place) (e) 4 (one decimal place) (f) 2

PROBLEM 2.5 Do the following calculations, and round off your answers to the proper number of significant figures.

(a) $190.6 \times 2.3 = 438.38$

The value 438.38 was obtained with a hand calculator. The answer should have two significant figures, because 2.3, the number with the fewest significant figures, has only two significant figures. The answer must, therefore, be expressed in scientific notation.

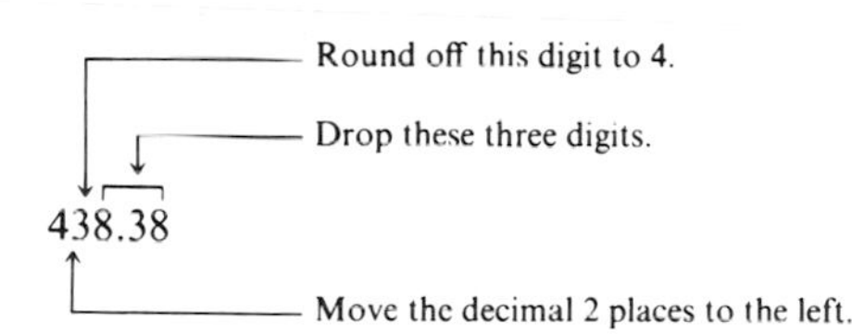

The correct answer is 4.4×10^2.

(b) $\frac{13.59 \times 6.3}{12} = 7.13475$

The value 7.13475 was obtained with a hand calculator. The answer should contain two significant figures because 6.3 and 12 have only two significant figures.

Drop these four digits.

7.13475

This digit remains the same.

The correct answer is 7.1.

(c) $\frac{1.039 - 1.020}{1.039} = 0.018286814$

The value 0.018286814 was obtained with a hand calculator. When the subtraction in the numerator is done,

$$1.039 - 1.020 = 0.019$$

the number of significant figures changes from four to two. Therefore the answer should contain two significant figures after the division is carried out:

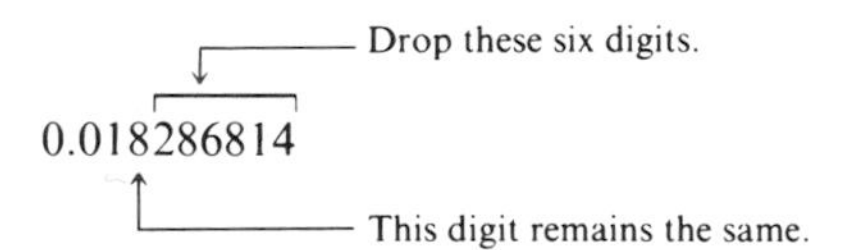
Drop these six digits.
0.018286814
This digit remains the same.

The correct answer is 0.018 or 1.8×10^{-2}.

Additional material on mathematical operations is given in the Mathematical Review in Appendix I. You are urged to review Appendix I and to study carefully any portions that are not familiar to you. This study may be done at various times during the course as the need for additional knowledge of mathematical operations arises.

2.6 The Metric System

metric system or SI

The **metric system**, or **International System** (**SI**, from *Système International*), is a decimal system of units for measurements of mass, length, time, and other physical quantities. It is built around a set of base units and uses factors of 10 to express larger or smaller numbers of these units. To express quantities that are larger or smaller than the base units, prefixes are added to the names of the units. These prefixes represent multiples of 10, making the metric system a decimal system of measurements. Table 2.1 shows the names, symbols, and numerical values of the prefixes. These are also shown in Appendix III. Some of the more commonly used prefixes are

mega	One million (1,000,000) times the unit expressed
kilo	One thousand (1000) times the unit expressed
deci	One-tenth (0.1) of the unit expressed

Table 2.1 Prefixes used in the metric system and their numerical values (The more commonly used prefixes are in color.)

Prefix	Symbol	Numerical value		
tera	T	1,000,000,000,000	or	10^{12}
giga	G	1,000,000,000	or	10^{9}
mega	M	1,000,000	or	10^{6}
kilo	k	1,000	or	10^{3}
hecto	h	100	or	10^{2}
deka	da	10	or	10^{1}
deci	d	0.1	or	10^{-1}
centi	c	0.01	or	10^{-2}
milli	m	0.001	or	10^{-3}
micro	μ	0.000001	or	10^{-6}
nano	n	0.000000001	or	10^{-9}
pico	p	0.000000000001	or	10^{-12}
femto	f	0.000000000000001	or	10^{-15}
atto	a	0.000000000000000001	or	10^{-18}

centi One-hundredth (0.01) of the unit expressed
milli One-thousandth (0.001) of the unit expressed
micro One-millionth (0.000001) of the unit expressed
nano One-billionth (0.000000001) of the unit expressed

Examples are

1 kilometer = 1000 meters
1 kilogram = 1000 grams
1 microsecond = 0.000001 second

The seven base units in the International System, their abbreviations, and the quantities they measure are given in Table 2.2. Other units are derived from these base units.

Table 2.2 International System base units of measurement

Quantity	Name of unit	Abbreviation
Length	Meter	m
Mass	Kilogram	kg
Temperature	Kelvin	K
Time	Second	s
Amount of substance	Mole	mol
Electric current	Ampere	A
Luminous intensity	Candela	cd

The metric system, or International System, is currently used by most of the countries in the world, not only for scientific and technical work, but also in commerce and industry. The United States is currently in the process of changing to the metric system of weights and measurements.

2.7 Measurement of Length

Standards for the measurement of length have an interesting historical development. The Old Testament mentions such units as the *cubit* (the distance from a man's elbow to the tip of his outstretched hand). In ancient Scotland the inch was once defined as a distance equal to the width of a man's thumb.

meter

Reference standards of measurements have undergone continuous improvements in precision. The standard unit of length in the metric system is the **meter**. When the metric system was first introduced in the 1790s, the meter was defined as one ten-millionth of the distance from the equator to the North Pole measured along the meridian passing through Dunkirk, France. In 1889 the meter was redefined as the distance between two engraved lines on a platinum–iridium alloy bar maintained at 0° Celsius. This international meter bar is stored in a vault at Sèvres near Paris. Duplicate meter bars have been made and are used as standards by many nations.

By the 1950s length could be measured with such precision that a new

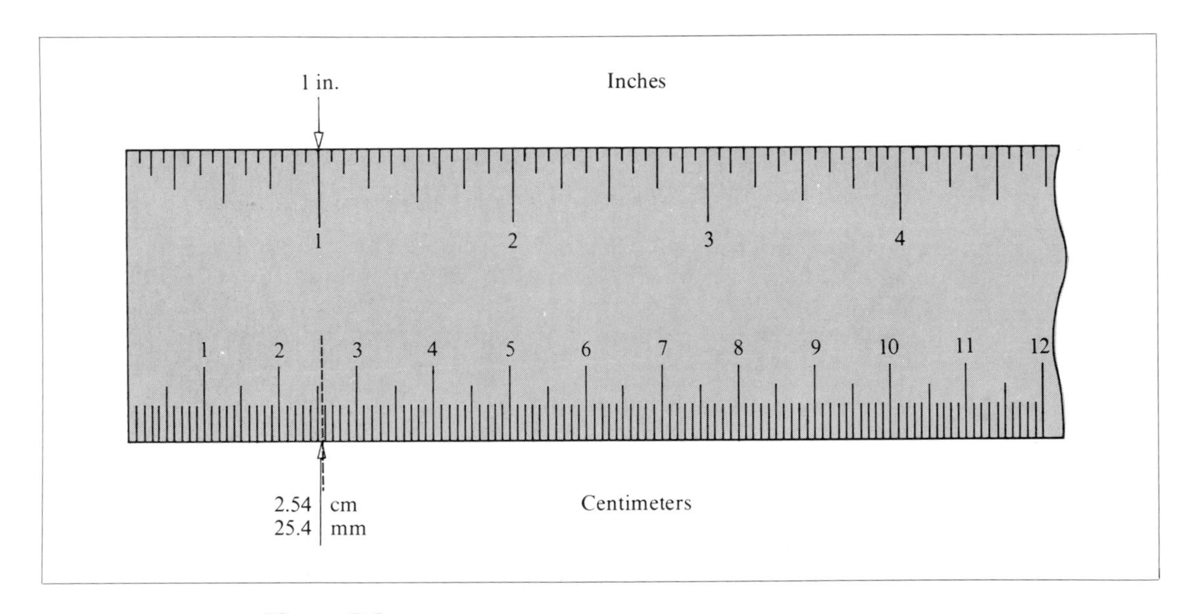

Figure 2.2

Comparison of the metric and American systems of length measurement: 2.54 cm = 1 in.

Table 2.3 Units of length

Unit	Abbreviation	Meter equivalent	Exponential equivalent
Kilometer	km	1000 m	10^3 m
Meter	m	1 m	10^0 m
Decimeter	dm	0.1 m	10^{-1} m
Centimeter	cm	0.01 m	10^{-2} m
Millimeter	mm	0.001 m	10^{-3} m
Micrometer	μm	0.000001 m	10^{-6} m
Nanometer	nm	0.000000001 m	10^{-9} m
Angstrom	Å	0.0000000001 m	10^{-10} m

standard was needed. Accordingly, the length of the meter was redefined in 1960 and again in 1983. The latest definition is: A meter is the distance that light travels in a vacuum during 1/299,792,458 of a second.

A meter is 39.37 inches, a little longer than 1 yard. One meter contains 10 decimeters, 100 centimeters, or 1000 millimeters (see Figure 2.2). A kilometer contains 1000 meters. Table 2.3 shows the relationships of these units.

The angstrom unit (10^{-10} m, abbreviated Å) is used extensively in expressing the wavelength of light and in atomic dimensions. Other important relationships are

$$1 \text{ m} = 100 \text{ cm} = 1000 \text{ mm} = 10^6 \ \mu\text{m} = 10^{10} \text{ Å}$$

$$1 \text{ cm} = 10 \text{ mm} = 0.01 \text{ m}$$

$$1 \text{ in.} = 2.54 \text{ cm}$$

$$1 \text{ mile} = 1.61 \text{ km}$$

2.8 Problem Solving

Many chemical principles are illustrated by mathematical concepts. Learning how to set up and solve numerical problems in a systematic fashion is essential in the study of chemistry. This skill, once acquired, is also very rewarding in other study areas. An electronic calculator will save you much time in computation.

Usually a problem can be solved by several methods. But in all methods it is best, especially for beginners, to use a systematic, orderly approach. The dimensional analysis, or factor-label, method is stressed in this book because

1. It provides a systematic, straightforward way to set up problems.
2. It gives a clear understanding of the principles involved.
3. It helps in learning to organize and evaluate data.

4 It helps to identify errors because unwanted units are not eliminated if the setup of the problem is incorrect.

The basic steps for solving problems are

1 Read the problem very carefully to determine what is to be solved for, and write it down.
2 Tabulate the data given in the problem. Even in tabulating data it is important to label all factors and measurements with the proper units.
3 Determine which principles are involved and which unit relationships are needed to solve the problem. Sometimes it is necessary to refer to tables for needed data.
4 Set up the problem in a neat, organized, and logical fashion, making sure that unwanted units cancel. Use sample problems in the text as guides for making setups.
5 Proceed with the necessary mathematical operations. Make certain that the answer contains the proper number of significant figures.
6 Check the answer to see if it is reasonable.

Label all factors with the proper units.

Just a few more words about problem solving. Don't allow any formal method of problem solving to limit your use of common sense and intuition. If a problem is clear to you and its solution seems simpler by another method, by all means use it. But in the long run you should be able to solve many otherwise difficult problems by using the dimensional analysis method.

The dimensional analysis method of problem solving converts one unit to another unit by the use of conversion factors.

$$\text{unit}_1 \times \text{conversion factor} = \text{unit}_2$$

If you want to know how many millimeters are in 2.5 meters, you need to convert meters (m) to millimeters (mm). Therefore, you start by writing

$$\text{m} \times \text{conversion factor} = \text{mm}$$

This conversion factor must accomplish two things. It must cancel, or eliminate, meters; and it must introduce millimeters, the unit wanted in the answer. Such a conversion factor will be in fractional form and have meters in the denominator and millimeters in the numerator:

$$\cancel{\text{m}} \times \frac{\text{mm}}{\cancel{\text{m}}} = \text{mm}$$

We know that 1 m = 1000 mm. From this relationship we can write two factors, 1 m per 1000 mm and 1000 mm per 1 m:

$$\frac{1\ \text{m}}{1000\ \text{mm}} \quad \text{and} \quad \frac{1000\ \text{mm}}{1\ \text{m}}$$

Using the factor 1000 mm/1 m, we can set up the calculation for the conversion of 2.5 m to millimeters,

$$2.5\ \cancel{\text{m}} \times \frac{1000\ \text{mm}}{1\ \cancel{\text{m}}} = 2500\ \text{mm} \qquad \text{or} \qquad 2.5 \times 10^3\ \text{mm}$$

(two significant figures)

Note that, in making this calculation, units are treated as numbers; meters in the numerator are canceled by meters in the denominator.

Now suppose you need to change 215 centimeters to meters. First you must determine that you need to convert centimeters to meters. We start with

$$\text{cm} \times \text{conversion factor} = \text{m}$$

The conversion factor must have centimeters in the denominator and meters in the numerator:

$$\cancel{\text{cm}} \times \frac{\text{m}}{\cancel{\text{cm}}} = \text{m}$$

From the relationship 100 cm = 1 m, we can write a factor that will accomplish this conversion:

$$\frac{1\ \text{m}}{100\ \text{cm}}$$

Now set up the calculation using all the data given,

$$215\ \cancel{\text{cm}} \times \frac{1\ \text{m}}{100\ \cancel{\text{cm}}} = \frac{215\ \text{m}}{100} = 2.15\ \text{m}$$

Some problems may require a series of conversions to reach the correct units in the answer. For example, suppose we want to know the number of seconds in 1 day. We need to go from the unit of days to seconds in this manner:

$$\text{day} \longrightarrow \text{hours} \longrightarrow \text{minutes} \longrightarrow \text{seconds}$$

This series requires three conversion factors, one for each step. We convert days to hours (hr), hours to minutes (min), and minutes to seconds (s). The conversions can be done individually or in a continuous sequence

$$\cancel{\text{day}} \times \frac{\text{hr}}{\cancel{\text{day}}} \qquad \cancel{\text{hr}} \times \frac{\text{min}}{\cancel{\text{hr}}} \qquad \cancel{\text{min}} \times \frac{\text{s}}{\cancel{\text{min}}}$$

$$\cancel{\text{day}} \times \frac{\cancel{\text{hr}}}{\cancel{\text{day}}} \times \frac{\cancel{\text{min}}}{\cancel{\text{hr}}} \times \frac{\text{s}}{\cancel{\text{min}}} = \text{s}$$

Inserting the proper factors we calculate the number of seconds in 1 day to be

$$1\ \cancel{\text{day}} \times \frac{24\ \cancel{\text{hr}}}{1\ \cancel{\text{day}}} \times \frac{60\ \cancel{\text{min}}}{1\ \cancel{\text{hr}}} \times \frac{60\ \text{s}}{1\ \cancel{\text{min}}} = 86{,}400.\ \text{s}$$

All five digits in 86,400 are significant, since all the factors in the calculation are exact numbers.

The dimensional analysis, or factor-label, method used in the preceding work shows how unit conversion factors are derived and used in calculations. After you become more proficient with the terms, you can save steps by writing the factors directly in the calculation. The problems that follow give examples of the conversion from American to metric units.

PROBLEM 2.6

How many centimeters are in 2.00 ft?

The stepwise conversion of units from feet to centimeters may be done in this manner: Convert feet to inches; then convert inches to centimeters.

$$\text{ft} \longrightarrow \text{in.} \longrightarrow \text{cm}$$

The conversion factors needed are

$$\frac{12\ \text{in.}}{1\ \text{ft}} \quad \text{and} \quad \frac{2.54\ \text{cm}}{1\ \text{in.}}$$

$$2.00\ \cancel{\text{ft}} \times \frac{12\ \text{in.}}{1\ \cancel{\text{ft}}} = 24.0\ \text{in.}$$

$$24.0\ \cancel{\text{in.}} \times \frac{2.54\ \text{cm}}{1\ \cancel{\text{in.}}} = 61.0\ \text{cm} \quad \text{(Answer)}$$

Since 1 ft and 12 in. are exact numbers, the number of significant figures allowed in the answer is three, based on the number 2.00.

PROBLEM 2.7

How many meters are in a 100 yd football field? The stepwise conversion of units from yards to meters may be done in this manner, using the proper conversion factors.

$$\text{yd} \longrightarrow \text{ft} \longrightarrow \text{in.} \longrightarrow \text{cm} \longrightarrow \text{m}$$

$$100\ \cancel{\text{yd}} \times \frac{3\ \text{ft}}{1\ \cancel{\text{yd}}} = 300\ \text{ft} \qquad (3\ \text{ft/yd})$$

$$300\ \cancel{\text{ft}} \times \frac{12\ \text{in.}}{1\ \cancel{\text{ft}}} = 3600\ \text{in.} \qquad (12\ \text{in./ft})$$

$$3600\ \cancel{\text{in.}} \times \frac{2.54\ \text{cm}}{1\ \cancel{\text{in.}}} = 9144\ \text{cm} \qquad (2.54\ \text{cm/in.})$$

$$9144\ \cancel{\text{cm}} \times \frac{1\ \text{m}}{100\ \cancel{\text{cm}}} = 91.4\ \text{m} \qquad (1\ \text{m}/100\ \text{cm}) \qquad \text{(three significant figures)}$$

Problems 2.6 and 2.7 may be solved using a running linear expression and writing down conversion factors in succession. This method often saves one or two calculation steps and allows numerical values to be reduced to simpler terms, leading to simpler calculations. The single linear expressions for Problems 2.6 and 2.7 are

$$2.00\,\cancel{\text{ft}} \times \frac{12\,\cancel{\text{in.}}}{1\,\cancel{\text{ft}}} \times \frac{2.54\text{ cm}}{1\,\cancel{\text{in.}}} = 61.0\text{ cm}$$

$$100\,\cancel{\text{yd}} \times \frac{3\,\cancel{\text{ft}}}{1\,\cancel{\text{yd}}} \times \frac{12\,\cancel{\text{in.}}}{1\,\cancel{\text{ft}}} \times \frac{2.54\,\cancel{\text{cm}}}{1\,\cancel{\text{in.}}} \times \frac{1\text{ m}}{100\,\cancel{\text{cm}}} = 91.4\text{ m}$$

Using the units alone (Problem 2.7), we see that the stepwise cancellation proceeds in succession until the desired unit is reached.

$$\cancel{\text{yd}} \times \frac{\cancel{\text{ft}}}{\cancel{\text{yd}}} \times \frac{\cancel{\text{in.}}}{\cancel{\text{ft}}} \times \frac{\cancel{\text{cm}}}{\cancel{\text{in.}}} \times \frac{\text{m}}{\cancel{\text{cm}}} = \text{m}$$

PROBLEM 2.8

How many cubic centimeters (cm^3) are in a box that measures 2.20 in. by 4.00 in. by 6.00 in.? First we need to determine the volume of the box in cubic inches ($in.^3$) by multiplying together the length times the width times the height.

$$2.20\text{ in.} \times 4.00\text{ in.} \times 6.00\text{ in.} = 52.8\text{ in.}^3$$

Now we need to convert $in.^3$ to cm^3, which can be done by using the inches and centimeters relationship three times.

$$\cancel{\text{in.}^3} \times \frac{\text{cm}}{\cancel{\text{in.}}} \times \frac{\text{cm}}{\cancel{\text{in.}}} \times \frac{\text{cm}}{\cancel{\text{in.}}} = \text{cm}^3$$

$$52.8\,\cancel{\text{in.}^3} \times \frac{2.54\text{ cm}}{1\,\cancel{\text{in.}}} \times \frac{2.54\text{ cm}}{1\,\cancel{\text{in.}}} \times \frac{2.54\text{ cm}}{1\,\cancel{\text{in.}}} = 865\text{ cm}^3$$

2.9 Measurement of Mass

kilogram

The gram is used as a unit of mass measurement, but it is a tiny amount of mass; for instance, a nickel has a mass of about 5 grams. Therefore the *standard unit* of mass in the SI system is the **kilogram** (equal to 1000 g). The amount of mass in a kilogram is defined by international agreement as exactly equal to the mass of a platinum–iridium weight (international prototype kilogram) kept in a vault at Sèvres, France. Comparing this unit of mass to 1 pound (16 ounces), we find that a kilogram is equal to 2.2 pounds. A pound is equal to 454 grams (0.454 kilogram). The same prefixes used in length measurement are used to indicate larger and smaller gram units (see Table 2.4).

Table 2.4 Metric units of mass

Unit	Abbreviation	Gram equivalent	Exponential equivalent
Kilogram	kg	1000 g	10^3 g
Gram	g	1 g	10^0 g
Decigram	dg	0.1 g	10^{-1} g
Centigram	cg	0.01 g	10^{-2} g
Milligram	mg	0.001 g	10^{-3} g
Microgram	μg	0.000001 g	10^{-6} g

It is convenient to remember that

1 g = 1000 mg

1 kg = 1000 g

1 kg = 2.2 lb

1 lb = 454 g

To change grams to milligrams, multiply grams by the conversion factor 1000 mg/g. The setup for converting 25 g to milligrams is

$$25\ \cancel{g} \times \frac{1000\ \text{mg}}{1\ \cancel{g}} = 25{,}000\ \text{mg} \qquad (2.5 \times 10^4\ \text{mg}) \quad \text{(Answer)}$$

Note that multiplying a number by 1000 is the same as multiplying the number by 10^3 and can be done simply by moving the decimal point three places to the right

$$6.428 \times 1000 = 6428 \qquad (6.\underset{3}{\underbrace{428}})$$

To change milligrams to grams, multiply milligrams by the conversion factor 1 g/1000 mg. For example, to convert 155 mg to grams:

$$155\ \cancel{\text{mg}} \times \frac{1\ \text{g}}{1000\ \cancel{\text{mg}}} = 0.155\ \text{g} \quad \text{(Answer)}$$

Examples of converting weights from American to metric units are shown in Problems 2.9 and 2.10.

PROBLEM 2.9 A 1.50 lb package of sodium bicarbonate costs 80 cents. How many grams of this substance are in this package?

We are solving for the number of grams equivalent to 1.50 lb. Since 1 lb = 454 g, the factor to convert pounds to grams is 454 g/lb.

$$1.50\,\cancel{lb} \times \frac{454\text{ g}}{1\,\cancel{lb}} = 681\text{ g} \quad \text{(Answer)}$$

Note: The cost of the sodium bicarbonate has no bearing on the question asked in this problem.

PROBLEM 2.10

Suppose four ostrich feathers weigh 1.00 lb. Assuming that each feather is equal in weight, how many milligrams does a single feather weigh? The unit conversion in this problem is from 1 lb/4 feathers to milligrams per feather. Since the unit *feathers* occurs in the denominator of both the starting unit and the desired unit, the unit conversions needed are

$$\text{lb} \longrightarrow \text{g} \longrightarrow \text{mg}$$

$$\frac{1.00\,\cancel{lb}}{4\text{ feathers}} \times \frac{454\,\cancel{g}}{1\,\cancel{lb}} \times \frac{1000\text{ mg}}{1\,\cancel{g}} = \frac{113{,}500\text{ mg}}{1\text{ feather}} \qquad (1.14 \times 10^5\text{ mg/feather})$$

2.10 Measurement of Volume

volume

Volume, as used here, is the amount of space occupied by matter. The SI unit of volume is the cubic meter (m^3). However, the liter (pronounced *leeter* and abbreviated L) and the milliliter (abbreviated mL) are the standard units of volume used in most chemical laboratories.

One liter is a little larger than one U.S. quart (see Figure 2.3). One liter is also equal to 1000 milliliters, and one milliliter is equal to one cubic centimeter (1 cm^3).

It is convenient to remember that

$$1\text{ L} = 1000\text{ mL} = 1000\text{ cm}^3$$
$$1\text{ mL} = 1\text{ cm}^3$$
$$1\text{ L} = 1.057\text{ qt}$$
$$946\text{ mL} = 1\text{ qt}$$

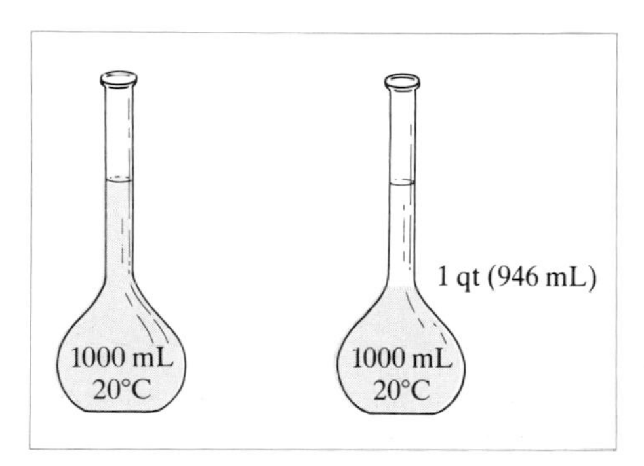

Figure 2.3

Comparison of the volume of a liter (1000 mL) and a quart (946 mL). The flasks shown are called volumetric flasks.

The volume of a cubic or rectangular container can be determined by multiplying its length times width times height. Thus a 10 cm square box has a volume of 10 cm × 10 cm × 10 cm = 1000 cm^3.

Examples of volume conversions are shown below.

PROBLEM 2.11 How many milliliters are contained in 3.5 liters?

The conversion factor to change liters to milliliters is 1000 mL/L.

$$3.5\,\cancel{L} \times \frac{1000\text{ mL}}{\cancel{L}} = 3500\text{ mL} \qquad (3.5 \times 10^3\text{ mL})$$

Liters may be changed to milliliters by moving the decimal point three places to the right and changing the units to milliliters.

$$1.500\text{ L} = 1500.\text{ mL}$$

PROBLEM 2.12 How many cubic centimeters are in a cube that is 11.1 inches on a side?

First change inches to centimeters. The conversion factor is 2.54 cm/in.

$$11.1\,\cancel{\text{in.}} \times \frac{2.54\text{ cm}}{1\,\cancel{\text{in.}}} = 28.2\text{ cm on a side}$$

Then change to cubic volume (length × width × height).

$$28.2\text{ cm} \times 28.2\text{ cm} \times 28.2\text{ cm} = 22{,}426\text{ cm}^3 \qquad (2.24 \times 10^4\text{ cm}^3)$$

2.11 Heat and Temperature Scales

heat

temperature

Heat is a form of energy associated with the motion of small particles of matter. The term *heat* refers to the quantity of energy within a system or to a quantity of energy added to or taken away from a system. *System* as used here simply refers to the entity that is being heated or cooled. Depending on the amount of heat energy present, a given system is said to be hot or cold. **Temperature** is a measure of the intensity of heat, or how hot a system is, regardless of its size. Heat always flows from a region of higher temperature to one of lower temperature. The SI unit of temperature is the Kelvin.

The temperature of a system can be expressed by several different scales. Three commonly used temperature scales are the Celsius scale (pronounced *sell-see-us*), the Kelvin (absolute) scale, and the Fahrenheit scale. The unit of temperature on the Celsius and Fahrenheit scales is called a *degree*, but the size of the Celsius and the Fahrenheit degree is not the same. The symbol for the Celsius and Fahrenheit degrees is °, and it is placed as a superscript after the number and before the symbol for the scales. Thus, 100°C means 100 *degrees Celsius*. The degree sign is not used with Kelvin temperatures.

degrees Celsius = °C

Kelvin (absolute) = K

degrees Fahrenheit = °F

On the Celsius scale the interval between the freezing and boiling temperatures of water is divided into 100 equal parts, or degrees. The freezing point of water is assigned a temperature of 0°C and the boiling point of water a temperature of 100°C. The Kelvin temperature scale is also known as the absolute temperature scale, because 0 K is the lowest temperature theoretically attainable. The Kelvin zero is 273.16 degrees below the Celsius zero. A Kelvin is equal in size to a Celsius degree. The freezing point of water on the Kelvin scale is 273.16 K (usually rounded to 273 K). The Fahrenheit scale has 180 degrees between the freezing and boiling temperatures of water. On this scale the freezing point of water is 32°F and the boiling point is 212°F.

$$0°\text{C} \cong 273\ \text{K} \cong 32°\text{F}$$

The three scales are compared in Figure 2.4. Although absolute zero (0 K) is the lower limit of temperature on these scales, temperature has no known upper limit. (Temperatures of several million degrees are known to exist in the sun and in other stars.)

By examining Figure 2.4 we can see that there are 100 Celsius degrees and 100 Kelvins between the freezing and boiling points of water. But there are 180 Fahrenheit degrees between these two temperatures. Hence, the size of a degree on the Celsius scale is the same as the size of one Kelvin, but one Celsius degree corresponds to 1.8 degrees on the Fahrenheit scale.

$$1°\text{C} = 1\ \text{K} = 1.8°\text{F}$$

From these data, mathematical formulas have been derived to convert a temperature on one scale to the corresponding temperature on another scale. These formulas are

$$\text{K} = °\text{C} + 273 \qquad (1)$$

$$°\text{F} = (1.8 \times °\text{C}) + 32 \qquad (2)$$

$$°\text{C} = \frac{(°\text{F} - 32)}{1.8} \qquad (3)$$

Interpretation: Formula (1) states that the addition of 273 to the degrees Celsius converts the temperature to Kelvins. Formula (2) states that, to obtain the Fahrenheit temperature corresponding to a given Celsius temperature, we multiply the degrees Celsius by 1.8 and then add 32. Formula (3) states that, to obtain the corresponding Celsius temperature, we subtract 32 from the degrees Fahrenheit and then divide this answer by 1.8. Examples of temperature conversions follow.

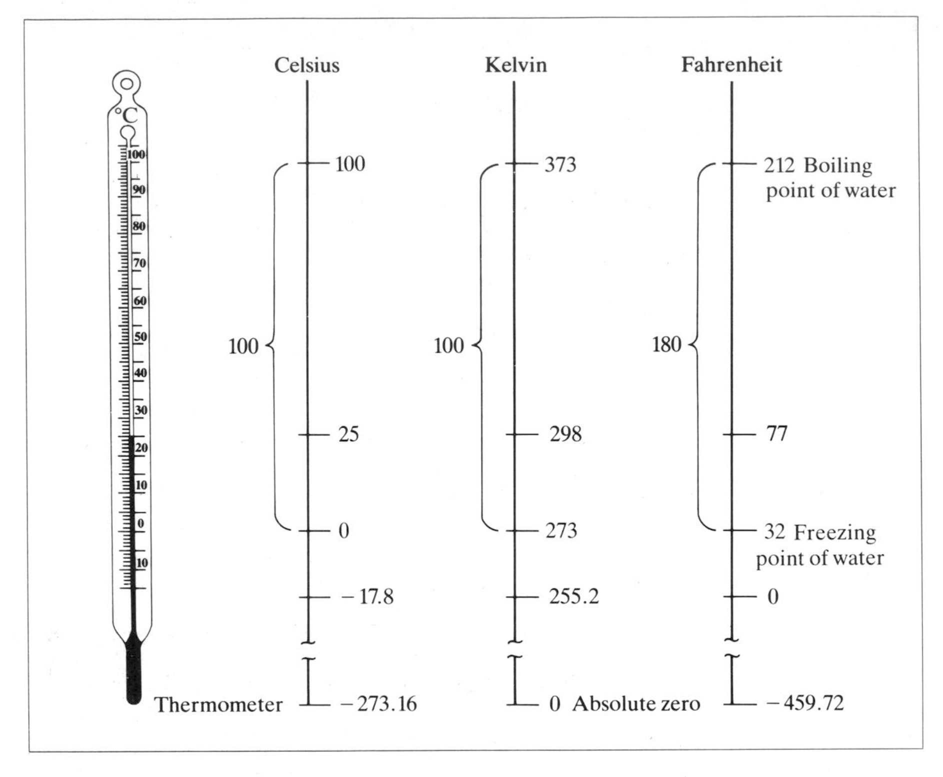

Figure 2.4
Comparison of Celsius, Kelvin, and Fahrenheit temperature scales

---

PROBLEM 2.13 The temperature at which table salt (sodium chloride) melts is 800°C. What is this temperature on the Kelvin and Fahrenheit scales?

We need to calculate K from °C, so we use formula (1) above. We also need to calculate °F from °C; for this calculation we use formula (2).

$$K = °C + 273$$

$$K = 800°C + 273 = 1073 \text{ K}$$

$$°F = (1.8 \times °C) + 32$$

$$°F = (1.8 \times 800°C) + 32$$

$$°F = 1440 + 32 = 1472°F$$

$$800°C = 1073 \text{ K} = 1472°F$$

PROBLEM 2.14 The temperature for December 1 was 110°F, a new record. Calculate this temperature in °C.

Formula (3) applies here.

$$°C = \frac{(°F - 32)}{1.8}$$

$$°C = \frac{(110 - 32)}{1.8} = \frac{78}{1.8} = 43°C$$

PROBLEM 2.15 What temperature on the Fahrenheit scale corresponds to −8.0°C? (Be alert to the presence of the minus sign in this problem.)

$$°F = (1.8 \times °C) + 32$$

$$°F = [1.8 \times (-8.0)] + 32 = -14.4 + 32$$

$$°F = 18 \quad \text{(2 significant figures)}$$

Temperatures used in this book are in degrees Celsius (°C) unless specified otherwise.

2.12 Heat: Quantitative Measurement

The SI-derived unit for heat is the joule (pronounced *jool* and abbreviated J). Another unit for heat, which has been used for many years, is the caloric (abbreviated cal). The relationship between joules and calories is

$$4.184 \text{ J} = 1 \text{ cal} \quad \text{(exactly)}$$

joule

calorie

To give you some idea of the magnitude of these heat units, 4.184 **J** or 1 **cal** is the quantity of heat energy required to change the temperature of 1 g of water by 1°C, usually measured from 14.5°C to 15.5°C.

Since joule and calorie are rather small units, the use of kilojoules (kJ) and kilocalories (kcal) are used to express heat energy in many chemical processes. The kilocalorie is also known as the nutritional or large Calorie (spelled with a capital *C* and abbreviated Cal). In this book heat energy will be expressed in joules with parenthetical values in calories.

$$1 \text{ kJ} = 1000 \text{ J}$$

$$1 \text{ kcal} = 1000 \text{ cal}$$

The difference in the meanings of the terms *heat* and *temperature* can be seen by this example: Visualize two beakers, A and B. Beaker A contains 100 g of water at 20°C, and beaker B contains 200 g of water also at 20°C. The beakers are heated until the temperature of the water in each reaches 30°C. The *temperature* of the water in the beakers was raised by exactly the same amount, 10°C. But twice as much *heat* (8368 J or 2000 cal) was required to raise the temperature of the water in beaker B as was required in beaker A (4184 J or 1000 cal).

Table 2.5 Specific heat of selected substances

Substance	Specific heat J/g°C	Specific heat cal/g°C
Water	4.184	1.00
Ethyl alcohol	2.138	0.511
Ice	2.059	0.492
Aluminum	0.900	0.215
Iron	0.473	0.113
Copper	0.385	0.0921
Gold	0.131	0.0312
Lead	0.128	0.0305

In the middle of the 18th century Joseph Black, a Scottish chemist, heated and cooled equal masses of iron and lead through the same temperature range. Black noted that much more heat was needed for the iron than for the lead. He had discovered a fundamental property of matter; namely, that every substance has a characteristic heat capacity. Heat capacities may be compared in terms of specific heats. The **specific heat** of a substance is the quantity of heat (lost or gained) required to change the temperature of 1 g of that substance by 1°C. It follows then that the specific heat of water is 4.184 J/g°C (1 cal/g°C). The specific heat of water is high compared with that of most substances. Aluminum and copper, for example, have specific heats of 0.900 and 0.385 J/g°C, respectively (see Table 2.5). The relation of mass, specific heat, temperature change (Δt), and quantity of heat lost or gained by a system is expressed by this general equation:

specific heat

$$\begin{pmatrix}\text{grams of}\\ \text{substance}\end{pmatrix} \times \begin{pmatrix}\text{specific heat}\\ \text{of substance}\end{pmatrix} \times \Delta t = \text{joules or calories} \qquad (4)$$

Thus, the amount of heat needed to raise the temperature of 200 g of water by 10°C can be calculated as follows:

$$200\ \cancel{g} \times \frac{4.184\ \text{J}}{\cancel{g}\cancel{^\circ C}} \times 10\cancel{^\circ C} = 8368\ \text{J}\ (2000\ \text{cal})$$

Examples of specific-heat problems follow.

PROBLEM 2.16 Calculate the specific heat of a solid in J/g°C and cal/g°C if 1638 J raise the temperature of 125 g of the solid from 25.0°C to 52.6°C.

First solve equation (4) to obtain an equation for specific heat.

$$\text{specific heat} = \frac{\text{J}}{\text{g} \times \Delta t}$$

Now substitute in the data:

$J = 1638; \quad g = 125; \quad \Delta t = 52.6 - 25.0°C = 27.6°C$

$$\text{specific heat} = \frac{1638\ \text{J}}{125\ \text{g} \times 27.6°\text{C}} = 0.475\ \text{J/g°C}$$

Now convert joules to calories using 1 cal/4.184 J:

$$\text{specific heat} = \frac{0.475\ \cancel{\text{J}}}{\text{g°C}} \times \frac{1\ \text{cal}}{4.184\ \cancel{\text{J}}} = 0.114\ \text{cal/g°C}$$

PROBLEM 2.17 A sample of a metal weighing 212 g is heated to 125.0°C and then dropped into 375 g of water at 24.0°C. If the final temperature of the water is 34.2°C, what is the specific heat of the metal? (Assume no heat losses to the surroundings.)

When the metal enters the water it begins to cool, losing heat to the water. At the same time the temperature of the water rises. This process continues until the temperature of the metal and the temperature of the water are equal, at which point (34.2°C) no net flow of heat occurs.

The heat lost or gained by a system is given by equation (4). We use this equation first to calculate the heat gained by the water and then to calculate the specific heat of the metal.

$$\text{temperature rise of the water } (\Delta t) = 34.2°\text{C} - 24.0°\text{C} = 10.2°\text{C}$$

$$\text{heat gained by the water} = 375\ \text{g} \times \frac{4.184\ \text{J}}{\text{g°C}} \times 10.2°\text{C} = 1.60 \times 10^4\ \text{J}$$

The metal dropped into the water must have a final temperature the same as the water (34.2°C).

$$\text{temperature drop by the metal } (\Delta t) = 125.0°\text{C} - 34.2°\text{C} = 90.8°\text{C}$$

$$\text{heat lost by the metal} = \text{heat gained by the water} = 1.60 \times 10^4\ \text{J}$$

Rearranging equation (4) we get

$$\text{specific heat} = \frac{\text{J}}{\text{g} \times \Delta t}$$

$$\text{specific heat of the metal} = \frac{1.60 \times 10^4\ \text{J}}{212\ \text{g} \times 90.8°\text{C}} = 0.831\ \text{J/g°C}$$

2.13 Tools for Measurement

Common measuring instruments used in chemical laboratories are illustrated in Figures 2.5 and 2.6. A balance is used to measure mass. Some balances will weigh objects to the nearest microgram. The choice of the balance depends on the accuracy required and the amount of material being weighed. Three standard balances are shown in Figure 2.5: a quadruple-beam balance with precision up to 0.01 g; a single-pan, top-loading balance with a precision of 0.001 g (1 mg); a

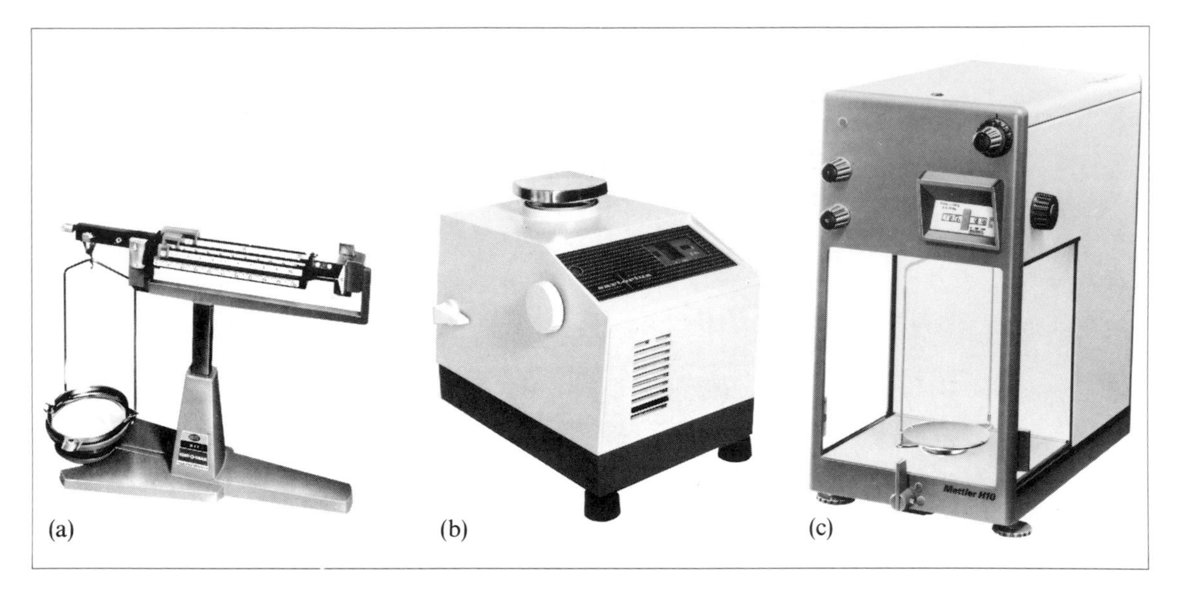

Figure 2.5

(a) A Cent-o-Gram R311 balance has four calibrated horizontal beams, each fitted with a movable weight. The weight of the object placed on the pan is determined by moving the weights along the beams until the swinging beam is in balance, as shown by the indicator on the right. (*Courtesy Ohaus Scale Corporation.*) (b) A single-pan, top-loading, rapid-weighing balance with direct readout to the nearest milligram (for example, 125.456 g). (*Courtesy Sartorius Balance Div., Brinkmann Instruments Inc.*) (c) A single-pan analytical balance for high-precision weighing. The precision of this balance is 0.1 mg (0.0001 g). (*Courtesy Mettler Instrument Corporation, Hightstown, N.J.*)

single-pan analytical balance with a precision up to 0.0001 g. Automatic-recording and electronic balances are also available.

The most common instruments or equipment for measuring liquids are the graduated cylinder, volumetric flask, buret, pipet, and syringe, which are shown in Figure 2.6. These calibrated pieces are usually made of glass and are available in various sizes.

The common laboratory instrument for measuring temperature is a thermometer (see Figure 2.4).

2.14 Density

density

Density (d) is the ratio of the mass of a substance to the volume occupied by that mass; it is the mass per unit of volume and is given by the equation

$$d = \frac{\text{mass}}{\text{volume}} \tag{5}$$

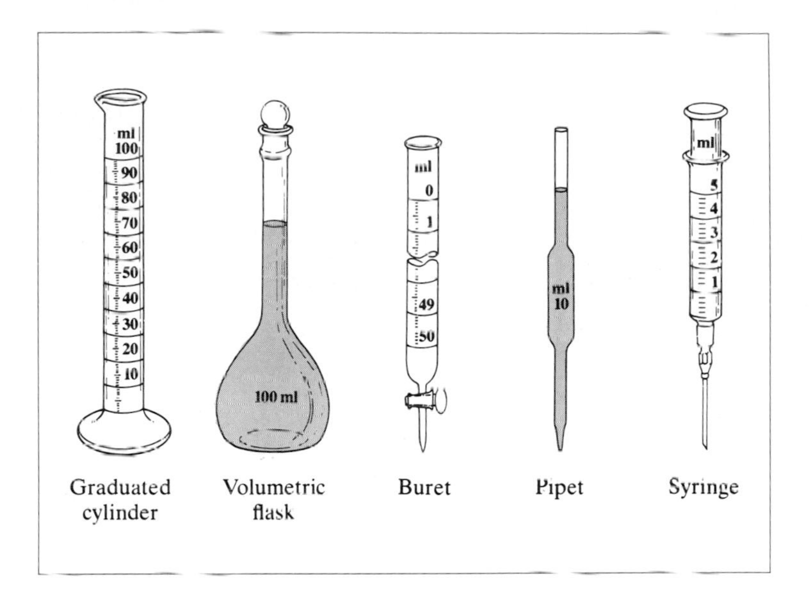

Figure 2.6
Calibrated glassware for measuring the volume of liquids

Density is a physical characteristic of a substance and may be used as an aid to its identification. When the density of a solid or a liquid is given, the mass is usually expressed in grams and the volume in milliliters or cubic centimeters.

$$d = \frac{\text{mass}}{\text{volume}} = \frac{\text{g}}{\text{mL}} \quad \text{or} \quad d = \frac{g}{\text{cm}^3}$$

Since the volume of a substance (especially liquids and gases) varies with temperature, it is important to state the temperature along with the density. For example, the volume of 1.0000 g water at 4°C is 1.0000 mL, at 20°C it is 1.0018 mL, and at 80°C it is 1.0290 mL. Density, therefore, also varies with temperature.

The density of water at 4°C is 1.0000 g/mL, but at 80°C the density of water is 0.9718 g/mL.

$$d^{4^\circ\text{C}} = \frac{1.0000\ \text{g}}{1.0000\ \text{mL}} = 1.0000\ \text{g/mL}$$

$$d^{80^\circ\text{C}} = \frac{1.0000\ \text{g}}{1.0290\ \text{mL}} = 0.9718\ \text{g/mL}$$

The density of iron at 20°C is 7.86 g/mL.

$$d^{20^\circ\text{C}} = \frac{7.86\ \text{g}}{1.00\ \text{mL}} = 7.86\ \text{g/mL}$$

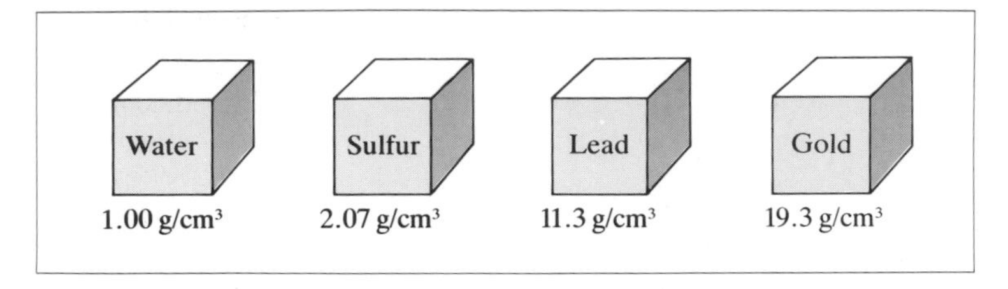

Figure 2.7

Comparison of the masses of equal volumes (1.00 cm^3) of water, sulfur, lead, and gold. The mass of each substance per cubic centimeter is its density. (Water is at 4°C; the three solids, at 20°C.)

The density of iron indicates that it is about eight times as heavy as water per unit of volume. The densities of water, sulfur, lead, and gold are compared in Figure 2.7.

Densities for liquids and solids are usually represented in terms of grams per milliliter (g/mL) or grams per cubic centimeter (g/cm^3). The density of gases, however, is expressed in terms of grams per liter (g/L). Unless otherwise stated, gas densities are given for 0°C and 1 atmosphere pressure (discussed further in Chapter 12). Table 2.6 lists the densities of a number of common materials.

Suppose that water, carbon tetrachloride, and cottonseed oil are successively poured into a graduated cylinder. The result is a layered three-liquid system

Table 2.6 Densities of some selected materials. For comparing densities the density of water is the reference for solids and liquids; air is the reference for gases.

Liquids and solids		Gases	
Substance	Density (g/mL at 20°C)	Substance	Density (g/L at 0°C)
Wood (Douglas fir)	0.512	Hydrogen	0.090
Ethyl alcohol	0.789	Helium	0.178
Cottonseed oil	0.926	Methane	0.714
Water (4°C)	**1.0000**	Ammonia	0.771
Sugar	1.59	Neon	0.90
Carbon tetrachloride	1.595	Carbon monoxide	1.25
Magnesium	1.74	Nitrogen	1.251
Sulfuric acid	1.84	**Air**	**1.293**
Sulfur	2.07	Oxygen	1.429
Salt	2.16	Hydrogen chloride	1.63
Aluminum	2.70	Argon	1.78
Silver	10.5	Carbon dioxide	1.963
Lead	11.34	Chlorine	3.17
Mercury	13.55		
Gold	19.3		

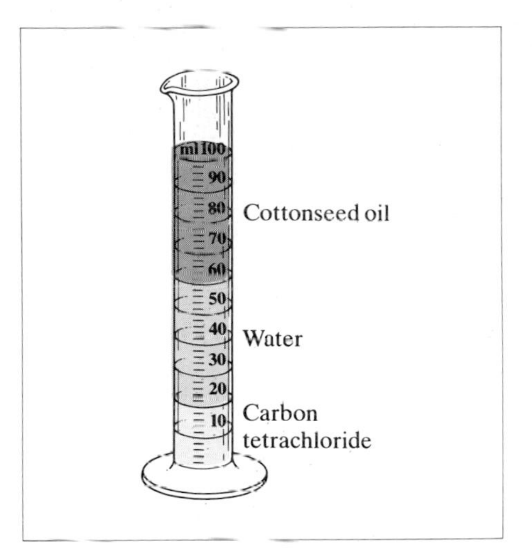

Figure 2.8

Relative density of liquids. When three immiscible (not capable of mixing) liquids are poured together, the liquid with the highest density will be the bottom layer. In the case of cottonseed oil, water, and carbon tetrachloride, cottonseed oil is the top layer.

(Figure 2.8). Can we predict the order of the liquid layers? Yes, by looking up the densities in Table 2.6. Carbon tetrachloride has the greatest density (1.595 g/mL), and cottonseed oil has the lowest density (0.926 g/mL). Carbon tetrachloride will be the bottom layer, and cottonseed oil will be the top layer. Water, with a density between the other two liquids, will form the middle layer. This information can also be determined by experiment. Add a few milliliters of carbon tetrachloride to a beaker of water. The carbon tetrachloride, being more dense than the water, will sink. Cottonseed oil, being less dense than water, will float when added to the beaker. Direct comparisons of density in this manner can be made only with liquids that are *immiscible* (do not mix with or dissolve in one another).

The density of air at 0°C is approximately 1.293 g/L. Gases with densities less than this value are said to be "lighter than air." A helium-filled balloon will rise rapidly in air, because the density of helium is only 0.178 g/L.

When an insoluble solid object is dropped into water, it will sink or float, depending on its density. If the object is less dense than water, it will float, displacing a *mass* of water equal to the mass of the object. If the object is more dense than water, it will sink, displacing a *volume* of water equal to the volume of the object. This information can be utilized to determine the volume (and density) of irregularly shaped objects.

Sample calculations of density problems follow.

PROBLEM 2.18

What is the density of a mineral if 427 g of the mineral occupy a volume of 35.0 mL? We need to solve for density, so we start by writing the formula for calculating density.

$$d = \frac{\text{mass}}{\text{volume}}$$

Then we substitute the data given in the problem into the equation and solve.

$$\text{mass} = 427 \text{ g} \qquad \text{volume} = 35.0 \text{ mL}$$

$$d = \frac{\text{mass}}{\text{volume}} = \frac{427 \text{ g}}{35.0 \text{ mL}} = 12.2 \text{ g/mL} \quad \text{(Answer)}$$

PROBLEM 2.19 The density of gold is 19.3 g/mL. What is the mass of 25.0 mL of gold?

Two ways to solve this problem are: (1) Solve the density equation for mass, then substitute the density and volume data into the new equation and calculate; (2) Solve by dimensional analysis.

Method 1 (a) Solve the density equation for mass:

$$d = \frac{\text{mass}}{\text{volume}} \qquad \text{mass} = d \times \text{volume}$$

(b) Substitute the data and calculate:

$$\text{mass} = \frac{19.3 \text{ g}}{\cancel{\text{mL}}} \times 25.0 \cancel{\text{mL}} = 482 \text{ g} \quad \text{(Answer)}$$

Method 2 Dimensional analysis: Use density as a conversion factor, converting mL ⟶ g.

The conversion of units is

$$\text{mL} \times \frac{\text{g}}{\text{mL}} = \text{g}$$

$$25.0 \cancel{\text{mL}} \times \frac{19.3 \text{ g}}{\cancel{\text{mL}}} = 482 \text{ g} \quad \text{(Mass of gold)}$$

PROBLEM 2.20 Calculate the volume (in mL) of 100 g of ethyl alcohol.

From Table 2.6 we see that the density of ethyl alcohol is 0.789 g/mL. This density also means that 1 mL of the alcohol weighs 0.789 g (1 mL/0.789 g). For a conversion factor, we can use either

$$\frac{\text{g}}{\text{mL}} \quad \text{or} \quad \frac{\text{mL}}{\text{g}}$$

In this case the conversion is from g ⟶ mL, so we use mL/g. Substituting the data, we get

$$100 \cancel{\text{g}} \times \frac{1 \text{ mL}}{0.789 \cancel{\text{g}}} = 127 \text{ mL of ethyl alcohol}$$

This problem may also be done by solving the density equation for volume and then substituting the data in the new equation.

PROBLEM 2.21

The water level in a graduated cylinder stands at 20.0 mL before and at 26.2 mL after a 16.74 g metal bolt is submerged in the water. (a) What is the volume of the bolt? (b) What is the density of the bolt?

(a) The bolt will displace a volume equal to its volume. Thus the increase in volume is the volume of the bolt.

$$\begin{array}{rl} 26.2\ \text{mL} & = \text{volume of water plus bolt} \\ -20.0\ \text{mL} & = \text{volume of water} \\ \hline 6.2\ \text{mL} & = \text{volume of bolt} \quad \text{(Answer)} \end{array}$$

(b) $d = \dfrac{\text{mass of bolt}}{\text{volume of bolt}} = \dfrac{16.74\ \text{g}}{6.2\ \text{mL}} = 2.7\ \text{g/mL}$ (Answer)

2.15 Specific Gravity

specific gravity

The **specific gravity** (sp gr) of a substance is a ratio of the density of that substance to the density of another substance. Water is usually used as the reference standard for solids and liquids. Air is usually used as the reference standard for gases. Specific gravity has no units because the density units cancel. The specific gravity tells us how many times as heavy a liquid, a solid, or a gas is as compared to the reference material.

$$\text{sp gr} = \frac{\text{density of a liquid or solid}}{\text{density of water}} \quad \text{or} \quad \frac{\text{density of a gas}}{\text{density of air}} \tag{6}$$

PROBLEM 2.22

What is the specific gravity of mercury with respect to water at 4°C? (Density of water at 4°C is 1.000 g/mL.)

$$\text{sp gr} = \frac{\text{density of mercury}}{\text{density of water}} = \frac{13.55\ \text{g/mL}}{1.000\ \text{g/mL}}$$

$$\text{sp gr of mercury} = 13.55$$

The value for the specific gravity of mercury (13.55) tells us that, per unit volume, mercury is 13.55 times as heavy as water. Do you think that you could readily lift a liter of mercury?

hydrometer

A **hydrometer** consists of a weighted bulb at the end of a sealed, calibrated tube. This instrument is used to measure the specific gravity of a liquid (see Figure 2.9). When a hydrometer is floated in a liquid, the specific gravity is indicated on the scale at the surface of the liquid.

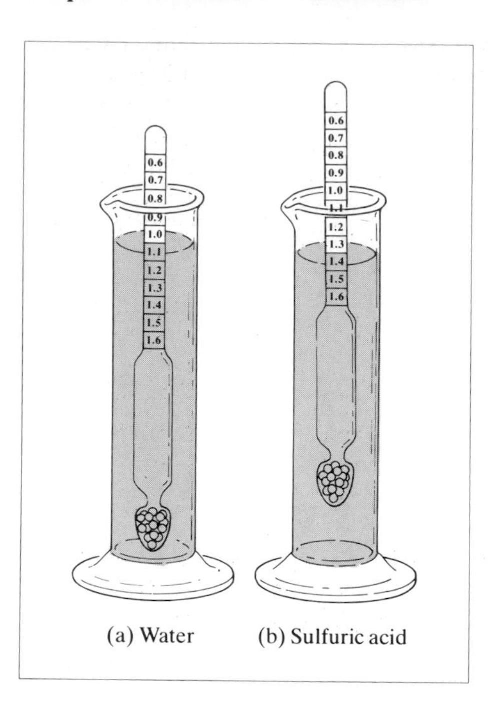

Figure 2.9
Specific gravity determination using hydrometers. The hydrometer in (a) is floating in water, showing a specific gravity of 1.0. The hydrometer in (b) is floating in dilute sulfuric acid (battery acid), showing a specific gravity of 1.3.

QUESTIONS

An asterisk indicates a more challenging question or problem.

A. *Review the meanings of the new terms introduced in this chapter.*

The terms listed in Section A of each set of Questions are new terms defined in the chapter. They appear in boldface type and occur in the chapter in the order listed in Question A.

1. Mass
2. Weight
3. Significant figures
4. Rounding off numbers
5. Scientific notation
6. Metric system, or SI
7. Meter
8. Kilogram
9. Volume
10. Heat
11. Temperature
12. Joule
13. A calorie
14. Specific heat
15. Density
16. Specific gravity
17. Hydrometer

B. *Information useful in answering these questions will be found in the tables and figures.*

1. Use Table 2.3 to determine how many decimeters make up 1 km.
2. What is the temperature difference in Fahrenheit degrees between 25°C and 100°C? (See Figure 2.4.)
3. Which metal will become warmer when 10 joules of energy are absorbed by one gram each of aluminum and iron? (See Table 2.5.)
4. Why do you suppose the neck of a 100 mL volumetric flask is narrower than the top of a 100 mL graduated cylinder? (See Figure 2.6.)
5. Refer to Table 2.6 and describe the arrangement that would be seen if these three immiscible substances were placed in a 100 mL graduated

cylinder: 25 mL mercury, 25 mL carbon tetrachloride, and a cube of magnesium measuring 2.0 cm on an edge.

6. Arrange these materials in order of increasing density: salt, cottonseed oil, sulfur, aluminum, and ethyl alcohol.
7. Will an argon-filled balloon rise or sink in a methane atmosphere? Explain.
8. Ice floats in cottonseed oil and sinks in ethyl alcohol. The density of ice must therefore lie between what numerical values?

C. *Review questions*

1. What are some of the important advantages of the metric system over the American system of weights and measurements?
2. What are the abbreviations for the following?
   (a) Gram (g) Micrometer
   (b) Kilogram (h) Angstrom
   (c) Milligram (i) Milliliter
   (d) Microgram (j) Microliter
   (e) Centimeter (k) Liter
   (f) Millimeter
3. For the following numbers tell whether the zeros are significant, are not significant, or may or may not be significant.
   (a) 503 (c) 4200 (e) 100.00
   (b) 0.007 (d) 3.0030 (f) 8.00×10^2
4. State the rules for rounding off numbers.
5. Distinguish between heat and temperature.
6. Will water or ethyl alcohol become hotter when 100 joules of energy are added to 25 g samples of these substances initially at 20°C?
7. Distinguish between density and specific gravity.
8. Which of the following statements are correct?
   (a) The prefix *micro* indicates one-millionth of the unit expressed.
   (b) The quantity 10 cm is equal to 1000 mm.
   (c) The number 383.263 rounded to four significant figures becomes 383.3.
   (d) The number of significant figures in the number 29,004 is five.
   (e) The number 0.00723 contains three significant figures.
   (f) The sum of 32.276 + 2.134 should contain four significant figures.
   (g) The product of 18.42 cm × 3.40 cm should contain three significant figures.
   (h) One microsecond is 10^{-6} second.
   (i) One thousand meters is a longer distance than 1000 yards.
   (j) One liter is a larger volume than one quart.
   (k) One centimeter is longer than one inch.
   (l) One cubic centimeter (cm^3) is equal to one milliliter.
   (m) The number 0.0002983 in exponential notation is 2.983×10^{-3}.
   (n) $3.0 \times 10^4 \times 6.0 \times 10^6 = 1.8 \times 10^{11}$
   (o) The direction of heat flow is from hot to cold.
   (p) A joule is a unit of temperature.
   (q) Temperature is a form of energy.
   (r) The density of water at 4°C is 1.00 g/mL.
   (s) A pipet is a more accurate instrument for measuring 10.0 mL of water than is a graduated cylinder.
   (t) A hydrometer is an instrument used to measure the specific heat of liquids.

D. *Review problems*

Significant Figures, Rounding, Exponential Notation, and Mathematical Review

1. How many significant figures are in each of the following numbers?
   (a) 0.025 (d) 0.0081 (g) 5.50×10^3
   (b) 40.0 (e) 0.0404 (h) 4.090×10^{-3}
   (c) 22.4 (f) 129,042
2. Round off the following numbers to three significant figures:
   (a) 93.246 (e) 4.644
   (b) 8.8726 (f) 129.509
   (c) 0.02854 (g) 34.250
   (d) 21.25 (h) 1.995×10^6
3. Express each of the following numbers in exponential notation:
   (a) 2,900,000 (d) 4082.2 (g) 12,000,000
   (b) 0.0456 (e) 0.00840 (h) 0.0000055
   (c) 0.58 (f) 40.30
4. Solve the following mathematical problems, stating answers to the proper number of significant figures:
   (a) $12.62 + 1.5 + 0.25 =$
   (b) $4.68 \times 12.5 =$
   (c) $2.25 \times 10^3 \times 4.80 \times 10^4 =$
   (d) $\dfrac{182.6}{4.6} =$
   (e) $\dfrac{452 \times 6.2}{14.3} =$

(f) $1986 + 23.48 + 0.012 =$
(g) $0.0394 \times 12.8 =$
(h) $2.92 \times 10^{-3} \times 6.14 \times 10^{5} =$
(i) $\dfrac{0.4278}{59.6} =$
(j) $\dfrac{29.3}{284 \times 415} =$

5. Change these fractions to decimals. Express each answer to three significant figures.
(a) $\frac{5}{6}$ (b) $\frac{3}{7}$ (c) $\frac{12}{16}$ (d) $\frac{9}{18}$

6. Solve each equation for X:
(a) $3.42X = 6.5$ (d) $0.298X = 15.3$
(b) $\dfrac{X}{12.3} = 7.05$ (e) $\dfrac{X}{0.819} = 10.9$
(c) $\dfrac{0.525}{X} = 0.25$ (f) $\dfrac{8.4}{X} = 282$

7. Solve each equation for the unknown:
(a) $°C = \dfrac{212 - 32}{1.8}$ (c) $K = 25 + 293$
(b) $°F = 1.8(22) + 32$ (d) $\dfrac{8.9\ g}{mL} = \dfrac{40.90\ g}{volume}$

Unit Conversions

8. Make the following conversions, showing mathematical setups:
(a) 28.0 cm to m
(b) 1000. m to km
(c) 9.28 cm to mm
(d) 150 mm to km
(e) 0.606 cm to km
(f) 4.5 cm to Å
(g) 6.5×10^{-7} m to Å
(h) 12.1 m to cm
(i) 8.0 km to m
(j) 315 mm to cm
(k) 25 km to mm
(l) 12 nm to cm
(m) 0.520 km to cm
(n) 3.884 Å to nm
(o) 42.2 in. to cm
(p) 0.64 mile to in.
(q) 504 miles to km
(r) 2.00 in.^{2} to cm^2
(s) 35.6 m to ft
(t) 16.5 km to miles
(u) 4.5 in.^{3} to mm^3
(v) 3.00 mile^3 to mm^3

9. Make the following conversions, showing mathematical setups:
(a) 10.68 g to mg
(b) 6.8×10^4 mg to kg
(c) 8.54 g to kg
(d) 42.8 kg to lb
(e) 164 mg to g
(f) 0.65 kg to mg
(g) 5.5 kg to g
(h) 95 lb to g

10. Make the following conversions, showing mathematical setups:
(a) 25.0 mL to L
(b) 22.4 L to mL
(c) 3.5 qt to mL
(d) 4.5×10^4 ft^3 to m^3
(e) 0.468 L to mL
(f) 35.6 L to gal
(g) 9.0 μL to mL
(h) 20.0 gal to L

11. An automobile traveling at 55 miles per hour is moving at what speed in (a) kilometers per hour and (b) feet per second?

12. A sprinter in the Olympic Games ran the 100 m dash in 9.82 s. What was his speed in (a) feet per second and (b) miles per hour?

13. A lab experiment requires each student to use 6.55 g of sodium chloride. The instructor opens a new 1.00 lb jar of the salt. If 24 students each take exactly the assigned amount of salt, how much should be left in the bottle at the end of the lab period?

14. When the space satellite Voyager I, which gave us new data on the planet Saturn, reaches the planet Neptune in 1989 it will be traveling at an average speed of 13 miles per second. What will be its speed in (a) miles per hour and (b) kilometers per hour?

15. How many kilograms does a 170 lb man weigh?

16. The usual aspirin tablet contains 5.0 grains of aspirin. How many grams of aspirin are in one tablet (1 grain = $\frac{1}{7000}$ lb)?

17. The sun is approximately 93 million miles from the earth. How many seconds will it take light to travel from the sun to the earth if the velocity of light is 3.00×10^{10} cm per second?

18. The average mass of the heart of a human baby is about 1 ounce. What is the mass in milligrams?

19. How much would 1.0 kg of potatoes cost if the price is $1.78 for 10 pounds?

20. The price of gold varies greatly and has been as high as $800 per ounce. What is the value of 227 g of gold at $345 per ounce? Gold is priced per troy ounce [1 lb (avoirdupois) = 14.58 oz (troy)].

21. An adult ruby-throated hummingbird has an average weight of 3.2 g, whereas an adult California condor may attain a weight of 21 lb. How many times heavier than the hummingbird is the condor?

22. At 35 cents per liter how much will it cost to fill a 15.8 gal tank with gasoline?

23. How many liters of gasoline will be used to drive 500 miles in a car that averages 34 miles per gallon?

24. Calculate the volume, in liters, of a box 75 cm long by 55 cm wide by 55 cm high.

25. Assuming that there are 20 drops in 1.0 mL, how many drops are in one gallon?

26. How many liters of oil are in a 42 gallon barrel of oil?
27. How many milliliters will be delivered by a filled 5.0 μL syringe?
*28. Calculate the number of milliliters of water in 1.00 cubic foot of water.
*29. Oil spreads in a thin layer on water and is called an "oil slick." How much area in square meters (m^2) will 200 cm^3 of oil cover if it forms a layer 5 Å in thickness?

Temperature Conversions and Specific Heat

30. Make the following conversions, showing mathematical setups:
    (a) 162°F to °C
    (b) 0.0°F to °C
    (c) 0.0°F to K
    (d) −18°C to °F
    (e) 32°C to °F
    (f) −8.6°F to °C
    (g) 273°C to K
    (h) 212 K to °C
31. Normal body temperature for humans is 98.6°F. What is this temperature on the Celsius scale?
32. Which is colder, −100°C or −138°F?
*33. (a) At what temperature are the Fahrenheit and Celsius temperatures exactly equal?
    (b) At what temperature are they numerically equal but opposite in sign?
34. How many joules of heat are required to raise the temperature of 80 g of water from 20°C to 70°C?
35. How many joules are required to raise the temperature of 80 g of iron from 20°C to 70°C? How many calories?
36. A 250. g metal bar requires 5.866 kJ to change its temperature from 22°C to 100°C. What is the specific heat of the metal in J/g°C?
*37. A 20.0 g piece of a metal at 203°C is dropped into 100. g of water at 25.0°C. The water temperature rises to 29.0°C. Calculate the specific heat (J/g°C) of the metal. Assume that all the heat lost by the metal is transferred to the water and no heat is lost to the surroundings.
*38. Assuming no heat losses by the system, what will be the final temperature when 50 g of water at 10°C are mixed with 10 g of water at 50°C?
*39. A 325 g piece of gold at 427°C is dropped into 200 mL of water at 22.0°C. The specific heat of gold is 0.131 J/g°C. Calculate the final temperature of the mixture. (Assume no heat losses to the surroundings.)

Density and Specific Gravity

40. Calculate the density of a liquid if 50.00 mL of the liquid weigh 78.26 g.
41. A 12.8 mL sample of bromine weighs 39.9 g. What is the density of bromine?
42. When a 32.7 g piece of chromium metal was placed into a graduated cylinder containing 25.0 mL of water, the water level rose to 29.6 mL. Calculate the density of the chromium.
43. Concentrated hydrochloric acid has a density of 1.19 g/mL. Calculate the weight, in grams, of 500. mL of this acid.
44. An empty graduated cylinder weighs 42.817 g. When filled with 50.0 mL of an unknown liquid it weighs 106.773 g. What is the density of the liquid?
45. What weight of mercury (density 13.6 g/mL) will occupy a volume of 25.0 mL?
46. A 35.0 mL sample of ethyl alcohol (density 0.789 g/mL) is added to a graduated cylinder that weighs 49.28 g. What will be the weight of the cylinder plus the alcohol?
47. You are given three cubes, A, B, and C; one is magnesium, one is aluminum, and the third is silver. All three cubes weigh the same, but cube A has a volume of 25.9 mL, cube B has a volume of 16.7 mL, and cube C has a volume of 4.29 mL. Identify cubes A, B, and C.
*48. A cube of aluminum weighs 500 g. What will be the weight of a cube of gold of the same dimensions?
49. A 25.0 mL sample of water of 90°C weighs 24.12 g. Calculate the density of water at this temperature.
50. Calculate (a) the density and (b) the specific gravity of a solid that weighs 136 g and has a volume of 50.0 mL.
51. The mass of an empty container is 88.25 g. The mass of the container when filled with a liquid (d = 1.25 g/mL) is 150.50 g. What is the volume of the container?
52. Which liquid will occupy the greater volume, 50 g of water or 50 g of ethyl alcohol? Explain.
53. A gold bullion dealer advertised a bar of pure gold for sale. The gold bar weighed 3300 g and measured 2.00 cm by 15.0 cm by 6.00 cm. Was the gold bar pure gold? Show evidence for your answer.

54. The largest nugget of gold on record was found in 1872 in New South Wales, Australia, and weighed 93.3 kg. Assuming the nugget is pure gold, what is its volume in cubic centimeters?

***55.** Forgetful Freddie placed 25.0 mL of a liquid in a graduated cylinder that weighed 89.450 g when empty. When Freddie placed a metal slug weighing 15.434 g into the cylinder, the volume rose to 30.7 mL. Freddie was asked to calculate the density of the liquid and of the metal slug from his data, but he forgot to obtain the weight of the liquid. He was told that if he weighed the cylinder containing the liquid and the slug, he would have enough data for the calculations. He did so and found it weighed 125.934 g. Calculate the density of the liquid and of the metal slug.

***56.** A solution made by adding 143 g of sulfuric acid to 500 mL of water had a volume of 554 mL.
(a) What value will a hydrometer read when placed in this solution?
(b) What volume of concentrated sulfuric acid ($d = 1.84$ g/mL) was added?

CHAPTER 3

Properties of Matter

After studying Chapter 3, you should be able to

1. Understand the new terms listed in Question A at the end of the chapter.
2. Identify the three physical states of matter and list the physical properties that characterize each state.
3. Distinguish between the physical and chemical properties of matter.
4. Classify the changes undergone by matter as either physical or chemical.
5. Distinguish between substances and mixtures.
6. Distinguish between kinetic and potential energy.
7. State the Law of Conservation of Mass.
8. State the Law of Conservation of Energy.
9. Explain why the laws dealing with the conservation of mass and energy may be combined into a single more accurate general statement.
10. Calculate the percent composition of compounds from the weights of the elements involved in a chemical reaction and vice versa.

3.1 Matter Defined

matter

The entire universe consists of matter and energy. Every day we come into contact with countless kinds of matter. Air, food, water, rocks, soil, glass, and this book are all different types of matter. Broadly defined, **matter** is *anything* that has mass and occupies space.

Matter may be quite invisible. If an apparently empty test tube is submerged mouth downward in a beaker of water, the water rises only slightly into the tube. The water cannot rise further because the tube is filled with invisible matter: air (see Figure 3.1).

To the eye matter appears to be continuous and unbroken. However, it is actually discontinuous and is composed of discrete, tiny particles called *atoms*. The particulate nature of matter will become evident when we study atomic structure and the properties of gases.

Figure 3.1

An apparently empty test tube is submerged, mouth downward, in water. Only a small volume of water rises into the tube, which is actually filled with air. This experiment proves that air, which is matter, occupies space.

3.2 Physical States of Matter

solid

Matter exists in three physical states: solid, liquid, and gas. A **solid** has a definite shape and volume, with particles that cohere rigidly to one another. The shape of a solid can be independent of its container. For example, a crystal of sulfur has the same shape and volume whether it is placed in a beaker or simply laid on a glass plate.

amorphous

Most commonly occurring solids, such as salt, sugar, quartz, and metals, are *crystalline*. Crystalline materials exist in regular, repeating, three-dimensional, geometric patterns. Solids such as plastics, glass, and gels, because they do not have any particular regular internal geometric pattern, are called **amorphous**

Figure 3.2

These three naturally occurring substances are examples of regular geometric formations that are characteristic of crystalline solids: (a) halite (salt), (b) quartz, and (c) gypsum.

solids. (*Amorphous* means without shape or form.) Figure 3.2 illustrates three crystalline solids: salt, quartz, and gypsum.

liquid

A **liquid** has a definite volume but not a definite shape, with particles that cohere firmly but not rigidly. Although the particles are held together by strong attractive forces and are in close contact with one another, they are able to move freely. Particle mobility gives a liquid fluidity and causes it to take the shape of the container in which it is stored. Figure 3.3 shows equal amounts of liquid in differently shaped containers.

gas

A **gas** has indefinite volume and no fixed shape, with particles that are moving independently of one another. Particles in the gaseous state have gained enough energy to overcome the attractive forces that held them together as liquids or solids. A gas presses continuously in all directions on the walls of any container. Because of this quality a gas completely fills a container. The particles

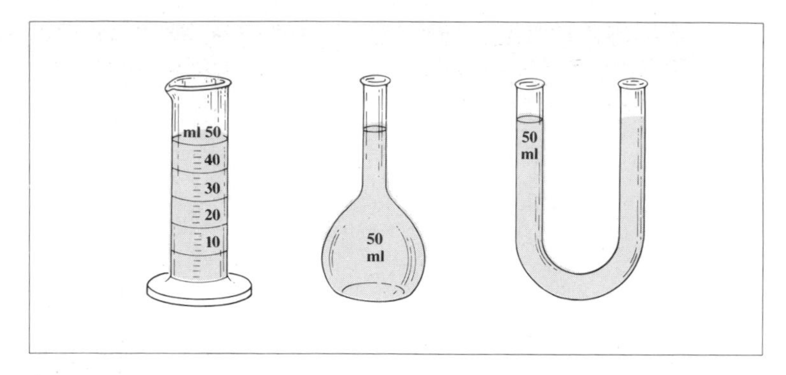

Figure 3.3

Liquids have the property of fluidity and assume the shape of their container, as illustrated in each of the three different calibrated containers.

of a gas are relatively far apart compared with those of solids and liquids. The actual volume of the gas particles is usually very small in comparison with the volume of the space occupied by the gas. A gas therefore may be compressed into a very small volume or expanded almost indefinitely. Liquids cannot be compressed to any great extent, and solids are even less compressible than liquids.

When a bottle of ammonia solution is opened in one corner of the laboratory, one can soon smell its familiar odor in all parts of the room. The ammonia gas escaping from the solution demonstrates that gaseous particles move freely and rapidly and tend to permeate the entire area into which they are released.

Although matter is discontinuous, attractive forces exist that hold the particles together and give matter its appearance of continuity. These attractive forces are strongest in solids, giving them rigidity; they are weaker in liquids but still strong enough to hold liquids to definite volumes. In gases the attractive

Table 3.1 Common materials in the solid, liquid, and gaseous states of matter

Solids	Liquids	Gases
Aluminum	Alcohol	Acetylene
Copper	Blood	Air
Gold	Gasoline	Butane
Polyethylene	Honey	Carbon dioxide
Salt	Mercury	Chlorine
Sand	Oil	Helium
Steel	Vinegar	Methane
Sulfur	Water	Oxygen

Table 3.2 Physical properties of solids, liquids, and gases

State	Shape	Volume	Particles	Compressibility
Solid	Definite	Definite	Rigidly cohering; tightly packed	Very slight
Liquid	Indefinite	Definite	Mobile; cohering	Slight
Gas	Indefinite	Indefinite	Independent of each other and relatively far apart	High

forces are so weak that the particles of a gas are practically independent of one another. Table 3.1 lists a number of common materials that exist as solids, liquids, and gases. Table 3.2 summarizes comparative properties of solids, liquids, and gases.

3.3 Substances and Mixtures

The term *matter* refers to all materials or material things that make up the universe. Many thousands of different and distinct kinds of matter or substances exist. A **substance** is a particular kind of matter with a definite, fixed composition. A substance, sometimes known as a *pure substance*, is either an element or a compound. Familiar examples of elements are copper, gold, and oxygen. Familiar compounds are salt, sugar, and water. Elements and compounds are discussed in more detail in Chapter 4.

substance

We can classify a sample of matter as either *homogeneous* or *heterogeneous* by examining it. **Homogeneous** matter is uniform in appearance and has the same properties throughout. Matter consisting of two or more physically distinct phases is **heterogeneous.** A **phase** is a homogeneous part of a system separated from other parts by physical boundaries. A system is simply the body of matter under consideration. Whenever we have a system in which visible boundaries exist between the parts or components, that system has more than one phase and is heterogeneous. It does not matter whether the components are in the solid, liquid, or gaseous states.

homogeneous

heterogeneous

phase

An important fact to keep in mind is that a pure substance, an element or compound, is always *homogeneous* in composition. However, a pure substance may exist as different phases in a heterogeneous system. Ice floating in water, for example, is a two-phase system made up of solid water and liquid water. The water in each phase is *homogeneous* in composition; but because two phases are present, the system is *heterogeneous.*

mixture

A **mixture** is a material containing two or more substances and can be either heterogeneous or homogeneous. Mixtures are variable in composition. If we add

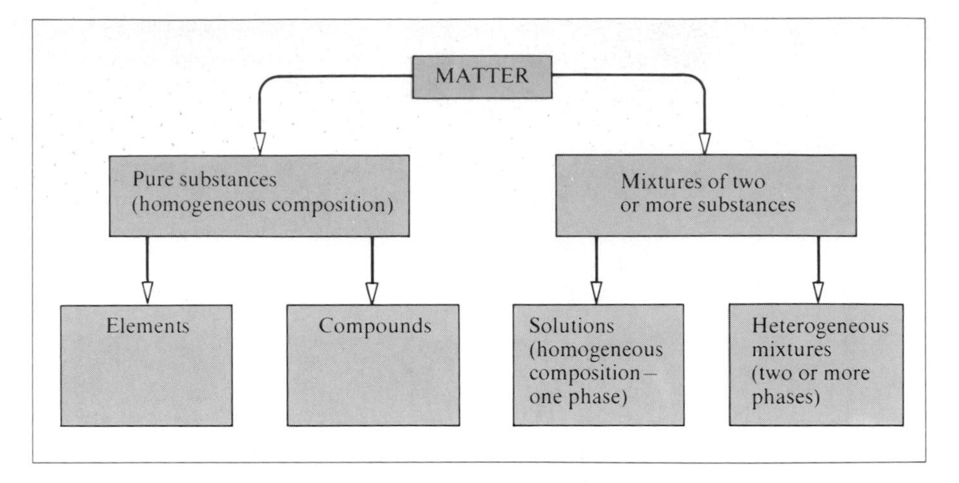

Figure 3.4

Classification of matter. A pure substance is always homogeneous in composition, whereas a mixture always contains two or more substances and may be either homogeneous (a solution) or heterogeneous.

a tablespoonful of sugar to a glass of water, a heterogeneous mixture is formed immediately. The two phases are a solid (sugar) and a liquid (water). But upon stirring the sugar dissolves to form a homogeneous mixture or solution. Both substances are still present: All parts of the solution are sweet and wet. The proportions of sugar and water can be varied simply by adding more sugar and stirring to dissolve.

Many substances do not form homogeneous mixtures. If we mix sugar and fine white sand, a heterogeneous mixture is formed. Careful examination may be needed to decide that the mixture is heterogeneous because the two phases (sugar and sand) are both white solids. Ordinary matter exists mostly as mixtures. If we examine soil, granite, iron ore, or other naturally occurring mineral deposits, we find them to be heterogeneous mixtures. Seawater is a homogeneous mixture (solution) containing many substances. Air is a homogeneous mixture (solution) of several gases. We shall consider mixtures further in Chapter 4. Figure 3.4 illustrates the relationships of substances and mixtures.

3.4 Properties of Substances

properties

How do we recognize substances? Each substance has a set of **properties** that is characteristic of that substance and gives it a unique identity. Properties are the personality traits of substances and are classified as either physical or chemical.

physical properties

Physical properties are the inherent characteristics of a substance that can be determined without altering its composition; they are associated with its physical existence. Common physical properties are color, taste, odor, state of

Table 3.3 Physical properties of chlorine, water, sugar, and acetic acid

Substance	Color	Odor	Taste	Physical state	Boiling point (°C)	Melting point (°C)
Chlorine	Yellowish-green	Sharp, suffocating	Sharp, sour	Gas	−34.6	−101.6
Water	Colorless	Odorless	Tasteless	Liquid	100.0	0.0
Sugar	White	Odorless	Sweet	Solid	Decomposes 170–186	—
Acetic acid	Colorless	Like vinegar	Sour	Liquid	118.0	16.7

chemical properties

matter (solid, liquid, or gas), density, melting point, and boiling point. **Chemical properties** describe the ability of a substance to form new substances, either by reaction with other substances or by decomposition.

We can select a few of the physical and chemical properties of chlorine as an example. Physically, chlorine is a gas about 2.4 times heavier than air. It is yellowish-green in color and has a disagreeable odor. Chemically, chlorine will not burn but will support the combustion of certain other substances. It can be used as a bleaching agent, as a disinfectant for water, and in many chlorinated substances such as refrigerants and insecticides. When chlorine combines with the metal sodium, it forms a salt called sodium chloride. These properties, among others, help to characterize and identify chlorine.

Substances, then, are recognized and differentiated by their properties. Table 3.3 lists four substances and tabulates several of their common physical properties. Information about common physical properties, such as that given in Table 3.3, is available in handbooks of chemistry and physics. Scientists do not pretend to know all the answers or to remember voluminous amounts of data, but it is important for them to know where to look for data in the literature. Handbooks are one of the most widely used resources for scientific data.*

No two substances will have identical physical and chemical properties.

3.5 Physical Changes

physical change

Matter can undergo two types of changes, physical and chemical. **Physical changes** are changes in physical properties (such as size, shape, and density) or changes in state of matter without an accompanying change in composition. The

* Robert C. Weast, ed., *Handbook of Chemistry and Physics*, 66th ed. (Cleveland: Chemical Rubber Company, 1985).

* Norbert A. Lange, comp., *Handbook of Chemistry*, 13th ed. (New York: McGraw-Hill, 1985).

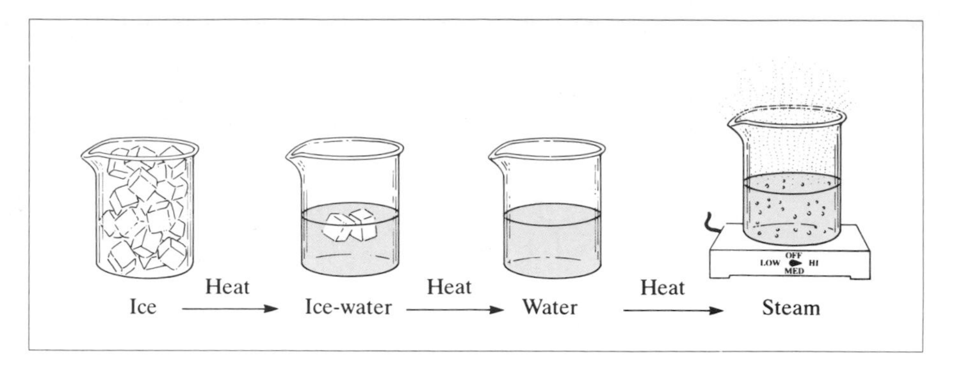

Figure 3.5

Physical changes in the appearance and state of water.

changing of ice into water and water into steam are physical changes from one state of matter into another. No new substances are formed in these physical changes (see Figure 3.5).

When a clean platinum wire is heated in a burner flame, the appearance of the platinum changes from silvery metallic to glowing red. This change is physical because the platinum can be restored to its original metallic appearance by cooling and, more importantly, because the composition of the platinum is not changed by heating and cooling.

3.6 Chemical Changes

chemical change

In a **chemical change**, new substances are formed that have different properties and composition from the original material. The new substances need not in any way resemble the initial material.

When a clean copper wire is heated in a burner flame, the appearance of the copper changes from coppery metallic to glowing red. Unlike the platinum previously mentioned, the copper is not restored to its original appearance by cooling but has become a black material. This black material is a new substance called copper(II) oxide. It was formed by chemical change when copper combined with oxygen in the air during the heating process. The unheated wire was essentially 100% copper, but the copper(II) oxide is 79.9% copper and 20.1% oxygen. One gram of copper will yield 1.251 g of copper(II) oxide (see Figure 3.6). The platinum was changed only physically when heated, but the copper was changed both physically and chemically when heated.

When 1.00 g of copper reacts with oxygen to yield 1.251 g of copper(II) oxide, the copper must have combined with 0.251 g of oxygen. The percentage of

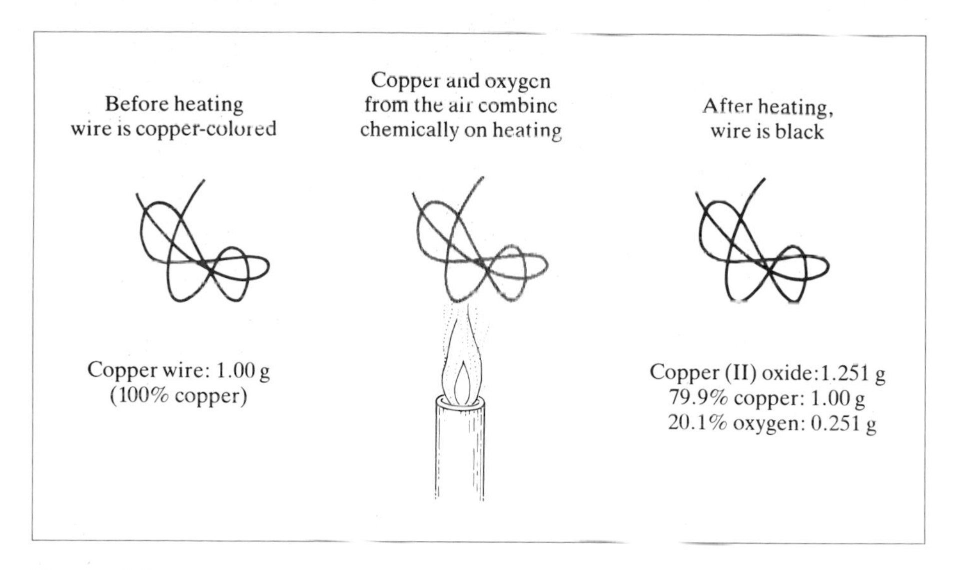
Before heating
wire is copper-colored
Copper and oxygen
from the air combine
chemically on heating
After heating,
wire is black
Copper wire: 1.00 g
(100% copper)
Copper (II) oxide:1.251 g
79.9% copper: 1.00 g
20.1% oxygen: 0.251 g

Figure 3.6
Chemical change: formation of copper(II) oxide from copper and oxygen.

copper and oxygen can be calculated from this data, the copper and oxygen each being a percent of the total weight of copper(II) oxide.

$$1.00 \text{ g copper} + 0.251 \text{ g oxygen} \longrightarrow 1.251 \text{ g copper(II) oxide}$$

$$\frac{1.00 \text{ g copper}}{1.251 \text{ g copper(II) oxide}} \times 100\% = 79.9\% \text{ copper}$$

$$\frac{0.251 \text{ g oxygen}}{1.251 \text{ g copper(II) oxide}} \times 100\% = 20.1\% \text{ oxygen}$$

Mercury(II) oxide is an orange-red powder that, when subjected to high temperature (500–600°C), decomposes into a colorless gas (oxygen) and a silvery, liquid metal (mercury). The composition and physical appearance of each product are noticeably different from those of the starting compound. When mercury(II) oxide is heated in a test tube (see Figure 3.7), small globules of mercury are observed collecting on the cooler part of the tube. Evidence that oxygen forms is observed when a glowing wood splint, lowered into the tube, bursts into flame. Oxygen supports and intensifies the combustion of the wood. From these observations we conclude that a chemical change has taken place.

Chemists have devised *chemical equations* as a shorthand method for expressing chemical changes. The two previous examples of chemical changes can be represented by the following word equations:

$$\text{copper} + \text{oxygen} \xrightarrow{\Delta} \text{copper(II) oxide} \quad \text{(cupric oxide)} \qquad (1)$$

$$\text{mercury(II) oxide} \xrightarrow{\Delta} \text{mercury} + \text{oxygen} \qquad (2)$$

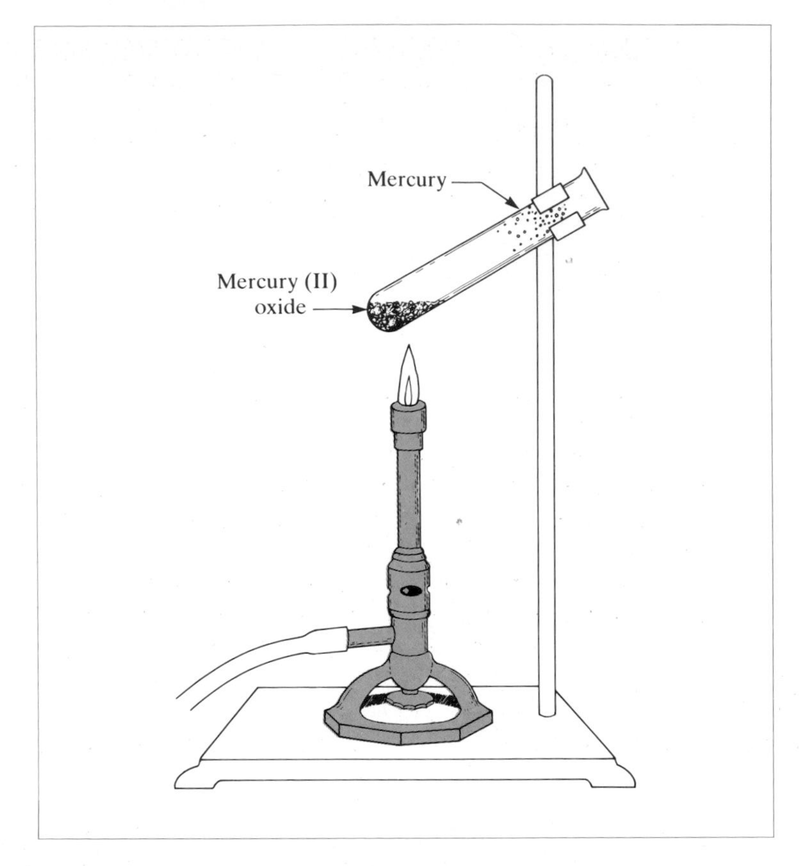

Figure 3.7

Heating mercury(II) oxide causes it to decompose into mercury and oxygen. Observation of the mercury and oxygen with properties different from those of mercury(II) oxide is evidence that a chemical change has occurred.

Equation (1) states: Copper plus oxygen when heated produce copper(II) oxide. Equation (2) states: Mercury(II) oxide when heated produces mercury plus oxygen. The arrow means "produces"; it points to the products. The Greek letter delta (Δ) represents heat. The starting substances (copper, oxygen, and mercury(II) oxide) are called the *reactants*, and the substances produced (copper(II) oxide, mercury, and oxygen) are called the *products*. In later chapters equations are presented in a still more abbreviated form, with symbols to represent substances.

Physical change usually accompanies a chemical change. Table 3.4 lists some common physical and chemical changes. In the examples given in the table, you will note that wherever a chemical change occurs, a physical change occurs also. However, wherever a physical change is listed, only a physical change occurs.

Table 3.4 Examples of processes involving physical or chemical changes

Process taking place	Type of change	Accompanying physical changes
Rusting of iron	Chemical	Shiny, bright metal changes to reddish-brown rust.
Boiling of water	Physical	Liquid changes to vapor.
Burning of sulfur in air	Chemical	Yellow solid sulfur changes to gaseous, choking sulfur dioxide.
Boiling an egg	Chemical	Liquid white and yolk change to solids.
Combustion of gasoline	Chemical	Liquid gasoline burns to gaseous carbon monoxide, carbon dioxide, and water.
Digesting food	Chemical	Food changes to liquid nutrients and partially solid wastes.
Sawing of wood	Physical	Smaller pieces of wood and sawdust are made from a larger piece of wood.
Burning of wood	Chemical	Wood burns to ashes, gaseous carbon dioxide, and water.
Heating of glass	Physical	Solid becomes pliable during heating, and the glass may change its shape.

3.7 Conservation of Mass

Law of Conservation of Mass

The **Law of Conservation of Mass** states that no detectable change is observed in the total mass of the substances involved in a chemical change. This law, tested by extensive laboratory experimentation, is the basis for the quantitative weight relationships among reactants and products.

The decomposition of mercury(II) oxide into mercury and oxygen illustrates this law. One hundred grams of mercury(II) oxide decomposes into 92.6 g of mercury and 7.39 g of oxygen.

$$\underset{100.\ \text{g}}{\text{mercury(II) oxide}} \longrightarrow \underset{92.6\ \text{g}}{\text{mercury}} + \underset{7.39\ \text{g}}{\text{oxygen}}$$

100. g Reactant	100. g Products

Operation of the ordinary photographic flashbulb also illustrates the law. Sealed within the bulb are fine wires of magnesium (a metal) and oxygen (a gas). When these reactants are energized, they combine chemically, producing

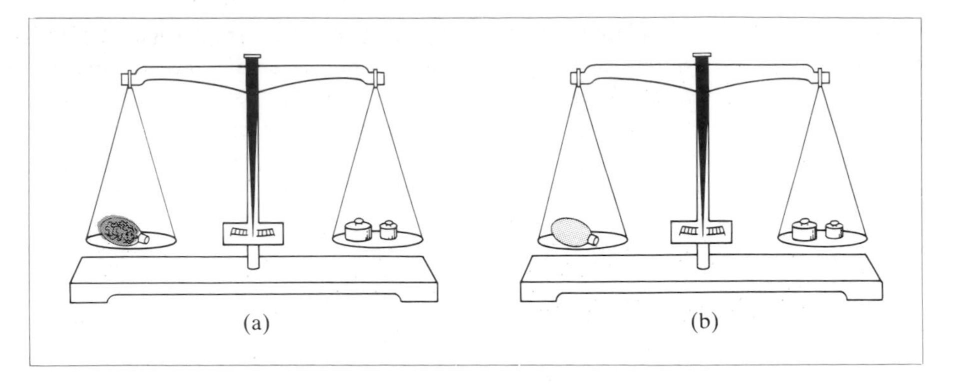

Figure 3.8

The flashbulb, containing magnesium and oxygen, weighs the same (a) before and (b) after the bulb is flashed. When the bulb is flashed, a chemical change occurs, and the original substances change into the white powder magnesium oxide.

magnesium oxide, a blinding white light, and considerable heat. The chemical change may be represented by this equation

$$\text{magnesium} + \text{oxygen} \longrightarrow \text{magnesium oxide} + \text{heat} + \text{light}$$

When weighed before and after the chemical change, as illustrated in Figure 3.8, the bulb shows no increase or decrease in weight.

mass of reactants = mass of products

3.8 Energy

energy

From the prehistoric discovery that fire could be used to warm shelters and cook food to the modern-day discovery that nuclear reactors can be used to produce vast amounts of controlled energy, man's technical progress has been directed by the ability to produce, harness, and utilize energy. **Energy** is the capacity of matter to do work. Energy exists in several forms; some of the more familiar forms are mechanical, chemical, electrical, heat, nuclear, and radiant or light energy. Matter can have both potential and kinetic energy.

potential energy

Potential energy is stored energy, or energy an object possesses due to its relative position. For example, a ball located 20 ft above the ground has more potential energy than when located 10 ft above the ground and will bounce higher when allowed to fall. Water backed up behind a dam represents potential energy that can be converted into useful work in the form of electrical or mechanical energy. Gasoline is a source of chemical potential energy. When gasoline burns (combines with oxygen), the heat released is associated with a decrease in potential energy. The new substances formed by burning have less chemical potential energy than the gasoline and oxygen did.

kinetic energy

Kinetic energy is the energy that matter possesses due to its motion. When the water behind the dam is released and allowed to flow, its potential energy is changed into kinetic energy, which can be used to drive generators and produce electricity. All moving bodies possess kinetic energy. The pressure exerted by a confined gas is due to the kinetic energy of rapidly moving gas particles. We all know the results when two moving vehicles collide: Their kinetic energy is expended in the crash that occurs.

Energy can be converted from one form to another form. Some kinds of energy can be converted to other forms easily and efficiently. For example, mechanical energy can be converted to electrical energy with an electric generator at better than 90% efficiency. On the other hand, solar energy has thus far been directly converted to electrical energy at an efficiency of only about 15%.

3.9 Energy in Chemical Changes

In all chemical changes matter either absorbs or releases energy. Chemical changes can produce different forms of energy. Electrical energy to start automobiles is produced by chemical changes in the lead storage battery. Light energy for photographic purposes occurs as a flash during the chemical change in the magnesium flashbulb. Heat and light energies are released from the combustion of fuels. All the energy needed for our life processes—breathing, muscle contraction, blood circulation, and so on—is produced by chemical changes occurring within the cells of our bodies.

Conversely, energy is used to cause chemical changes. For example, a chemical change occurs in the electroplating of metals when electrical energy is passed through a salt solution in which the metal is submerged. A chemical change also occurs when radiant energy from the sun is used by green plants in the process of photosynthesis. And, as we saw, a chemical change occurs when heat causes mercury(II) oxide to decompose into mercury and oxygen. Chemical changes are often used primarily to produce energy rather than to produce new substances. The heat or thrust generated by the combustion of fuels is more important than the new substances formed.

3.10 Conservation of Energy

An energy transformation occurs whenever a chemical change occurs. If energy is absorbed during the change, the products will have more chemical potential energy than the reactants. Conversely, if energy is given off in a chemical change, the products will have less chemical potential energy than the reactants. Water, for example, can be decomposed in an electrolytic cell. Electrical energy is absorbed in the decomposition, and the products, hydrogen and oxygen, have a

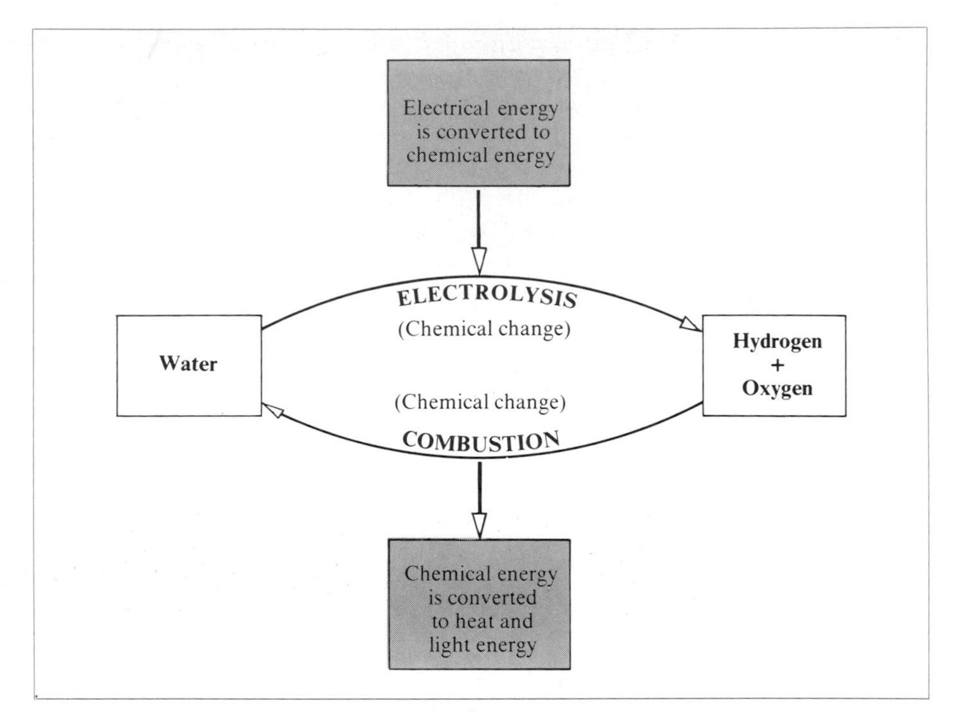

Figure 3.9

Energy transformations during the electrolysis of water and the combustion of hydrogen and oxygen. Electrical energy is converted to chemical energy in the electrolysis, and chemical energy is converted to heat and light energy in the combustion.

greater chemical potential energy level than that of water. This potential energy is released in the form of heat and light when the hydrogen and oxygen are burned to form water again (see Figure 3.9). Thus, energy can be changed from one form to another or from one substance to another and therefore is not lost.

The energy changes occurring in many systems have been thoroughly studied by many investigators. No system has been found to acquire energy except at the expense of energy possessed by another system. This principle is stated in other words as the **Law of Conservation of Energy**: Energy can be neither created nor destroyed though it can be transformed from one form to another.

Law of Conservation of Energy

3.11 Interchangeability of Matter and Energy

Sections 3.7–3.10 dealt with matter and energy, which are clearly related; any attempt to deal with one inevitably involves the other. The nature of this relationship eluded the most able scientists until the beginning of the 20th

Figure 3.10
Albert Einstein (1879–1955), world-renowned physicist and author of the theory of relativity and the interrelationship between matter and energy: $E = mc^2$. (*Courtesy of The Bettmann Archives.*)

century. Then, in 1905, Albert Einstein (Figure 3.10) presented one of the most original scientific concepts ever devised.

Einstein stated that the quantity of energy E equivalent to the mass m could be calculated by the equation $E = mc^2$. In this equation the energy units of E are ergs (1 erg $= 1 \times 10^{-7}$ J), m is in grams, and c is the velocity of light (3.0×10^{10} cm/s). According to Einstein's equation, whenever energy is absorbed or released by a substance, mass is lost or gained. Although the energy changes in chemical reactions are measurable and may appear to be large, the amounts are relatively small. The accompanying difference in mass between reactants and products in chemical changes is so small that it cannot be detected by available measuring instruments. According to Einstein's equation 9.2×10^7 J (9.2×10^{14} ergs) of energy are equivalent to 0.0000010 g (1.0 μg) of mass.

$$1\ \mu\text{g mass} = 9.2 \times 10^7\ \text{J energy}\ (9.2 \times 10^{14}\ \text{ergs})$$

In a more practical sense, when 2.8×10^3 g of carbon are burned to carbon dioxide, 9.2×10^7 J of energy are released. Of this very large amount of carbon, only about one-millionth of a gram, which is 3.6×10^{-8} % of the starting mass, is converted to energy. Therefore, in actual practice we may treat the reactants and products of chemical changes as having constant mass. However, because mass and energy are interchangeable, the two laws dealing with the conservation of matter may be combined into a single and generally more accurate statement:

The total amount of mass and energy remains constant during chemical change.

QUESTIONS

A. *Review the meanings of the new terms introduced in this chapter.*

1. Matter
2. Solid
3. Amorphous
4. Liquid
5. Gas
6. Substance
7. Homogeneous
8. Heterogeneous
9. Phase
10. Mixture
11. Properties
12. Physical properties
13. Chemical properties
14. Physical change
15. Chemical change
16. Law of Conservation of Mass
17. Energy
18. Potential energy
19. Kinetic energy
20. Law of Conservation of Energy

B. *Information useful in answering these questions will be found in the tables and figures.*

1. What evidence can you find in Figure 3.1 that gases occupy space?
2. Which three liquids listed in Table 3.1 are not mixtures?
3. Which of the gases listed in Table 3.1 is not a pure substance?
4. In what physical state does acetic acid exist at 10°C? (See Table 3.3.)
5. In what physical state does chlorine exist at 102 K? (See Table 3.3.)
6. What evidence of chemical change is visible when mercury(II) oxide is heated as shown in Figure 3.7?
7. What chemical changes occur to the matter in the flashbulb of Figure 3.8 when the bulb is flashed?
8. What physical changes occur during the electrolysis of water? (See Figure 3.9.)

C. *Review Questions*

1. List four different substances in each of the three states of matter.
2. In terms of the properties of the ultimate particles of a substance, explain
   (a) Why a solid has a definite shape but a liquid does not.
   (b) Why a liquid has a definite volume but a gas does not.
   (c) Why a gas can be compressed rather easily but a solid cannot be compressed appreciably.
3. When the stopper is removed from a partly filled bottle containing solid and liquid acetic acid at 16.7°C, a strong vinegarlike odor is noticeable immediately. How many acetic acid phases must be present in the bottle? Explain.
4. Is the system enclosed in the bottle of Question 3 homogeneous or heterogeneous? Explain.
5. Is a system that contains only one substance necessarily homogeneous? Explain.
6. Is a system that contains two or more substances necessarily heterogeneous? Explain.
7. Distinguish between physical and chemical properties.
8. What is the fundamental difference between a chemical change and a physical change?
9. Classify the following as being primarily physical or primarily chemical changes:
   (a) Formation of a snowflake
   (b) Freezing ice cream
   (c) Boiling water
   (d) Boiling an egg
   (e) Churning cream to make butter
   (f) Souring milk
10. Classify the following as being primarily physical or chemical changes:
    (a) Lighting a candle
    (b) Stirring cake batter
    (c) Dissolving sugar in water
    (d) Decomposition of limestone by heat
    (e) A leaf turning yellow
    (f) Gas escaping from a freshly opened bottle of soda pop
11. Cite the evidence that indicated that only physical changes occurred when a platinum wire was heated in a burner flame.
12. Cite the evidence that indicated that both physical and chemical changes occurred when a copper wire was heated in a burner flame.
13. Cite the evidence that heating mercury(II) oxide brings about a chemical change.
14. Identify the reactants and products for each of the following:
    (a) Heating a copper wire in a burner flame
    (b) Heating mercury(II) oxide as shown in Figure 3.7
    (c) Flashing the flashbulb shown in Figure 3.8

15. In a chemical change why can we consider that mass is neither lost nor gained (for practical purposes)?
16. Distinguish between potential and kinetic energy.
17. What happens to the kinetic energy of a speeding automobile when the automobile is braked to a stop?
18. What energy transformation is responsible for the fiery reentry of a returning space vehicle?
19. When the flashbulb of Figure 3.8 is flashed, energy is given off to the surroundings. Explain why the mass of the bulb appears to be the same after flashing as it was before, even though (according to Einstein) energy is equivalent to mass.
20. Which of the following statements are correct? (Try to answer this question without referring to the text.)
    (a) Liquids are the least compact state of matter.
    (b) Liquids have a definite volume and a definite shape.
    (c) Matter in the solid state is discontinuous; that is, it is made up of discrete particles.
    (d) Wood is homogeneous.
    (e) Wood is a substance.
    (f) Dirt is a mixture.
    (g) Seawater, although homogeneous, is a mixture.
    (h) Any system made up of only one substance is homogeneous.
    (i) Any system containing two or more substances is heterogeneous.
    (j) A solution, although it contains dissolved material, is homogeneous.
    (k) Boiling water represents a chemical change, because a change of state occurs.
    (l) All the following represent chemical changes: baking a cake, frying an egg, leaves changing color, iron changing to rust.
    (m) All of the following represent physical changes: breaking a stick, melting wax, folding a napkin, burning hydrogen to form water.
    (n) Chemical changes can produce electrical energy.
    (o) Electrical energy can produce chemical changes.
    (p) A stretched rubber band possesses kinetic energy.
    (q) An automobile rolling down a hill possesses both kinetic and potential energy.
    (r) When heated in the air, a platinum wire gains weight.
    (s) When heated in the air, a copper wire loses weight.
    (t) The two kinds of pure substances are elements and compounds.

D. *Review Problems*

1. Calculate the boiling point of acetic acid in (a) Kelvins and (b) degrees Fahrenheit. (See Table 3.3.)
2. What is the percentage of iron in a mixture that contains 15.0 g of iron, 16.0 g of sulfur, and 18.5 g of sand?
3. How many grams of copper(II) oxide can be obtained from 1.80 g of copper? (See Figure 3.6.)
4. How many grams of copper will combine with 5.50 g of oxygen to form copper(II) oxide?
5. How many grams of mercury can be obtained from 65.0 g of mercury(II) oxide?
6. If 40.0 g of a meat sample contains 3.40 g of fat, what percentage fat is present?
7. When 10.5 g of magnesium was heated in air, 17.4 g of magnesium oxide was produced. Given the chemical reaction

$$\text{magnesium} + \text{oxygen} \longrightarrow \text{magnesium oxide}$$

    (a) What weight of oxygen has combined with the magnesium?
    (b) What percentage of the magnesium oxide is magnesium?
8. When aluminum combines with chlorine gas, they produce the substance aluminum chloride. If 4.94 g of aluminum chloride is formed from 1.00 g of aluminum, how many grams of chlorine will combine with 5.50 g of aluminum?
9. If a U.S. 25-cent coin weighs about 5.5 g,
    (a) How many joules would be released by the complete conversion of a 25-cent coin to energy? How many calories?
    (b) The energy calculated in (a) could heat how many gallons of water from room temperature to the boiling point if 1.27×10^6 J are needed to heat a gallon of water from room temperature (20°C) to boiling (100°C)?

CHAPTER 4

Elements and Compounds

After studying Chapter 4, you should be able to

1. Understand the terms listed in Question A at the end of the chapter.
2. List in order of abundance the five most abundant elements in the earth's crust, seawater, and atmosphere.
3. List in order of abundance the six most abundant elements in the human body.
4. Classify common materials as elements, compounds, or mixtures.
5. Write the symbols when given the names or write the names when given the symbols of the common elements listed in Table 4.3.
6. State the Law of Definite Composition.
7. Understand how symbols, including subscripts and parentheses, are used to write chemical formulas.
8. Differentiate among atoms, molecules, and ions.
9. List the characteristics of metals and nonmetals.
10. Name binary compounds from their formulas.
11. Balance simple chemical equations when the formulas are given.
12. List the elements that occur as diatomic molecules.

4.1 Elements

All the words in the English dictionary are formed from an alphabet consisting of only 26 letters. All known substances on earth—and most probably in the universe, too—are formed from a sort of "chemical alphabet" consisting of 108 presently known elements. An **element** is a fundamental or elementary substance that cannot be broken down by chemical means to simpler substances. Elements are the building blocks of all substances. The elements are numbered in order of increasing complexity beginning with hydrogen, number 1. Of the first 92 elements, 88 are known to occur in nature. The other four—technetium (43), promethium (61), astatine (85), and francium (87)—either do not occur in nature or have only transitory existences during radioactive decay. With the exception of number 94, plutonium, elements above number 92 are not known to occur naturally but have been synthesized, usually in very small quantities, in laboratories. The discovery of trace amounts of element 94 (plutonium) in nature has been reported recently. The syntheses of elements 107 and 109 were reported in 1981 and 1982; element 108 has not been reported. No elements other than those on the earth have been detected on other bodies in the universe.

element

Most substances can be decomposed into two or more simpler substances. We have seen that mercury(II) oxide can be decomposed into mercury and oxygen and that water can be decomposed into hydrogen and oxygen. Sugar can be decomposed into carbon, hydrogen, and oxygen. Table salt is easily decomposed into sodium and chlorine. An element, however, cannot be decomposed into simpler substances by ordinary chemical changes.

If we could take a small piece of an element, say copper, and divide it and subdivide it into smaller and smaller particles, we finally would come to a single unit of copper that we could no longer divide and still have copper. This ultimate particle, the smallest particle of an element that can exist, is called an **atom**. An atom is also the smallest unit of an element that can enter into a chemical reaction. Atoms are made up of still smaller subatomic particles. But these subatomic particles (described in Chapter 5) do not have the properties of elements.

atom

4.2 Distribution of Elements

Elements are distributed very unequally in nature. At normal room temperature two of the elements, bromine and mercury, are liquids; eleven elements, hydrogen, nitrogen, oxygen, fluorine, chlorine, helium, neon, argon, krypton, xenon, and radon, are gases; all the other elements are solids.

Ten elements make up about 99% of the weight of the earth's crust, seawater, and atmosphere. Oxygen, the most abundant of these, constitutes about 50% of this mass. The distribution of the elements shown in Table 4.1 includes the earth's crust to a depth of about 10 miles, the oceans, fresh water, and the atmosphere but

Table 4.1 Distribution of the elements in the earth's crust, seawater, and atmosphere

Element	Weight percent	Element	Weight percent
Oxygen	49.20	Chlorine	0.19
Silicon	25.67	Phosphorus	0.11
Aluminum	7.50	Manganese	0.09
Iron	4.71	Carbon	0.08
Calcium	3.39	Sulfur	0.06
Sodium	2.63	Barium	0.04
Potassium	2.40	Nitrogen	0.03
Magnesium	1.93	Fluorine	0.03
Hydrogen	0.87		
Titanium	0.58	All others	0.47

Table 4.2 Average elemental composition of the human body

Element	Weight percent
Oxygen	65.0
Carbon	18.0
Hydrogen	10.0
Nitrogen	3.0
Calcium	2.0
Phosphorus	1.0
Traces of several other elements	1.0

does not include the mantle and core of the earth, which are believed to consist of metallic iron and nickel. Because the atmosphere contains relatively little matter, its inclusion has almost no effect on the distribution shown in Table 4.1. But the inclusion of fresh and salt water does have an appreciable effect since water contains about 11.1% hydrogen. Nearly all of the 0.87% hydrogen shown is from water.

The average distribution of the elements in the human body is shown in Table 4.2. Note again the high percentage of oxygen.

4.3 Names of the Elements

The names of the elements came to us from various sources. Many are derived from early Greek, Latin, or German words that generally described some property of the element. For example, iodine is taken from the Greek word *iodes*, meaning violetlike. Iodine, indeed, is violet in the vapor state. The name of the metal bismuth had its origin from the German words *weisse masse*, which means white mass. Miners called it *wismat*; it was later changed to *bismat*, and finally to bismuth. Some elements are named for the location of their discovery—for example, germanium, discovered in 1886 by Winkler, a German chemist. Others are named in commemoration of famous scientists, such as einsteinium and curium, named for Albert Einstein and Marie Curie, respectively.

4.4 Symbols of the Elements

We all recognize Mr., N.Y., and Ave. as abbreviations for mister, New York, and avenue. In like manner chemists have assigned an abbreviation to each element; these are called **symbols** of the elements. Fourteen of the elements have a single letter as their symbol, five have three-letter symbols, and the rest have two letters. A symbol stands for the element itself, for one atom of the element, and (as we shall see later) for a particular quantity of the element.

symbol

Rules governing symbols of elements are as follows:

1. Symbols are composed of one, two, or three letters.
2. If one letter is used, it is capitalized.
3. If two or three letters are used, the first is capitalized and the others are lowercase letters.

Examples: Sulfur S Barium Ba

The symbols and names of all the elements are given in the table on the inside back cover of this book. Table 4.3 lists the more commonly used symbols. If we examine this table carefully, we note that most of the symbols start with the same letter as the name of the element that is represented. A number of symbols, however, appear to have no connection with the names of the elements they represent (see Table 4.4). These symbols have been carried over from earlier names (usually in Latin) of the elements and are so firmly implanted in the literature that their use is continued today.

Special care must be used in writing symbols. Begin each with a capital letter and use a lowercase second letter if needed. For example, consider Co, the symbol for the element cobalt. If through error CO (capital C and capital O) is written, the two elements carbon and oxygen (the *formula* for carbon monoxide) are represented instead of the single element cobalt. Another example of the need for care in writing symbols is the symbol Ca for calcium versus Co for cobalt. The letters must be distinct or else the symbol for the element may be misinterpreted.

Knowledge of symbols is essential for writing chemical formulas and equations. You should begin to learn the symbols immediately because they will be used extensively in the remainder of this book and in any future chemistry courses you may take. One way to learn the symbols is to practice a few minutes a day by making side-by-side lists of names and symbols and then covering each list alternately and writing the corresponding name or symbol. Initially it is a

Table 4.3 Symbols of the most common elements

Element	Symbol	Element	Symbol	Element	Symbol
Aluminum	Al	Fluorine	F	Phosphorus	P
Antimony	Sb	Gold	Au	Platinum	Pt
Argon	Ar	Helium	He	Potassium	K
Arsenic	As	Hydrogen	H	Radium	Ra
Barium	Ba	Iodine	I	Silicon	Si
Bismuth	Bi	Iron	Fe	Silver	Ag
Boron	B	Lead	Pb	Sodium	Na
Bromine	Br	Lithium	Li	Strontium	Sr
Cadmium	Cd	Magnesium	Mg	Sulfur	S
Calcium	Ca	Manganese	Mn	Tin	Sn
Carbon	C	Mercury	Hg	Titanium	Ti
Chlorine	Cl	Neon	Ne	Tungsten	W
Chromium	Cr	Nickel	Ni	Uranium	U
Cobalt	Co	Nitrogen	N	Zinc	Zn
Copper	Cu	Oxygen	O		

Table 4.4 Symbols of the elements derived from early names. These symbols are in use today even though they do not correspond to the current name of the element

Present name	Symbol	Former name
Antimony	Sb	Stibium
Copper	Cu	Cuprum
Gold	Au	Aurum
Iron	Fe	Ferrum
Lead	Pb	Plumbum
Mercury	Hg	Hydrargyrum
Potassium	K	Kalium
Silver	Ag	Argentum
Sodium	Na	Natrium
Tin	Sn	Stannum
Tungsten	W	Wolfram

good plan to learn the symbols of the most common elements shown in Table 4.3.

The experiments of alchemists paved the way for the development of chemistry. Alchemists surrounded their work with mysticism, partly by devising a system of symbols known only to practitioners of alchemy (see Figure 4.1). The symbol ℞ (from the Latin *recipe*) is still used in medicine and was established during this time. In the early 1800s the Swedish chemist J. J. Berzelius (1779–1848) made a great contribution to chemistry by devising the present system of symbols using letters of the alphabet.

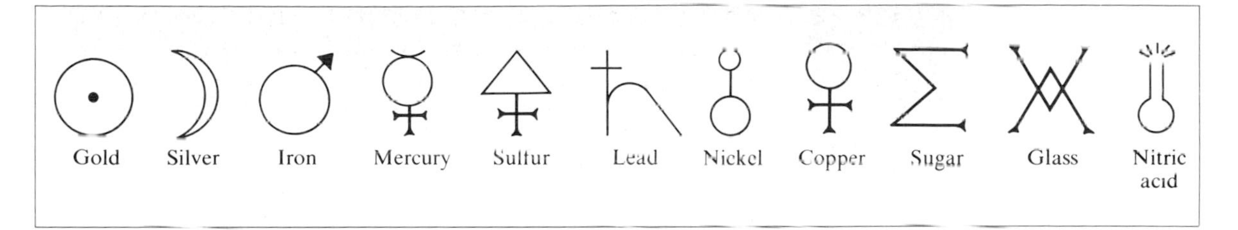

Figure 4.1
Some typical alchemists' symbols

4.5 Compounds

compound

A **compound** is a distinct substance containing two or more elements chemically combined in definite proportions by weight. Compounds, unlike elements, can be decomposed chemically into simpler substances—that is, into simpler compounds and/or elements. Atoms of the elements in a compound are combined in whole-number ratios, never as fractional parts of atoms. Compounds fall into two general types, *molecular* and *ionic*.

molecule

A **molecule** is the smallest uncharged individual unit of a compound formed by the union of two or more atoms. Water is a typical molecular compound. If we divide a drop of water into smaller and smaller particles, we finally obtain a single molecule of water consisting of two hydrogen atoms bonded to one oxygen atom. This molecule is the ultimate particle of water; it cannot be further subdivided without destroying the water and forming hydrogen and oxygen.

ion
cation
anion

An **ion** is a positively or negatively charged atom or group of atoms. An ionic compound is held together by attractive forces that exist between positively and negatively charged ions. A positively charged ion is called a **cation** (pronounced *cat-eye-on*); a negatively charged ion is called an **anion** (pronounced *an-eye-on*).

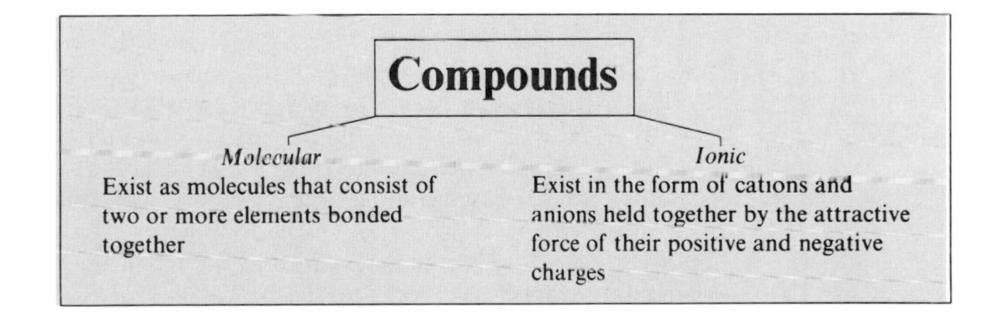

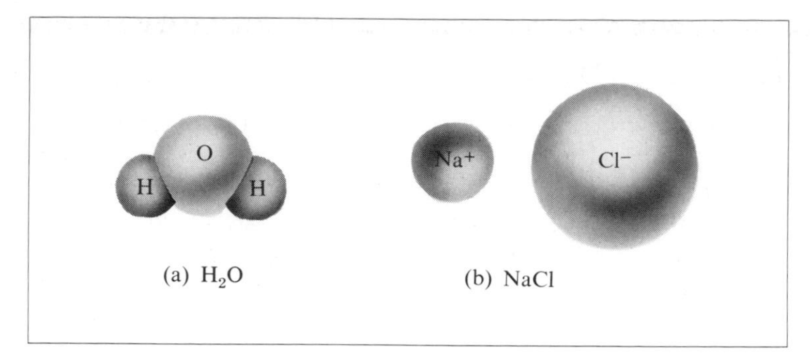

Figure 4.2

Representation of molecular and ionic (nonmolecular) compounds. (a) Two hydrogen atoms combined with an oxygen atom to form a molecule of water. (b) A positively charged sodium ion and a negatively charged chloride ion form the compound sodium chloride.

Sodium chloride is a typical ionic compound. The ultimate particles of sodium chloride are positively charged sodium ions and negatively charged chloride ions. Sodium chloride is held together in a crystalline structure by the attractive forces existing between these oppositely charged ions. Although ionic compounds consist of large aggregates of cations and anions, their formulas are normally represented by the simplest possible ratio of the atoms in the compound. For example in sodium chloride the ratio is one sodium ion to one chlorine ion, and the formula is NaCl (see Section 4.7). The two types of compounds, molecular and ionic, are illustrated in Figure 4.2.

There are more than 6 million known registered compounds, with no end in sight as to the number that will be prepared in the future. Each compound is unique and has characteristic physical and chemical properties. Let us consider two compounds, water and mercury(II) oxide in some detail. Water is a colorless, odorless, tasteless liquid that can be changed to a solid (ice) at 0°C and to a gas (steam) at 100°C. Composed of two atoms of hydrogen and one atom of oxygen per molecule, water is 11.2% hydrogen and 88.8% oxygen by weight. Water reacts chemically with sodium to produce hydrogen gas and sodium hydroxide, with lime to produce calcium hydroxide, and with sulfur trioxide to produce sulfuric acid. No other compound has all these exact physical and chemical properties; they are characteristic of water alone.

Mercury(II) oxide is a dense, orange-red powder having a ratio of one atom of mercury to one atom of oxygen. Its composition by weight is 92.6% mercury and 7.4% oxygen. When it is heated to temperatures greater than 360°C, a colorless gas, oxygen, and a silvery, liquid metal, mercury, are produced. These specific physical and chemical properties belong to mercury(II) oxide and to no other substance. Thus, a compound may be identified and distinguished from all other compounds by its characteristic properties.

4.6 Law of Definite Composition of Compounds

A large number of experiments extending over a long period of time have established the fact that a particular compound always contains the same elements in the same proportions by weight. For example, water will always contain 11.2% hydrogen and 88.8% oxygen by weight. The fact that water contains hydrogen and oxygen in this particular ratio does not mean that hydrogen and oxygen cannot combine in some other ratio. However, a compound with a different ratio would not be water. In fact, hydrogen peroxide is made up of two atoms of hydrogen and two atoms of oxygen per molecule and contains 5.9% hydrogen and 94.1% oxygen by weight; its properties are markedly different from those of water.

	Water	Hydrogen peroxide
Percent H	11.2	5.9
Percent O	88.8	94.1
Atomic composition	2 H + 1 O	2 H + 2 O

Law of Definite Composition

The **Law of Definite Composition** states: A compound always contains two or more elements combined in a definite proportion by weight. The reliability of this law, which in essence states that the composition of a substance will always be the same no matter what its origin or how it is formed, is the cornerstone of the science of chemistry.

4.7 Chemical Formulas

chemical formula

Chemical formulas are used as abbreviations for compounds. A **chemical formula** shows the symbols and the ratio of the atoms of the elements in a compound. Sodium chloride contains one atom of sodium per atom of chlorine; its formula is NaCl. The formula for water is H_2O; it shows that a molecule of water contains two atoms of hydrogen and one atom of oxygen.

The formula of a compound tells us which elements it is composed of and how many atoms of each element are present in a formula unit. For example, a molecule of sulfuric acid is composed of two atoms of hydrogen, one atom of sulfur, and four atoms of oxygen. We could express this compound as HHSOOOO, but the usual formula for writing sulfuric acid is H_2SO_4. The formula may be expressed verbally as "H-two-S-O-four." Numbers that appear partially below the line and to the right of a symbol of an element are called *subscripts.* Thus the 2 and the 4 in H_2SO_4 are subscripts. Characteristics of chemical formulas are

1 The formula of a compound contains the symbols of all the elements in the compound.
2 When the formula contains one atom of an element, the symbol of that element represents that one atom. The number one (1) is not used as a subscript to indicate one atom of an element.
3 When the formula contains more than one atom of an element, the number of atoms is indicated by a subscript written to the right of the symbol of that atom. For example, the two (2) in H_2O indicates two atoms of H in the formula.
4 When the formula contains more than one of a group of atoms that occurs as a unit, parentheses are placed around the group, and the number of units of the group are indicated by a subscript placed to the right of the parentheses. Consider the nitrate group, NO_3^-. The formula for sodium nitrate, $NaNO_3$, has only one nitrate group; therefore no parentheses are needed. Calcium nitrate, $Ca(NO_3)_2$, has two nitrate groups, as indicated by the use of parentheses and the subscript 2. $Ca(NO_3)_2$ has a total of nine atoms: one Ca, two N, and six O atoms. The formula $Ca(NO_3)_2$ is read as "C-A [pause] N-O-three-taken twice."
5 Formulas written as H_2O, H_2SO_4, $Ca(NO_3)_2$, and $C_{12}H_{22}O_{11}$ show only the number and kind of each atom contained in the compound; they do not show the arrangement of the atoms in the compound or how they are chemically bonded to one another.

Figure 4.3 illustrates how symbols and numbers are used in chemical formulas.

PROBLEM 4.1 Write formulas for the following compounds, the atom composition of which is given. (a) Hydrogen chloride: 1 atom hydrogen + 1 atom chlorine; (b) Methane: 1 atom carbon + 4 atoms hydrogen; (c) Glucose: 6 atoms carbon + 12 atoms hydrogen + 6 atoms oxygen.

(a) First write the symbols of the atoms in the formula: H Cl Since the ratio of atoms is one to one, we merely bring the symbols together to give the formula for hydrogen chloride as HCl.
(b) Write the symbols of the atoms: C H Now bring the symbols together and place a subscript 4 after the hydrogen atom. The formula is CH_4.
(c) Write the symbols of the atoms: C H O Now write the formula, bringing together the symbols followed by the correct subscripts according to the data given (six C, twelve H, six O). The formula is $C_6H_{12}O_6$.

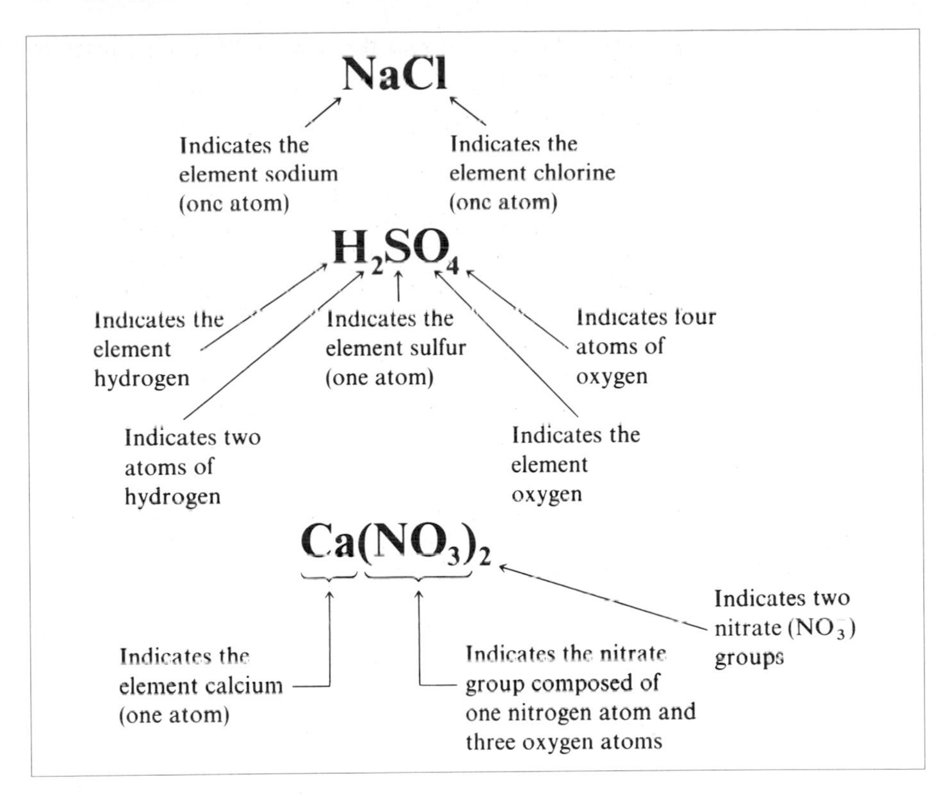

Figure 4.3
Explanation of the formulas NaCl, H_2SO_4, and $Ca(NO_3)_2$

4.8 Mixtures

Single substances—elements or compounds—seldom occur naturally in the pure state. Air is a mixture of gases; seawater is a mixture containing a variety of dissolved minerals; ordinary soil is a complex mixture of minerals and various organic materials.

How is a mixture distinguished from a pure substance? A mixture (see Section 3.3) always contains two or more substances that can be present in varying concentrations. Let us consider an example of a homogeneous mixture and an example of a heterogeneous mixture. Homogeneous mixtures (solutions) containing either 5% or 10% salt in water can be prepared simply by mixing the correct amounts of salt and water. These mixtures can be separated by boiling away the water, leaving the salt as a residue. The composition of a heterogeneous mixture of sulfur crystals and iron filings can be varied by merely blending in either more sulfur or more iron filings. This mixture can be separated physically by using a magnet to attract the iron or by adding carbon disulfide to dissolve the sulfur.

Iron(II) sulfide (FeS) contains 63.5% Fe and 36.5% S by weight. If we mix iron and sulfur in this proportion, do we have iron(II) sulfide? No, it is still a mixture; the iron is still attracted by a magnet. But if this mixture is heated strongly, a chemical change (reaction) occurs in which the reacting substances, iron and sulfur, form a new substance, iron(II) sulfide. Iron(II) sulfide, FeS, is a compound of iron and sulfur and has properties that are different from those of either iron or sulfur: It is neither attracted by a magnet nor dissolved by carbon disulfide. The differences between the iron and sulfur *mixture* and the iron(II) sulfide *compound* are as follows:

Mixture of iron and sulfur	Compound of iron and sulfur
Formula: mixture has no definite formula; consists of Fe and S.	Formula: FeS
Composition: mixture contains Fe and S in any proportion by weight.	Composition: compound contains 63.5% Fe and 36.5% S by weight.
Separation: Fe and S can be separated by physical means.	Separation: Fe and S can be separated only by chemical changes.

The general characteristics of mixtures and compounds are compared in Table 4.5.

Table 4.5 Comparison of mixtures and compounds

	Mixture	Compound
Composition	May be composed of elements, compounds, or both in variable composition	Composed of two or more elements in a definite, fixed proportion by weight
Separation of components	Separation may be made by physical or mechanical means.	Elements can be separated by chemical changes only.
Identification of components	Components do not lose their identity.	A compound does not resemble the elements from which it is formed.

4.9 Metals, Nonmetals, and Metalloids

metal
nonmetal
metalloid

Three primary classifications of the elements are **metals, nonmetals,** and **metalloids.** Most of the elements are metals. We are familiar with metals because of their widespread use in tools, materials of construction, automobiles, and so on. But nonmetals are equally useful in our everyday life as major components of such items as clothing, food, fuel, glass, plastics, and wood.

The metallic elements are solids at room temperature (mercury is an exception). They have high luster, are good conductors of heat and electricity, are *malleable* (can be rolled or hammered into sheets), and are *ductile* (can be drawn into wires). Most metals have a high melting point and high density. Familiar metals are aluminum, chromium, copper, gold, iron, lead, magnesium, mercury, nickel, platinum, silver, tin, and zinc. Less familiar but still important metals are calcium, cobalt, potassium, sodium, uranium, and titanium.

Metals have little tendency to combine with each other to form compounds. But many metals readily combine with nonmetals such as chlorine, oxygen, and sulfur to form mainly ionic compounds such as metallic chlorides, oxides, and sulfides. In nature the more active metals are found combined with other elements as minerals. A few of the less active ones such as copper, gold, and silver are sometimes found in a native, or free, state.

Nonmetals, unlike metals, are not lustrous, have relatively low melting points and densities, and are generally poor conductors of heat and electricity. Carbon, phosphorus, sulfur, selenium, and iodine are solids; bromine is a liquid; the rest of the nonmetals are gases. Common nonmetals found uncombined in nature are carbon (graphite and diamond), nitrogen, oxygen, sulfur, and the noble gases (helium, neon, argon, krypton, xenon, and radon).

Nonmetals combine with one another to form molecular compounds such as carbon dioxide (CO_2), methane (CH_4), butane (C_4H_{10}), and sulfur dioxide (SO_2). Fluorine, the most reactive nonmetal, combines readily with almost all the other elements.

Table 4.6 Classification of the elements into metals, metalloids, and nonmetals

Key
- Metals
- Metalloids
- Nonmetals

1 H																	2 He
3 Li	4 Be											5 B	6 C	7 N	8 O	9 F	10 Ne
11 Na	12 Mg											13 Al	14 Si	15 P	16 S	17 Cl	18 Ar
19 K	20 Ca	21 Sc	22 Ti	23 V	24 Cr	25 Mn	26 Fe	27 Co	28 Ni	29 Cu	30 Zn	31 Ga	32 Ge	33 As	34 Se	35 Br	36 Kr
37 Rb	38 Sr	39 Y	40 Zr	41 Nb	42 Mo	43 Tc	44 Ru	45 Rh	46 Pd	47 Ag	48 Cd	49 In	50 Sn	51 Sb	52 Te	53 I	54 Xe
55 Cs	56 Ba	57 La	72 Hf	73 Ta	74 W	75 Re	76 Os	77 Ir	78 Pt	79 Au	80 Hg	81 Tl	82 Pb	83 Bi	84 Po	85 At	86 Rn
87 Fr	88 Ra	89 Ac	104 Unq	105 Unp	106 Unh	107 —		109 —									

58 Ce	59 Pr	60 Nd	61 Pm	62 Sm	63 Eu	64 Gd	65 Tb	66 Dy	67 Ho	68 Er	69 Tm	70 Yb	71 Lu
90 Th	91 Pa	92 U	93 Np	94 Pu	95 Am	96 Cm	97 Bk	98 Cf	99 Es	100 Fm	101 Md	102 No	103 Lr

Several elements (boron, silicon, germanium, arsenic, antimony, tellurium, and polonium) are classified as *metalloids* and have properties that are intermediate between those of metals and those of nonmetals. The intermediate position of these elements is shown in Table 4.6, which lists and classifies all the elements as metals, nonmetals, or metalloids. Certain metalloids, such as boron, silicon, and germanium, are the raw materials for the semiconductor devices that make our modern electronics industry possible.

4.10 Elements that Exist as Diatomic Molecules

Seven of the elements (all nonmetals) occur as *diatomic molecules*. These elements and their symbols, formulas, and brief descriptions are listed in Table 4.7. Whether found free in nature or prepared in the laboratory, the molecules of these elements always contain two atoms. The formulas of the free elements are therefore always written to show this molecular composition: H_2, N_2, O_2, F_2, Cl_2, Br_2, and I_2.

It is important to see how symbols are used to designate either an atom or a molecule of an element. Consider hydrogen and oxygen. Hydrogen gas is present in volcanic gases and can be prepared by many chemical reactions. Regardless of their source, all samples of free hydrogen gas consist of diatomic molecules. Free hydrogen is designated and its composition is expressed by the formula H_2. Oxygen makes up about 21% by volume of the air that we breathe. This free oxygen is constantly being replenished by photosynthesis; it can also be prepared in the laboratory by several reactions. All free oxygen is diatomic and is designated by the formula O_2. Now consider water, a compound designated by the formula H_2O (sometimes HOH). Water contains neither free hydrogen (H_2) nor free oxygen (O_2). The H_2 part of the formula H_2O simply indicates that two atoms of hydrogen are combined with one atom of oxygen to form water. Thus symbols are used to designate elements, show the composition of molecules of elements, and give the elemental composition of compounds.

Table 4.7 Elements that exist as diatomic molecules

Element	Symbol	Molecular formula	Normal state
Hydrogen	H	H_2	Colorless gas
Nitrogen	N	N_2	Colorless gas
Oxygen	O	O_2	Colorless gas
Fluorine	F	F_2	Pale yellow gas
Chlorine	Cl	Cl_2	Yellow-green gas
Bromine	Br	Br_2	Reddish-brown liquid
Iodine	I	I_2	Bluish-black solid

4.11 Nomenclature and Chemical Equations

Knowledge of chemical names and of the writing and balancing of chemical equations is vital to the study of chemistry. This section serves only as an introduction to the naming of compounds and the writing of equations. More complete details of the systematic methods of naming inorganic compounds are given in Chapter 8, and a more detailed explanation of chemical equations is given in Chapter 10. Refer to these two chapters often, as needed. Neither chapter is intended to be studied only in the sequence given in the text; rather, they are common depositories of information on chemical nomenclature and equations.

Nomenclature

We have already used such names as hydrogen chloride (HCl), mercury(II) oxide (HgO), magnesium oxide (MgO), and sodium chloride (NaCl). Note that all four names end in *ide*. This *ide* ending is characteristic of the names of *binary* compounds—that is, compounds composed of atoms of two different elements. Some compounds contain several atoms of the same element (for example, CCl_4, carbon tetrachloride), but as long as only two different kinds of atoms are present, the compound is considered to be binary.

When a compound consisting of a metal and a nonmetal is named, the name of the metal is given first, followed by the name of the nonmetal, which is modified to end in *ide* (see Table 4.8).

Table 4.8 Names of selected nonmetals modified to end in *ide*, for use in naming binary compounds

Nonmetal	Modified name used in binary compounds
Oxygen	Oxide
Fluorine	Fluoride
Chlorine	Chloride
Bromine	Bromide
Iodine	Iodide
Sulfur	Sulfide
Nitrogen	Nitride

There are exceptions to the rule that all names ending in *ide* are for binary compounds. Some compounds containing more than two elements also have names that end in *ide* (for example, NH_4Cl, ammonium chloride; NaOH, sodium hydroxide). Refer to Section 8.7 for more details on naming binary compounds. Examples of binary compounds with names ending in *ide* are

$NaCl$	Sodium chloride	H_2S	Hydrogen sulfide
$BaBr_2$	Barium bromide	$AlBr_3$	Aluminum bromide
NaI	Sodium iodide	K_2S	Potassium sulfide
CaF_2	Calcium fluoride	Mg_3N_2	Magnesium nitride

Other names and formulas that you should become familiar with early in your study of chemistry are the commonly used acids and bases shown below. Methods used for naming these compounds are discussed in Chapter 8.

HCl	Hydrochloric acid	$NaOH$	Sodium hydroxide
H_2SO_4	Sulfuric acid	KOH	Potassium hydroxide
HNO_3	Nitric acid	$Ca(OH)_2$	Calcium hydroxide
$HC_2H_3O_2$	Acetic acid	NH_4OH	Ammonium hydroxide
H_3PO_4	Phosphoric acid		

PROBLEM 4.2 Name the binary compounds having the formulas NaCl and MgO.

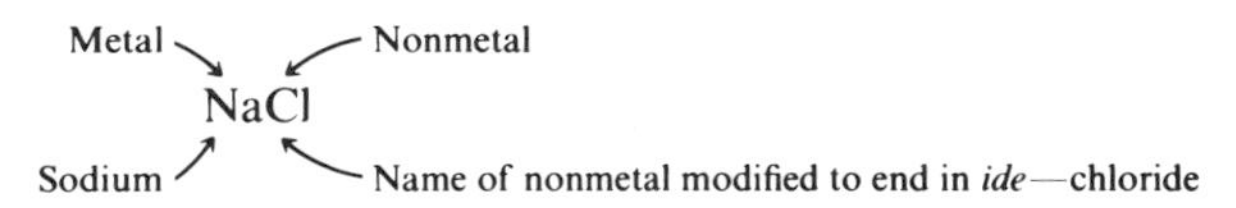

The name is sodium chloride.

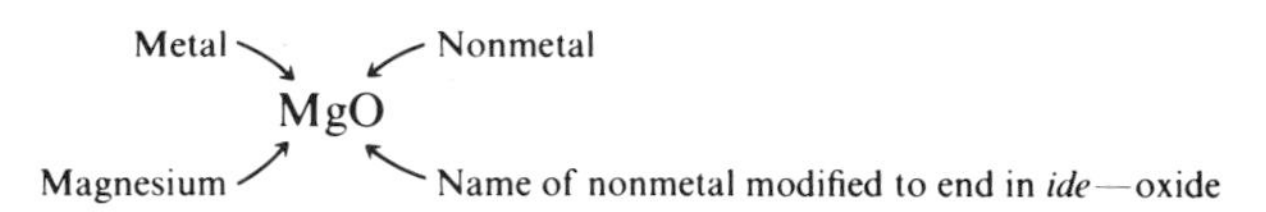

The name is magnesium oxide.

Chemical Equations

chemical equation

Chemical changes, or reactions, form substances with compositions that are different from those of the starting substances. A **chemical equation** is a shorthand expression for a chemical reaction. Substances in the reaction are represented by their symbols or formulas in the equation. The equation indicates both the reactants (starting substances) and the products and often shows the conditions necessary to facilitate the chemical change. The reactants are written on the left side and the products on the right side of the equation. An arrow (→) pointing to the products separates the reactants from the products. A plus sign (+) is used to separate one reactant (or product) from another.

Reactants ⟶ Products

We shall see in Chapter 5 that every atom has a specific mass. In an equation the symbols or formulas that represent a substance also represent a specific mass of that substance. Because no detectable change in mass results from a chemical change, the mass of the products must equal the mass of the reactants. When a chemical change is represented by an equation, this conservation of mass is attained by balancing the equation. After the correct formulas for the reactants and products are established, an equation is balanced by placing integral numbers (as needed) in front of the formulas of the substances in the equation. We use these numbers to obtain an equation that has the same number of atoms of each kind of element on each side of the equation.

A balanced chemical equation contains the same number of atoms of each kind of element on each side of the equation.

Consider, again, the reaction of metallic copper heated in air. The chemical change may be represented by the following equations:

$$\text{copper} + \text{oxygen} \xrightarrow{\Delta} \text{copper(II) oxide} \quad (1)$$

$$Cu + O_2 \longrightarrow CuO \quad \text{(unbalanced)} \quad (2)$$

Copper and oxygen are the reactants, and copper(II) oxide is the product. Equation (2) as written is not balanced because two oxygen atoms appear on the left side and only one on the right side. We place a 2 in front of Cu and a 2 in front of CuO to obtain the balanced equation:

$$2\,Cu + O_2 \xrightarrow{\Delta} 2\,CuO \quad \text{(balanced)} \quad (3)$$

This balanced equation contains 2 Cu atoms and 2 O atoms on each side of the equation.

A very important factor to remember when balancing equations is that you must not change the subscripts of a correct formula for the convenience of balancing the equation. In the unbalanced equation (2) we cannot change the formula of CuO to CuO_2 to balance the equation, even though by doing so we balance the number of atoms of each element on each side of the equation. The formula CuO_2 is not the correct formula for the true product. It is also important to be aware that a number in front of a formula multiplies every atom in that formula by that number. Thus,

2 CuO means 2 Cu atoms and 2 O atoms
$3\,H_2O$ means 6 H atoms and 3 O atoms
$4\,H_2SO_4$ means 8 H atoms, 4 S atoms, and 16 O atoms

Once a correct formula is written, it must not be changed during the balancing of an equation.

PROBLEM 4.3 How many atoms of each element are in each of these expressions?
(a) CH_4 (b) $2\ CH_4$ (c) $3\ Ca(OH)_2$

(a) CH_4

1 atom C ↗ ↖ 4 atoms H

Each molecule of CH_4 contains one C atom and four H atoms.

(b) $2\ CH_4$ means two molecules of CH_4. Since one molecule of CH_4 contains one C atom and four H atoms, two CH_4 molecules contain two C atoms and eight H atoms.

(c) $Ca(OH)_2$

1 atom Ca ↗ ↖ 2 units of OH

Each unit of OH contains one atom of O and one atom of H. Therefore $(OH)_2$ contains two O atoms and two H atoms. One formula unit of $Ca(OH)_2$ contains one Ca atom, two O atoms, and two H atoms; $3\ Ca(OH)_2$ indicates three formula units of $Ca(OH)_2$. Therefore $3\ Ca(OH)_2$ contains three Ca atoms, six O atoms, and six H atoms.

PROBLEM 4.4 When methane gas (CH_4) is burned in air, it reacts with oxygen to yield carbon dioxide and water. The equation for the reaction is

$$CH_4 + O_2 \longrightarrow CO_2 + H_2O \qquad \text{(unbalanced)}$$

Balance the equation for this reaction.

Inspecting each side of the equation we observe

Left side:	1 C atom	Right side:	1 C atom
	4 H atoms		2 H atoms
	2 O atoms		3 O atoms

The numbers of H and O atoms on each side of the equation are not equal, so the equation is unbalanced. To balance the equation, first place a 2 in front of the H_2O. This balances the H atoms, giving four H atoms on each side:

$$CH_4 + O_2 \longrightarrow CO_2 + 2\ H_2O \qquad \text{(unbalanced)}$$

The oxygen atoms are still unbalanced, with two O atoms on the left and four O atoms on the right. Place a 2 in front of the O_2, and the equation is balanced:

$$CH_4 + 2\ O_2 \longrightarrow CO_2 + 2\ H_2O \qquad \text{(balanced)}$$

The balanced equation contains one C, four H, and four O atoms on each side.

PROBLEM 4.5 Balance the following equation:

$$Al + Cl_2 \longrightarrow AlCl_3 \qquad \text{(unbalanced)}$$

In this equation the Cl atoms are not in balance (two Cl atoms on the left and three Cl atoms on the right side of the equation). Placing integral numbers in front of Cl_2 will always result in an even number of Cl atoms on the left side of the equation. Therefore the number in front of $AlCl_3$ will have to be an even number (2, 4, 6, ...). The number of Cl atoms needed on each side of the equation is 6, the lowest common multiple of 2 and 3. Placing a 3 in front of Cl_2 and a 2 in front of $AlCl_3$ will give 6 Cl atoms on each side of the equation:

$$Al + 3\ Cl_2 \longrightarrow 2\ AlCl_3 \quad \text{(unbalanced)}$$

By balancing the Cl atoms we have unbalanced the Al atoms. But we can correct this imbalance by placing a 2 in front of Al to give the balanced equation:

$$2\ Al + 3\ Cl_2 \longrightarrow 2\ AlCl_3 \quad \text{(balanced)}$$

Each side of the balanced equation has two Al and six Cl atoms.

QUESTIONS

A. *Review the meanings of the new terms introduced in this chapter.*

1. Element
2. Atom
3. Symbol
4. Compound
5. Molecule
6. Ion
7. Cation
8. Anion
9. Law of Definite Composition
10. Chemical formula
11. Metal
12. Nonmetal
13. Metalloid
14. Chemical equation

B. *Information useful in answering these questions will be found in the tables and figures.*

1. Of the ten most abundant elements in the earth's crust, seawater, and atmosphere, how many are metals? Nonmetals? Metalloids?
2. Of the six most abundant elements in the human body, how many are metals? Nonmetals? Metalloids?
3. Why is the symbol for gold Au rather than G or Go?
4. Give the names of (a) the solid diatomic nonmetal and (b) the liquid diatomic nonmetal.
5. What would be the modified names of the two elements of Question 4 when they are combined with metals in binary compounds?
6. Are there more atoms of silicon or hydrogen in the earth's crust, seawater, and atmosphere? Use Table 4.1 and the fact that the mass of the silicon atom is about 28 times that of a hydrogen atom.
7. How many metals are there? Nonmetals? Metalloids? (See Table 4.6.)

C. *Review Questions*

1. What does the symbol of an element stand for?
2. Write down what you believe to be the symbols for the elements phosphorus, aluminum, hydrogen, potassium, magnesium, sodium, nitrogen, nickel, silver, and plutonium. Now look up the correct symbols and rewrite them, comparing the two sets.
3. Interpret the difference in meanings for each of these pairs:
   (a) Si and SI (b) Pb and PB (c) 4 P and P_4
4. List six elements and their symbols in which the first letter of the symbol is different from that of the name.
5. Write the names and symbols for the fourteen elements that have only one letter as their symbol. (See table on inside back cover.)
6. Distinguish between an element and a compound.
7. Explain why the Law of Definite Composition does not pertain to mixtures.

8. What are the two general types of compounds? How do they differ from each other?
9. What is the basis for distinguishing one compound from another?
10. Given the following list of compounds and their formulas, what elements are present in each compound?
    (a) Potassium iodide KI
    (b) Sodium carbonate Na_2CO_3
    (c) Aluminum oxide Al_2O_3
    (d) Calcium bromide $CaBr_2$
    (e) Carbon tetrachloride CCl_4
    (f) Magnesium bromide $MgBr_2$
    (g) Nitric acid HNO_3
    (h) Barium sulfate $BaSO_4$
    (i) Aluminum phosphate $AlPO_4$
    (j) Acetic acid $HC_2H_3O_2$
11. Write the formula for each of the following compounds, the composition of which is given after each name:
    (a) Zinc oxide 1 atom Zn, 1 atom O
    (b) Potassium chlorate 1 atom K, 1 atom Cl, 3 atoms O
    (c) Sodium hydroxide 1 atom Na, 1 atom O, 1 atom H
    (d) Aluminum bromide 1 atom Al, 3 atoms Br
    (e) Calcium fluoride 1 atom Ca, 2 atoms F
    (f) Lead(II) chromate 1 atom Pb, 1 atom Cr, 4 atoms O
    (g) Ethyl alcohol 2 atoms C, 6 atoms H, 1 atom O
    (h) Benzene 6 atoms C, 6 atoms H
12. Explain the meaning of each symbol and number in the following formulas:
    (a) H_2O (d) $Ni(NO_3)_2$
    (b) $AlBr_3$ (e) $C_{12}H_{22}O_{11}$ (sucrose)
    (c) Na_2SO_4
13. How many atoms are represented in each of these formulas?
    (a) KF (d) $Ba(ClO_3)_2$
    (b) $CaCO_3$ (e) $K_2Cr_2O_7$
    (c) N_2 (f) $NaC_2H_3O_2$
    (g) CCl_2F_2 (Freon) (i) $(NH_4)_2C_2O_4$
    (h) $Al_2(SO_4)_3$
14. How many atoms are contained in (a) one molecule of hydrogen, (b) one molecule of water, and (c) one molecule of sulfuric acid?
15. What is the major difference between a cation and an anion?
16. Write the names and formulas of the elements that exist as diatomic molecules.
17. How many atoms of oxygen are represented in each expression?
    (a) $4\ H_2O$ (c) H_2O_2 (e) $Al(ClO_3)_3$
    (b) $3\ CuSO_4$ (d) $3\ Fe(OH)_3$
18. How many atoms of hydrogen are represented in each expression?
    (a) $5\ H_2$ (c) $2\ C_6H_{12}O_6$
    (b) $2\ Ba(C_2H_3O_2)_2$ (d) $2\ HC_2H_3O_2$
19. Distinguish between homogeneous and heterogeneous mixtures.
20. Classify each of the following materials as an element, compound, or mixture:
    (a) Air (e) Wine
    (b) Oxygen (f) Iodine
    (c) Sodium chloride (g) Sulfuric acid
    (d) Platinum (h) Crude oil
21. Classify each of the following materials as an element, compound, or mixture:
    (a) Paint (e) Sulfuric acid
    (b) Salt (f) Silver
    (c) Copper (g) Milk
    (d) Beer (h) Sodium hydroxide
22. A white solid, on heating, formed a colorless gas and a yellow solid. Assuming that there was no reaction with the air, is the original solid an element or a compound? Explain.
23. Tabulate the properties that characterize metals and nonmetals.
24. Which of the following are diatomic molecules?
    (a) H_2 (c) HCl (e) NO (g) $MgCl_2$
    (b) SO_2 (d) H_2O (f) NO_2
25. Name the following binary compounds. Refer to Chapter 8 if necessary.
    (a) $AgBr$ (e) NaF (h) BN
    (b) HI (f) K_2O (i) $BaCl_2$
    (c) $MgBr_2$ (g) $LiCl$ (j) Al_2S_3
    (d) CaS
    What is common to the names of these binary compounds?
26. An atom of silver is represented by the symbol Ag; a hydrogen molecule by the formula H_2; a

water molecule by H_2O. Write the expressions to represent:
(a) Five silver atoms
(b) Four hydrogen molecules
(c) Three water molecules

27. Balance these equations (all formulas are correct as written):
(a) $H_2 + Cl_2 \longrightarrow HCl$
(b) $Zn + CuSO_4 \longrightarrow Cu + ZnSO_4$
(c) $HCl + NaOH \longrightarrow NaCl + H_2O$
(d) $Ca + O_2 \longrightarrow CaO$
(e) $Fe + HCl \longrightarrow FeCl_2 + H_2$
(f) $P + I_2 \longrightarrow PI_3$
(g) $MgO + HCl \longrightarrow MgCl_2 + H_2O$
(h) $HNO_3 + Ba(OH)_2 \longrightarrow Ba(NO_3)_2 + H_2O$
(i) $BiCl_3 + H_2S \longrightarrow Bi_2S_3 + HCl$
(j) $Mg_3N_2 + H_2O \longrightarrow Mg(OH)_2 + NH_3$

28. Balance the following equations, each of which represents a method of preparing oxygen gas:
(a) $H_2O_2 \longrightarrow H_2O + O_2$
(b) $KClO_3 \xrightarrow{\Delta} KCl + O_2$
(c) $KNO_3 \xrightarrow{\Delta} KNO_2 + O_2$
(d) $Na_2O_2 + H_2O \longrightarrow NaOH + O_2$
(e) $H_2O \xrightarrow[H_2SO_4]{\text{Electrical energy}} H_2 + O_2$

29. Balance the following equations, each of which represents a method of preparing hydrogen gas:
(a) $Zn + HCl \longrightarrow ZnCl_2 + H_2$
(b) $Al + H_2SO_4 \longrightarrow Al_2(SO_4)_3 + H_2$
(c) $Na + H_2O \longrightarrow NaOH + H_2$
(d) $C + H_2O$ (steam) $\longrightarrow CO + H_2$
(e) $Fe + H_2O$ (steam) $\longrightarrow Fe_3O_4 + H_2$

30. Which of the following statements are correct?
(a) The smallest unit of an element that can exist and enter into a chemical reaction is called a molecule.
(b) The basic building blocks of all substances, which cannot be decomposed into simpler substances by ordinary chemical change, are compounds.
(c) The most abundant element in the earth's crust, seawater, and atmosphere by weight is oxygen.
(d) The most abundant element in the human body, by weight, is carbon.
(e) Most of the elements are represented by symbols consisting of one or two letters.
(f) The symbol for copper is Co.
(g) The symbol for sodium is Na.
(h) The symbol for potassium is P.
(i) The symbol for lead is Le.
(j) Early names for some elements led to unlikely symbols, such as Fe for iron.
(k) A compound is a distinct substance that contains two or more elements combined in a definite proportion by weight.
(l) The smallest uncharged individual unit of a compound formed by the union of two or more atoms is called a substance.
(m) An ion is a positive or negative electrically charged atom or group of atoms.
(n) The Law of Definite Composition states that a compound always contains two or more elements combined in a definite proportion by weight.
(o) A chemical formula is a shorthand expression for a chemical reaction.
(p) The formula Na_2CO_3 indicates a total of six atoms, including three oxygen atoms.
(q) A general property of nonmetals is that they are good conductors of heat and electricity.
(r) Metals have the properties of ductility and malleability.
(s) *Malleable* means that when struck a hard blow the substance will shatter.
(t) Elements that have properties intermediate between metals and nonmetals are called mixtures.
(u) More of the elements are metals than nonmetals.
(v) In a balanced chemical equation, the mass of the products is equal to the mass of the reactants.
(w) Bromine is an element that occurs as a diatomic molecule, Br_2.
(x) The binary compound CaS is called calcium sulfate.
(y) The substances on the left side of the arrow in a chemical equation are called reactants and those on the right side are called products.
(z) The equation $2\ C_2H_2 + 4\ O_2 \longrightarrow 4\ CO_2 + 2\ H_2O$ is balanced.

D. *Review Problems*

1. Common table salt, NaCl, contains 39.3% sodium and 60.7% chlorine. What weight of sodium is present in 35.0 g of salt?

2. Red brass is a homogeneous mixture of 90% copper and 10% zinc. If 40 g of zinc is added to 100 g of red brass, what will be the new composition?

3. Calcium oxide, CaO, contains 71.5% calcium. What size sample of CaO would contain 12.0 g of calcium?

***4.** What would be the density of a solution made by mixing 2.50 mL of carbon tetrachloride (CCl_4, $d = 1.595$ g/mL) and 3.50 mL of carbon tetrabromide (CBr_4, $d = 3.420$ g/mL)? Assume that the volume of the mixed liquids is the sum of the two volumes used.

***5.** Only two elements, Br_2 ($d = 3.12$ g/mL) and Hg ($d = 13.6$ g/mL), are liquids at room temperature. How many milliliters of Br_2 will have the same mass as 12.5 mL Hg?

***6.** When 12.0 g of calcium and 12.0 g of sulfur were mixed and reacted to give the compound calcium sulfide (CaS), 2.40 g of sulfur remained unreacted.
(a) What percentage of the compound is sulfur?
(b) An atom of which element, Ca or S, has the greater mass? Explain.
(c) How many grams of sulfur will combine with 30.0 g of calcium?

7. Pure gold is too soft a metal for many uses, so it is alloyed to give it more mechanical strength. One particular alloy is made by mixing 60 g of gold, 8.0 g of silver, and 12 g of copper. What carat gold is this alloy if pure gold is considered to be 24 carat?

***8.** Methane, the chief component of natural gas, has the formula CH_4. Each atom of carbon weighs 12 times as much as an atom of hydrogen. Calculate the weight percent of carbon in methane.

9. White gold is a homogeneous solution of 90% gold and 10% palladium. How much gold is present in a bar of white gold weighing 8420 g?

***10.** The metal used to make the U.S. nickel coin is an alloy of 75% copper and 25% nickel. What maximum weight of alloy could be produced if only 450 kg of nickel and 1180 kg of copper were on hand?

Review Exercises for Chapters 1–4

CHAPTER 1 INTRODUCTION

True–False. *Answer the following as either true or false.*

1. Chemistry is the science that deals with the composition of substances and the transformations they undergo.
2. Scientific laws are simple statements of natural phenomena to which no exceptions are known.
3. From 1803 to 1810, John Dalton advanced his atomic theory.
4. A key feature of the scientific method is to plan and do additional experiments to test a hypothesis.
5. An explanation of many observations that have been proven by many tests is called a hypothesis.
6. The early Greek philosophers believed that all matter was derived from four elements: earth, air, fire, and water.
7. One of the principal goals of the alchemists was to change metals such as iron into gold.
8. Oxygen was discovered by Robert Boyle in 1774.
9. The use of the chemical balance revolutionized quantitative measurements in chemical reactions.
10. The two main branches of chemistry are organic and inorganic chemistry.

CHAPTER 2 STANDARDS FOR MEASUREMENT

True–False. *Answer the following as either true or false.*

1. A milligram is 0.001 g.
2. A centimeter is longer than a millimeter.
3. If 1 mL = 0.001 liter, then we can use the factor 10^3 mL/liter to convert liters to milliliters.
4. The density of water at 4°C is 1.00 g/mL.
5. As a metric prefix, *kilo* means 1000, or 10^3.
6. One milliliter equals 1 cm^3 exactly.
7. The joule is a unit of temperature.
8. The measurement 12.200 g contains three significant figures.
9. The answer to 25.2×0.1465 should contain three significant figures.
10. The number 14.0667 rounded off to four digits is 14.07.
11. The answer to $16.215 - 2.32$ should contain three digits.
12. A liter contains 100 mL.
13. 90°C is hotter than 210°F.
14. The units of specific gravity are g/mL.
15. The mass of an object is fixed and independent of its location.
16. The prefix *milli* means one-hundredth of.
17. The number 0.002040 written in scientific notation is 2.04×10^{-3}.
18. Two cubes of the same size but different masses will have different densities.
19. The density of liquid A is 2.20 g/mL, and that of liquid B is 1.44 g/mL. When equal volumes of these two immiscible liquids are mixed, liquid A will float on liquid B.
20. A hydrometer is used to measure the density of liquids.

Multiple Choice. *Choose the correct answer to each of the following.*

1. 1.00 cm is equal to how many meters?
   (a) 2.54 (b) 100 (c) 10 (d) 0.01
2. 1.00 cm is equal to how many inches?
   (a) 0.394 (b) 0.10 (c) 12 (d) 2.54
3. 4.50 ft is how many centimeters?
   (a) 11.4 (b) 21.3 (c) 454 (d) 137

4. The number 0.0048 contains how many significant figures?
   (a) 1 (b) 2 (c) 3 (d) 4
5. Express 0.00382 in scientific notation.
   (a) 3.82×10^3 (c) 3.82×10^{-2}
   (b) 3.8×10^{-3} (d) 3.82×10^{-3}
6. 42.0°C is equivalent to:
   (a) 273 K (b) 5.55°F (c) 108°F (d) 53.3°F
7. 267°F is equivalent to:
   (a) 404 K (b) 116°C (c) 540 K (d) 389 K
8. To heat 30 g of water from 20°C to 50°C will require:
   (a) 30 cal (c) 3.8×10^3 J
   (b) 50 cal (d) 6.3×10^3 J
9. An object has a mass of 62 g and a volume of 4.6 mL. Its density is:
   (a) 0.074 mL/g (c) 7.4 g/mL
   (b) 285 g/mL (d) 13 g/mL
10. The mass of a block is 9.43 g and its density is 2.35 g/mL. The volume of the block is:
    (a) 4.01 mL (c) 22.2 mL
    (b) 0.249 mL (d) 2.49 mL
11. The density of copper is 8.92 g/mL. The mass of a piece of copper that has a volume of 9.5 mL is:
    (a) 2.58 g (b) 85 g (c) 0.94 g (d) 1.07 g
12. An empty graduated cylinder weighs 54.772 g. When filled with 50.0 mL of an unknown liquid it weighs 101.074 g. The density of the liquid is:
    (a) 0.926 g/mL (c) 2.02 g/mL
    (b) 1.00 g/mL (d) 1.845 g/mL
13. The conversion factor to change grams to milligrams is:
    (a) $\frac{100 \text{ mg}}{1 \text{ g}}$ (c) $\frac{1 \text{ g}}{1000 \text{ mg}}$
    (b) $\frac{1 \text{ g}}{100 \text{ g}}$ (d) $\frac{1000 \text{ mg}}{1 \text{ g}}$
14. What Fahrenheit temperature is twice the Celsius temperature?
    (a) 64°F (b) 320°F (c) 200°F (d) 746°F
15. The specific heat of aluminum is 0.891 J/g°C. How many joules of energy are required to raise the temperature of 20.0 g of Al from 10.0°C to 15.0°C?
    (a) 79 J (b) 89 J (c) 100 J (d) 112 J
16. A gold alloy has a density of 12.41 g/mL and contains 75.0% gold by weight. The volume of this alloy that can be made from 255 g of pure gold is
    (a) 4.22×10^3 mL (c) 27.4 mL
    (b) 2.37×10^3 mL (d) 15.4 mL

CHAPTER 3 PROPERTIES OF MATTER

True–False. *Answer the following as either true or false.*

1. A substance is homogeneous but does not have a fixed composition.
2. A system having more than one phase is heterogeneous.
3. In a chemical change, substances are formed that are entirely different, having different properties and composition from the original material.
4. The Law of Conservation of Energy says that, because of the energy shortage, anyone wasting energy can be arrested.
5. The Law of Conservation of Mass states that no detectable change is observed in the total mass of the substances involved in a chemical change.
6. The starting substances in a chemical reaction are called the reactants.
7. Plastics, glass, and gels are examples of amorphous solids.
8. Matter that has identical properties throughout is homogeneous.
9. A gas is the least compact of the three states of matter.
10. When a clean copper wire is heated in a burner flame it gains weight.
11. The energy released when hydrogen and oxygen react to form water was stored in the hydrogen and oxygen as chemical or kinetic energy.
12. In a physical change the composition of matter does not change.
13. A mixture is a combination of two or more substances in which the substances retain their identity.
14. A liquid has both a definite shape and a definite volume.
15. Pure substances occur in two forms, elements and ions.

Multiple Choice. *Choose the correct answer to each of the following.*

1. Which of the following is not a physical property?
   (a) Boiling point (c) Bleaching action
   (b) Physical state (d) Color
2. Which of the following is a physical change?
   (a) A piece of sulfur is burned.
   (b) A firecracker explodes.
   (c) A rubber band is stretched.
   (d) A nail rusts.
3. Which of the following is a chemical change?
   (a) Water evaporates.
   (b) Ice melts.
   (c) Rocks are ground to sand.
   (d) A penny tarnishes.
4. Which of the following is a mixture?
   (a) Water (c) Sugar solution
   (b) Mercury(II) oxide (d) Copper(II) oxide
5. When 9.44 g of calcium are heated in air, 13.22 g of calcium oxide are formed. The percentage by weight of oxygen in the compound is:
   (a) 28.6% (b) 40.0% (c) 71.4% (d) 13.2%
6. Mercury(II) sulfide, HgS, contains 86.2% mercury by weight. The grams of HgS that can be made from 30.0 g of mercury are:
   (a) 2586 g (b) 2.87 g (c) 25.9 g (d) 34.8 g
7. Which is the most compact state of matter?
   (a) Solid (c) Gas
   (b) Liquid (d) Amorphous
8. Which is not characteristic of a solution?
   (a) A homogeneous mixture
   (b) A heterogeneous mixture
   (c) Contains two or more substances
   (d) Has a variable composition
9. The changing of liquid water to ice is known as a
   (a) Chemical change
   (b) Heterogeneous change
   (c) Homogeneous change
   (d) Physical change
10. Which of the following does not represent a chemical change?
    (a) Heating of copper in air
    (b) Combustion of gasoline
    (c) Cooling of red-hot iron
    (d) Digestion of food

CHAPTER 4 ELEMENTS AND COMPOUNDS

True–False. *Answer the following as either true or false.*

1. The basic building blocks of all substances, which cannot be decomposed into simpler substances by ordinary chemical change, are compounds.
2. The smallest particle of an element that can exist and still retain the properties of the element is called an atom.
3. The symbol for silver is Ag.
4. The symbol for nitrogen is Ni.
5. The Law of Definite Composition states that a compound contains two or more elements combined in a definite proportion by weight.
6. The name of $ZnBr_2$ is zinc bromide.
7. Elements that have properties resembling both metals and nonmetals are called mixtures.
8. A molecule is a small, uncharged individual unit of a compound formed by the union of two or more atoms.
9. An ion is a positive or negative electrically charged atom or group of atoms.
10. A chemical formula is a shorthand expression for a chemical reaction.
11. The most abundant element in the earth's crust, seawater, and atmosphere is nitrogen.
12. The symbol for cobalt can be written as Co or CO.
13. Metalloids are elements that have properties intermediate between those of metals and nonmetals.
14. Compounds exist as either molecules or ions.
15. The main characteristic of a mixture is that it has a definite composition.
16. The characteristic name ending for a binary compound is *ide*.
17. In a balanced chemical equation, the mass of the products is equal to the mass of the reactants.
18. The symbols of the elements have two letters.
19. The present system of symbols for the elements was devised by J. J. Berzelius in the early 1800s.

20. The smallest uncharged unit of a compound is an atom.

Multiple Choice. *Choose the correct answer to each of the following.*

1. Which of the following is not one of the five most abundant elements by weight in the earth's crust, seawater, and atmosphere?
   (a) Oxygen (c) Silicon
   (b) Hydrogen (d) Aluminum
2. Which of the following is a compound?
   (a) Lead (c) Potassium
   (b) Wood (d) Water
3. Which of the following is a mixture?
   (a) Water (c) Wood
   (b) Chromium (d) Sulfur
4. How many atoms are represented in the formula Na_2CrO_4?
   (a) 3 (b) (5) (c) (7) (d) 8
5. Which of the following is a characteristic of metals?
   (a) Ductile (c) Extremely strong
   (b) Easily shattered (d) Dull
6. Which of the following is a characteristic of nonmetals?
   (a) Always a gas
   (b) Poor conductor of electricity
   (c) Shiny
   (d) Combine only with metals
7. When the equation $Al + O_2 \longrightarrow Al_2O_3$ is properly balanced, which of the following terms appears?
   (a) 2 Al (b) 2 Al_2O_3 (c) 3 Al (d) 2 O_2
8. Which of the following does not occur as a diatomic molecule?
   (a) O_2 (b) I_2 (c) H_2 (d) Na_2
9. Barium iodide, BaI_2, contains 35.1% barium by weight. An 8.50 g sample of barium iodide contains what weight of iodine?
   (a) 5.52 g (b) 2.98 g (c) 3.51 g (d) 6.49 g
10. Which equation is incorrectly balanced?
    (a) $2\,KNO_3 \xrightarrow{\Delta} 2\,KNO_2 + O_2$
    (b) $H_2O_2 \longrightarrow H_2O + O_2$
    (c) $2\,Na_2O_2 + 2\,H_2O \longrightarrow 4\,NaOH + O_2$
    (d) $2\,H_2O \xrightarrow[H_2SO_4]{\text{Electrical energy}} 2\,H_2 + O_2$
11. Which of the following is not a binary compound?
    (a) NaI (b) HClO (c) K_2S (d) Mg_3N_2
12. Which of the following is the formula for sodium bromide?
    (a) SBr (b) SoBr (c) NaBr (d) NaB
13. When a pure substance was analyzed, it was found to contain carbon and chlorine. This substance must be classified as:
    (a) An element
    (b) A mixture
    (c) A compound
    (d) Both a mixture and a compound
14. Chromium, fluorine, and magnesium have the symbols
    (a) Ch, F, Ma (c) Cr, F, Mg
    (b) Cr, Fl, Mg (d) Cr, F, Ma
15. Sodium, carbon, and sulfur have the symbols
    (a) Na, C, S (c) Na, Ca, Su
    (b) So, C, Su (d) So, Ca, Su
16. Coffee is an example of
    (a) An element
    (b) A compound
    (c) A homogeneous mixture
    (d) A heterogeneous mixture
17. The number of oxygen atoms in $Al(C_2H_3O_2)_3$ is
    (a) 2 (b) 3 (c) 5 (d) 6

CHAPTER 5

Atomic Theory and Structure

After studying Chapter 5, you should be able to

1. Understand the terms listed in Question A at the end of the chapter.
2. State the major provisions of Dalton's atomic theory.
3. Give the names, symbols, charges, and relative masses of the three principal subatomic particles.
4. Describe the atom as conceived by Ernest Rutherford after his alpha scattering experiments.
5. Describe the atom as conceived by Niels Bohr.
6. Discuss the contributions to atomic theory made by Dalton, Thomson, Rutherford, Bohr, Chadwick, and Schrödinger.
7. Describe what is meant by an electron orbital.
8. Determine the maximum number of electrons that can exist in the principal energy levels and sublevels.
9. Determine the atomic number, mass number, or number of neutrons of any isotope when given the values of any two of these three items.
10. Draw the diagram of any isotope of the first 56 elements, showing the composition of the nucleus and the numbers of electrons in the main energy levels.
11. Write the electron structure ($1s^2 2s^2 2p^6 \ldots$) for any of the first 56 elements.
12. Explain what is represented by the Lewis-dot (electron-dot) structure of an element.
13. Write Lewis-dot (electron-dot) symbols for the first 20 elements.
14. Understand the basis for the Octet Rule.
15. Name and distinguish among the three isotopes of hydrogen.
16. Calculate the average atomic mass (weight) of an element, given the isotopic masses and abundance of its isotopes.

5.1 Early Thoughts

The structure of matter has long intrigued and engaged the minds of people. The seed of modern atomic theory was sown during the time of the ancient Greek philosophers. About 440 B.C. Empedocles stated that all matter was composed of four "elements"—earth, air, water, and fire. Democritus (about 470–370 B.C.), one of the early atomistic philosophers, thought that all forms of matter were finitely divisible into invisible particles, which he called atoms. He held that atoms were in constant motion and that they combined with one another in various ways. This purely speculative hypothesis was not based on scientific observations. Shortly thereafter, Aristotle (384–322 B.C.) opposed the theory of Democritus and endorsed and advanced the Empedoclean theory. So strong was the influence of Aristotle that his theory dominated the thinking of scientists and philosophers until the beginning of the 17th century. The term *atom* is derived from the Greek word *atomos*, meaning indivisible.

5.2 Dalton's Atomic Theory

More than 2000 years after Democritus, the English schoolmaster John Dalton (1766–1844) revived the concept of atoms and proposed an atomic theory based on facts and experimental evidence. This theory, described in a series of papers published during the period 1803–1810, rested on the idea of a different kind of atom for each element. The essence of **Dalton's atomic theory** may be summed up as follows:

Dalton's atomic theory

1 Elements are composed of minute, indivisible particles called atoms.
2 Atoms of the same element are alike in mass and size.
3 Atoms of different elements have different masses and sizes.
4 Chemical compounds are formed by the union of two or more atoms of different elements.
5 Atoms combine to form compounds in simple numerical ratios such as one to one, two to one, two to three, and so on.
6 Atoms of two elements may combine in different ratios to form more than one compound.

Dalton's atomic theory stands as a landmark in the development of chemistry. The major premises of his theory are still valid today. However, some of the statements must be modified or qualified because investigations since Dalton's time have shown that (1) atoms are composed of subatomic particles; (2) not all the atoms of a specific element have the same mass; and (3) atoms, under special circumstances, can be decomposed.

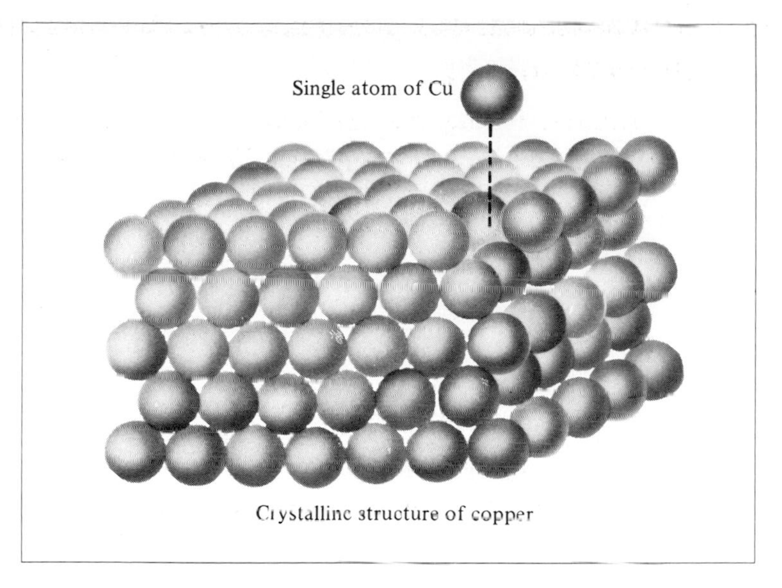

Figure 5.1
A single atom of copper compared with copper as it occurs in its regular crystalline lattice structure. Billions of atoms are present in a cross section of even the smallest strand of copper wire.

5.3 Subatomic Parts of the Atom

The concept of the atom—a particle so small that it could not be seen even with the most powerful microscope—and the subsequent determination of its structure stand among the very greatest creative intellectual human achievements.

Any visible quantity of an element contains a vast number of identical atoms. But when we refer to an atom of an element, we isolate a single atom from the multitude in order to present that element in its simplest form. Figure 5.1 illustrates the hypothetical isolation of a single copper atom from its crystal lattice.

Let us examine this tiny particle we call the atom. The diameter of a single atom ranges from 0.1 to 0.5 nanometers (1 nm = 1×10^{-9} m). Hydrogen, the smallest atom, has a diameter of about 0.1 nm. To arrive at some idea of how small an atom is, consider this dot (•), which has a diameter of about 1 mm, or 1×10^{6} nm. It would take 10 million hydrogen atoms to form a line of atoms across this dot. As inconceivably small as atoms are, they contain even smaller particles, the **subatomic particles**, such as electrons, protons, and neutrons.

subatomic particles

The experimental discovery of the electron (e^-) was made in 1897 by J. J. Thomson (1856–1940). The **electron** is a particle with a negative electrical charge

electron

Table 5.1 Electrical charge and relative mass of electrons, protons, and neutrons

Particle	Symbol	Relative electrical charge	Relative mass (amu)	Actual mass (g)
Electron	e^-	-1	$\frac{1}{1837}$	9.110×10^{-28}
Proton	p	$+1$	1	1.673×10^{-24}
Neutron	n	0	1	1.675×10^{-24}

and a mass of 9.110×10^{-28} g. This mass is 1/1837 the mass of a hydrogen atom and corresponds to 0.0005486 atomic mass unit (amu) (defined in Section 5.16). One atomic mass unit has a mass of 1.661×10^{-24} g. Although the actual electrical charge of an electron is known, its value is too cumbersome for practical use. Therefore, the electron has been assigned a relative electrical charge of -1. The size of an electron has not been determined exactly, but its diameter is believed to be less than 10^{-12} cm.

Protons were first observed by E. Goldstein (1850–1930) in 1886. However, it was J. J. Thomson who discovered the nature of the proton. He showed that the proton is a particle, and he calculated its mass to be about 1837 times that of an electron. The **proton** (p) is a particle with a relative mass of 1 amu and an actual mass of 1.673×10^{-24} g. Its relative charge ($+1$) is equal in magnitude but of opposite sign to the charge on the electron. The mass of a proton is only very slightly less than that of a hydrogen atom.

proton

The third major subatomic particle was discovered in 1932 by James Chadwick (1891–1974). This particle, the **neutron** (n), bears neither a positive nor a negative charge and has a relative mass of about 1 amu. Its actual mass (1.675×10^{-24} g) is only very slightly greater than that of a proton. The properties of these three subatomic particles are summarized in Table 5.1.

neutron

Nearly all the ordinary chemical properties of matter can be explained in terms of atoms consisting of electrons, protons, and neutrons. The discussion of atomic structure that follows is based on the assumption that atoms contain only these principal subatomic particles. Many other subatomic particles such as mesons, positrons, neutrinos, and antiprotons have been discovered. At this time it is not clear whether all these particles are actually present in the atom or whether they are produced by reactions occurring within the nucleus. The fields of atomic and particle or high-energy physics are fascinating and have attracted many young scientists in recent years. This interest has resulted in a great deal of research that is producing a long list of subatomic particles. Descriptions of the properties of many of these particles are to be found in recent physics textbooks and in various articles appearing in *Scientific American* over the past several years.

Figure 5.2

Ernest Rutherford (1871–1937), British physicist who identified two of the three principal rays emanating from radioactive substances. His experiments with alpha particles led to the first laboratory transmutation of an element and to his formulation of the nuclear atom. Rutherford was awarded the Nobel prize in 1908 for his work on transmutation. (*Courtesy Rutherford Museum, McGill University.*)

5.4 The Nuclear Atom

The discovery that positively charged particles were present in atoms came soon after the discovery of radioactivity by Henri Becquerel in 1896.

Ernest Rutherford (Figure 5.2) had, by 1907, established that the positively charged alpha particles emitted by certain radioactive elements were ions of the element helium. Rutherford used these alpha particles to establish the nuclear nature of atoms. In some experiments performed in 1911, he directed a stream of positively charged helium ions (alpha particles) at a very thin sheet of gold foil (about 1000 atoms thick). He observed that most of the alpha particles passed through the foil with little or no deflection; but a few of the particles were

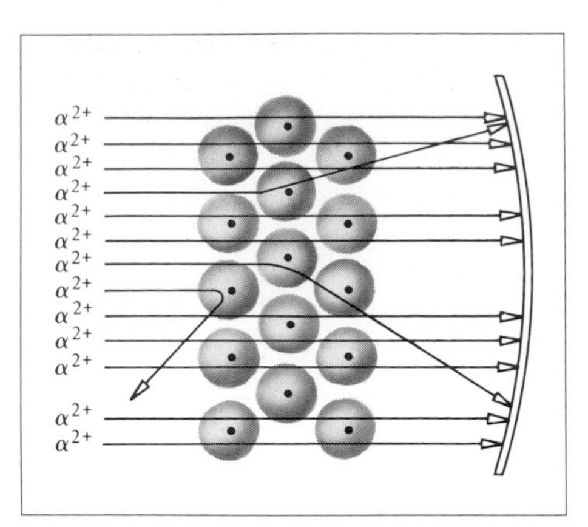

Figure 5.3

Diagram representing Rutherford's experiment on alpha particle scattering. Positive alpha particles (α^{2+}), emanating from a radioactive source, were directed at a thin metal foil. Diagram illustrates the deflection and repulsion of the positive alpha particles by the positive nuclei of the metal atoms.

deflected at large angles, and occasionally one even bounced back from the foil (see Figure 5.3). It was known that like charges repel each other and that an electron with a mass of 1/1837 amu could not possibly have an appreciable effect on the path of a 4 amu alpha particle, which is about 7350 times more massive than an electron. Rutherford therefore reasoned that each gold atom must contain a positively charged mass occupying a relatively tiny volume and that, when an alpha particle approached close enough to this positive mass, it was deflected. Rutherford spoke of this positively charged mass as the *nucleus* of the atom. Because alpha particles have relatively high masses, the extent of the deflections (some actually bounced back) indicated to Rutherford that the nucleus is relatively very heavy and dense. (The density of the nucleus of a hydrogen atom is about 10^{12} g/cm^3, about one trillion times the density of water.) Because most of the alpha particles passed through the thousand or so gold atoms without any apparent deflection, he further concluded that most of an atom consists of empty space.

When we speak of the mass of an atom, we are, for practical purposes, referring primarily to the mass of the nucleus. The nucleus contains all the protons and neutrons, which represent more than 99.9% of the total mass of any atom (see Table 5.1). By way of illustration, the largest number of electrons known to exist in an atom is 109. The mass of even 109 electrons is only about 1/17 of the mass of a single proton or neutron. The mass of an atom, therefore, is primarily determined by the combined masses of its protons and neutrons.

5.5 General Arrangement of Subatomic Particles

The alpha particle scattering experiments of Rutherford established that the atom contains a dense, positively charged nucleus. The later work of Chadwick demonstrated that the atom contains neutrons, which are particles with mass but no charge. Light, negatively charged electrons are also present and offset the positive charges in the nucleus. Based on this experimental evidence, a general description of the atom and the location of its subatomic particles was devised.

nucleus

Each atom consists of a **nucleus** surrounded by electrons. The nucleus contains protons and neutrons but not electrons. In a neutral atom the positive charge of the nucleus (due to protons) is exactly offset by the negative electrons. Because the charge of an electron is equal but of opposite sign to the charge of a proton, a neutral atom must contain exactly the same number of electrons as protons. However, this generalized picture of atomic structure provides no information on the arrangement of electrons within the atom.

A neutral atom contains the same number of protons and electrons.

5.6 The Bohr Atom

At high temperatures or when subjected to high voltages, elements in the gaseous state give off colored light. Brightly colored neon signs illustrate this property of matter very well. When passed through the prism or the grating of a spectroscope, the light emitted by a gas appears as a set of brightly colored lines and is called a line spectrum. These colored lines indicate that the light is being emitted only at certain wavelengths, or frequencies, that correspond to specific colors. Each element possesses a unique set of these spectral lines that is different from the sets of all the other elements.

In 1912–1913, while studying the line spectra of hydrogen, Niels Bohr (1885–1962), a Danish physicist, made a significant contribution to the rapidly growing knowledge of atomic structure. His research led him to believe that electrons in an atom exist in specific regions at various distances from the nucleus. He also visualized the electrons as rotating in orbits around the nucleus, like planets rotating around the sun.

Bohr's first paper in this field dealt with the hydrogen atom, which he described as a single electron rotating in an orbit about a relatively heavy nucleus. He applied the concept of energy quanta, proposed in 1900 by the German physicist Max Planck (1858–1947), to the observed line spectra of hydrogen. Planck stated that energy is never emitted in a continuous stream but only in small discrete packets called quanta (Latin, *quantus*, how much). Bohr theorized that electrons have several possible orbits at different distances from the nucleus and that an electron had to be in one specific orbit (energy level) or another; it could not exist between orbits. In other words, the energy of the electron is said to be quantized. Bohr also stated that, when a hydrogen atom absorbed one or more quanta of energy, its electron would "jump" to an orbit at a greater distance from the nucleus.

Bohr was able to account for spectral lines this way. A number of orbits are available, each corresponding to a different energy level. The orbit closest to the

Figure 5.4

Line spectrum of hydrogen. Each line corresponds to the wavelength of the energy emitted when the electron of a hydrogen atom, which has absorbed energy, falls back to a lower energy level.

nucleus is the lowest, or ground state, energy level; orbits at increasing distances are the second, third, fourth, etc., energy levels. When an electron falls from a high-energy orbit to a lower one (say, from the fourth to the second), a quantum of energy is emitted as light at a specific frequency, or wavelength. This light corresponds to one of the lines visible in the hydrogen spectrum (see Figure 5.4). Several lines are visible in this spectrum. Each line corresponds to a specific electron energy-level shift within the hydrogen atom.

The chemical properties of an element and its position in the periodic table (Chapter 6) depend on electron behavior within the atoms. In turn much of our knowledge of the behavior of electrons within atoms is based on spectroscopy. Niels Bohr contributed a great deal to our knowledge of atomic structure by (1) suggesting quantized energy levels for electrons and (2) showing that spectral lines result from the radiation of small increments of energy (Planck's quanta) when electrons shift from one energy level to another. Bohr's calculations succeeded very well in correlating the experimentally observed spectral lines with electron energy levels for the hydrogen atom. However, Bohr's methods of calculation did not succeed for heavier atoms. More theoretical work on atomic structure was needed.

In 1924 the French physicist, Louis de Broglie, suggested that moving electrons had properties of waves as well as mass. In 1926 Erwin Schrödinger, an Austrian physicist, introduced a new method of calculation—quantum mechanics, or wave mechanics. By Schrödinger's method electrons are described in mathematical terms as having dual characteristics; that is, some electron properties are best described in terms of waves (like light) and others in terms of particles having mass.

Since the late 1920's quantum mechanical concepts have generally replaced Bohr's ideas in theoretical considerations of atomic structure. One important difference between the Bohr and the quantum mechanics concepts is in the treatment of electrons. Quantum mechanics retains the concept of electrons being in specific energy levels. But the electrons are treated not as revolving about

orbital

the nucleus in *orbits* but as being located in *orbitals*. An **orbital** is simply a region in space about the nucleus where there is a high probability of finding a given electron.

5.7 Energy Levels of Electrons

energy levels of electrons

electron shells

Not all the electrons in an atom are located the same distance from the nucleus. As pointed out by both the Bohr theory and quantum mechanics, the probability of finding electrons is greatest at certain specified distances from the nucleus, called **energy levels**. Energy levels are also referred to as **electron shells** and may contain only a limited number of electrons. The main or principal energy levels (**n**) are numbered, starting with **n** = 1 as the energy level nearest to the nucleus and going to **n** = 7, for the known elements. (Theoretically, the number of energy levels is infinite.) Each succeeding energy level is located farther from the nucleus. The electrons in energy levels at increasing distances from the nucleus have increasingly higher energies. The order of energy for the principal energy levels **n** is

Principal energy levels: $1 < 2 < 3 < 4 < 5 < 6 < 7$

The number of electrons that can exist in each energy level is limited. The maximum number of electrons for a specific energy level can be calculated from the formula $2\mathbf{n}^2$, where **n** is the number of the principal energy level. For example, energy level 1 (**n** = 1) can have a maximum of two electrons ($2 \times 1^2 = 2$); energy level 2 (**n** = 2) can have a maximum of eight electrons ($2 \times 2^2 = 8$), and so on. Table 5.2 shows the maximum number of electrons that can exist in each of the first five energy levels.

Table 5.2 Maximum number of electrons that can occupy each principal energy level

Principal energy level, n	Maximum number of electrons in each energy level, $2n^2$
1	$2 \times 1^2 = 2$
2	$2 \times 2^2 = 8$
3	$2 \times 3^2 = 18$
4	$2 \times 4^2 = 32$
5	$2 \times 5^2 = 50$[a]

[a] The theoretical value of 50 electrons in energy level 5 has never been attained in any element known to date.

5.8 Energy Sublevels

The principal energy levels contain sublevels designated by the letters *s*, *p*, *d*, and *f*. These orbitals are the ones in which electrons are located. The *s* sublevel consists of one orbital; the *p* sublevel consists of three orbitals; the *d* sublevel consists of five orbitals; and the *f* sublevel consists of seven orbitals. An electron spins on its own axis in one of only two directions, clockwise or counterclockwise. As a result, only two electrons can occupy the same orbital, one spinning clockwise and the other spinning counterclockwise. When an orbital contains two electrons, the electrons are said to be paired. Because no more than two electrons can exist in an orbital, the maximum numbers of electrons that can exist in the sublevels are 2 in the *s* orbital, 6 in the three *p* orbitals, 10 in the five *d* orbitals, and 14 in the seven *f* orbitals.

Type of sublevel	Number of orbitals possible	Number of electrons possible
s	1	2
p	3	6
d	5	10
f	7	14

The order of energy of the sublevels within a principal energy level is the following: *s* electrons are lower in energy than *p* electrons, which are lower than *d* electrons, which are lower than *f* electrons. This order may be expressed in the following manner:

Sublevel energy: $s < p < d < f$

Not all principal energy levels contain each and every type of sublevel. To determine what types of sublevels occur in any given energy level, we need to know the maximum number of electrons possible in that energy level (see Table 5.2), and we need to use these three rules:

1. No more than two electrons can occupy one orbital.
2. Electrons occupy the lowest possible energy sublevels; they enter a higher sublevel only when the lower sublevels are filled.
3. Orbitals in a given sublevel of equal energy are each occupied by a single electron before a second electron enters them. For example, all three *p* orbitals must contain one electron before a second electron enters a *p* orbital.

The maximum number of electrons in the first energy level is two; both are *s* orbital electrons, designated as $1s^2$. (The *s* orbital in the second energy level ($\mathbf{n} = 2$) is written as $2s$, in the third energy level as $3s$, and so on.) The second energy level, with a maximum of eight electrons, contains only *s* and *p* electrons—

namely, a maximum of two s and six p electrons, designated as $2s^2 2p^6$. The following diagram shows how to read these electron designations:

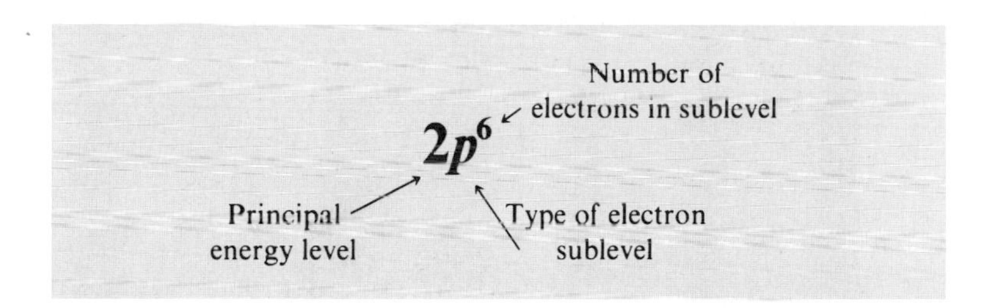

If each orbital contains two electrons, the second energy level can have four orbitals (8 electrons): one s orbital and three individual p orbitals. These three p orbitals are energetically equivalent to each other and are labeled $2p_x$, $2p_y$, and $2p_z$ to indicate their orientation in space (see Figure 5.5). The symbols $3s^2$, $3p^6$, and $3d^{10}$ show the sublevel breakdown of electrons in the third energy level. From this line of reasoning, we can see that, if an atom has sufficient electrons, f electrons first appear in the fourth energy level. Table 5.3 shows the type of sublevel electron orbitals and the maximum number of orbitals and electrons in each energy level. No elements in the fifth, sixth and seventh energy levels contain the calculated maximum number of electrons.

Since the *spdf* atomic orbitals have definite orientations in space, they are represented by particular spatial shapes. At this time we will consider only the s and p orbitals. The s orbitals are spherically symmetrical about the nucleus, as illustrated in Figure 5.5. A $2s$ orbital is a larger sphere than a $1s$ orbital. The p orbitals (p_x, p_y, p_z) are dumbbell-shaped and are oriented at right angles to each other along the x, y, and z axes in space. An electron has an equal probability of

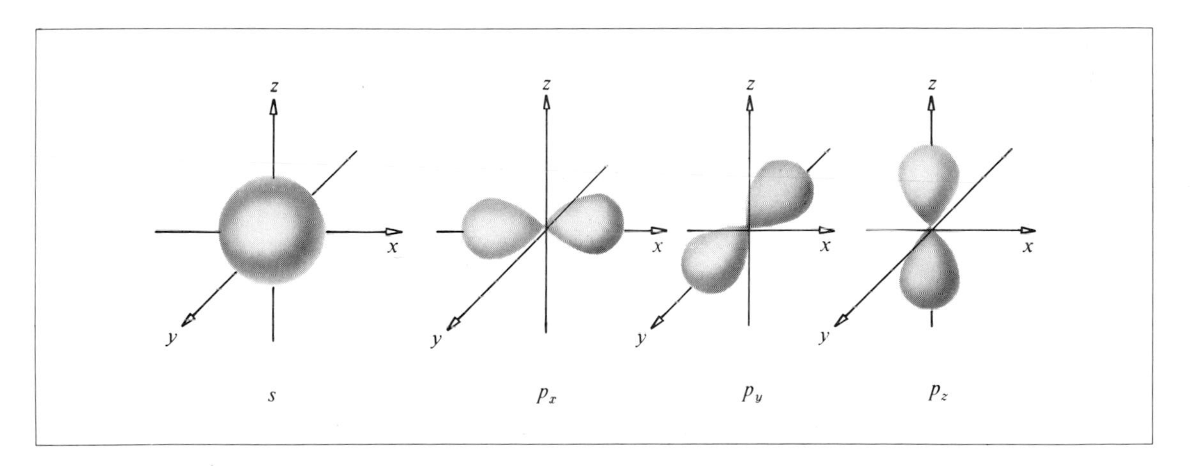

Figure 5.5

Perspective representation of the s, p_x, p_y, and p_z atomic orbitals

Table 5.3 Sublevel electron orbitals in each principal energy level and the maximum number of orbitals and electrons in each energy level

Principal energy level	Sublevel electron	Maximum number of orbitals	Maximum number of electrons
1	*s*	1	2
2	*s*, *p*	4	8
3	*s*, *p*, *d*	9	18
4	*s*, *p*, *d*, *f*	16	32
5	*s*, *p*, *d*, *f*	Incomplete[a]	(50)[a]
6	*s*, *p*, *d*	Incomplete[a]	(72)[a]
7	*s*	Incomplete[a]	(96)[a]

[a] Insufficient electrons to complete the shell.

being located in either lobe of the *p* orbital. In illustrations such as Figure 5.5, the boundaries of the orbitals enclose the region of the greatest probability (about a 90% chance) of finding an electron. In the ground state, or lowest energy level, of a hydrogen atom, this region is a sphere having a radius of 0.053 nm.

5.9 Atomic Numbers of the Elements

atomic number

The **atomic number** of an element is the number of protons in the nucleus of an atom of that element. The atomic number determines the identity of an atom. For example, every atom with an atomic number of 1 is a hydrogen atom; it contains one proton in its nucleus. Every atom with an atomic number of 8 is an oxygen atom; it contains 8 protons in its nucleus. Every atom with an atomic number of 92 is a uranium atom; it contains 92 protons in its nucleus. The atomic number tells us not only the number of positive charges in the nucleus but also the number of electrons in the neutral atom.

atomic number = number of protons in the nucleus

5.10 The Simplest Atom: Hydrogen

The common hydrogen atom, consisting of a nucleus containing one proton and an electron orbital containing one electron, is the simplest known atom. (Some hydrogen atoms are known to contain one or two neutrons in their nucleus. See

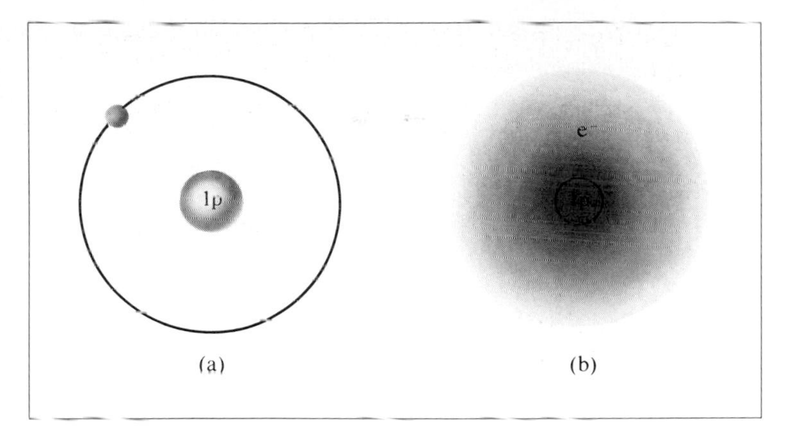

Figure 5.6

The hydrogen atom. (a) Illustration of the Bohr description, indicating a discrete electron moving around its nucleus of one proton.
(b) Illustration of the modern concept of a hydrogen atom consisting of an electron in an *s* orbital as a cloud of negative charge surrounding the proton in the nucleus.

Section 5.16 on isotopes.) The electron in hydrogen occupies an *s* orbital in the first energy level. This electron does not move in any definite path but rather in a rapid random motion throughout its entire orbital, forming an electron "cloud" about the nucleus. The diameter of the nucleus is believed to be about 10^{-13} cm, and the diameter of the electron orbital to be about 10^{-8} cm. Hence, the diameter of the electron orbital of a hydrogen atom is about 100,000 times greater than the diameter of the nucleus.

What we have, then, is a positive nucleus surrounded by an electron cloud formed by an electron in an *s* orbital. The net electrical charge on the hydrogen atom is zero; it is called a *neutral atom.* Figure 5.6 shows two methods of representing a hydrogen atom.

5.11 Atomic Structures of the First Twenty Elements

Starting with hydrogen and progressing in order of increasing atomic number to helium, lithium, beryllium, and so on, the atoms of each successive element contain one more proton and one more electron than do the atoms of the preceding element. This sequence continues, without exception, through the entire list of known elements. It is one of the most impressive examples of order in nature.

The number of neutrons also increases as we progress through the list of elements. But this number, unlike the number of protons and electrons, does not increase in a perfectly uniform manner. Furthermore, atoms of the same element may contain different numbers of neutrons; such atoms are called *isotopes*. For example, three different hydrogen isotopes are described in Section 5.16. The most abundant helium isotope (element number 2) contains two neutrons, but two other helium isotopes exist, containing one and four neutrons, respectively.

The ground-state electron structures of the first 20 elements fall into a regular pattern. The one hydrogen electron is in the first energy level as are both helium electrons. The electron structures for hydrogen and helium are written $1s^1$ and $1s^2$, respectively. The maximum number of electrons in the first energy level is two ($2\mathbf{n}^2 = 2 \times 1^2 = 2$; see Section 5.7), so the two electrons fill the first energy level of helium.

An atom with three electrons will have its third electron in the second energy level because the first level can contain only two electrons. Thus in lithium (atomic number 3) the third electron is in the $2s$ sublevel of the second energy level. Lithium has the electron structure $1s^2 2s^1$.

In succession, the atoms of beryllium (4), boron (5), carbon (6), nitrogen (7), oxygen (8), fluorine (9), and neon (10) have one more proton and one more electron than the preceding element until, in neon, both the first and second energy levels are filled to capacity, with 2 and 8 electrons, respectively.

H	$1s^1$	C	$1s^2 2s^2 2p^2$
He	$1s^2$	N	$1s^2 2s^2 2p^3$
Li	$1s^2 2s^1$	O	$1s^2 2s^2 2p^4$
Be	$1s^2 2s^2$	F	$1s^2 2s^2 2p^5$
B	$1s^2 2s^2 2p^1$	Ne	$1s^2 2s^2 2p^6$

Element 11, sodium (Na), has two electrons in the first energy level and eight electrons in the second energy level, with the remaining electron occupying the $3s$ orbital in the third energy level. The electron structure of sodium is $1s^2 2s^2 2p^6 3s^1$. Magnesium (12), aluminum (13), silicon (14), phosphorus (15), sulfur (16), chlorine (17), and argon (18) follow in order, each adding one electron to the third energy level up to argon, which has eight electrons in the third energy level.

Up through the $3p$ level the sequence of filling the sublevels is exactly as expected, based on the increasing principal and sublevel energy levels. However, after the $3p$ level is filled, variations occur. The third energy level might logically be expected to fill to its capacity of 18 electrons with $3d$ electrons before electrons enter the $4s$ sublevel. However, this order of filling the third energy level does not occur because the $4s$ sublevel is at a lower energy than the $3d$ sublevel (see Figure 5.7). Consequently, because the sublevels fill in order of increasing energy, the last electron in potassium (19) and the last two electrons in calcium (20) are in the $4s$ sublevel. The electron structures for potassium and calcium are

K $1s^2 2s^2 2p^6 3s^2 3p^6 4s^1$ Ca $1s^2 2s^2 2p^6 3s^2 3p^6 4s^2$

This break in sequence does not invalidate the formula $2\mathbf{n}^2$, which prescribes the maximum number of electrons that each shell can contain but not the order in

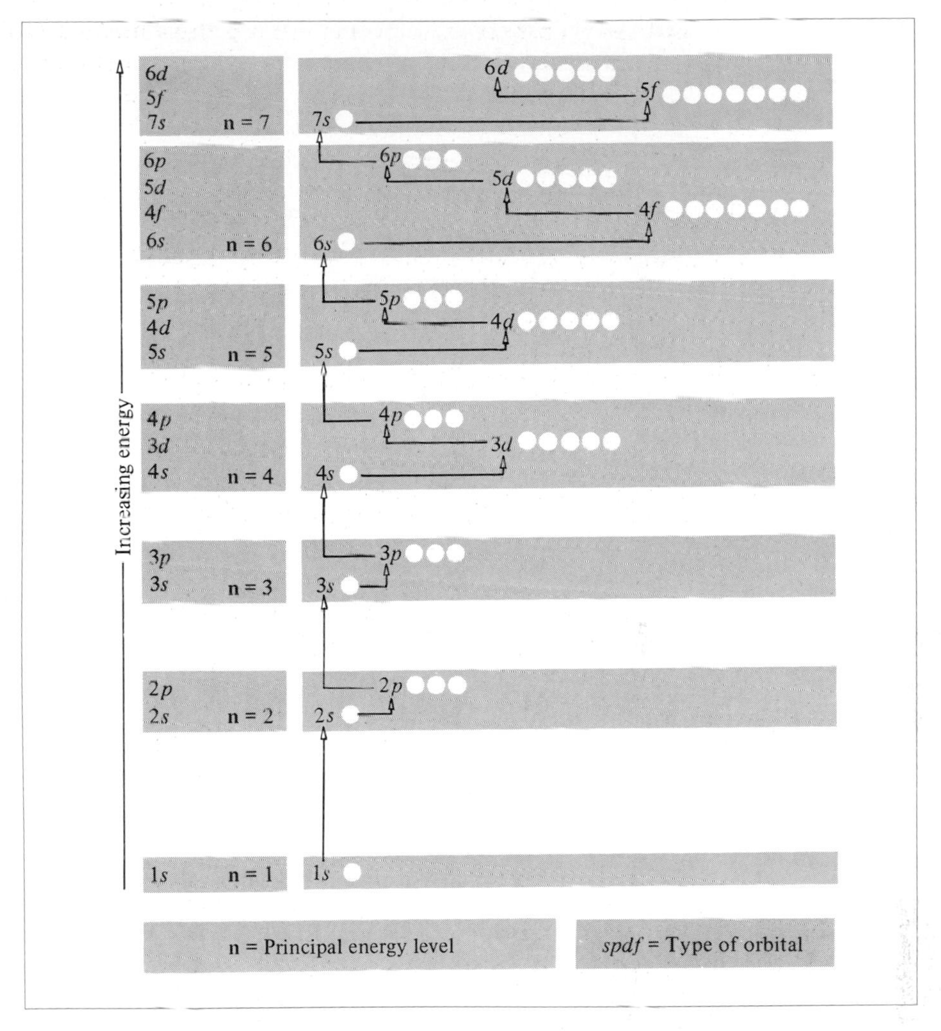

Figure 5.7

Order of filling electron orbitals. Each circle represents an orbital, which can contain two electrons. The progression of electrons filling the energy sublevels is shown, starting with 1*s* and going to 6*d*. All lower sublevels must be filled before an electron enters the next higher sublevel. For example, all three 2*p* orbitals must be filled before an electron enters the 3*s* sublevel. (Some exceptions to this order are known.)

which the shells are filled. Table 5.4 shows the electron structure of the first 20 elements.

The relative energies of the electron orbitals are shown in Figure 5.7. The order given can be used to determine the electron distribution in the atoms of the elements, although some exceptions to the pattern are known. Suppose we wish to determine the electron structure of a chlorine atom (atomic number 17), which

Table 5.4 Electron structure of the first twenty elements

Element	Number of protons (atomic number)	Number of electrons	Electron structure
H	1	1	$1s^1$
He	2	2	$1s^2$
Li	3	3	$1s^22s^1$
Be	4	4	$1s^22s^2$
B	5	5	$1s^22s^22p^1$
C	6	6	$1s^22s^22p^2$
N	7	7	$1s^22s^22p^3$
O	8	8	$1s^22s^22p^4$
F	9	9	$1s^22s^22p^5$
Ne	10	10	$1s^22s^22p^6$
Na	11	11	$1s^22s^22p^63s^1$
Mg	12	12	$1s^22s^22p^63s^2$
Al	13	13	$1s^22s^22p^63s^23p^1$
Si	14	14	$1s^22s^22p^63s^23p^2$
P	15	15	$1s^22s^22p^63s^23p^3$
S	16	16	$1s^22s^22p^63s^23p^4$
Cl	17	17	$1s^22s^22p^63s^23p^5$
Ar	18	18	$1s^22s^22p^63s^23p^6$
K	19	19	$1s^22s^22p^63s^23p^64s^1$
Ca	20	20	$1s^22s^22p^63s^23p^64s^2$

has 17 electrons. Following the order in Figure 5.7, we begin by placing two electrons in the 1*s* orbital, then two electrons in the 2*s* orbital, and then six electrons in the 2*p* orbitals. We now have used ten electrons.

$$1s^22s^22p^6$$

Finally we place the next two electrons in the 3*s* orbital and the remaining five electrons in the 3*p* orbitals, which uses all 17 electrons, giving the electron structure for a chlorine atom as $1s^22s^22p^63s^23p^5$. The sum of the superscripts equals 17, the number of electrons in the atom. This procedure is summarized below.

Order of orbitals to be filled: $1s2s2p3s3p$

Distribution of the 17 electrons in a chlorine atom: $1s^22s^22p^63s^23p^5$

PROBLEM 5.1 What is the electron distribution in a phosphorus atom?

First determine the number of electrons contained in a phosphorus atom. The atomic number of phosphorus is 15; therefore, each atom contains 15 protons and 15 electrons. Now tabulate the number of electrons in each principal and subenergy level until all 15 electrons are assigned.

Sublevel	Number of e^-	Total e^-
1*s* orbital	$2e^-$	2
2*s* orbital	$2e^-$	4
2*p* orbital	$6e^-$	10
3*s* orbital	$2e^-$	12
3*p* orbital	$3e^-$	15

Therefore the electron distribution in phosphorus is $1s^2 2s^2 2p^6 3s^2 3p^3$.

5.12 Electron Structures of the Elements beyond Calcium

The elements following calcium have a less regular pattern of adding electrons. The lowest energy level available for the 21st electron is the 3*d* sublevel. Scandium (21) has one more electron than calcium (20). Its electron structure will be the same as calcium plus one electron in the 3*d* sublevel. The electron structure for scandium is $1s^2 2s^2 2p^6 3s^2 3p^6 4s^2 3d^1$. The elements following scandium, titanium (22) through copper (29), continue to add *d* electrons until the third energy level has its maximum of 18. Two exceptions in the orderly electron addition are chromium (24) and copper (29), the structures of which are given in Table 5.5. The third energy level of electrons is first completed in the element copper. Table 5.5 shows the order of filling of the electron orbitals and the electron configuration of all the known elements.

PROBLEM 5.2 What is the electron structure for a sulfur atom (atomic number 16) and for an iron atom (atomic number 26)?

Look in Table 5.5 for the element with atomic number 16 and write down its structure, [Ne] $3s^2 3p^4$. [Ne] is an abbreviated structure for neon, which is $1s^2 2s^2 2p^6$. Therefore, the electron structure for a sulfur atom is $1s^2 2s^2 2p^6 3s^2 3p^4$.

For an iron atom, Table 5.5 shows a structure of [Ar] $4s^2 3d^6$. This notation means that the electron structure of iron consists of the electron structure for argon plus $4s^2 3d^6$. The table shows that the structure for [Ar] is [Ne] $3s^2 3p^6$, which is equal to $1s^2 2s^2 2p^6 3s^2 3p^6$. Therefore, the electron structure for iron is $1s^2 2s^2 2p^6 3s^2 3p^6 4s^2 3d^6$.

If Table 5.5 is not available, the structure can be determined by tabulating electrons as in Problem 5.1 and using Figure 5.7. Structures for all the noble gases, He, Ne, Ar, and so on, are also given in Table 5.6.

Table 5.5 Electron structures of the elements

Element	Atomic number	Electron structure	Element	Atomic number	Electron structure
H	1	$1s^1$	Zn	30	[Ar] $4s^2 3d^{10}$
He	2	$1s^2$	Ga	31	[Ar] $4s^2 3d^{10} 4p^1$
Li	3	$1s^2 2s^1$	Ge	32	[Ar] $4s^2 3d^{10} 4p^2$
Be	4	$1s^2 2s^2$	As	33	[Ar] $4s^2 3d^{10} 4p^3$
B	5	$1s^2 2s^2 2p^1$	Se	34	[Ar] $4s^2 3d^{10} 4p^4$
C	6	$1s^2 2s^2 2p^2$	Br	35	[Ar] $4s^2 3d^{10} 4p^5$
N	7	$1s^2 2s^2 2p^3$	Kr	36	[Ar] $4s^2 3d^{10} 4p^6$
O	8	$1s^2 2s^2 2p^4$	Rb	37	[Kr] $5s^1$
F	9	$1s^2 2s^2 2p^5$	Sr	38	[Kr] $5s^2$
Ne	10	$1s^2 2s^2 2p^6$	Y	39	[Kr] $5s^2 4d^1$
Na	11	[Ne] $3s^1$	Zr	40	[Kr] $5s^2 4d^2$
Mg	12	[Ne] $3s^2$	Nb	41	[Kr] $5s^1 4d^4$
Al	13	[Ne] $3s^2 3p^1$	Mo	42	[Kr] $5s^1 4d^5$
Si	14	[Ne] $3s^2 3p^2$	Tc	43	[Kr] $5s^2 4d^5$
P	15	[Ne] $3s^2 3p^3$	Ru	44	[Kr] $5s^1 4d^7$
S	16	[Ne] $3s^2 3p^4$	Rh	45	[Kr] $5s^1 4d^8$
Cl	17	[Ne] $3s^2 3p^5$	Pd	46	[Kr] $4d^{10}$
Ar	18	[Ne] $3s^2 3p^6$	Ag	47	[Kr] $5s^1 4d^{10}$
K	19	[Ar] $4s^1$	Cd	48	[Kr] $5s^2 4d^{10}$
Ca	20	[Ar] $4s^2$	In	49	[Kr] $5s^2 4d^{10} 5p^1$
Sc	21	[Ar] $4s^2 3d^1$	Sn	50	[Kr] $5s^2 4d^{10} 5p^2$
Ti	22	[Ar] $4s^2 3d^2$	Sb	51	[Kr] $5s^2 4d^{10} 5p^3$
V	23	[Ar] $4s^2 3d^3$	Te	52	[Kr] $5s^2 4d^{10} 5p^4$
Cr	24	[Ar] $4s^1 3d^5$	I	53	[Kr] $5s^2 4d^{10} 5p^5$
Mn	25	[Ar] $4s^2 3d^5$	Xe	54	[Kr] $5s^2 4d^{10} 5p^6$
Fe	26	[Ar] $4s^2 3d^6$	Cs	55	[Xe] $6s^1$
Co	27	[Ar] $4s^2 3d^7$	Ba	56	[Xe] $6s^2$
Ni	28	[Ar] $4s^2 3d^8$	La	57	[Xe] $6s^2 5d^1$
Cu	29	[Ar] $4s^1 3d^{10}$	Ce	58	[Xe] $6s^2 4f^1 5d^1$

5.13 Diagramming Atomic Structures

We can use several methods to diagram the atomic structures of atoms, depending on what we are trying to illustrate. When we want to show both the nuclear makeup and the electron structure of each energy level (without orbital detail), we can use a diagram such as Figure 5.8.

A method of diagramming subenergy levels is shown in Figure 5.9. Each orbital is represented by a square □. When the orbital contains one electron, an arrow (↑) is placed in the square. A second arrow, pointing downward (↓), indicates the second electron in that orbital.

The diagram for hydrogen is [↑]. Helium, with two electrons, is drawn as [↑↓]; both electrons are $1s$ electrons. The diagram for lithium shows three electrons in two energy levels, $1s^2 2s^1$. All four electrons of beryllium are s electrons, $1s^2 2s^2$. Boron has the first p electron, which is located in the $2p_x$ orbital. Because it is

Table 5.5 (*continued*)

Element	Atomic number	Electron structure	Element	Atomic number	Electron structure
Pr	59	[Xe] $6s^24f^3$	Bi	83	[Xe] $6s^24f^{14}5d^{10}6p^3$
Nd	60	[Xe] $6s^24f^4$	Po	84	[Xe] $6s^24f^{14}5d^{10}6p^4$
Pm	61	[Xe] $6s^24f^5$	At	85	[Xe] $6s^24f^{14}5d^{10}6p^5$
Sm	62	[Xe] $6s^24f^6$	Rn	86	[Xe] $6s^24f^{14}5d^{10}6p^6$
Eu	63	[Xe] $6s^24f^7$	Fr	87	[Rn] $7s^1$
Gd	64	[Xe] $6s^24f^75d^1$	Ra	88	[Rn] $7s^2$
Tb	65	[Xe] $6s^24f^9$	Ac	89	[Rn] $7s^26d^1$
Dy	66	[Xe] $6s^24f^{10}$	Th	90	[Rn] $7s^26d^2$
Ho	67	[Xe] $6s^24f^{11}$	Pa	91	[Rn] $7s^25f^26d^1$
Er	68	[Xe] $6s^24f^{12}$	U	92	[Rn] $7s^25f^36d^1$
Tm	69	[Xe] $6s^24f^{13}$	Np	93	[Rn] $7s^25f^46d^1$
Yb	70	[Xe] $6s^24f^{14}$	Pu	94	[Rn] $7s^25f^6$
Lu	71	[Xe] $6s^24f^{14}5d^1$	Am	95	[Rn] $7s^25f^7$
Hf	72	[Xe] $6s^24f^{14}5d^2$	Cm	96	[Rn] $7s^25f^76d^1$
Ta	73	[Xe] $6s^24f^{14}5d^3$	Bk	97	[Rn] $7s^25f^9$
W	74	[Xe] $6s^24f^{14}5d^4$	Cf	98	[Rn] $7s^25f^{10}$
Re	75	[Xe] $6s^24f^{14}5d^5$	Es	99	[Rn] $7s^25f^{11}$
Os	76	[Xe] $6s^24f^{14}5d^6$	Fm	100	[Rn] $7s^25f^{12}$
Ir	77	[Xe] $6s^24f^{14}5d^7$	Md	101	[Rn] $7s^25f^{13}$
Pt	78	[Xe] $6s^14f^{14}5d^9$	No	102	[Rn] $7s^25f^{14}$
Au	79	[Xe] $6s^14f^{14}5d^{10}$	Lr	103	[Rn] $7s^25f^{14}6d^1$
Hg	80	[Xe] $6s^24f^{14}5d^{10}$	Unq	104	[Rn] $7s^25f^{14}6d^2$
Tl	81	[Xe] $6s^24f^{14}5d^{10}6p^1$	Unp	105	[Rn] $7s^25f^{14}6d^3$
Pb	82	[Xe] $6s^24f^{14}5d^{10}6p^2$	Unh	106	[Rn] $7s^25f^{14}6d^4$

Note: For simplicity of expression, symbols of the chemically stable noble gases are used as a portion of the electron structure for the elements beyond neon. For example, the electron structure of a sodium atom, Na, consists of ten electrons, as in neon [Ne], plus a $3s^1$ electron. Detailed electron structures for the noble gases are given in Table 5.6.

energetically more difficult for the next p electron to pair up with the electron in the p_x orbital than to occupy a second p orbital, the second p electron in carbon is located in the $2p_y$ orbital. The third p electron in nitrogen is still unpaired and is found in the $2p_z$ orbital. The next three electrons pair with each of the $2p$ electrons through the element neon. Also shown in Figure 5.9 are the equivalent linear expressions for these orbital electron structures.

The electrons in successive elements are found in sublevels of increasing energy. The general sequence of increasing energy of sublevels and the order of filling sublevels with electrons is

$$1s\ 2s\ 2p\ 3s\ 3p\ 4s\ 3d\ 4p\ 5s\ 4d\ 5p\ 6s\ 4f\ 5d\ 6p\ 7s\ 5f\ 6d$$

Minor variations from the electron structure predicted by the foregoing general sequence are found in a number of atoms. Table 5.5 shows the accepted ground-state electron structure for the elements.

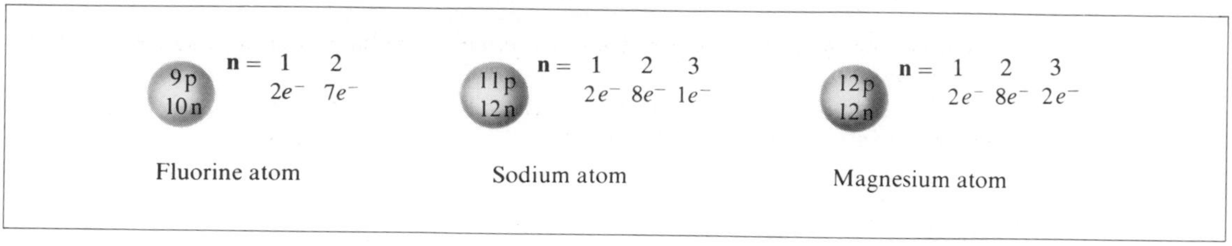

9p 10n n = 1 2 2e^- 7e^-
Fluorine atom

11p 12n n = 1 2 3 2e^- 8e^- 1e^-
Sodium atom

12p 12n n = 1 2 3 2e^- 8e^- 2e^-
Magnesium atom

Figure 5.8

Atomic structure diagrams of fluorine, sodium, and magnesium atoms. The numbers of protons and neutrons are shown in the nucleus. The number of electrons is shown in each principal energy level outside the nucleus.

Element	Orbital electron structure						Linear expression of electron structure
	1s	2s	2p_x	2p_y	2p_z	3s	
H	↑						$1s^1$
He	↑↓						$1s^2$
Li	↑↓	↑					$1s^22s^1$
Be	↑↓	↑↓					$1s^22s^2$
B	↑↓	↑↓	↑				$1s^22s^22p_x^1$
C	↑↓	↑↓	↑	↑			$1s^22s^22p_x^12p_y^1$
N	↑↓	↑↓	↑	↑	↑		$1s^22s^22p_x^12p_y^12p_z^1$
O	↑↓	↑↓	↑↓	↑	↑		$1s^22s^22p_x^22p_y^12p_z^1$
F	↑↓	↑↓	↑↓	↑↓	↑		$1s^22s^22p_x^22p_y^22p_z^1$
Ne	↑↓	↑↓	↑↓	↑↓	↑↓		$1s^22s^22p_x^22p_y^22p_z^2$
Na	↑↓	↑↓	↑↓	↑↓	↑↓	↑	$1s^22s^22p_x^22p_y^22p_z^23s^1$

Figure 5.9

Subenergy-level electron structure of hydrogen through sodium atoms. Each electron is indicated by an arrow placed in the square, which represents the orbital.

PROBLEM 5.3 Diagram the electron structure of a zinc atom and a rubidium atom. Use the $1s^2 2s^2 2p^6$, etc., method.

The atomic number of zinc is 30; therefore it has 30 protons and 30 electrons in a neutral atom. Using Figure 5.7 tabulate the 30 electrons as follows:

Orbital	Number of e⁻	Total e⁻
1*s*	2e⁻	2
2*s*	2e⁻	4
2*p*	6e⁻	10
3*s*	2e⁻	12
3*p*	6e⁻	18
4*s*	2e⁻	20
3*d*	10e⁻	30

The electron structure of a zinc atom is $1s^2 2s^2 2p^6 3s^2 3p^6 4s^2 3d^{10}$. Check by adding the superscripts, which should equal 30.

The atomic number of rubidium is 37; therefore it has 37 protons and 37 electrons in a neutral atom. With a little practice, and using Figure 5.7, the electron structure may be written directly in the linear form. The electron structure of a rubidium atom is $1s^2 2s^2 2p^6 3s^2 3p^6 4s^2 3d^{10} 4p^6 5s^1$. Check by adding the superscripts, which should equal 37.

5.14 Lewis-Dot Representation of Atoms

The Lewis-dot (or electron-dot) method of representing atoms, proposed by the American chemist G. N. Lewis (1875–1946), uses the symbol for the element and dots for electrons. The number of dots placed around the symbol equals the number of *s* and *p* electrons in the outermost energy level of the atom. Paired dots represent paired electrons; unpaired dots represent unpaired electrons. For example, **H·** is the Lewis symbol for a hydrogen atom, $1s^1$; **:Ḃ** is the Lewis symbol for a boron atom, $1s^2 2s^2 2p^1$; **:Ï:·** is an iodine atom, which has seven electrons in its outermost energy level. In the case of boron, the symbol **B** represents the boron nucleus and the $1s^2$ electrons; the dots represent only the $2s^2 2p^1$ electrons.

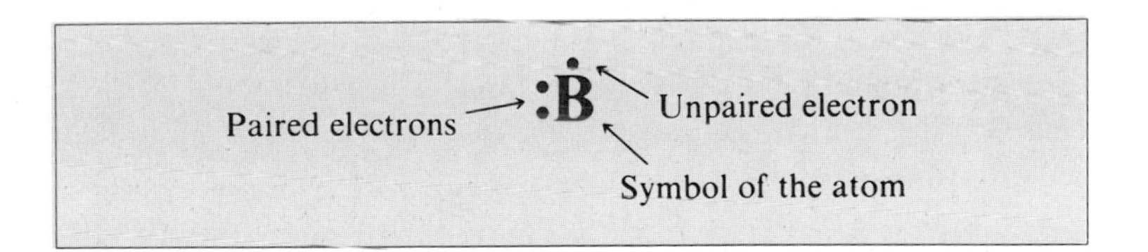

The Lewis-dot method is often used, not only because of its simplicity of expression, but also because much of the chemistry of the atom is directly

H·   He: Li·   Be: :B   :C·   :N·   ·O: :F: :Ne:

Na·   Mg: :Al   :Si·   :P·   ·S: :Cl: :Ar:

K·   Ca:

Figure 5.10

Lewis-dot diagrams of the first twenty elements. Dots represent electrons in the outermost energy level only.

associated with the electrons in the outermost energy level. This association is especially true for the first 20 elements and the remaining Group A elements of the periodic table (see Chapter 6). Figure 5.10 shows Lewis-dot diagrams for the elements hydrogen through calcium.

PROBLEM 5.4 Write the Lewis-dot structure for a phosphorus atom.

First establish the electron structure for a phosphorus atom. It is $1s^22s^22p^63s^23p^3$. Note that there are five electrons in the outermost principal energy level; they are $3s^23p^3$. Write the symbol for phosphorus and place the five electrons as dots around it:

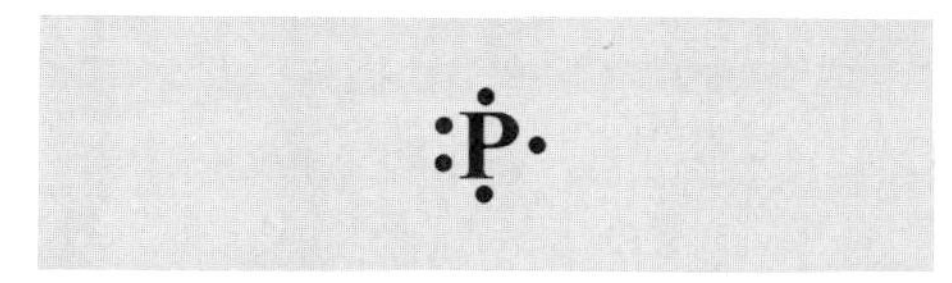

The $3s^2$ electrons are paired and are represented by the paired dots. The $3p^3$ electrons, which are unpaired, are represented by the single dots.

5.15 The Octet Rule

noble gases

The family of elements consisting of helium, neon, argon, krypton, xenon, and radon is known as the **noble gases**. These elements, formerly called *inert gases*, have almost no chemical reactivity; in fact, no compounds of any of them were known before 1962.

Each of these elements, except helium, has an outer shell of eight electrons, two *s* and six *p* (see Table 5.6). The only shell of helium is an *s* orbital filled with two electrons. The electron structure of the noble gases is such that the outer shell *s* and *p* orbitals are filled with paired electrons. This arrangement is very stable and makes the atoms of the noble gases chemically unreactive. Recognition of the

Table 5.6 Arrangement of electrons in the noble gases. Each gas except helium has eight electrons in its outermost energy level.

		Electron structure					
Noble gas	Symbol	n = 1	2	3	4	5	6
Helium	He	$1s^2$					
Neon	Ne	$1s^2$	$2s^22p^6$				
Argon	Ar	$1s^2$	$2s^22p^6$	$3s^23p^6$			
Krypton	Kr	$1s^2$	$2s^22p^6$	$3s^23p^63d^{10}$	$4s^24p^6$		
Xenon	Xe	$1s^2$	$2s^22p^6$	$3s^23p^63d^{10}$	$4s^24p^64d^{10}$	$5s^25p^6$	
Radon	Rn	$1s^2$	$2s^22p^6$	$3s^23p^63d^{10}$	$4s^24p^64d^{10}4f^{14}$	$5s^25p^65d^{10}$	$6s^26p^6$

octet rule

extraordinary stability of this structure led to the **octet rule**: Through chemical changes many of the elements tend to attain an electron structure of eight electrons in their outermost energy level, identical to that of the chemically stable noble gases. Although the octet rule is useful and applies to the behavior of many elements and compounds, it is not universally applicable; some elements do not obey this rule. Applications of the octet rule are given in Chapter 7.

5.16 Isotopes of the Elements

Shortly after Rutherford's conception of the nuclear atom, experiments were performed to determine the masses of individual atoms. These experiments showed that the masses of nearly all atoms were greater than could be accounted for by simply adding up the masses of all the protons and electrons that were known to be present in an atom. This fact led to the concept of the neutron, a particle with no charge but with a mass about the same as that of a proton. Because this particle has no charge, it was very difficult to detect, and the existence of the neutron was not proven experimentally until 1932. All atomic nuclei except that of the simplest hydrogen atom are now believed to contain neutrons.

All atoms of a given element have the same number of protons, but experimental evidence has shown that, in most cases, all atoms of a given element do not have identical masses because atoms of the same element may have different numbers of neutrons in their nuclei.

isotopes

Atoms of an element having the same atomic number but different atomic masses are called **isotopes** of that element. Atoms of the various isotopes of an element, therefore, have the same number of protons and electrons but different numbers of neutrons.

Three isotopes of hydrogen (atomic number 1) are known. Each has one proton in the nucleus and one electron in the first energy level. The first isotope (protium), without a neutron, has a mass of 1; the second isotope (deuterium),

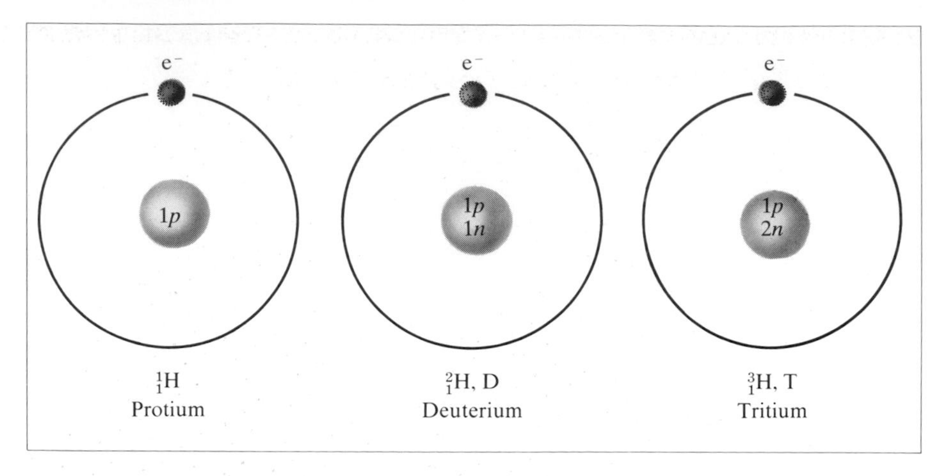

Figure 5.11
Schematic diagram of the isotopes of hydrogen. The number of protons (p) and neutrons (n) are shown within the nucleus. The electron, e^-, exists in an orbital outside the nucleus.

with one neutron in the nucleus, has a mass of 2; the third isotope (tritium), with two neutrons, has a mass of 3 (see Figure 5.11).

The three isotopes of hydrogen may be represented by the symbols ^1_1H, ^2_1H, ^3_1H, indicating an atomic number of 1 and mass numbers of 1, 2, and 3, respectively. This method of representing atoms is called *isotopic notation.* The subscript (Z) is the atomic number; the superscript (A) is the **mass number**, which is the sum of the number of protons and the number of neutrons in the nucleus. The hydrogen isotopes may also be referred to as hydrogen-1, hydrogen-2, and hydrogen-3.

mass number

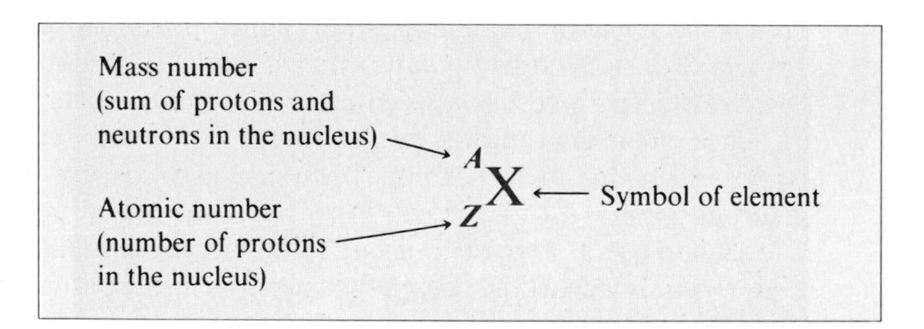

Most of the elements occur in nature as mixtures of isotopes. However, not all isotopes are stable; some are radioactive and are continuously decomposing to form other elements. For example, of the seven known isotopes of carbon, only two, carbon-12 and carbon-13, are stable. Of the seven known isotopes of oxygen, only three, $^{16}_8O$, $^{17}_8O$, and $^{18}_8O$, are stable. Of the fifteen known isotopes of arsenic, $^{75}_{33}As$ is the only one that is stable.

5.17 Atomic Weight (Atomic Mass)

Single atoms are far too small to weigh individually on a balance. But fairly precise determinations of the masses of individual atoms can be made with an instrument called a *mass spectrometer*. The mass of a single hydrogen atom is 1.6736×10^{-24} g. However, it is neither convenient nor practical to compare the actual weights of atoms expressed in grams; therefore a table of relative atomic weights using *atomic mass units* was devised. (The term *atomic mass* is often used instead of atomic weight.) The carbon isotope having six protons and six neutrons and designated carbon-12, or $^{12}_{6}C$, was chosen as the standard for atomic weights. This reference isotope was assigned a value of exactly 12 atomic mass units (amu). Thus one **atomic mass unit** is defined as equal to exactly 1/12 of the mass of a carbon-12 atom. The actual mass of a carbon-12 atom is 1.9927×10^{-23} g, and that of one atomic mass unit is 1.6606×10^{-24} g. In the table of atomic weights all elements then have values that are relative to the mass assigned to the reference isotope carbon-12.

atomic mass unit

A table of atomic weights is given on the inside back cover of this book. Hydrogen atoms, with a mass of about 1/12 that of a carbon atom, have an average atomic mass of 1.00797 amu on this relative scale. Magnesium atoms, which are about twice as heavy as carbon, have an average mass of 24.305 amu. The average atomic mass of oxygen is 15.9994 amu (usually rounded off to 16.0 for calculations).

Since most elements occur as mixtures of isotopes with different masses, the atomic weight determined for an element represents the average relative mass of all the naturally occurring isotopes of that element. The atomic weights of the individual isotopes are approximately whole numbers, because the relative masses of the protons and neutrons are approximately 1.0 amu each. Yet we find that the atomic weights given for many of the elements deviate considerably from whole numbers. For example, the atomic weight of rubidium is 85.4678 amu, that of copper is 63.546 amu, and that of magnesium is 24.305 amu. The deviation of an atomic weight from a whole number is due mainly to the unequal occurrence of the various isotopes of an element. It is also due partly to the difference between the mass of a free proton or neutron and the mass of these same particles in the nucleus. For example, the two principal isotopes of copper are $^{63}_{29}Cu$ and $^{65}_{29}Cu$. It is apparent that copper-63 atoms are the more abundant isotope, since the atomic weight of copper, 63.546 amu, is closer to 63 than to 65 amu. The actual values of the copper isotopes observed by mass spectra determination are shown in the following table:

Isotope	Isotopic mass (amu)	Abundance (%)	Average atomic mass (amu)
$^{63}_{29}Cu$	62.9298	69.09	63.55
$^{65}_{29}Cu$	64.9278	30.91	

The average atomic mass can be calculated by multiplying the atomic mass of each isotope by the fraction of each isotope present and adding the results. The calculation for copper is

$$\begin{aligned} 62.9298 \text{ amu} \times 0.6909 &= 43.48 \text{ amu} \\ 64.9278 \text{ amu} \times 0.3091 &= \underline{20.07} \text{ amu} \\ & 63.55 \text{ amu} \end{aligned}$$

atomic weight

The **atomic weight** of an element is the average relative mass of the isotopes of that element referred to the atomic mass of carbon-12 (exactly 12.0000 amu).

The relationship between mass number and atomic number is such that, if we subtract the atomic number from the mass number of a given isotope, we obtain the number of neutrons in the nucleus of an atom of that isotope. Table 5.7 shows the application of this method of determining the number of neutrons. For example, the fluorine atom ($^{19}_{9}F$), atomic number 9, having a mass of 19 amu, contains 10 neutrons:

$$\begin{array}{ccccc} \text{Mass number} & - & \text{Atomic number} & = & \text{Number of neutrons} \\ 19 & - & 9 & = & 10 \end{array}$$

The atomic weights given in the table on the inside back cover of this book are values accepted by international agreement. You need not memorize atomic weights. In most of the calculations needed in this book, use of atomic weights to the first decimal place will give results of sufficient accuracy.

Table 5.7 Determination of the number of neutrons in an atom by subtracting the atomic number from the mass number

	Hydrogen ($^{1}_{1}H$)	Oxygen ($^{16}_{8}O$)	Sulfur ($^{32}_{16}S$)	Fluorine ($^{19}_{9}F$)	Iron ($^{56}_{26}Fe$)
Mass number	1	16	32	19	56
Atomic number	−1	−8	−16	−9	−26
Number of neutrons	0	8	16	10	30

QUESTIONS

A. *Review the meanings of the new terms introduced in this chapter.*

1. Dalton's atomic theory
2. Subatomic particles
3. Electron
4. Proton
5. Neutron
6. Nucleus
7. Orbital
8. Energy levels of electrons
9. Electron shells
10. Atomic number
11. Noble gases
12. Octet rule
13. Isotopes
14. Mass number
15. Atomic mass unit
16. Atomic weight

B. *Information useful in answering these questions will be found in the tables and figures.*

1. What are the atomic numbers of (a) copper, (b) nitrogen, (c) phosphorus, (d) radium, and (e) zinc?
2. A neutron is approximately how many times heavier than an electron?
3. Explain why, in Rutherford's experiments, some alpha particles were scattered at large angles by the gold foil or even bounced back from it?
4. How many electrons can be present in the fourth energy level?
5. How many orbitals can exist in the third energy level? What are they?
6. Sketch the s, p_x, p_y, and p_z orbitals.
7. Diagram the atomic structures of the following atoms:
   (a) $^{14}_{7}N$ (b) $^{35}_{17}Cl$ (c) $^{65}_{30}Zn$ (d) $^{91}_{40}Zr$
   (e) $^{127}_{53}I$
8. Show the Lewis-dot structures for C, Mg, Al, Cl, and K.
9. In the designation $3d^7$, give the significance of the 3, the d, and the 7.
10. Using the method shown in Figure 5.9, show the orbital electron structure for an atom of:
    (a) Si (b) S (c) Ar (d) V
11. What electron structure do the noble gases have in common?

C. *Review Questions*

Atomic Structure

1. From the point of view of a chemist what are the essential differences among a proton, a neutron, and an electron?
2. Describe the general arrangement of subatomic particles in the atom.
3. What part of the atom contains practically all its mass?
4. What experimental evidence led Rutherford to conclude each of the following?
   (a) The nucleus of the atom contains most of the atomic mass.
   (b) The nucleus of the atom is positively charged.
   (c) The atom consists of mostly empty space.
5. What contribution did each of the following scientists make to the atomic theory?
   (a) Dalton (c) Rutherford (e) Bohr
   (b) Thomson (d) Chadwick
6. What is the major difference between an orbital and a Bohr orbit?
7. Explain how the spectral lines of hydrogen occur.
8. Which of the following statements are correct?
   (a) John Dalton developed an important atomic theory in the early 1800s.
   (b) Dalton said that elements are composed of minute indivisible particles called *atoms*.
   (c) Dalton said that when atoms combine to form compounds, they do so in simple numerical ratios.
   (d) Dalton said that atoms are composed of protons, neutrons, and electrons.
   (e) All of Dalton's theory is still considered valid today.
   (f) Hydrogen is the smallest atom.
   (g) A proton is about 1837 times as heavy as an electron.
   (h) The nucleus of an atom contains protons, neutrons, and electrons.
   (i) The Bohr theory proposed that electrons move around the nucleus in circular orbits.
   (j) Bohr concluded from his experiment that the positive charge and almost all the mass were concentrated in a very small nucleus.

Electron Structure

9. What is an electron orbital?
10. Under which conditions can a second electron enter an orbital already containing one electron?
11. What is meant when we say that the electron structure of an atom is in its ground state?
12. How do 1s and 2s orbitals differ? How are they alike?
13. What letters are used to designate the energy sublevels?
14. List the following electron sublevels in order of increasing energy: 2s, 2p, 4s, 1s, 3d, 3p, 4p, 3s.
15. How many s electrons, p electrons, and d electrons are possible in any electron shell?
16. How many protons are in the nucleus of an atom of each of these elements: H, B, F, Sc, Ag, U, Br, Sb, and Pb?
17. Give the electron structure ($1s^22s^22p^6\ldots$) for B, Ti, Zn, Br, and Sr.
18. Why is the eleventh electron of the sodium atom located in the third energy level rather than in the second energy level?

19. Why is the last electron in potassium located in the fourth energy level rather than in the third energy level?

20. Which atoms have the following electron structures?
(a) $1s^22s^22p^63s^2$
(b) $1s^22s^22p^5$
(c) $1s^22s^22p^63s^23p^64s^23d^8$
(d) $1s^22s^22p^63s^23p^64s^23d^5$
(e) $1s^22s^22p^63s^23p^64s^23d^{10}4p^65s^14d^5$

21. Show the electron structures ($1s^22s^22p^6$ ...) for elements of atomic numbers 8, 11, 17, 23, 28, and 34.

22. Using only Table 5.5 show the electron structures ($1s^22s^22p^6$ ...) for the elements having the following numbers of electrons:
(a) 9 (b) 26 (c) 31 (d) 39 (e) 52

23. Which elements have the following electron structures?
(a) $[Ar]4s^23d^1$ (c) $[Kr]5s^24d^{10}5p^2$
(b) $[Ar]4s^23d^{10}4p^6$ (d) $[Xe]6s^1$

24. Identify these atoms from their atomic structure diagrams:
(a) (16p 16n) $2e^-$ $8e^-$ $6e^-$
(b) (28p 32n) $2e^-$ $8e^-$ $16e^-$ $2e^-$

25. Diagram the atomic structures (as in Question 24) for these atoms:
(a) $^{27}_{13}Al$ (b) $^{51}_{23}V$

26. State the octet rule and its relationship to the noble gases.

27. Write Lewis-dot symbols for these atoms: He, B, O, Na, Si, Ar, Ga, Ca, Br, and Kr.

28. Which of the following statements are correct?
(a) In the ground state, electrons tend to occupy orbitals having the lowest possible energy.
(b) The maximum number of *p* electrons in the first energy level is six.
(c) A 2*s* electron is in a lower energy state than a 2*p* electron.
(d) The electron structure for a carbon atom is $1s^22s^22p^2$.
(e) The $2p_x$, $2p_y$, and $2p_z$ electron orbitals are all in the same energy state.
(f) The energy level of a 3*d* electron is higher than that of a 4*s* electron.
(g) The electron structure for a calcium atom is $1s^22s^22p^63s^23p^63d^2$.
(h) There are seven principal energy levels for the known elements.
(i) The third energy level can have a maximum of 18 electrons.
(j) The number of possible *d* electrons in the third energy level is ten.
(k) The first *f* electron occurs in the fourth principal energy level.
(l) The Lewis-dot symbol for nitrogen is **:Ṅ·** (with a dot below)
(m) The Lewis-dot symbol for potassium is **P·**
(n) Atoms of all the noble gases (except helium) have eight electrons in their outermost energy level.
(o) A *p* orbital is spherically symmetrical around the nucleus.
(p) An atom of nitrogen has two electrons in a 1*s* orbital, two electrons in a 2*s* orbital, and one electron in each of three different 2*p* orbitals.
(q) The maximum number of electrons that can occupy a specific energy level **n** is given by $2\mathbf{n}^2$.
(r) $^{12}_{6}C$ is an isotope of carbon that is used as the reference standard for the atomic mass system.
(s) The Lewis-dot symbol for the noble gas helium is **:He:** (with dots above and below)
(t) When an orbital contains two electrons, the electrons have parallel spins.

Isotopes, Isotopic Notation

29. In what ways are isotopes alike? In what ways are they different?

30. What special names are given to the isotopes of hydrogen?

31. List the similarities and differences in the three isotopes of hydrogen.

32. What is the symbol and name of the element that has an atomic number of 24 and a mass number of 52.

33. An atom of an element has a mass number of 201 and has 121 neutrons in its nucleus.
(a) What is the electrical charge of the nucleus?
(b) What is the symbol and name of the element?

34. What is the nuclear composition of the six naturally occurring isotopes of calcium having mass numbers of 40, 42, 43, 44, 46, and 48?

35. What are the numbers of protons, neutrons, and electrons in each of the following?
(a) $^{79}_{35}Br$ (b) $^{131}_{56}Ba$ (c) $^{238}_{92}U$ (d) $^{56}_{26}Fe$
36. What letters are used to designate atomic number and mass number in isotopic notation of atoms?
37. Write isotopic notation symbols for the following.
(a) $Z = 26, A = 55$ (d) $Z = 14, A = 29$
(b) $Z = 12, A = 26$ (e) $Z = 79, A = 188$
(c) $Z = 3, A = 6$
38. Give the isotopic notation ($^{73}_{32}Ge$, for example) for:
(a) An atom containing 27 protons, 32 neutrons, and 27 electrons.
(b) An atom containing 110 neutrons, 74 electrons, and 74 protons.
39. Which of the following statements are correct?
(a) An element with an atomic number of 29 has 29 protons, 29 neutrons, and 29 electrons.
(b) An atom of the isotope $^{60}_{26}Fe$ has 34 neutrons in its nucleus.
(c) $^{2}_{1}H$ is a symbol for the isotope deuterium.
(d) An atom of $^{31}_{15}P$ contains 15 protons, 16 neutrons, and 31 electrons.
(e) In the isotope $^{6}_{3}Li$, $Z = 3$ and $A = 3$.
(f) Isotopes of a given element have the same number of protons but differ in the number of neutrons.
(g) The three isotopes of hydrogen are called protium, deuterium, and tritium.
(h) $^{23}_{11}Na$ and $^{24}_{11}Na$ are isotopes.
(i) $^{24}_{11}Na$ has one more electron than $^{23}_{11}Na$.
(j) $^{24}_{11}Na$ has one more proton than $^{23}_{11}Na$.
(k) $^{24}_{11}Na$ has one more neutron than $^{23}_{11}Na$.
(l) Only a few of the elements exist in nature as mixtures of isotopes.

Atomic weight (Atomic mass)

40. Explain why the atomic weights of elements are not whole numbers.
41. Is the isotopic mass of a given isotope ever an exact whole number? Is it always? In answering, consider the masses of $^{12}_{6}C$ and $^{63}_{29}Cu$.
42. Which of the isotopes of calcium in Question 34 is the most abundant isotope? Can you be sure? Explain your choice.
43. Which of the following statements are correct?
(a) One atomic mass unit weighs 12 times as much as one carbon-12 atom.
(b) The atomic masses of protium and deuterium differ by about 100%.
(c) $^{23}_{11}Na$ and $^{24}_{11}Na$ have the same atomic masses.
(d) The atomic weight of an element represents the average relative atomic mass of all the naturally occurring isotopes of that element.

D. *Review problems*

1. Change the following to powers of 10 (scientific notation):
(a) 510,000 (c) $(0.001)^2$
(b) 0.000274 (d) $(8.0)^3$
2. Using the formula $2\mathbf{n}^2$, calculate the number of electrons that can exist in principal energy levels: $\mathbf{n} = 1, 2, 3, 4, 5$, and 6.
3. Complete the following table with the appropriate data for each isotope given:

Atomic number	Mass number	Symbol of element	Number of protons	Number of neutrons
(a) 8	16			
(b)		Ni		30
(c)	199		80	

4. The actual mass of one atom of an unknown isotope is 2.18×10^{-22} amu. Calculate the atomic mass of this isotope.
5. Naturally occurring silver exists as two stable isotopes, ^{107}Ag with a mass of 106.9041 amu (51.82%) and ^{109}Ag with a mass of 108.9047 amu (48.18%). Calculate the average atomic weight of silver.
6. Naturally occurring magnesium consists of three stable isotopes: ^{24}Mg, 23.985 amu (78.99%); ^{25}Mg, 24.986 amu (10.00%); and ^{26}Mg, 25.983 amu (11.01%). Calculate the average atomic weight of magnesium.

CHAPTER 6

The Periodic Arrangement of the Elements

After studying Chapter 6, you should be able to

1 Understand the terms listed in Question A at the end of the chapter.
2 Describe briefly the contributions of Döbereiner, Newlands, Mendeleev, Meyer, and Moseley to the development of the periodic law.
3 State the periodic law in its modern form.
4 Indicate the locations of the metals, nonmetals, metalloids, and noble gases in the periodic table.
5 Indicate in the periodic table the areas where the *s*, *p*, *d*, and *f* orbitals are being filled.
6 Describe how atomic radii vary (a) from left to right in a period and (b) from top to bottom in a group.
7 Distinguish between representative elements and transition elements.
8 Identify groups of elements by their special names.
9 Describe the changes in outer-level electron structure when (a) moving from left to right in a period and (b) going from top to bottom in a group.
10 Explain the relationship between group number and the number of outer-shell electrons for the representative elements.
11 List the general characteristics of group properties.
12 Write Lewis-dot symbols for the representative elements from their position in the periodic table.
13 Predict formulas of simple compounds formed between the representative (Group A) elements using the periodic table.

14 Point out how the change in electron structure between adjacent transition elements differs from that between representative elements.

6.1 Early Attempts to Classify the Elements

Chemists of the early 19th century had sufficient knowledge of the properties of elements to recognize similarities among groups of elements. As early as 1817 J. W. Döbereiner (1780–1849), professor at the University of Jena in Germany, observed the existence of *triads* of similarly behaving elements, in which the middle element had an atomic weight approximating the average of the other two elements. He also noted that for many other properties the value for the central element was approximately the average of the values for the other two elements. Table 6.1 presents comparative data on atomic weight and density for two sets of Döbereiner's triads.

In 1864, J. A. R. Newlands (1837–1898), an English chemist, reported his *Law of Octaves*. In his studies Newlands observed that, when the elements were arranged according to increasing atomic weights, every eighth element had similar properties. (The noble gases were not yet discovered at that time.) Newlands' theory was ridiculed by his contemporaries in the Royal Chemical Society, and they refused to publish his work. Many years later, however, Newlands was awarded the highest honor of the society for this important contribution to the development of the periodic law.

In 1869 Dmitri Ivanovitch Mendeleev (1834–1907) of Russia and Lothar Meyer (1830–1895) of Germany independently published their periodic arrangements of the elements that were based on increasing atomic weights. Because his arrangement was published slightly earlier and was in a somewhat more useful form than that of Meyer, Mendeleev's name is usually associated with the modern periodic table.

Table 6.1 Döbereiner's triads

Triads	Atomic weight	Density (g/mL at 4°C)
Chlorine	35.5	1.56[a]
Bromine	79.9	3.12
Iodine	126.9	4.95
Average of chlorine and iodine	81.2	3.26
Calcium	40.1	1.55
Strontium	87.6	2.6
Barium	137.4	3.5
Average of calcium and barium	88.8	2.52

[a] Density at −34°C (liquid)

6.2 The Periodic Law

Only about 63 elements were known when Mendeleev constructed his table. He arranged these elements so that those with similar chemical properties fitted into columns to form family groups. The arrangement left many gaps between elements, and Mendeleev predicted that these spaces would be filled as new elements were discovered. For example, spaces for undiscovered elements were left after calcium, under aluminum, and under silicon. He called these unknown elements eka-boron, eka-aluminum, and eka-silicon. The term *eka* comes from Sanskrit meaning "one" and was used to indicate that the missing element was one place away in the table from the element indicated. Mendeleev even went so far as to predict with high accuracy the physical and chemical properties of these undiscovered elements. The three elements, scandium (atomic number 21), gallium (31), and germanium (32), were in fact discovered during Mendeleev's lifetime and were found to have properties agreeing very closely with the predictions that he had made for eka-boron, eka-aluminum, and eka-silicon. The amazing way in which Mendeleev's predictions were fulfilled is illustrated in Table 6.2, which compares the predicted properties of eka-silicon with those of germanium, discovered by the German chemist C. Winkler in 1886.

Two major additions have been made to the periodic table since Mendeleev's time: (1) A new family of elements, the noble gases, was discovered and added; and (2) elements having atomic numbers greater than 92 have been discovered and fitted into the table.

The term *periodic* means recurring at regular intervals. The original periodic tables were based on the premise that the properties of the elements are periodic functions of their atomic weights. However, this basic premise had some disturbing discrepancies. For example, the atomic weight for argon is greater than that of potassium, yet argon had to be placed before potassium, because argon is certainly a noble gas, and potassium behaves like the other alkali metals. These discrepancies were resolved by the work of the British physicist, H. G. J. Moseley (1887–1915) and by the discovery of the existence of isotopes. Moseley noted that the X-ray emission frequencies of the elements increased in a regular, stepwise fashion each time the nuclear charge (atomic number) increased by one unit. This observation meant that the periodic table must be based on a revised premise—namely, that the properties of the elements are a periodic function of their *atomic numbers*. Under this revised premise argon (18) properly comes before potassium (19), even though the atomic weight of argon is greater than the atomic weight of potassium. The current statement of the **periodic law** is

periodic law

The properties of the chemical elements recur periodically when the elements are arranged in increasing order of their atomic numbers.

As the format of the periodic table is studied, it becomes evident that the periodicity of the properties of the elements is due to the recurring similarities of their electron structures.

Table 6.2 Comparison of the properties of eka-silicon predicted by Mendeleev with the properties of germanium

Property	Mendeleev's predictions in 1871 for eka-silicon (Es)	Observed properties for germanium (Ge)
Atomic weight	72	72.6
Color of metal	Dirty gray	Grayish-white
Density	5.5 g/mL	5.35 g/mL
Oxide formula	EsO_2	GeO_2
Oxide density	4.7 g/mL	4.70 g/mL
Chloride formula	$EsCl_4$	$GeCl_4$
Chloride density	1.9 g/mL	1.87 g/mL
Boiling temperature of chloride	Under 100°C	86°C

6.3 Arrangement of the Periodic Table

periodic table

periods of elements

groups or families of elements

The modern **periodic table** is shown in Table 6.3 and on the inside front cover of this book. In this table the elements are arranged horizontally in numerical sequence according to their atomic numbers; the result is seven horizontal rows called **periods**. Each period, with the exception of the first, starts with an alkali metal and ends with a noble gas. This arrangement forms vertical columns of elements that have identical or similar outer-shell electron structures and thus similar chemical properties. These columns are known as **groups** or **families of elements**.

The heavy zigzag line starting at boron and running diagonally down the table separates the elements into metals and nonmetals. The elements to the right of the line are nonmetallic, and those to the left are metallic. The elements bordering the zigzag line are the metalloids, which show both metallic and nonmetallic properties. With some exceptions the characteristic electron arrangement of metals is that their atoms have one, two, or three electrons in the outer energy level, whereas nonmetals have five, six, or seven electrons in the outer energy level.

In this periodic arrangement the elements also fall into blocks according to the sublevel of electrons that is being filled. The grouping of the elements into *spdf* blocks is shown in Table 6.4. The *s* and *p* blocks are made up of Group A elements and noble gases. The *s* block consists of the Group IA and IIA elements and helium; each element in this block has one or two *s* electrons in its outer energy level. The *p* block contains the Group IIIA through VIIA elements and the noble gases (except helium). In the *p* block, electrons are filling the *p* sublevel orbitals. Each *p* block element has two *s* electrons and from one to six *p* electrons in its outermost energy level.

The *d* block includes the transition elements of Groups IB through VIIB and Group VIII (see Section 6.7). In this block electrons are filling the *d* sublevel

Table 6.3 Periodic table of the elements

Atomic number — 11
Name — Sodium
Symbol — Na
Electron structure — 2 8 1
Atomic weight — 22.98977

[a]Mass number of most stable or best-known isotope

[b]Mass of the isotope of longest half-life

Transition elements (IIIB – VIII)

Period	Group IA	IIA	IIIB	IVB	VB	VIB	VIIB	VIII	VIII
1	1 Hydrogen **H** 1.0079 1								
2	3 Lithium **Li** 6.941 2 1	4 Beryllium **Be** 9.01218 2 2							
3	11 Sodium **Na** 22.98977 2 8 1	12 Mag-nesium **Mg** 24.305 2 8 2							
4	19 Potas-sium **K** 39.098 2 8 8 1	20 Calcium **Ca** 40.08 2 8 8 2	21 Scandium **Sc** 44.9559 2 8 9 2	22 Titanium **Ti** 47.90 2 8 10 2	23 Vanadium **V** 50.9414 2 8 11 2	24 Chro-mium **Cr** 51.996 2 8 13 1	25 Manga-nese **Mn** 54.9380 2 8 13 2	26 Iron **Fe** 55.847 2 8 14 2	27 Cobalt **Co** 58.9332 2 8 15 2
5	37 Rubid-ium **Rb** 85.4678 2 8 18 8 1	38 Strontium **Sr** 87.62 2 8 18 8 2	39 Yttrium **Y** 88.9059 2 8 18 9 2	40 Zirco-nium **Zr** 91.22 2 8 18 10 2	41 Niobium **Nb** 92.9064 2 8 18 12 1	42 Molyb-denum **Mo** 95.94 2 8 18 13 1	43 Techne-tium **Tc** 98.9062[b] 2 8 18 14 1	44 Ruthe-nium **Ru** 101.07 2 8 18 15 1	45 Rhodium **Rh** 102.9055 2 8 18 16 1
6	55 Cesium **Cs** 132.9054 2 8 18 18 8 1	56 Barium **Ba** 137.34 2 8 18 18 8 2	57 Lan-thanum **La*** 138.9055 2 8 18 18 9 2	72 Hafnium **Hf** 178.49 2 8 18 32 10 2	73 Tantalum **Ta** 180.9479 2 8 18 32 11 2	74 Wolfram (Tungsten) **W** 183.85 2 8 18 32 12 2	75 Rhenium **Re** 186.2 2 8 18 32 13 2	76 Osmium **Os** 190.2 2 8 18 32 14 2	77 Iridium **Ir** 192.22 2 8 18 32 17 0
7	87 Francium **Fr** (223)[a] 2 8 18 32 18 8 1	88 Radium **Ra** 226.0254[b] 2 8 18 32 18 8 2	89 Actinium **Ac**** (227)[a] 2 8 18 32 18 9 2	104 Unnilquad-ium **Unq** (261)[a] 2 8 18 32 32 10 2	105 Unnilpent-ium **Unp** (262)[a] 2 8 18 32 32 11 2	106 Unnilhex-ium **Unh** (263)[a] 2 8 18 32 32 12 2	107	—	109

* Lanthanide series 6	58 Cerium **Ce** 140.12 2 8 18 20 8 2	59 Praseo-dymium **Pr** 140.9077 2 8 18 21 8 2	60 Neo-dymium **Nd** 144.24 2 8 18 22 8 2	61 Pro-methium **Pm** (145)[a] 2 8 18 23 8 2	62 Samar-ium **Sm** 150.4 2 8 18 24 8 2
** Actinide series 7	90 Thorium **Th** 232.0381[b] 2 8 18 32 18 10 2	91 Protac-tinium **Pa** 231.0359[b] 2 8 18 32 20 9 2	92 Uranium **U** 238.029 2 8 18 32 21 9 2	93 Nep-tunium **Np** 237.0482 2 8 18 32 22 9 2	94 Plu-tonium **Pu** (242)[a] 2 8 18 32 23 9 2

Atomic weights are based on carbon-12. Atomic weights in parentheses indicate the most stable or best-known isotope. Slight disagreement exists as to the exact electronic configuration of several of the high-atomic-number elements. Names and symbols for elements 104, 105, and 106 are unofficial.

	IB	IIB	IIIA	IVA	VA	VIA	VIIA	Noble gases
								2 Helium **He** 4.00260 2
			5 Boron **B** 10.81 2 3	6 Carbon **C** 12.011 2 4	7 Nitrogen **N** 14.0067 2 5	8 Oxygen **O** 15.9994 2 6	9 Fluorine **F** 18.99840 2 7	10 Neon **Ne** 20.179 2 8
			13 Aluminum **Al** 26.98154 2 8 3	14 Silicon **Si** 28.086 2 8 4	15 Phosphorus **P** 30.97376 2 8 5	16 Sulfur **S** 32.06 2 8 6	17 Chlorine **Cl** 35.453 2 8 7	18 Argon **Ar** 39.948 2 8 8
28 Nickel **Ni** 58.71 2 8 16 2	29 Copper **Cu** 63.546 2 8 18 1	30 Zinc **Zn** 65.38 2 8 18 2	31 Gallium **Ga** 69.72 2 8 18 3	32 Germanium **Ge** 72.59 2 8 18 4	33 Arsenic **As** 74.9216 2 8 18 5	34 Selenium **Se** 78.96 2 8 18 6	35 Bromine **Br** 79.904 2 8 18 7	36 Krypton **Kr** 83.80 2 8 18 8
46 Palladium **Pd** 106.4 2 8 18 18 0	47 Silver **Ag** 107.868 2 8 18 18 1	48 Cadmium **Cd** 112.40 2 8 18 18 2	49 Indium **In** 114.82 2 8 18 18 3	50 Tin **Sn** 118.69 2 8 18 18 4	51 Antimony **Sb** 121.75 2 8 18 18 5	52 Telurium **Te** 127.60 2 8 18 18 6	53 Iodine **I** 126.9045 2 8 18 18 7	54 Xenon **Xe** 131.30 2 8 18 18 8
78 Platinum **Pt** 195.09 2 8 18 32 17 1	79 Gold **Au** 196.9665 2 8 18 32 18 1	80 Mercury **Hg** 200.59 2 8 18 32 18 2	81 Thallium **Tl** 204.37 2 8 18 32 18 3	82 Lead **Pb** 207.2 2 8 18 32 18 4	83 Bismuth **Bi** 208.9804 2 8 18 32 18 5	84 Polonium **Po** (210)[a] 2 8 18 32 18 6	85 Astatine **At** (210)[a] 2 8 18 32 18 7	86 Radon **Rn** (222)[a] 2 8 18 32 18 8

Inner transition elements

63 Europium **Eu** 151.96 2 8 18 25 8 2	64 Gadolinium **Gd** 157.25 2 8 18 25 9 2	65 Terbium **Tb** 158.9254 2 8 18 27 8 2	66 Dysprosium **Dy** 162.50 2 8 18 28 8 2	67 Holmium **Ho** 164.9304 2 8 18 29 8 2	68 Erbium **Er** 167.26 2 8 18 30 8 2	69 Thulium **Tm** 168.9342 2 8 18 31 8 2	70 Ytterbium **Yb** 173.04 2 8 18 32 8 2	71 Lutetium **Lu** 174.97 2 8 18 32 9 2
95 Americium **Am** (243)[a] 2 8 18 32 25 8 2	96 Curium **Cm** (247)[a] 2 8 18 32 25 9 2	97 Berkelium **Bk** (249)[a] 2 8 18 32 26 9 2	98 Californium **Cf** (251)[a] 2 8 18 32 27 9 2	99 Einsteinium **Es** (254)[a] 2 8 18 32 28 9 2	100 Fermium **Fm** (253)[a] 2 8 18 32 29 9 2	101 Mendelevium **Md** (256)[a] 2 8 18 32 30 9 2	102 Nobelium **No** (254)[a] 2 8 18 32 31 9 2	103 Lawrencium **Lr** (257)[a] 2 8 18 32 32 9 2

Table 6.4 Arrangement of the elements into blocks according to the sublevel of electrons being filled in their atomic structure

s block; *d* block, Transition elements; *p* block; Noble gases

Period	IA	IIA	IIIB	IVB	VB	VIB	VIIB	VIII	VIII	VIII	IB	IIB	IIIA	IVA	VA	VIA	VIIA	Noble gases
1	1*s* H																	1*s* He
2	2*s* → Li	Be											2*p* → B	C	N	O	F	Ne
3	3*s* → Na	Mg											3*p* → Al	Si	P	S	Cl	Ar
4	4*s* → K	Ca	3*d* → Sc	Ti	V	Cr	Mn	Fe	Co	Ni	Cu	Zn	4*p* → Ga	Ge	As	Se	Br	Kr
5	5*s* → Rb	Sr	4*d* → Y	Zr	Nb	Mo	Tc	Ru	Rh	Pd	Ag	Cd	5*p* → In	Sn	Sb	Te	I	Xe
6	6*s* → Cs	Ba	5*d* → La	Hf	Ta	W	Re	Os	Ir	Pt	Au	Hg	6*p* → Tl	Pb	Bi	Po	At	Rn
7	7*s* → Fr	Ra	6*d* → Ac	Unq	Unp	Unh												

f block
Inner transition elements

4*f* → Ce	Pr	Nd	Pm	Sm	Eu	Gd	Tb	Dy	Ho	Er	Tm	Yb	Lu
5*f* → Th	Pa	U	Np	Pu	Am	Cm	Bk	Cf	Es	Fm	Md	No	Lr

s block
p block
d block
f block

orbitals. Each period beginning with period 4 can have ten *d* block elements. Periods 4, 5, and 6 each have ten; period 7 is incomplete (see Table 6.4).

The *f* block consists of the inner transition elements and includes two series of 14 elements in periods 6 and 7. The first series, found in period 6, is the lanthanide series (Ce to Lu); the second, found in period 7, is the actinide series (Th to Lr). Electrons are filling the 4*f* sublevel orbitals in the lanthanide series and the 5*f* orbitals in the actinide series.

6.4 Periods of Elements

Of the seven periods of elements (see Tables 6.4 and 6.5), the first three are known as *short periods*, and the remaining four as *long periods*. The first period contains 2 elements, hydrogen and helium. Period 2 contains 8 elements, starting with lithium and ending with neon. Period 3 also contains 8 elements, sodium to argon. Periods 4 and 5 each contain 18 elements; period 6 has 32 elements; and period 7, which is incomplete, contains the remaining elements.

The number of each period corresponds to the number of the outermost energy level that contains electrons of the elements in that period. For example, the period 1 elements contain electrons in the first energy level only; period 2

Table 6.5 The number of elements in each period

Period number	Number of elements	Electron orbitals being filled in each period
1	2	1*s*
2	8	2*s*2*p*
3	8	3*s*3*p*
4	18	4*s*3*d*4*p*
5	18	5*s*4*d*5*p*
6	32	6*s*4*f*5*d*6*p*
7	22	7*s*5*f*6*d*

elements contain electrons in the first and second energy levels; period 3 elements contain electrons in the first, second, and third levels; and so on.

From the standpoint of properties, we find that the two elements in the first period are gases; hydrogen (1) is reactive, and helium (2) is a noble gas. Thereafter, as we move from left to right across the successive periods from 2 to 6, we find that the elements vary from strongly metallic at the beginning to nonmetallic at the end of each period. Period 7 is incomplete but begins with a strongly metallic element, francium (87).

Starting with the third element of the long periods, 4, 5, 6, and 7 (scandium, Sc; yttrium, Y; lanthanum, La; actinium, Ac), the inner shells of *d* and *f* orbital electrons begin to fill in, forming the transition and inner transition elements (*d* and *f* blocks). Both the transition and inner transition elements are metallic.

In general the atomic radii of the elements within a period decrease with increasing nuclear charge. This decrease occurs because, as the positive charge on the nucleus increases, it exerts a greater attractive force on the electrons, causing the atom to become smaller. Therefore the size of the atoms becomes progressively smaller from left to right within each period (see Figure 6.1). Because the noble gases do not readily combine with other elements to form compounds, the radii of their atoms are not determined in the same comparative manner as are those of the other elements. However, calculations have shown that the radii

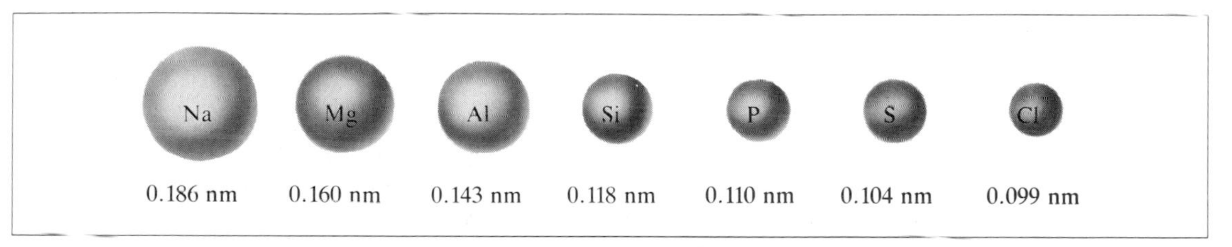

Figure 6.1

Radii of period 3 elements. In general the size of the atoms in a period decreases with increasing nuclear charge.

of the noble gas atoms are about the same as, or slightly smaller than, the element immediately preceding them. Slight deviations in atomic radii occur in the middle of the long periods of the elements.

6.5 Groups or Families of Elements

representative elements

transition elements

The groups, or families, of elements are numbered IA through VIIA, IB through VIIB, VIII, and noble gases. Group A elements are often referred to as the **representative**, or *main*, groups of **elements**. They are filling in electrons in the *s* and *p* orbitals. Group B and Group VIII elements are known as the **transition elements**. They are filling electrons in the *d* and *f* orbitals.

The elements comprising each family have similar outer energy-level electron structures. In the atoms of Group A elements the number of electrons in the outer energy level is identical to the group number. Group IA is known as the *alkali metals*. Each atom of this family of elements has one *s* electron in its outer energy level. Group IIA atoms, known as the *alkaline earth metals*, each have two *s* electrons in their outer energy level. Group VIIA is known as the halogens; their atoms each have seven electrons (two *s* and five *p*) in their outer energy level. The noble gases (except helium) have eight electrons in their outer energy level.

Each alkali metal starts a new period of elements in which the *s* electron occupies a principal energy level one greater than in the previous period. As a result the size of the atoms of this and other families of elements increases from the top to the bottom of the family. (This generality has some exceptions.) The relative size of the alkali metal atoms is illustrated in Figure 6.2.

One major distinction between groups is the energy level to which the last electron is added. In the elements of Groups IA through VIIA, IB, IIB, and the noble gases, the last electron is added either to an *s* or to a *p* orbital located in the outermost energy level. (This rule has some exceptions; see Table 5.5.) In the elements of Groups IIIB through VIIB and VIII, the last electron goes to a *d* or to an *f* orbital located in an inner energy level. For examples see Figure 6.3.

The general characteristics of group properties are as follows:

1. The number of electrons in the outer energy level of Groups IA through VIIA, IB, and IIB elements is the same as the group number. The other B groups and Group VIII do not show this characteristic. Each noble gas except helium has eight electrons in its outer energy level.
2. The groups on the left and in the middle sections of the table tend to be metallic in nature. The groups on the right tend to be nonmetallic.
3. The radii of the elements increase from top to bottom within a particular group (for example, from lithium to francium).
4. Elements at the bottom of a group tend to be more metallic in their properties than those at the top. This tendency is especially noticeable in Groups IVA through VIIA.

5 Elements within an A group have the same number of electrons in their outer shell and show closely related chemical properties.
6 Elements within a B group have some similarity in electron structure and also show some similarities in chemical properties.

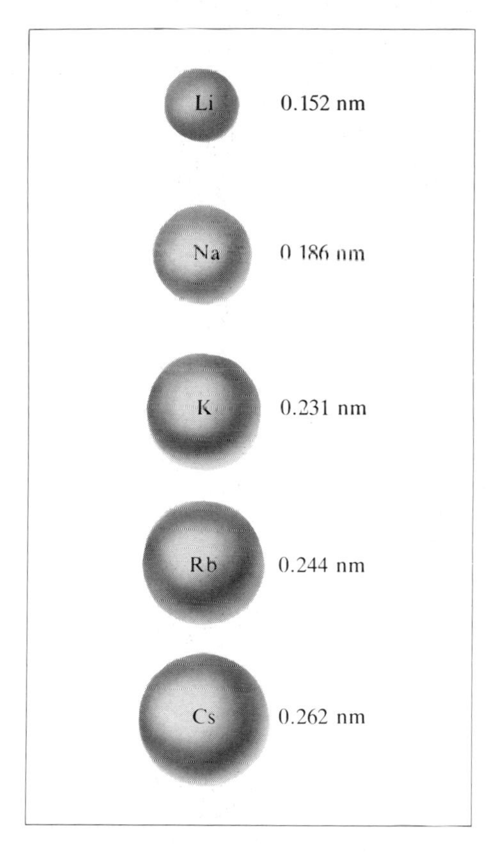

Figure 6.2

Radii of the atoms of the alkali metal family. The size of the atoms in a family increases from top to bottom, because each atom progressively contains electrons in a higher principal energy level.

Group	Noble gas	IA	IIA	IIIB	IVB
Element	Ar	K	Ca	Sc	Ti
Electron structure	2,8,8	2,8,8,1	2,8,8,2	2,8,9,2	2,8,10,2
Energy level	1,2,3	1,2,3,4	1,2,3,4	1,2,3,4	1,2,3,4
Last electron added	↑	↑	↑	↑	↑

Figure 6.3

Comparison of the placement of the last electron in Group A and Group B elements. The energy level to which the last electron is added is indicated by the arrow, an outer level for Group A elements and an inner level for Group B elements.

A new format for numbering the groups in the periodic table has been recommended by the American Chemical Society Committee on Nomenclature and by the International Union of Pure and Applied Chemistry (IUPAC) Commission on the Nomenclature of Inorganic Chemistry. The new format numbers the groups from 1 to 18 going from left to right. A periodic table using this numbering method is given in Appendix VII.

PROBLEM 6.1 Write Lewis-dot symbols for the representative elements Na, K, Sr, Se, and Br.

First we need to locate these elements in the periodic table. They are all Group A elements. Then, knowing that Group A numbers are the same as the number of outer-shell electrons and that Lewis-dot symbols show only the outer-shell electrons, we can proceed to write the Lewis-dot symbols.

Na, K	Group IA	$(1\,e^-)$	Na· K·
Sr	Group IIA	$(2\,e^-)$	Sr:
Se	Group VIA	$(6\,e^-)$	:S̤̈e·
Br	Group VIIA	$(7\,e^-)$	:B̤̈r·

6.6 Predicting Formulas by Use of the Periodic Table

The periodic table can be used to predict the formulas of simple compounds. As we shall see in Chapter 7, the chemical properties of the elements are dependent on their electrons. The representative (Group A) elements ordinarily form compounds using only the electrons in their outer energy level. If we examine Group IA, the alkali metals, we see that all of them have one electron in their outer energy level; all follow a noble gas in the table; and all, except lithium, have eight electrons in their next inner shell. These likenesses suggest that these metals should have a great deal of similarity in their chemistry, since their chemical properties are vested primarily in their outer-shell electron. And they do have similarity: All readily lose their outer electron and attain a noble gas electron structure. In doing so they form compounds with similar atomic compositions. For example, all the monoxides of Group IA contain two atoms of the alkali metal to one atom of oxygen. Their formulas are Li_2O, Na_2O, K_2O, Rb_2O, Cs_2O, and Fr_2O.

How can we use the table to predict formulas of other compounds? Because of similar electron structures, the elements in a family generally form compounds with the same atomic ratios, as was shown in the preceding paragraph for the oxides of the Group IA metals. In general, if we know the atomic ratio of a particular compound, say sodium chloride (NaCl), we can predict the atomic

Table 6.6 Elements in Group A families have the same outer-shell electron structure and are likely to form compounds with the same atomic ratios. Shown, for example, are the monoxides, chlorides, bromides, and sulfates of the alkali metals.

Lewis-dot structure	Monoxides	Chlorides	Bromides	Sulfates
Li·	Li_2O	LiCl	LiBr	Li_2SO_4
Na·	Na_2O	NaCl	NaBr	Na_2SO_4
K·	K_2O	KCl	KBr	K_2SO_4
Rb·	Rb_2O	RbCl	RbBr	Rb_2SO_4
Cs·	Cs_2O	CsCl	CsBr	Cs_2SO_4

ratios and formulas of the other alkali metal chlorides. These formulas are LiCl, KCl, RbCl, CsCl, and FrCl (see Table 6.6).

In a similar way, if we know that the formula of the oxide of hydrogen is H_2O, we can predict that the formula of the sulfide will be H_2S, because sulfur has the same outer-shell electron structure as oxygen. It must be recognized, however, that these are only predictions; it does not necessarily follow that every element in a group will behave like the others or even that a predicted compound will actually exist. Knowing the formulas for potassium chlorate, bromate, and iodate to be $KClO_3$, $KBrO_3$, and KIO_3, we can correctly predict the corresponding sodium compounds to have the formulas $NaClO_3$, $NaBrO_3$, and $NaIO_3$. Fluorine belongs to the same family of elements (Group VIIA) as chlorine, bromine, and iodine. So we can predict that potassium and sodium fluorates will have the formulas KFO_3 and $NaFO_3$. But this prediction would not be correct, because potassium and sodium fluorates are not known to exist. However, if they did exist, the formulas could very well be correct, for these predictions are based on comparisons with known formulas and similar electron structures.

Predicting formulas using this type of analogy is not as reliable for the transition (Group B) elements.

PROBLEM 6.2

The formula for calcium sulfate is $CaSO_4$ and that for lithium carbonate is Li_2CO_3. Predict formulas for (a) magnesium sulfate, (b) potassium carbonate, and (c) magnesium selenate.

(a) Look in the periodic table for calcium and magnesium. They are both in Group IIA. Since the formula for calcium sulfate is $CaSO_4$, it is reasonable to predict that the formula for magnesium sulfate is $MgSO_4$.

(b) Find lithium and potassium in the periodic table. They are in Group IA. Since the formula for lithium carbonate is Li_2CO_3, it is reasonable to predict that K_2CO_3 is the formula for potassium carbonate.

(c) Find selenium in the periodic table. It is in Group VIA just below sulfur. Therefore it is reasonable to assume that selenium forms selenate in the same way that sulfur forms sulfate. Since $MgSO_4$ was the predicted formula for magnesium sulfate in part (a), it is reasonable to assume that the formula for magnesium selenate is $MgSeO_4$.

6.7 Transition Elements

Elements in Groups IB, IIIB through VIIB, and VIII are known as the transition elements. Transition elements occur in periods 4, 5, 6, and 7. The transition elements are characterized by an increasing number of *d* or *f* electrons in an inner shell; they all have either one or two electrons in their outer shell. In period 4, electrons enter the 3*d* sublevel. In period 5, electrons enter the 4*d* sublevel. The transition elements in period 6 include the lanthanide series (or *rare earth elements*, Ce to Lu), in which electrons are entering the 4*f* sublevel. The 5*d* sublevel also fills up in the sixth period. The seventh period of elements is an incomplete period. It includes the actinide series (Th to Lr) in which electrons are entering the 5*f* and 6*d* sublevels.

All of the transition elements are metals. Their oxidation states may be variable; that is, in the formation of compounds electrons may come from more than one energy level. For this reason many of these metals form multiple series of compounds. These compounds often occur in the form of some of the most beautifully colored crystals to be found in nature.

6.8 Noble Gases

The noble gases are the last group in the periodic table and are sometimes called the zero group. These gases are characterized by extremely low chemical reactivity and were, in fact, formerly known as the inert gases because they were believed to be chemically inert. Their low reactivity is associated with a stable electron outer shell consisting of filled *s* orbitals and (except helium) filled *p* orbitals (see Section 5.15).

All the noble gases except radon are normally present in the atmosphere. Argon is the most abundant, about 1% by volume. The others are present only in trace amounts. Argon was discovered in 1894 by Lord Raleigh (1842–1919) and Sir William Ramsay (1852–1916). Helium was first observed in the spectrum of the sun during an eclipse in 1868. It was not until 1894 that Ramsay recognized that helium exists in the earth's atmosphere. In 1898 he and his co-worker, Morris W. Travers (1872–1961), announced the discovery of neon, krypton, and xenon, having isolated them from liquid air. Friedrich E. Dorn (1848–1916) first identified radon, the heaviest member of the noble gases, as a radioactive gas emanating from the element radium.

Because of its low density and nonflammability, helium has been used for filling balloons and dirigibles. Only hydrogen surpasses helium in lifting power. Helium mixed with oxygen is used by deep-sea divers for breathing. This mixture reduces the danger of acquiring the "bends" (caisson disease), pains and paralysis suffered by divers on returning from the ocean depths to normal atmospheric pressure. Helium is also used in heliarc welding, to supply an inert atmosphere for the welding of active metals such as magnesium. Helium is found in some natural gas wells in the southwestern United States. As a liquid it is used to study the properties of substances at very low temperatures. The boiling point of liquid helium, 4.2 K, is not far above absolute zero.

We are all familiar with the neon sign, in which a characteristic red color is produced when an electric discharge is passed through a tube filled with neon. This color may be modified by mixing the neon with other gases or by changing the color of the glass. Argon is used primarily in gas-filled electric light bulbs and other types of electronic tubes to provide an inert atmosphere for prolonging tube life. Argon is also used in some welding applications where an inert atmosphere is needed. Krypton and xenon have not been used extensively because of their limited availability. Radon is radioactive; it has been used medicinally in the treatment of cancer.

For many years it was believed that the noble gases could not be made to combine chemically with any other element. Then, in 1962, Neil Bartlett at the University of British Columbia, Vancouver, synthesized the first noble gas compound, xenon hexafluoroplatinate, $XePtF_6$. This outstanding discovery opened a new field in the techniques of preparing noble gas compounds and investigating their chemical bonding and properties. Other compounds of xenon and compounds of krypton and radon have been prepared. Some of these are XeF_2, XeF_4, XeF_6, $XeOF_4$, $Xe(OH)_6$, and KrF_2.

6.9 New Elements

Mendeleev allowed gaps in his orderly periodic table for elements whose discovery he predicted. These were actually discovered, as were all the elements up to atomic number 92 that occur naturally on the earth. Sixteen elements beyond uranium (atomic numbers 93–107 and 109) have been discovered or synthesized since 1939. All these elements have unstable nuclei and are radioactive. Beyond element 101 the isotopes synthesized thus far have such short lives that chemical identification has not been accomplished.

Intensive research is continuing on the synthesis of still heavier elements to extend the periodic table beyond the presently known elements. It is predicted that elements 110 to 118 will be very stable but still radioactive. Element 114 will lie below lead (82) in the periodic table and should be exceptionally stable. Element 118 should be a member of the noble gas family. Elements 119 and 120 should be in Groups IA and IIA, respectively, and have electrons in the 8*s* sublevel.

6.10 Summary

The arrangement of the elements into the modern periodic table has been and continues to be of great value to chemists and chemistry students. This value increases with your knowledge of chemistry. For a given element the following data can usually be obtained directly from the table: name, symbol, atomic number, atomic weight, electron configuration, group number, period number, and whether it is a metal, nonmetal, or metalloid. From the location of the element in the periodic table, one can estimate and compare many of its properties—such as ionization energy, density, atomic radius, atomic volume, oxidation states, electrical conductance, and electronegativity—with those of other elements.

The periodic table is still used as a guide in predicting the synthesis of possible new elements. It presents a very large amount of chemical information in compact form and correlates the properties and relationships of all the elements. The table is so useful that a copy hangs in nearly every chemistry lecture hall and laboratory in the world. Refer to it often.

QUESTIONS

A. *Review the meanings of the new terms introduced in this chapter.*

1. Periodic law
2. Periodic table
3. Periods of elements
4. Groups or families of elements
5. Representative elements
6. Transition elements

B. *Information useful in answering these questions will be found in the tables and figures.*

1. From the standpoint of electron structure, what do the elements in the *s* block have in common?
2. Write the symbols for the elements having atomic numbers 8, 16, 34, 52, and 84. What do these elements have in common?
3. Mendeleev described a then-undiscovered element as eka-silicon. When this element was discovered, (a) what was it named and (b) how did the actual density agree with that predicted by Mendeleev?
4. How does the size of the atoms of the elements in the third period vary from left to right?
5. Write the symbols of the family of elements that have seven electrons in their outer energy level.
6. Write the symbols of the alkali metal family in order of increasing atomic size.
7. What is the greatest number of elements to be found in any period? Which periods have this number?
8. From the standpoint of energy level, how does the placement of the last electron differ in Group A elements from that of Group B elements?

C. *Review questions*

1. What were Döbereiner's triads? In what way did they lead to later developments in periodicity?
2. What do you think is the basis for Newlands' Law of Octaves?
3. Given that the noble gases were not discovered at the time of Newlands' Law of Octaves, could his law be extended as far as the element bromine? Explain.
4. What is meant by the term *periodicity* as applied to the elements?

5. Why are some missing elements in Mendeleev's periodic table considered a victory for his table rather than a defeat?
6. How does our modern periodic table differ from Mendeleev's?
7. What additional understanding of periodic properties was added by the work of H. G. J. Moseley?
8. State the periodic law in its current form. How is this law different from the law as stated by Mendeleev?
9. Find the places in the modern periodic table where elements are not in proper sequence according to atomic weight.
10. How are elements in a period related to one another?
11. How are elements in a group related to one another?
12. What is common about the electron structures of the alkali metals?
13. Why would you expect the elements zinc, cadmium, and mercury to be in the same chemical family?
14. Draw the Lewis-dot symbols for Cs, Ba, Tl, Pb, Po, At, and Rn. How do these structures correlate with the group in which each element occurs?
15. Pick the electron structures below that represent elements in the same chemical family:
    (a) $1s^2 2s^1$
    (b) $1s^2 2s^2 2p^4$
    (c) $1s^2 2s^2 2p^2$
    (d) $1s^2 2s^2 2p^6 3s^2 3p^4$
    (e) $1s^2 2s^2 2p^6 3s^2 3p^6$
    (f) $1s^2 2s^2 2p^6 3s^2 3p^6 4s^2$
    (g) $1s^2 2s^2 2p^6 3s^2 3p^6 4s^1$
    (h) $1s^2 2s^2 2p^6 3s^2 3p^6 4s^2 3d^1$
16. Pick the electron structures below that represent elements in the same chemical family:
    (a) $[He]2s^2 2p^6$ (e) $[Ar]4s^1 3d^{10}$
    (b) $[Ne]3s^1$ (f) $[Ar]4s^2 3d^{10} 4p^6$
    (c) $[Ne]3s^2$ (g) $[Ar]4s^2 3d^5$
    (d) $[Ne]3s^2 3p^3$ (h) $[Kr]5s^1 4d^{10}$
17. In the periodic table, calcium, element 20, is surrounded by elements 12, 19, 21, and 38. Which of these have physical and chemical properties most resembling calcium?
18. Oxygen is a gas. Sulfur is a solid. What is it about their electron structures that causes them to be grouped in the same chemical family?
19. Classify each of the following elements as metals, nonmetals, or metalloids:
    (a) Potassium (e) Iodine
    (b) Plutonium (f) Tungsten
    (c) Sulfur (g) Molybdenum
    (d) Antimony (h) Germanium
20. In which period and group does an electron first appear in a *d* orbital?
21. How many electrons occur in the outer shell of Group IIIA and IIIB elements? Why are they different?
22. In which groups are transition elements located?
23. How do the electron structures of the transition elements differ from representative elements?
24. Which element in each of the following pairs has the larger atomic radius?
    (a) Na or K (c) O or F (e) Ti or Zr
    (b) Na or Mg (d) Br or I
25. Which element in each of Groups IA–VIIA has the smallest atomic radius?
26. Why does the atomic size increase in going down any family of the periodic table?
27. All the atoms within each Group A family of elements can be represented by the same Lewis-dot symbol. Complete the table, expressing the Lewis-dot symbol for each group. Use E to represent the elements.

Group	IA	IIA	IIIA	IVA	VA	VIA	VIIA
	E·						

28. Let E be any representative element. Following the pattern in the table, write the formulas for the hydrogen and oxygen compounds of
    (a) Na (c) Al (e) Sb (g) Cl
    (b) Ca (d) Sn (f) Se

Group IA	IIA	IIIA	IVA	VA	VIA	VIIA
EH	EH_2	EH_3	EH_4	EH_3	H_2E	HE
E_2O	EO	E_2O_3	EO_2	E_2O_5	EO_3	E_2O_7

29. Group IB elements have one electron in their outer shell, as do Group IA elements. Would you expect them to form compounds such as CuCl, AgCl, and AuCl? Explain.

30. The formula for lead(II) bromide is $PbBr_2$; predict formulas for tin(II) and germanium(II) bromides. (If needed, see Section 8.7, part (b), for the use of Roman numerals in naming compounds.)

31. The formula for sodium sulfate is Na_2SO_4. Write the names and formulas for the other alkali metal sulfates.

32. The formula for calcium bromide is $CaBr_2$. Write formulas for magnesium bromide, strontium bromide, and barium bromide.

33. Why should the discovery of the existence of isotopes have any bearing on the fact that the periodicity of the elements is a function of their atomic numbers and not their atomic weights?

Try to answer Questions 34–36 *without referring to the periodic table.*

34. The atomic numbers of the noble gases are 2, 10, 18, 36, 54, and 86. What are the atomic numbers for the elements with six electrons in their outer electron shells?

35. Element number 87 is in Group IA, period 7. Describe its outermost energy level. How many energy levels of electrons does it have?

36. If element 36 is a noble gas, in which groups would you expect elements 35 and 37 to occur?

37. Write a paragraph describing the general features of the periodic table.

38. Rank the following five elements according to the radii of their atoms, from smallest to largest: Na, Mg, Cl, K, and Rb.

39. Which of the following statements are correct?

(a) Properties of the elements are periodic functions of their atomic numbers.
(b) There are more nonmetallic elements than metallic elements.
(c) Metallic properties of the elements increase from left to right across a period.
(d) Metallic properties of the elements increase from top to bottom in a family of elements.
(e) Calcium is a member of the alkaline earth metal family.
(f) Iron belongs to the alkali metal family.
(g) Bromine belongs to the halogen family.
(h) Neon is a noble gas.
(i) Group A elements do not contain partially filled *d* or *f* sublevels.
(j) An atom of oxygen is larger than an atom of lithium.
(k) An atom of sulfur is larger than an atom of oxygen.
(l) An atom of aluminum (Group IIIA) has five electrons in its outer shell.
(m) If the formula for calcium iodide is CaI_2, then the formula for cesium iodide is CsI_2.
(n) If the formula for aluminum oxide is Al_2O_3, then the formula for gallium oxide is Ga_2O_3.
(o) Uranium is an inner transition element.
(p) The element $[Ar]4s^2 3d^{10} 4p^5$ is a halogen.
(q) The element $[Kr]5s^2$ is a nonmetal.
(r) The element with $Z = 12$ forms compounds similar to the element with $Z = 37$.
(s) Nitrogen, fluorine, neon, gallium, and bromine are all nonmetals.
(t) The Lewis-dot symbol for tin is $:\dot{\mathrm{Sn}}\cdot$
(u) The atom having an outer-shell electron structure of $5s^2 5p^2$ would be in period 6, Group IVA.
(v) The yet-to-be-discovered element with an atomic number of 118 would be a noble gas.
(w) The most metallic element of Group VIIA is iodine.

CHAPTER 7

Chemical Bonds: The Formation of Compounds from Atoms

After studying Chapter 7, you should be able to

1 Understand the terms listed in Question A at the end of the chapter.
2 Describe how the ionization energies of the elements vary with respect to (a) position in the periodic table and (b) the removal of successive electrons.
3 Determine the number of valence electrons in an atom for any Group A element.
4 Describe (a) the formation of ions by electron transfer and (b) the nature of the chemical bond formed by electron transfer.
5 Show by means of Lewis-dot structures the formation of an ionic compound from atoms.
6 Describe a crystal of sodium chloride.
7 Predict the formulas of the monatomic ions of Group A metals and nonmetals.
8 Predict the relative sizes of an atom and a monatomic ion for a given element.
9 Describe the covalent bond and predict whether a given covalent bond would be polar or nonpolar.
10 Draw Lewis-dot structures for the diatomic elements.
11 Identify single, double, and triple covalent bonds.
12 Describe the changes in electronegativity in (1) moving across a period and (2) moving down a group in the periodic table.
13 Describe the effect of electronegativity on the type of chemical bonds in a compound.
14 Draw Lewis-dot structures for (a) the molecules of covalent compounds and (b) polyatomic ions.

15 Describe the difference between polar and nonpolar bonds.
16 Distinguish clearly between ionic and molecular substances.
17 Predict whether the bonding in a compound will be primarily ionic or covalent.
18 Distinguish coordinate-covalent from covalent bonds in a Lewis-dot structure.

7.1 Chemical Bonds

Except in very rare instances matter does not fly apart spontaneously. It is prevented from doing so by forces acting at the ionic and molecular levels. Through chemical reactions atoms tend to attain more stable states at lower chemical potential energy levels. Atoms react chemically by losing, gaining, or sharing electrons. Forces arise from electron transferring and electron sharing interactions. Those forces that hold oppositely charged ions together or that bind atoms together in molecules are called **chemical bonds**. The two principal types of bonds are: the ionic bond and the covalent bond.

chemical bonds

7.2 Ionization Energy and Electron Affinity

According to Niels Bohr's concept, electrons exist at various discrete energy levels depending on the amount of energy that the atom has absorbed. If sufficient energy is applied to an atom, it is possible to remove completely (or "knock out") one or more electrons from its structure, thereby forming a positive ion:

$$\text{atom} + \text{energy} \longrightarrow \text{positive ion} + \text{electron } (e^-)$$

The amount of energy required to remove one electron from an atom or an ion is called the **ionization energy**. The *first* ionization energy is the amount of energy required to remove the first electron from an atom, the *second* ionization energy is the amount needed to remove the second electron from that atom, and so on.

ionization energy

Ionization energies may be expressed in terms of various energy units such as electron volts, kilojoules, or kilocalories per mole. The mole (abbreviated mol), which is discussed in detail in Chapter 9, is the unit in the International System for expressing an amount of substance. One mole of a substance is 6.022×10^{23} individual particles of that substance. Thus a mole of atoms is 6.022×10^{23} atoms, or a mole of oranges is 6.022×10^{23} oranges (and that's a lot of oranges). In the following discussion ionization energy is expressed as kilojoules per mole (kJ/mol) indicating the number of kilojoules required to remove one electron from each atom in a mole of atoms.

Na	+ 494 kJ ⟶	Na^+	+	e^-
1 mol sodium atoms		1 mol sodium ions		1 mol electrons

Table 7.1 gives the ionization energies for the removal of one to five electrons from several elements. Thus, 1314 kJ are required to remove one electron from a mole of hydrogen atoms, but only 494 kJ are needed to remove the first electron from a mole of sodium atoms. The table shows that increasingly higher amounts of energy are required to remove the second, third, fourth, and fifth electrons. This sequence is logical, because the removal of electrons does not decrease the number of protons, or the amount of charge, in the nucleus. Therefore the remaining electrons are held more tightly. The data of Table 7.1 also show that an extra-large ionization energy is needed whenever an electron is pulled away from a noble-gas electron structure, clearly demonstrating the high stability of this structure.

First ionization energies have been experimentally determined for most of the elements. Figure 7.1 is a graphic plot of the first ionization energies of the first 56 elements, H through Ba.

Certain periodic relationships are evident in Figure 7.1. In the Group A, or representative, elements, the first ionization energy generally decreases from top to bottom in groups or families. For example, in alkali metals (Group IA) the first ionization energy decreases from 520 kJ/mol for lithium to 376 kJ/mol for cesium. The two main reasons for this family trend are that (1) as we go from top to bottom in the family, the electron being removed is farther away from its nucleus, and (2) the electron is shielded from its nucleus by more inner shells of electrons. Furthermore all the alkali metals have relatively low first ionization energies, indicating that each has one electron that is easily removed.

From left to right within a period, the ionization energy gradually increases

Table 7.1 Ionization energies for selected elements. Values are expressed in kilojoules per mole, showing energies required to remove one to five electrons per atom. Color indicates the energy needed to remove an electron from a noble-gas electron structure.

	Required amounts of energy (kJ/mol)				
Element	1st e^-	2nd e^-	3rd e^-	4th e^-	5th e^-
H	1,314				
He	2,372	5,247			
Li	520	7,297	11,810		
Be	900	1,757	14,845	21,000	
B	800	2,430	3,659	25,020	32,810
C	1,088	2,352	4,619	6,222	37,800
Ne	2,080	3,962	6,276	9,376	12,190
Na	496	4,565	6,912	9,540	13,355

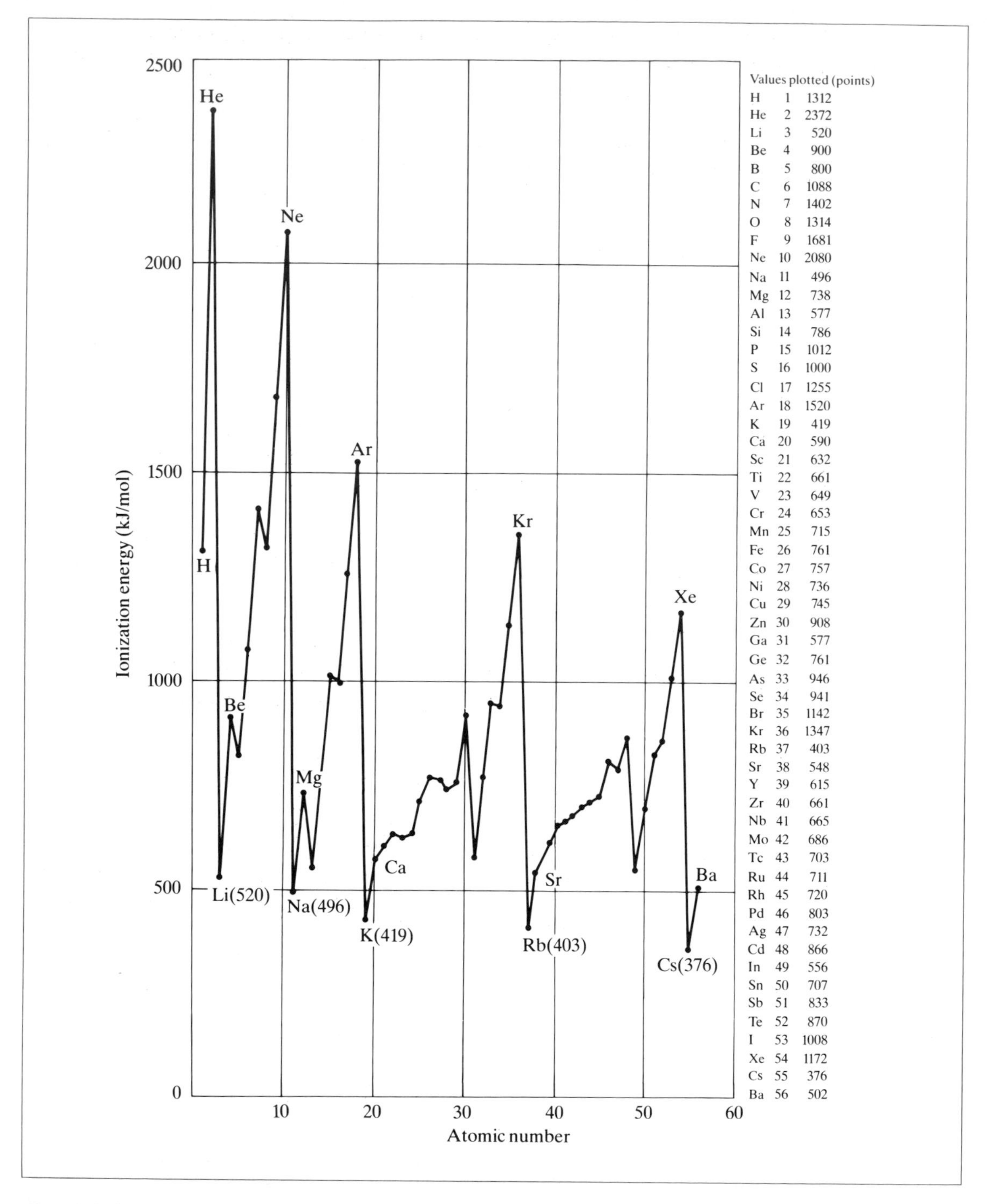

Element	Atomic number	Value
H	1	1312
He	2	2372
Li	3	520
Be	4	900
B	5	800
C	6	1088
N	7	1402
O	8	1314
F	9	1681
Ne	10	2080
Na	11	496
Mg	12	738
Al	13	577
Si	14	786
P	15	1012
S	16	1000
Cl	17	1255
Ar	18	1520
K	19	419
Ca	20	590
Sc	21	632
Ti	22	661
V	23	649
Cr	24	653
Mn	25	715
Fe	26	761
Co	27	757
Ni	28	736
Cu	29	745
Zn	30	908
Ga	31	577
Ge	32	761
As	33	946
Se	34	941
Br	35	1142
Kr	36	1347
Rb	37	403
Sr	38	548
Y	39	615
Zr	40	661
Nb	41	665
Mo	42	686
Tc	43	703
Ru	44	711
Rh	45	720
Pd	46	803
Ag	47	732
Cd	48	866
In	49	556
Sn	50	707
Sb	51	833
Te	52	870
I	53	1008
Xe	54	1172
Cs	55	376
Ba	56	502

Figure 7.1

Periodic relationship of the first ionization energy to the atomic number of the elements

despite some irregularities. The noble gases have relatively high values, confirming the nonreactivity of this family and the stability of an eight-electron structure in the outer energy level.

electron affinity

Atoms also have *electron affinity*—that is, the ability to attract electrons and form negative ions. **Electron affinity** may be defined as the amount of energy released or absorbed when an electron is added to an atom to form a negative ion. For most of the elements, heat is released when an atom adds an electron.

$$\text{atom} + \text{electron} \longrightarrow \text{negative ion} + \text{energy}$$

Electron affinity is a measure of the attraction of an atom for an electron—in other words, of the tendency to form a negative ion. Chlorine, for example, is a nonmetallic element and has a strong tendency to form negative ions. Consequently the electron affinity of chlorine is high.

$$\underset{\text{chlorine atoms}}{\underset{1\ \text{mol}}{Cl}} + \underset{\text{electrons}}{\underset{1\ \text{mol}}{e^-}} \longrightarrow \underset{\text{chloride ions}}{\underset{1\ \text{mol}}{Cl^-}} + 348\ \text{kJ}\ (83\ \text{kcal})$$

Electron affinity tends to be high for nonmetals (particularly for the halogens, oxygen, and sulfur) and low for metals. The general trend of electron affinity is to increase from left to right in any period and to decrease from top to bottom in a family of elements.

7.3 Electrons in the Outer Shell — Valence Electrons

One outstanding property of the elements is their tendency to form a stable outer-shell electron structure. For many elements this stable outer shell contains eight electrons (two *s* and six *p*) identical to the outer-shell electron structure of the noble gases. Atoms undergo rearrangements of electron structure to attain a state of greater stability. These rearrangements are accomplished by losing, gaining, or sharing electrons with other atoms. For example, a hydrogen atom has a tendency to accept another electron and thus attain an electron structure like that of the stable noble gas helium; a fluorine atom can acquire one more electron to attain a stable electron structure like neon; a sodium atom tends to lose one electron to attain a stable electron structure like neon.

$$Na + \text{energy} \longrightarrow Na^+ + 1\ e^-$$

valence electrons

The electrons in the outermost shell of an atom are responsible for most of this electron activity and are called the **valence electrons**. In Lewis-dot symbols of atoms, the dots represent the outer-shell electrons and thus also represent the valence electrons. For example, hydrogen has one valence electron; sodium, one;

aluminum, three; and oxygen, six. When a rearrangement of these electrons takes place between atoms, a chemical change occurs.

H·	$\dot{\ddot{\text{Al}}}$·	·$\ddot{\underset{\cdot}{\text{O}}}$:
One valence electron	Three valence electrons	Six valence electrons

7.4 The Ionic Bond: Transfer of Electrons from One Atom to Another

The chemistry of the elements, especially the representative ones, is to attain an outer-shell electron structure like that of the chemically stable noble gases. With the exception of helium, this stable structure consists of eight electrons in the outer shell (see Table 5.6).

Let us look at the electron structures of sodium and chlorine to see how each element can attain a structure of 8 electrons in its outer shell. A sodium atom has 11 electrons: 2 in the first energy level, 8 in the second energy level, and 1 in the third energy level. A chlorine atom has 17 electrons: 2 in the first energy level, 8 in the second energy level, and 7 in the third energy level. If a sodium atom transfers or loses its 3*s* electron, its third energy level becomes vacant, and it becomes a sodium ion with an electron configuration identical to that of the noble gas neon. This process absorbs energy.

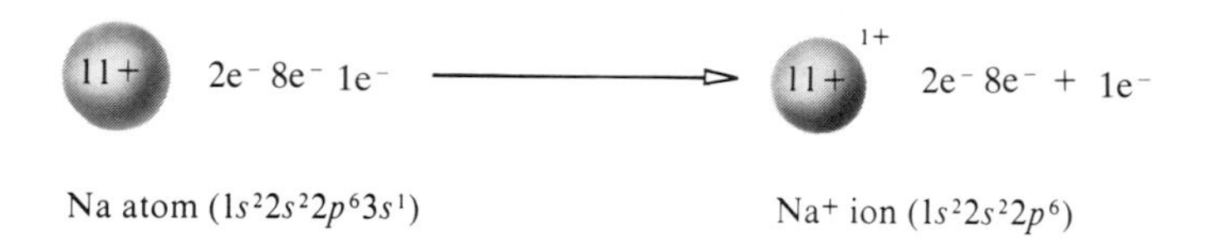

An atom that has lost or gained electrons will have a plus or minus electric charge, depending on which charged particles, protons or electrons, are in excess. Recall that a charged atom or group of atoms is called an *ion*.

By losing a negatively charged electron, the sodium atom becomes a positively charged particle known as a sodium ion. The charge, +1, results because the nucleus still contains 11 positively charged protons, and the electron orbitals contain only 10 negatively charged electrons. The charge is indicated by a plus sign (+) and is written as a superscript after the symbol of the element (Na^+).

A chlorine atom with 7 electrons in the third energy level needs 1 electron to pair up with its one unpaired 3*p* electron to attain the stable outer-shell electron structure of argon. By gaining 1 electron the chlorine atom becomes a chloride ion (Cl^-), a negatively charged particle containing 17 protons and 18 electrons. This process releases energy.

Consider the case in which sodium and chlorine atoms react with each other. The $3s$ electron from the sodium atom transfers to the half-filled $3p$ orbital in the chlorine atom to form a positive sodium ion and a negative chloride ion. The compound sodium chloride results because the Na^+ and Cl^- ions are strongly attracted to each other by their opposite electrostatic charges. The force holding the oppositely charged ions together is an ionic bond.

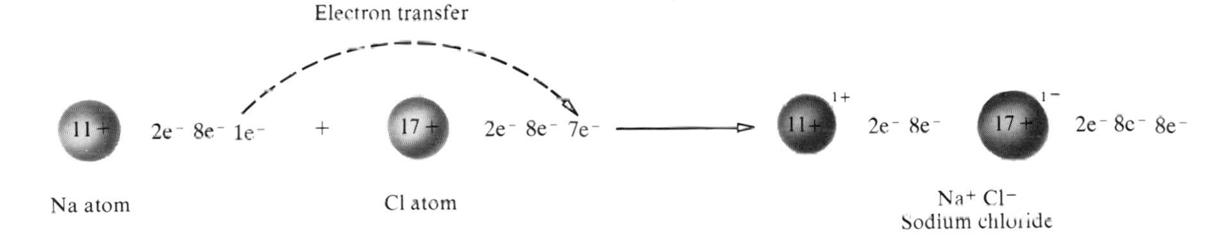

The Lewis-dot representation of sodium chloride formation is shown below.

$$Na\cdot + \cdot\ddot{\underset{\cdot\cdot}{Cl}}: \longrightarrow Na^+ :\ddot{\underset{\cdot\cdot}{Cl}}:^-$$

The chemical reaction between sodium and chlorine is a very vigorous one, producing considerable heat in addition to the salt formed. When energy is released in a chemical reaction, the products are more stable than the reactants. Note that in NaCl both atoms attained a noble-gas electron structure.

Sodium chloride is made up of cubic crystals in which each sodium ion is surrounded by six chloride ions and each chloride ion by six sodium ions, except at the crystal surface. A visible crystal is a regularly arranged aggregate of millions of these ions, but the ratio of sodium to chloride ions is one-to-one, hence the formula NaCl. The cubic crystalline lattice arrangement of sodium chloride is shown in Figure 7.2.

Figure 7.3 contrasts the relative sizes of sodium and chlorine atoms with those of their ions. The sodium ion is smaller than the atom due primarily to two factors: (1) The sodium atom has lost its outer shell of 1 electron, thereby reducing its size; and (2) the 10 remaining electrons are now attracted by 11 protons and are thus drawn closer to the nucleus. Conversely the chloride ion is larger than the atom because it has 18 electrons but only 17 protons; the nuclear attraction on each electron is thereby decreased, allowing the chlorine atom to expand as it forms an ion.

We have seen that when sodium reacts with chlorine, each atom becomes an electrically charged ion. Sodium chloride, like all ionic substances, is held

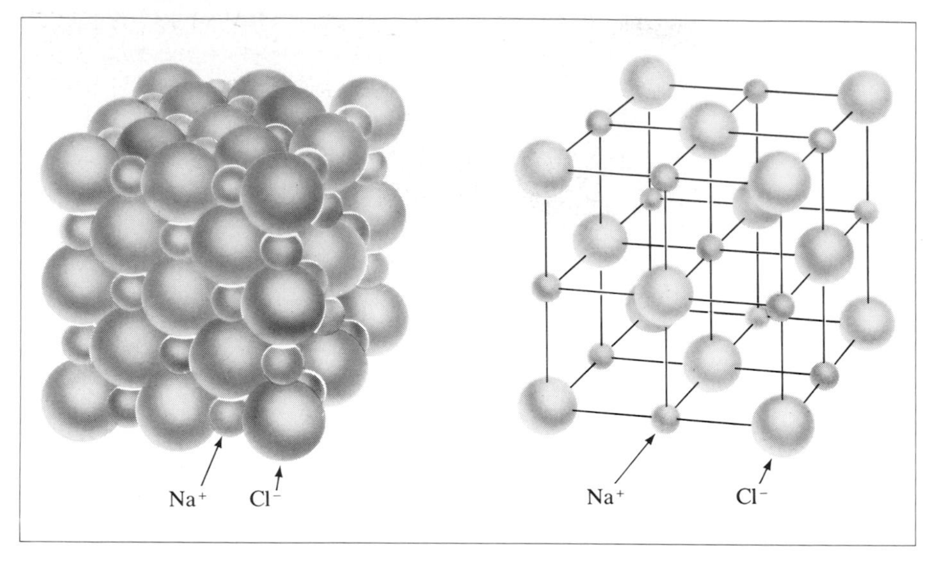

Figure 7.2

Sodium chloride crystal. Diagram represents a small fragment of sodium chloride, which forms cubic crystals. Each sodium ion is surrounded by six chloride ions, and each chloride ion is surrounded by six sodium ions.

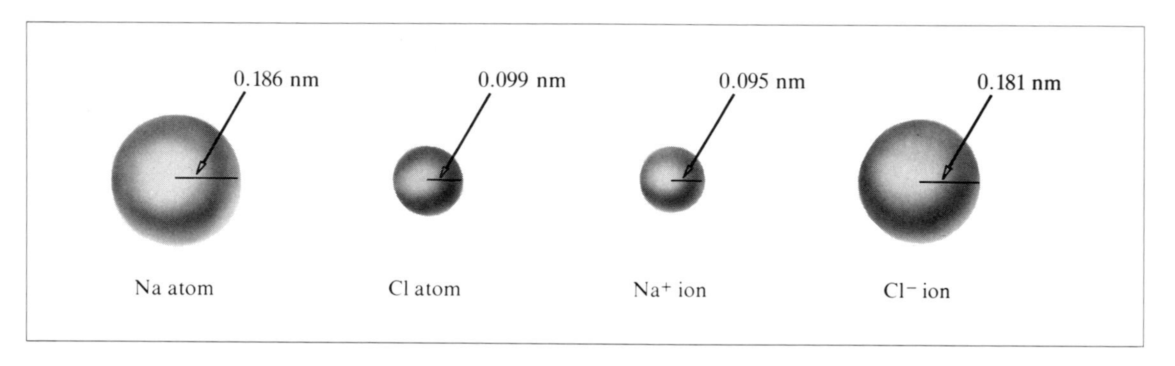

Figure 7.3

Relative sizes of sodium and chlorine atoms and their ions

ionic bond

together by the attraction existing between positive and negative charges. An **ionic bond** is the attraction between oppositely charged ions.

Ionic bonds are formed whenever one or more electrons are transferred from one atom to another. The metals, which have relatively little attraction for their valence electrons, tend to form ionic bonds when they combine with nonmetals.

Table 7.2 Change in atomic radii of selected metals and nonmetals. The metals shown lose electrons to become positive ions. The nonmetals gain electrons to become negative ions.

Atomic radius (nm)	Ionic radius (nm)	Atomic radius (nm)	Ionic radius (nm)
Li 0.152	Li^+ 0.060	F 0.071	F^- 0.136
Na 0.186	Na^+ 0.095	Cl 0.099	Cl^- 0.181
K 0.227	K^+ 0.133	Br 0.114	Br^- 0.195
Mg 0.160	Mg^{2+} 0.065	O 0.074	O^{2-} 0.140
Al 0.143	Al^{3+} 0.050	S 0.103	S^{2-} 0.184

It is important to recognize that substances with ionic bonds do not exist as molecules. In sodium chloride, for example, the bond does not exist solely between a single sodium ion and a single chloride ion. Each sodium ion in the crystal attracts six near-neighbor negative chloride ions; in turn, each negative chloride ion attracts six near-neighbor positive sodium ions (see Figure 7.2).

A metal will usually have one, two, or three electrons in its outer energy level. In reacting, metal atoms characteristically lose these electrons, attain the electron structure of a noble gas, and become positive ions. A nonmetal, on the other hand, is only a few electrons short of having a complete octet in its outer energy level and thus has a tendency to gain electrons (electron affinity). In reacting with metals, nonmetal atoms characteristically gain one, two, or three electrons, attain the electron structure of a noble gas, and become negative ions. The ions formed by loss of electrons are much smaller than the corresponding metal atoms; the ions formed by gaining electrons are larger than the corresponding nonmetal atoms. The actual dimensions of the atomic and ionic radii of several metals and nonmetals are given in Table 7.2.

Study the following examples. Note the loss and gain of electrons between atoms; also note that the ions in each compound have a noble-gas electron structure.

EXAMPLE 7.1

Formation of magnesium chloride, $MgCl_2$

A magnesium atom of electron structure $1s^22s^22p^63s^2$ must lose two electrons or gain six electrons to reach a stable electron structure. If magnesium reacts with chlorine and each chlorine atom can accept only one electron, two chlorine atoms will be needed for the two electrons from one magnesium atom. The compound formed will contain one magnesium ion and two chloride ions. The magnesium atom, having lost two electrons, becomes a magnesium ion with a $+2$ charge. Each chloride ion will have a -1 charge. The transfer of electrons from a magnesium atom to two chlorine atoms is shown in the following illustration.

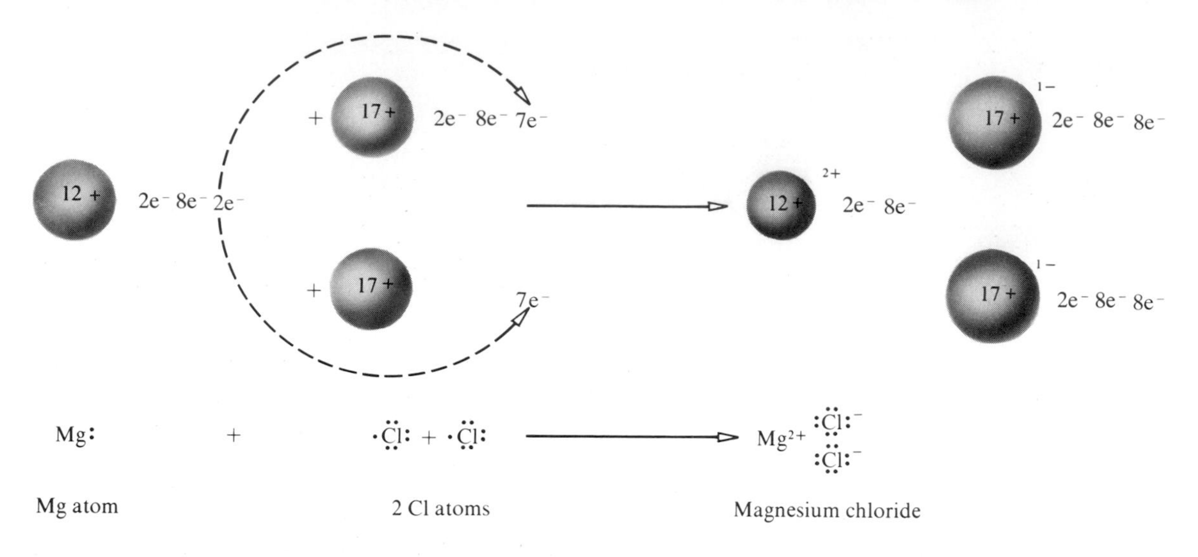

EXAMPLE 7.2 Formation of sodium fluoride, NaF

11+ 2e⁻ 8e⁻ 1e⁻ + 9+ 2e⁻ 7e⁻ → $^{1+}$ 11+ 2e⁻ 8e⁻ $^{1-}$ 9+ 2e⁻ 8e⁻

Na·	+	·F:	→	Na^{+} :F:⁻
Sodium atom		Fluorine atom		Sodium fluoride

The fluorine atom, with seven electrons in its outer shell, behaves similarly to the chlorine atom.

EXAMPLE 7.3 Formation of aluminum chloride, $AlCl_3$

$$\ddot{Al}\cdot + \begin{matrix}\cdot\ddot{Cl}: \\ \cdot\ddot{Cl}: \\ \cdot\ddot{Cl}:\end{matrix} \longrightarrow Al^{3+}\begin{matrix}:\ddot{Cl}:^{-} \\ :\ddot{Cl}:^{-} \\ :\ddot{Cl}:^{-}\end{matrix} \quad \text{or} \quad AlCl_3$$

Aluminum atom   Chlorine atoms   Aluminum chloride

Each chlorine atom can accept only one electron. Therefore three chlorine atoms are needed to combine with the three outer-shell electrons of one aluminum atom. The aluminum atom has lost three electrons to become an aluminum ion, Al^{3+}, with a +3 charge.

EXAMPLE 7.4 Formation of magnesium oxide, MgO

12+ 2e⁻ 8e⁻ 2e⁻ + 8+ 2e⁻ 6e⁻ ⟶ 12+ 2e⁻ 8e (2+) 8+ 2e⁻ 8e⁻ (2−)

$$\text{Mg}: \; + \; \cdot\ddot{\text{O}}: \longrightarrow \text{Mg}^{2+} \; :\ddot{\text{O}}:^{2-}$$

Magnesium atom | Oxygen atom | Magnesium oxide

The magnesium atom, with two electrons in the outer energy level, exactly fills the need of two electrons of one oxygen atom. The resulting compound has a ratio of one atom of magnesium to one atom of oxygen. The oxygen (oxide) ion has a -2 charge, having gained two electrons. In combining with oxygen, magnesium behaves the same way as when combining with chlorine; it loses two electrons.

EXAMPLE 7.5 Formation of sodium sulfide, Na_2S

$$\begin{matrix}\text{Na}\cdot \\ \text{Na}\cdot\end{matrix} + \cdot\ddot{\text{S}}: \longrightarrow \begin{matrix}\text{Na}^+ \\ \text{Na}^+\end{matrix} \; :\ddot{\text{S}}:^{2-} \quad \text{or} \quad \text{Na}_2\text{S}$$

Sodium atoms | Sulfur atom | Sodium sulfide

Two sodium atoms supply the electrons that one sulfur atom needs to make eight in its outer shell.

EXAMPLE 7.6 Formation of aluminum oxide, Al_2O_3

$$\begin{matrix}\dot{\text{Al}}\cdot \\ \\ \dot{\text{Al}}\cdot\end{matrix} + \begin{matrix}\cdot\ddot{\text{O}}: \\ \cdot\ddot{\text{O}}: \\ \cdot\ddot{\text{O}}:\end{matrix} \longrightarrow \begin{matrix} & :\ddot{\text{O}}:^{2-} \\ \text{Al}^{3+} & \\ & :\ddot{\text{O}}:^{2-} \\ \text{Al}^{3+} & \\ & :\ddot{\text{O}}:^{2-}\end{matrix} \quad \text{or} \quad \text{Al}_2\text{O}_3$$

Aluminum atoms | Oxygen atoms | Aluminum oxide

One oxygen atom, needing two electrons, cannot accommodate the three electrons from one aluminum atom. One aluminum atom falls one electron short of the four electrons needed by two oxygen atoms. A ratio of two atoms of aluminum to three atoms of oxygen, involving the transfer of six electrons (two to each oxygen atom), gives each atom a stable electron configuration.

Note that in each of the examples above outer shells containing eight electrons were formed in all the negative ions. This formation resulted from the pairing of all the *s* and *p* electrons in these outer shells.

Chemistry would be considerably simpler if all compounds were made by the direct formation of ions as outlined in the examples just given. However, this method is only one of the two general methods of compound formation. The second general method will be outlined in the sections that follow.

7.5 The Covalent Bond: Sharing Electrons

Some atoms do not transfer electrons from one atom to another to form ions. Instead they form a chemical bond by sharing pairs of electrons between them.

covalent bond

A **covalent bond** consists of a pair of electrons shared between two atoms. This bonding concept was introduced in 1916 by G. N. Lewis of the University of California at Berkeley. In the millions of compounds that are known, the covalent bond is the predominant chemical bond.

True molecules exist in substances in which the atoms are covalently bonded. Hence, it is proper to refer to molecules of such substances as hydrogen, chlorine, hydrogen chloride, carbon dioxide, water, or sugar. These substances contain only covalent bonds and exist as aggregates of molecules. We do not use the term *molecule* when talking about ionically bonded compounds such as sodium chloride, because such substances exist as large aggregates of positive and negative ions, not as molecules.

A study of the hydrogen molecule will give an insight into the nature of the covalent bond and its formation. The formation of a hydrogen molecule, H_2, involves the overlapping and pairing of 1*s* electron orbitals from two hydrogen atoms. This overlapping and pairing is shown in Figure 7.4. Each atom contributes one electron of the pair that is shared jointly by two hydrogen nuclei.

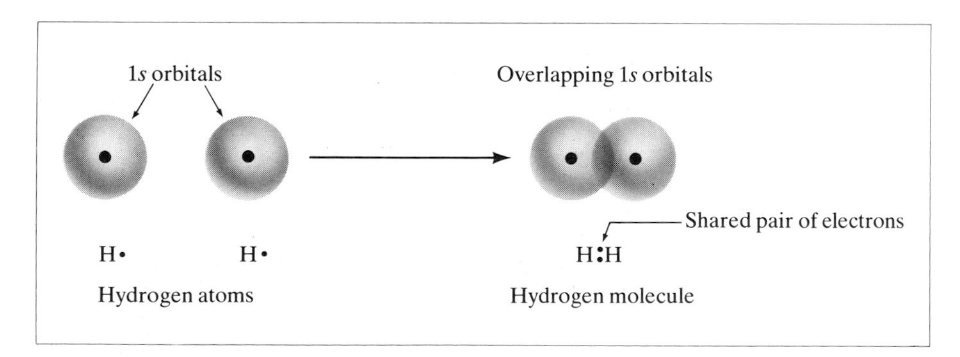

Figure 7.4

The formation of a hydrogen molecule from two hydrogen atoms. The two 1*s* orbitals overlap, forming the H_2 molecule. In this molecule the two electrons are shared between the atoms, forming a covalent bond.

The orbital of the electrons now includes both hydrogen nuclei, but probability factors show that the most likely place to find the electrons (the point of highest electron density) is between the two nuclei. The two nuclei are shielded from each other by the pair of electrons, allowing the two nuclei to be drawn very close to each other. (The average bond length between the hydrogen nuclei is 7.4×10^{-9} cm.)

The tendency for hydrogen atoms to form a molecule is very strong. In the molecule, each electron is attracted by two positive nuclei. This attraction gives the hydrogen molecule a more stable structure than the individual hydrogen atoms had. Energy is released when a bond forms between two atoms. Consequently, the same amount of energy is needed to break that bond. Experimental evidence of stability is shown by the fact that 436 kJ are needed to break the bonds between the hydrogen atoms in 1 mole of hydrogen molecules. The strength of a bond may be determined by the energy required to break it. The energy required to break a covalent bond is known as the *bond dissociation energy*. The following bond dissociation energies illustrate relative bond strengths. (All substances are considered to be in the gaseous state and to form neutral atoms.)

Reaction	Bond dissociation energy (kJ/mol)
$H_2 \longrightarrow 2\,H$	435
$N_2 \longrightarrow 2\,N$	946
$O_2 \longrightarrow 2\,O$	595
$F_2 \longrightarrow 2\,F$	153
$Cl_2 \longrightarrow 2\,Cl$	243
$Br_2 \longrightarrow 2\,Br$	193
$I_2 \longrightarrow 2\,I$	151

The formula of chlorine gas is Cl_2. When the two atoms of chlorine combine to form this molecule, the electrons must interact in a manner that is similar to that shown in the preceding example. Each chlorine atom would be more stable with eight electrons in its outer shell. But chlorine atoms are identical, and neither is able to pull an electron away from the other. What happens is this: The unpaired $3p$ electron orbital of one chlorine atom overlaps the unpaired $3p$ electron orbital of the other atom, resulting in a pair of electrons that are mutually shared between the two atoms. Each atom furnishes one of the pair of shared electrons. Thus, each atom attains a stable structure of eight electrons by sharing an electron pair with the other atom. The pairing of the p electrons and formation of a chlorine molecule are illustrated in Figure 7.5. Neither chlorine atom has a positive or negative charge, since both contain the same number of protons and have equal attraction for the pair of electrons being shared. Other examples of molecules in which electrons are equally shared between two atoms are hydrogen, H_2; oxygen, O_2; nitrogen, N_2; fluorine, F_2; bromine, Br_2; and iodine, I_2. Note that more than one pair of electrons may be shared between atoms.

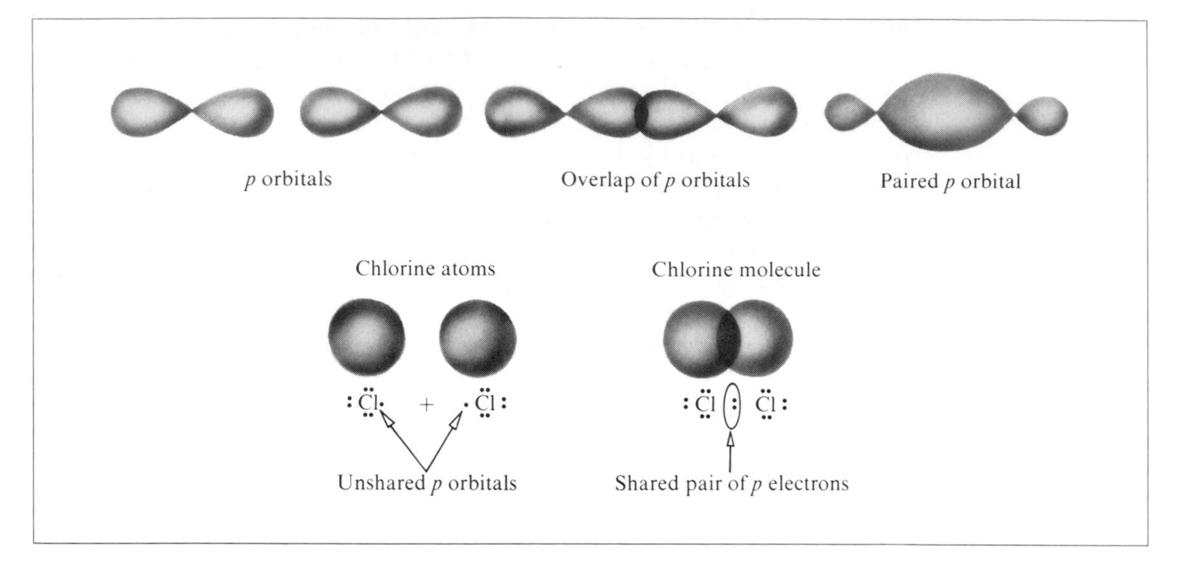

Figure 7.5
Pairing of *p* electrons in the formation of a chlorine molecule.

H:H	:F:F:	:Br:Br:	:I:I:	:O::O:	:N:::N:
Hydrogen	Fluorine	Bromine	Iodine	Oxygen	Nitrogen

The Lewis structure given for oxygen does not adequately account for all the properties of the oxygen molecule. Other theories explaining the bonding in oxygen molecules have been advanced; however, they are complex and beyond the scope of this book.

A common practice in writing structures is to replace the pair of dots used to represent a shared pair of electrons by a dash (—). But you must remember that a dash represents a shared pair of electrons. One dash represents a single bond; two dashes, a double bond; and three dashes, a triple bond. The six structures just shown may be written thus:

H—H :F—F: :Br—Br: :I—I: :O═O: :N≡N:

7.6 Electronegativity

When two different kinds of atoms share a pair of electrons, one atom assumes a partial positive charge and the other a partial negative charge with respect to each other. This difference in charge occurs because the two atoms exert unequal attraction for the pair of shared electrons. The attractive force that an atom of an element has for shared electrons in a molecule is known as its **electronegativity**. Elements differ in their electronegativities. For example, both hydrogen and

electro-negativity

chlorine need one electron to form stable electron configurations. They share a pair of electrons in hydrogen chloride, HCl. Chlorine is more electronegative and therefore has a greater attraction for the shared electrons than does hydrogen. As a result the pair of electrons is displaced toward the chlorine atom, giving it a partial negative charge and leaving the hydrogen atom with a partial positive charge. It should be understood that the electron is not transferred entirely to the chlorine atom, as in the case of sodium chloride, and no ions are formed. The entire molecule, HCl, is electrically neutral. A partial charge is usually indicated by the Greek letter delta, δ. Thus, a partial positive charge is represented by $\delta+$ and a partial negative charge by $\delta-$.

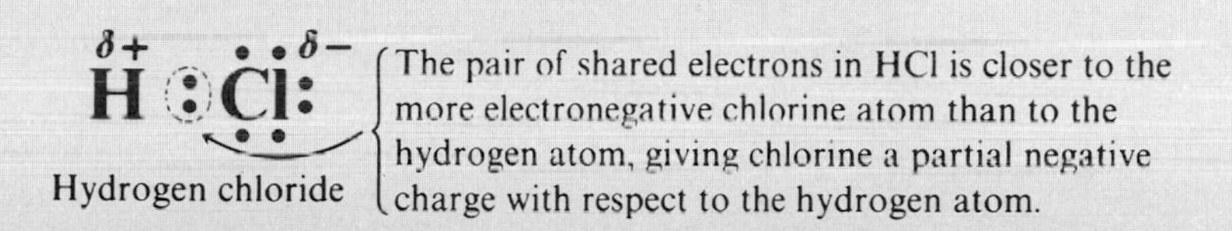

The pair of shared electrons in HCl is closer to the more electronegative chlorine atom than to the hydrogen atom, giving chlorine a partial negative charge with respect to the hydrogen atom.

The electronegativity, or ability of an atom to attract an electron, depends on several factors: (1) the charge on the nucleus, (2) the distance of the outer electrons from the nucleus, and (3) the amount the nucleus is shielded from the outer-shell electrons by intervening shells of electrons. A scale of relative electronegativities, in which the most electronegative element, fluorine, is assigned a value of 4.0, was developed by the Nobel laureate (1954 and 1962) Linus Pauling (b. 1901). Table 7.3 shows that the relative electronegativity of the nonmetals is high and

Table 7.3 Relative electronegativity of the elements. The electronegativity value is given below the symbol of each element.

Key: 9 — Atomic number; F — Symbol; 4.0 — Electronegativity

1 H 2.1																	2 He —
3 Li 1.0	4 Be 1.5											5 B 2.0	6 C 2.5	7 N 3.0	8 O 3.5	9 F 4.0	10 Ne —
11 Na 0.9	12 Mg 1.2											13 Al 1.5	14 Si 1.8	15 P 2.1	16 S 2.5	17 Cl 3.0	18 Ar —
19 K 0.8	20 Ca 1.0	21 Sc 1.3	22 Ti 1.4	23 V 1.6	24 Cr 1.6	25 Mn 1.5	26 Fe 1.8	27 Co 1.8	28 Ni 1.8	29 Cu 1.9	30 Zn 1.6	31 Ga 1.6	32 Ge 1.8	33 As 2.0	34 Se 2.4	35 Br 2.8	36 Kr —
37 Rb 0.8	38 Sr 1.0	39 Y 1.2	40 Zr 1.4	41 Nb 1.6	42 Mo 1.8	43 Tc 1.9	44 Ru 2.2	45 Rh 2.2	46 Pd 2.2	47 Ag 1.9	48 Cd 1.7	49 In 1.7	50 Sn 1.8	51 Sb 1.9	52 Te 2.1	53 I 2.5	54 Xe —
55 Cs 0.7	56 Ba 0.9	57–71 La-Lu 1.1–1.2	72 Hf 1.3	73 Ta 1.5	74 W 1.7	75 Re 1.9	76 Os 2.2	77 Ir 2.2	78 Pt 2.2	79 Au 2.4	80 Hg 1.9	81 Tl 1.8	82 Pb 1.8	83 Bi 1.9	84 Po 2.0	85 At 2.2	86 Rn —
87 Fr 0.7	88 Ra 0.9	89– Ac– 1.1–1.7	104 Unq	105 Unp	106 Unh												

that of the metals is low. These electronegativities indicate that atoms of metals have a greater tendency to lose electrons than do atoms of nonmetals and that nonmetals have a greater tendency to gain electrons than do metals. The higher the electronegativity value, the greater the attraction for electrons.

7.7 Writing Lewis Structures

As we have seen, Lewis structures are a convenient way of showing the covalent bonds in many molecules or ions of the representative elements. In writing Lewis structures the object is to connect the atoms in a molecule with covalent bonds by rearranging the valence electrons of the atoms so that each atom has eight outer-shell electrons around it. Exceptions to this rule are hydrogen, which requires only two electrons, and several other elements such as lithium, beryllium, and boron.

The most difficult part of writing Lewis structures is determining the arrangement of the atoms in a molecule or an ion. In simple molecules with more than two atoms, one atom will be the central atom surrounded by the other atoms. Thus Cl_2O has two possible arrangements, Cl—Cl—O or Cl—O—Cl. Usually, but not always, the single atom in the formula (except H) will be the central atom.

Although Lewis structures for many molecules and ions can be written by inspection, the following procedure will be helpful for learning to write these structures:

1. Obtain the total number of valence electrons to be used in the structure by adding the number of valence electrons in all of the atoms in the molecule or ion. If you are writing the structure of an ion, add one electron for each negative charge or subtract one electron for each positive charge on the ion. Remember, the number of valence electrons of Group A elements is the same as their group number in the periodic table.
2. Write down the skeletal arrangement of the atoms and connect them with a single covalent bond (two dots or one dash). Hydrogen, which contains only one bonding electron, can form only one covalent bond. Oxygen atoms are not normally bonded to each other, except in compounds known to be peroxides. Oxygen atoms normally have a maximum of two covalent bonds, two single bonds or one double bond.
3. Subtract two electrons for each single bond you used in Step 2 from the total number of electrons calculated in Step 1. This calculation gives you the net number of electrons available for completing the structure.
4. Distribute pairs of electrons (pairs of dots) around each atom (except hydrogen) to give each atom a total of eight electrons around it.
5. If there are not enough electrons to give these atoms eight electrons, change single bonds between atoms to double or triple bonds by shifting unbonded pairs of electrons as needed. Check to see that each atom (except H) has eight electrons around it. A double bond counts as four electrons for each atom to which it is bonded.

PROBLEM 7.1 How many valence electrons are in each of these atoms: Cl, H, C, O, N, S, P, I?

You can look in Table 5.5 for the electron structure, or, if the element is in an A Group of the periodic table, the number of valence electrons is equal to the group number.

Atom	Periodic group	Valence electrons
Cl	VIIA	7
H	IA	1
C	IVA	4
O	VIA	6
N	VA	5
S	VIA	6
P	VA	5
I	VIIA	7

PROBLEM 7.2 Write the Lewis structure for water, H_2O.

Step 1 The total number of valence electrons is 8, 2 from the two hydrogen atoms and 6 from the oxygen atom.

Step 2 The two hydrogen atoms are connected to the oxygen atom. Write the skeletal structure:

H O
  H

Place two dots between the hydrogen and oxygen atoms to form the covalent bonds:

H:O
  ··
  H

Step 3 Subtract the 4 electrons used in Step 2 from 8 to obtain 4 electrons yet to be used.

Step 4 Distribute the four electrons around the oxygen atom. Hydrogen atoms cannot accommodate any more electrons.

H:Ö: or H—Ö:
  ·· |
  H   H

This arrangement is the Lewis structure. Each atom has a noble-gas electron structure.

PROBLEM 7.3 Write Lewis structures for a molecule of (a) methane, CH_4, and (b) carbon tetrachloride, CCl_4.

(a) **Step 1** The total number of valence electrons is 8, 1 from each hydrogen atom and 4 from the carbon atom.

Step 2 The skeletal structure contains four H atoms around a central C atom. Place 2 electrons between the C and each H.

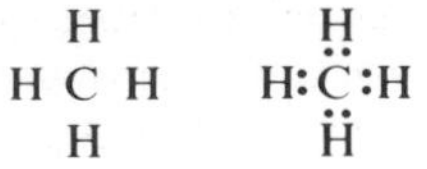

```
  H          H
                ··
H C H      H:C:H
                ··
  H          H
```

Step 3 Subtract the 8 electrons used in Step 2 from 8 to obtain zero electrons yet to be placed. Therefore, the Lewis structure must be as written in Step 2.

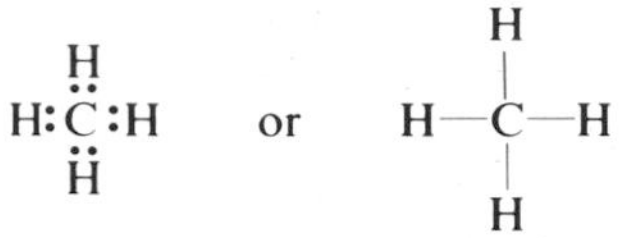

```
   H                    H
  ··                     |
H:C:H     or     H—C—H
  ··                     |
   H                    H
```

(b) **Step 1** The total number of valence electrons to be used is 32, 4 from the carbon atom and 7 from each of the four chlorine atoms.

Step 2 The skeletal structure contains the four Cl atoms around a central C atom. Place 2 electrons between the C and each Cl.

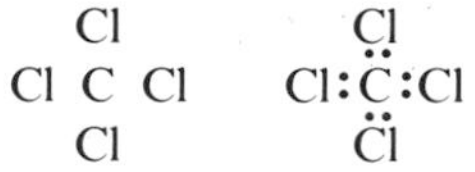

```
    Cl               Cl
                      ··
Cl C Cl       Cl:C:Cl
                      ··
    Cl               Cl
```

Step 3 Subtract the 8 electrons used in Step 2 from 32, to obtain 24 electrons yet to be placed.

Step 4 Distribute the 24 electrons (12 pairs) around the Cl atoms so that each Cl atom has 8 electrons around it.

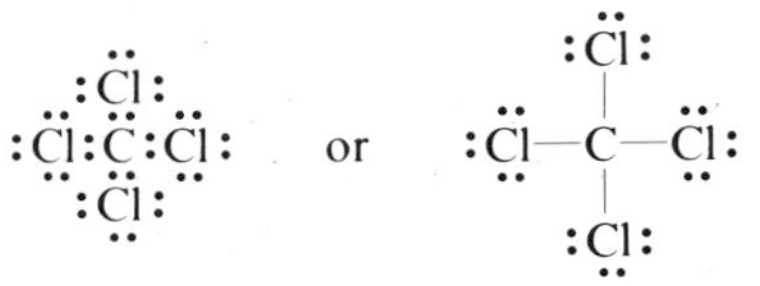

```
       ··                          ··
     :Cl:                        :Cl:
  ··  ··  ··                      |
:Cl:C:Cl:      or      :Cl—C—Cl:
  ··  ··  ··                      |
     :Cl:                        :Cl:
       ··                          ··
```

This arrangement is the Lewis structure; CCl_4 contains four covalent bonds.

PROBLEM 7.4 Write Lewis structures for (a) carbon dioxide, CO_2, and (b) nitric acid, HNO_3.

(a) **Step 1** The total number of valence electrons is 16, 4 from the carbon atom and 6 from each oxygen atom.

Step 2 The two O atoms are bonded to a central C atom. Write the skeletal structure and place 2 electrons between the C and each O atom.

O:C:O

Step 3 Subtract the 4 electrons used in Step 2 from 16 to obtain 12 electrons yet to be placed.

Step 4 Distribute the 12 electrons around the C and O atoms. Several possibilities exist:

```
  ··      ··          ··   ··          ·· ··
:O:C:O:        :O:C:O:         :O:C:O:
  ··      ··           ·· ··            ·· ··

     I                 II                III
```

Step 5 All the atoms do not have 8 electrons around them. Move one pair of unbonded electrons from each O atom in structure I and place one additional pair of electrons between each C and O atom, forming two double bonds:

```
 ..       ..           ..     ..
:O::C::O:     or     :O=C=O:
```

Each atom now has 8 electrons around it. Carbon is sharing four pairs of electrons, and each oxygen is sharing two pairs.

(b) **Step 1** The total number of valence electrons is 24, 1 from the hydrogen atom, 5 from the nitrogen atom, and 6 from each oxygen atom.

Step 2 The three O atoms are bonded to a central N atom. The H atom is bonded to one of the O atoms. Write the skeletal structure and place 2 electrons between each pair of atoms:

```
     O
     ..
H:O:N:O
```

Step 3 Subtract the 8 electrons used in Step 2 from 24 to obtain 16 electrons yet to be placed.

Step 4 Distribute the 16 electrons around the N and O atoms:

```
       ..
      :O <—— electron deficient
  ..   ..  ..
H:O:N:O:
  ..  ..  ..
```

Step 5 One pair of electrons is still needed to give all the N and O atoms 8 electrons. Move the unbonded pair of electrons from the N atom and place it between the N and the electron-deficient O atom, making a double bond:

```
      ..                      ..
     :O                      :O
  .. :: ..                    ||
H:O:N:O:       or     H—O—N—O:
  ..    ..               ..     ..
```

This arrangement is the Lewis structure.

PROBLEM 7.5 Write the Lewis structure for a sulfate ion, SO_4^{2-}.

Step 1 These five atoms have 30 valence electrons (6 in each atom) plus 2 electrons from the -2 charge, which makes 32 electrons to be placed.

Step 2 In the sulfate ion the sulfur is the central atom surrounded by the four oxygen atoms. Write the skeletal structure and place two electrons between each pair of atoms:

```
  O
  ..
O:S:O
  ..
  O
```

Step 3 Subtract the 8 electrons used in Step 2 from 32 to give 24 electrons yet to be placed.

Step 4 Distribute the 24 electrons around the four O atoms and indicate that the sulfate ion has a -2 charge:

```
      ..                    ..
     :O:                   :O:
  ..  ..  ..                |
 :O:S:O:2−     or    :O—S—O:2−
  ..  ..  ..          ..    ..
     :O:                    |
      ..                   :O:
                            ..
```

This arrangement is the Lewis structure. Each atom has 8 electrons around it.

Although many compounds follow the octet rule for covalent bonding, the rule has numerous exceptions. Sometimes it is impossible to write a structure in which each atom has eight electrons around it. For example, in BF_3 the boron atom has only 6 electrons around it, and in SF_6 the sulfur atom has 12 electrons around it.

7.8 Nonpolar and Polar Covalent Bonds

nonpolar covalent bond

We have considered bonds to be either ionic or covalent depending on whether electrons are transferred from one atom to another or are shared between two atoms. In a **nonpolar covalent bond** the shared pair of electrons is attracted equally by the two atoms. Nonpolar covalent bonds form between the same kind of atoms. For example, the covalent bond in a hydrogen molecule or in a chlorine molecule is nonpolar because the electronegativity difference between identical atoms is zero. (See Figure 7.6.)

polar covalent bond

A covalent bond between two different kinds of atoms has a partial ionic character and is known as a **polar covalent bond**. This bond is polar because the two atoms have different electronegativities. The polarity is due to unequal sharing of the electrons because the more electronegative atom has a greater attraction for the shared pair of electrons. The more electronegative atom thus acquires a partial negative charge, and the other atom acquires a partial positive charge. When a polar covalent bond is present in a diatomic molecule such as HCl, the molecule is polar. By *polar* we mean an unequal distribution of electrical charge. Overall, however, the HCl molecule is neutral because the partial negative charge on the chlorine atom is exactly balanced by the partial positive charge on the hydrogen atom.

dipole

A **dipole** is a molecule that is electrically unsymmetrical, causing it to be oppositely charged at two points. A dipole is often written as (+ −). A hydrogen chloride molecule is polar and behaves as a small dipole. The HCl dipole may be written as H ⟶ Cl. The arrow points toward the negative end of the dipole. Molecules of H_2O, HBr, and ICl are polar.

$$\overset{\delta+}{H} \longrightarrow \overset{\delta-}{Cl} \qquad \overset{\delta+}{H} \longrightarrow \overset{\delta-}{Br} \qquad \overset{\delta+}{I} \longrightarrow \overset{\delta-}{Cl} \qquad \overset{\delta+}{H} \nearrow \overset{\delta-}{O} \nwarrow \overset{\delta+}{H}$$

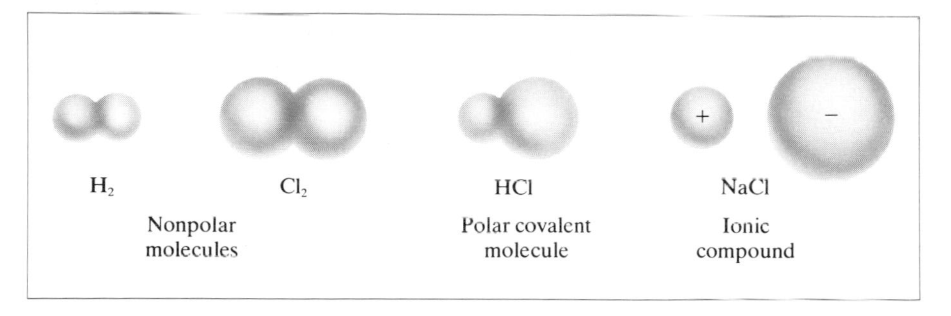

Figure 7.6
Nonpolar, polar covalent, and ionic compounds

How do we know whether a bond between two atoms is ionic or covalent? The difference in electronegativity between two atoms determines the character of the bond formed between them. As the difference in electronegativity increases, the polarity of the bond (or percentage ionic character) increases. As a rule, if the electronegativity difference between two bonded atoms is greater than 1.7–1.9, the bond will be more ionic than covalent. If the electronegativity difference is greater than 2.0, the bond is strongly ionic. If the electronegativity difference is less than 1.5, the bond is strongly covalent.

Care must be taken to distinguish between polar bonds and polar molecules. A covalent bond between different kinds of atoms is always polar. But a molecule containing different kinds of atoms may or may not be polar, depending on its shape or geometry. Molecules of HF, HCl, HBr, HI, and ICl are all polar because each contains a single polar bond. However, CO_2, CH_4, and CCl_4 are nonpolar molecules despite the fact that all three contain polar bonds. The carbon dioxide molecule, O=C=O, is nonpolar because the carbon-oxygen dipoles cancel each other by acting in opposite directions.

$$\overset{\longleftarrow\!\!\!+\quad +\!\!\!\longrightarrow}{O{=}C{=}O}$$

Dipoles in opposite directions

Methane (CH_4) and carbon tetrachloride (CCl_4) are nonpolar because the four C—H and C—Cl polar bonds are identical, and, because these bonds emanate from the center to the corners of a tetrahedron in the molecule, the effect of their polarities cancel one another (Figure 7.7).

We have said that water is a polar molecule. If the atoms in water were linear like those in carbon dioxide, the two O—H dipoles would cancel each other, and the molecule would be nonpolar. However, water is definitely polar and has a nonlinear (bent) structure with an angle of 105° between the two O—H bonds (see Figure 7.8).

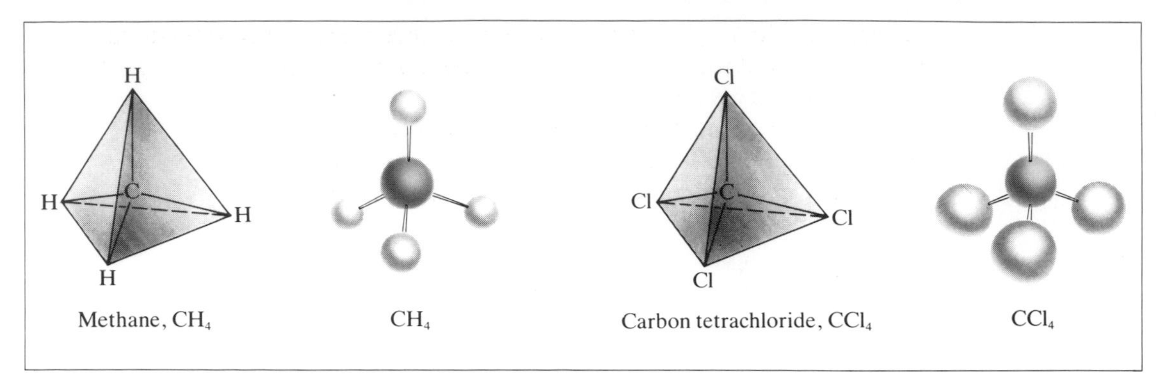

Figure 7.7

Ball-and-stick models of methane and carbon tetrachloride. Methane and carbon tetrachloride are nonpolar molecules because their polar bonds cancel each other in the tetrahedral arrangement of their atoms. The carbon atoms are located in the centers of the tetrahedrons.

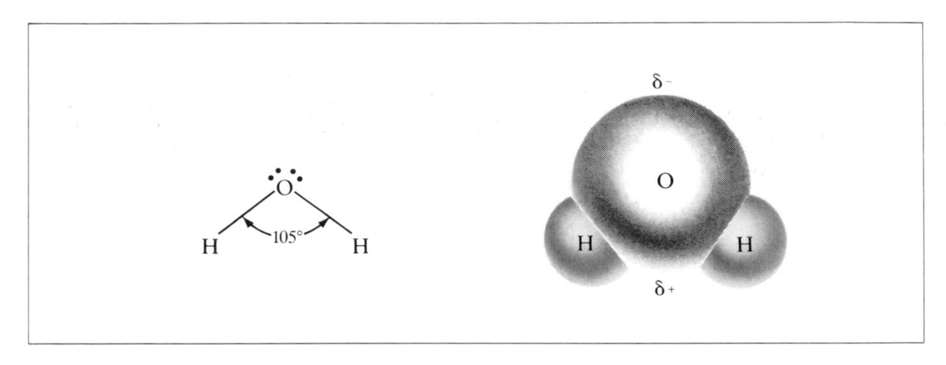

Figure 7.8

The polar water molecule, H_2O

7.9 Coordinate Covalent Bonds

In Section 7.5 we saw that a covalent bond was formed by the overlapping of electron orbitals between two atoms. The two atoms each furnish an electron to make a pair that is shared between them.

coordinate covalent bond

Covalent bonds can also be formed by one atom furnishing both electrons that are shared between the two atoms. The bond so formed is called a **coordinate covalent bond**. This bond is often designated by an arrow pointing away from the electron-pair donor (for example, A ⟶ B). Once formed, a coordinate covalent bond cannot be distinguished from any other covalent bond; it is simply a pair of electrons shared between two atoms.

The Lewis structures of sulfurous and sulfuric acids show coordinate covalent bonds between the sulfur and the oxygen atoms that are not bonded to hydrogen atoms. The colored dots indicate the electrons of the sulfur atom.

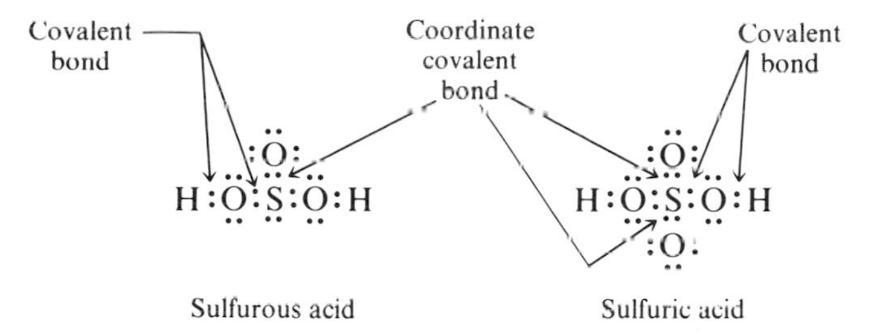

Sulfurous acid   Sulfuric acid

The lone (unbonded) pair of electrons on the sulfur atom in sulfurous acid allows room for another oxygen atom with six valence electrons to fit perfectly into its structure to form sulfuric acid. Other atoms with six valence electrons in their outer shell, such as sulfur, could also fit into this pattern. The coordinate covalent bond explains the formation of many complex molecules.

7.10 Polyatomic Ions

polyatomic ion A **polyatomic ion** is a stable group of atoms that has either a positive or a negative charge and behaves as a single unit in many chemical reactions. Sodium sulfate, Na_2SO_4, contains two sodium ions and a sulfate ion. The sulfate ion, SO_4^{2-}, is a polyatomic ion composed of one sulfur atom and four oxygen atoms and has a charge of −2. One sulfur and four oxygen atoms have a total of 30 electrons in their outer shells. The sulfate ion contains 32 outer-shell electrons and therefore has a charge of −2. In this case the two additional electrons come from the two sodium atoms, which are now sodium ions.

```
        [   :Ö:   ]2−                  [   :Ö:   ]2−
Na+     [ :Ö:S:Ö: ]    Na+             [ :Ö:S:Ö: ]
        [   :Ö:   ]                    [   :Ö:   ]
```

Sodium sulfate   Sulfate ion

Sodium sulfate has both ionic and covalent bonds. Ionic bonds exist between each of the sodium ions and the sulfate ion. Covalent bonds are present between the sulfur and oxygen atoms within the sulfate ion. One important difference between the ionic and covalent bonds in this compound can be demonstrated by dissolving sodium sulfate in water. It dissolves in water forming three charged particles, two sodium ions and one sulfate ion, per formula unit of sodium sulfate:

$$Na_2SO_4 \xrightarrow{\text{Water}} 2\,Na^+ + SO_4^{2-}$$

Sodium sulfate Sodium ions Sulfate ion

The SO_4^{2-} ion remains as a unit, held together by covalent bonds; whereas, where the bonds were ionic, dissociation of the ions took place. Do not think, however, that polyatomic ions are so stable that they cannot be altered. Chemical reactions by which polyatomic ions can be changed to other substances do exist.

The Lewis structures for several common polyatomic ions are shown below:

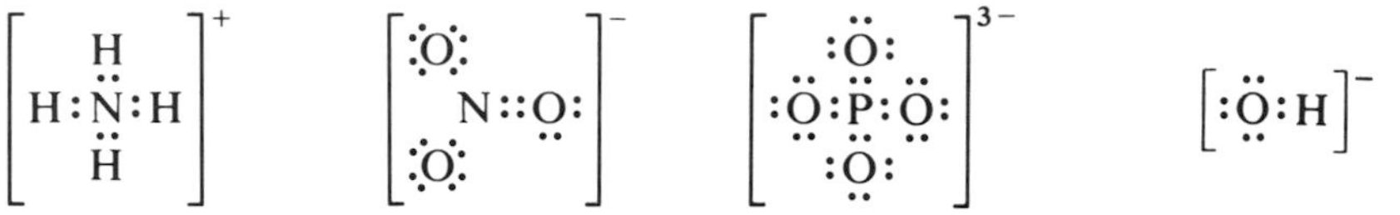

Ammonium ion, NH_4^+ Nitrate ion, NO_3^- Phosphate ion, PO_4^{3-} Hydroxide ion, OH^-

QUESTIONS

A. *Review the meanings of the new terms introduced in this chapter.*

1. Chemical bonds
2. Ionization energy
3. Electron affinity
4. Valence electrons
5. Ionic bond
6. Covalent bond
7. Electronegativity
8. Nonpolar covalent bond
9. Polar covalent bond
10. Dipole
11. Coordinate covalent bond
12. Polyatomic ion

B. *Information useful in answering these questions will be found in the tables and figures.*

1. Explain why much more ionization energy is required to remove the first electron from neon than from sodium.
2. Explain the large increase in ionization energy needed to remove the third electron from beryllium compared with that needed for the second electron.
3. Does the first ionization energy increase or decrease from top to bottom in the periodic table for the alkali metal family? Explain.
4. Does the first ionization energy increase or decrease from top to bottom in the periodic table for the noble gas family? Explain.
5. In which general areas of the periodic table are the elements with (a) the highest and (b) the lowest electronegativities found?
6. Which is larger, a magnesium atom or a magnesium ion? Explain.
7. Which is smaller, a bromine atom or a bromide ion? Explain.
8. Using the table of electronegativity values (Table 7.3), indicate which element is positive and which is negative in the following compounds:
   (a) H_2O (e) NO (i) CCl_4
   (b) NaF (f) CH_4 (j) IBr
   (c) NH_3 (g) HCl (k) MgH_2
   (d) PbS (h) LiH (l) OF_2
9. Classify the bond between the following pairs of elements as principally ionic or principally covalent (use Table 7.3):
   (a) Sodium and chlorine
   (b) Carbon and hydrogen
   (c) Chlorine and carbon
   (d) Calcium and oxygen
   (e) Hydrogen and sulfur
   (f) Barium and oxygen
   (g) Fluorine and fluorine
   (h) Potassium and fluorine

C. *Review Questions*

1. Explain why, in general, the ionization energy decreases from top to bottom in a family of elements.
2. Why does barium (Ba) have a lower ionization energy than beryllium (Be)?
3. Why is there such a large increase in the ionization energy required to remove the second electron from a sodium atom as opposed to the first?
4. Explain how electron affinity differs from ionization energy.
5. Explain what happens to the electron structures of Mg and Cl atoms when they react to form $MgCl_2$.
6. Write an equation representing (a) the change of a fluorine atom to a fluoride ion (F^-) and (b) the change of a calcium atom to a calcium ion (Ca^{2+}).
7. Use Lewis-dot symbols to show the electron transfer for the formation of the following ionic compounds from the atoms.
   (a) MgF_2 (b) K_2O (c) CaO (d) NaBr
8. What are valence electrons?
9. How many valence electrons are in each of these atoms? H, K, Mg, He, Al, Si, N, P, O, Cl
10. How many electrons must be gained or lost for each of the following to achieve a noble-gas electron structure?
    (a) A calcium atom (d) A chloride ion
    (b) A sulfur atom (e) A nitrogen atom
    (c) A helium atom (f) A potassium atom
11. Explain why potassium forms a K^+ ion but not a K^{2+} ion.
12. Why does an aluminum ion have a +3 charge?
13. Which is larger? Explain.
    (a) A potassium atom or a potassium ion
    (b) A bromine atom or a bromide ion
    (c) A magnesium ion or an aluminum ion
    (d) Fe^{2+} or Fe^{3+}
14. Write Lewis-dot structures for
    (a) Na (b) Br^- (c) O^{2-} (d) Ga (e) Ga^{3+}
15. Why is it not proper to speak of sodium chloride molecules?
16. What is a covalent bond? How does it differ from an ionic bond?
17. What is different about the formation of a coordinate covalent bond from that of an ordinary covalent bond?
18. Classify the bonding in each compound as ionic or covalent:
    (a) H_2O (c) MgO (e) HCl (g) NH_3
    (b) NaCl (d) Br_2 (f) $BaCl_2$ (h) SO_2
19. Predict the type of bond that would be formed between each of the following pairs of atoms:
    (a) Na and N (d) H and Si
    (b) N and S (e) O and F
    (c) Br and I
20. Draw Lewis structures for:
    (a) H_2 (b) N_2 (c) Cl_2 (d) O_2 (e) Br_2
21. Draw Lewis structures for:
    (a) NCl_3 (c) H_3PO_4 (e) H_2S
    (b) H_2CO_3 (d) C_2H_6 (f) CS_2
22. Briefly comment on the structure $Na{:}\underset{\cdot\cdot}{\ddot{O}}{:}Na$ for the compound Na_2O.
23. What are the four most electronegative elements?
24. Draw Lewis structures for:
    (a) Ba^{2+} (e) SO_4^{2-} (h) CO_3^{2-}
    (b) Al^{3+} (f) SO_3^{2-} (i) ClO_3^-
    (c) I^- (g) CN^- (j) NO_3^-
    (d) S^{2-}
25. The Lewis structure for chloric acid is

```
   ..  ..  ..
H :O :Cl :O :
   ..  ..  ..
      :O :
       ..
```

    Point out the covalent and coordinate covalent bonds in this structure.
26. Draw the Lewis structure for ammonia (NH_3). What type of bond is present? Can this molecule form coordinate covalent bonds? If so, how many?
27. Rank these elements from highest electronegativity to lowest: Mg, S, F, H, O, Cs.
28. Classify the following molecules as polar or nonpolar:
    (a) H_2O (c) CF_4 (e) CO_2
    (b) HBr (d) F_2 (f) NH_3
29. Is it possible for a molecule to be nonpolar even though it contains polar covalent bonds? Explain.
30. Why is CO_2 a nonpolar molecule, whereas CO is a polar molecule?
31. Which of these statements are correct? (Try to answer this question using only the periodic table.)

(a) The amount of energy required to remove one electron from an atom is known as the ionization energy.
(b) Metallic elements tend to have relatively low electronegativities.
(c) Elements with a high ionization energy tend to have very metallic properties.
(d) Sodium and chlorine react to form molecules of NaCl.
(e) A chlorine atom has fewer electrons than a chloride ion.
(f) The noble gases have a tendency to lose one electron to become positively charged ions.
(g) The chemical bonds in a water molecule are ionic.
(h) The chemical bonds in a water molecule are polar.
(i) Valence electrons are those electrons in the outermost shell of an atom.
(j) An atom with eight electrons in its outer shell has all its *s* and *p* orbitals filled.
(k) Fluorine has the lowest electronegativity of all the elements.
(l) Oxygen has a greater electronegativity than carbon.
(m) A cation is larger than its corresponding neutral atom.
(n) Cl_2 is more ionic in character than HCl.
(o) A neutral atom with eight electrons in its valence shell will likely be an atom of a noble gas.
(p) A nitrogen atom has four valence electrons.
(q) An aluminum atom must lose three electrons to become an aluminum ion, Al^{3+}.
(r) A stable group of atoms that has either a positive or a negative charge and behaves as a single unit in many chemical reactions is called a polyatomic ion.
(s) Sodium sulfate, Na_2SO_4, has covalent bonds between sulfur and the oxygen atoms, and ionic bonds between sodium ions and the sulfate ion.
(t) The water molecule is a dipole.
(u) In an ethylene molecule, C_2H_4,

```
H       H
 \     /
  C = C
 /     \
H       H
```

two pairs of electrons are shared between the carbon atoms.
(v) The octet rule is mainly useful for atoms where only *s* and *p* electrons enter into bonding.
(w) When electrons are transferred from one atom to another, the resulting compound contains ionic bonds.
(x) A phosphorus atom, $\cdot\ddot{\underset{\cdot}{P}}$, needs three additional electrons to attain a stable octet of electrons.
(y) The simplest compound between oxygen, $\cdot\ddot{\underset{\cdot\cdot}{O}}\cdot$, and fluorine, $:\ddot{\underset{\cdot\cdot}{F}}\cdot$, atoms is FO_2.
(z) The $H:\ddot{\underset{\cdot\cdot}{Cl}}:$ molecule has three unshared pairs of electrons.
(aa) The smaller the difference in electronegativity between two atoms, the more ionic the bond between them will be.
(bb) Lewis structures are mainly useful for the representative elements.
(cc) The correct Lewis structure for NH_3 is

$$H:\ddot{\underset{\cdot\cdot}{N}}:H$$
$$H$$

(dd) The correct Lewis structure for CO_2 is

$$:\ddot{\underset{\cdot\cdot}{O}}:C:\ddot{\underset{\cdot\cdot}{O}}:$$

(ee) The correct Lewis structure for SO_4^{2-} is

$$:\ddot{\underset{\cdot\cdot}{O}}:\ddot{\underset{\cdot\cdot}{O}}:\ddot{\underset{\cdot\cdot}{S}}:\ddot{\underset{\cdot\cdot}{O}}:\ddot{\underset{\cdot\cdot}{O}}:^{2-}$$

(ff) In period 4 of the periodic table, the element having the lowest ionization energy is Xe.
(gg) An atom having an electron structure of $1s^22s^22p^63s^23p^2$ has four valence electrons.
(hh) When an atom of bromine becomes a bromide ion, its size increases.
(ii) The structures that show that H_2O is a dipole and that CO_2 is not a dipole are $H—\ddot{\underset{\cdot\cdot}{O}}—H$ and $:\ddot{O}=C=\ddot{O}:$
(jj) The Cl^- and S^{2-} ions have the same electron structure.

Review Exercises for Chapters 5–7

CHAPTER 5 ATOMIC THEORY AND STRUCTURE

True–False. *Answer the following as either true or false.*

1. Dalton's atomic theory states that all atoms are composed of protons, neutrons, and electrons.
2. The electron was discovered by J. J. Thomson.
3. The proton was discovered by James Chadwick in 1932.
4. A *p* orbital is spherically symmetrical around the nucleus.
5. An atom of nitrogen has two electrons in a 1*s* orbital, two electrons in a 2*s* orbital, and one electron in each of three different 2*p* orbitals.
6. The Lewis-dot symbol for calcium is Ca:
7. An atom of $^{108}_{47}Ag$ contains 47 protons, 47 electrons, and 108 neutrons.
8. One atomic mass unit is defined as one-twelfth the mass of a carbon-12 atom.
9. The proton and the neutron have approximately equal charge.
10. In the ground state, electrons tend to occupy orbitals of the lowest possible energy.
11. The Lewis-dot symbol for potassium is $\cdot\ddot{\underset{\cdot}{P}}\cdot$
12. The fourth principal energy level can contain a total of 32 electrons.
13. The second principal energy level contains only *s* and *p* sublevels.
14. The first *f* sublevel electrons are in the fourth principal energy level.
15. The 4*s* energy sublevel fills before the 3*d* sublevel because the 4*s* is at a lower energy level.
16. A *p* type energy sublevel can contain six electrons.
17. A *d* type energy sublevel can contain five orbitals.
18. The listed atomic weight of an element represents the average relative mass of all the naturally occurring isotopes of that element.
19. In a Lewis-dot symbol representation of an element, the symbol of the element is shown surrounded by the number of electrons in the outermost energy level of the atom.
20. The atoms of two different elements must have different mass numbers.
21. $^{35}_{17}Cl$ and $^{37}_{17}Cl$ are isotopes of chlorine.
22. The element represented by $1s^2 2s^2 2p^6 3s^2 3p^4$ has an atomic number of 16.
23. The maximum number of electrons in an *f* orbital is 14.
24. An atom with eight electrons in its outer shell is a noble gas.
25. In the isotope $^{112}_{47}Ag$, $Z = 112$ and $A = 47$.
26. All the isotopes of an element have the same number of electrons.
27. An atom of $_{16}S$ has nine sublevel orbitals that contain one or more electrons.
28. The electron configuration of $_{22}Ti$ is $1s^2 2s^2 2p^6 3s^2 3p^6 3d^4$.
29. The reason the atomic weight of Mg is 24.305 rather than almostly exactly 24 is that protons and neutrons do not have exactly the same mass.
30. The lightest element is helium.

Multiple Choice. *Choose the correct answer to each of the following.*

1. The concept of the positive charge and most of the mass concentrated in a small nucleus surrounded by the electrons was the contribution of:
   (a) Dalton (c) Bohr
   (b) Rutherford (d) Schrödinger
2. The concept of electrons existing in specific orbits around the nucleus was the contribution of:
   (a) Thomson (c) Bohr
   (b) Rutherford (d) Schrödinger
3. The neutron was discovered in 1932 by
   (a) Bohr (c) Thomson
   (b) Rutherford (d) Chadwick

4. The correct electron structure for a fluorine atom, F, is:
   (a) $1s^22s^22p^5$ (c) $1s^22s^22p^43s^1$
   (b) $1s^22s^22p^23s^23p^1$ (d) $1s^22s^22p^3$
5. The correct electron structure for $_{48}$Cd is:
   (a) $1s^22s^22p^63s^23p^64s^23d^{10}$
   (b) $1s^22s^22p^63s^23p^64s^23d^{10}4p^65s^24d^{10}$
   (c) $1s^22s^22p^63s^23p^64s^23d^{10}4p^64d^4$
   (d) $1s^22s^22p^63s^23p^64s^24p^64d^{10}5s^25d^{10}$
6. The correct electron structure of $_{23}$V is:
   (a) [Ar] $4s^23d^3$ (c) [Ar] $4s^24d^3$
   (b) [Ar] $4s^24p^3$ (d) [Kr] $4s^23d^3$
7. Which of the following is the correct atomic structure for $^{48}_{22}$Ti?
   (a) (22p 26n) 2 8 10 2 (c) (26p 22n) 2 8 8 4
   (b) (22p 48n) 2 8 8 4 (d) (22p 26n) 2 8 8 4
8. The number of orbitals in a *d* sublevel is
   (a) 3 (c) 7
   (b) 5 (d) No correct answer given
9. An atom of atomic number 53 and mass number 127 contains how many neutrons?
   (a) 53 (b) 74 (c) 127 (d) 180
10. How many electrons are in an atom of $^{40}_{18}$Ar?
    (a) 20 (c) 40
    (b) 22 (d) No correct answer given
11. The number of neutrons in an atom of $^{139}_{56}$Ba is:
    (a) 56 (c) 139
    (b) 83 (d) No correct answer given
12. The number of electrons in the third principal energy level in an atom having the electron structure $1s^22s^22p^63s^23p^2$ is:
    (a) 2 (b) 4 (c) 6 (d) 8
13. The total number of orbitals that contain at least one electron in an atom having the structure $1s^22s^22p^63s^23p^2$ is:
    (a) 5 (c) 14
    (b) 8 (d) No correct answer given
14. The name of the isotope containing one proton and two neutrons is:
    (a) Protium (c) Deuterium
    (b) Tritium (d) Helium
15. Each atom of a specific element has the same
    (a) Number of protons
    (b) Atomic mass
    (c) Number of neutrons
    (d) No correct answer given
16. Which of these elements has two *s* and six *p* electrons in its outer energy level?
    (a) He (c) Ar
    (b) O (d) No correct answer given
17. Which is not a correct Lewis-dot symbol for phosphorus, $_{15}$P?
    (a) $:\underset{\cdot}{\dot{P}}\cdot$ (b) $:\ddot{P}\cdot$ (c) $\cdot\underset{\cdot}{\ddot{P}}\cdot$ (d) $\cdot\underset{\cdot}{\dot{P}}:$
18. Which pair of symbols represents isotopes?
    (a) $^{23}_{11}$Na and $^{23}_{12}$Na
    (b) $^{7}_{3}$Li and $^{6}_{3}$Li
    (c) $^{63}_{29}$Cu and $^{29}_{64}$Cu
    (d) $^{12}_{24}$Mg and $^{12}_{26}$Mg
19. Which element is not a noble gas?
    (a) Ra (b) Xe (c) He (d) Ar
20. Two naturally occurring isotopes of an element have masses and abundance as follows: 54.00 amu (20.00%) and 56.00 amu (80.00%). What is the relative atomic weight of the element?
    (a) 54.20 (b) 54.40 (c) 54.80 (d) 55.60

CHAPTER 6 THE PERIODIC ARRANGEMENT OF THE ELEMENTS

True–False. *Answer the following as either true or false.*

1. The periodic law states that the atomic weights of the elements are a periodic function of their atomic number.
2. The atomic radius of phosphorus, P, will be larger than that of sodium, Na.
3. The atomic radius of barium, Ba, will be larger than that of magnesium, Mg.
4. Elements at the bottom of a group in the periodic table tend to be more metallic in their properties than those at the top.
5. Elements within a group will have similar electron structures and will show some similarities in their chemical properties.
6. The transition elements are characterized by an increasing number of *d* or *f* electrons in an inner shell.

7. Based on the periodic table, the formula for aluminum oxide would be Al_2O_3.
8. J. A. R. Newlands observed that, when the elements were arranged according to increasing atomic weights, every eighth element had similar properties.
9. The horizontal rows of elements in the periodic table are called periods.
10. The *p* block of elements contains transition elements.
11. Group IIIA elements all have three electrons in their outer energy level.
12. Group B elements are referred to as the representative elements.
13. The element in period 3 having the largest radius is argon.
14. The element in period 3 that is classified as a metalloid is silicon.
15. The halogen family is located in Group VIIA.
16. There are seven periods of elements, two short and five long periods.
17. The representative elements do not contain *d* or *f* electrons.
18. Group VIA elements are called the alkaline earth metals.
19. The Lewis-dot symbol for $_{84}Po$ is $\cdot\underset{\cdot}{\ddot{\text{P}}}\text{o}:$
20. Oxygen has the smallest radius of the elements in Group VIA.

Multiple Choice. *Choose the correct answer to each of the following.*

1. The scientist who noticed the existence of triads of elements with similar properties was:
   (a) Döbereiner (c) Mendeleev
   (b) Newlands (d) Meyer
2. The chemist who developed a periodic table with the known elements arranged by atomic weight, who left spaces where elements seemed to be missing, and who predicted the properties of the missing elements was:
   (a) Newlands (c) Meyer
   (b) Mendeleev (d) Moseley
3. The physicist who, through his work in X-ray emission frequencies of elements, established that atomic number was the correct basis for arranging the elements was:
   (a) Döbereiner (c) Meyer
   (b) Mendeleev (d) Moseley
4. The German scientist who in 1869 independently published a periodic arrangement of elements based on increasing atomic weights similar to Mendeleev's arrangement was:
   (a) Döbereiner (c) Newlands
   (b) Meyer (d) Moseley
5. Periods IIIA through VIIA plus the noble gases form the area of the periodic table where the electron sublevels being filled are:
   (a) *p* sublevels (c) *d* sublevels
   (b) *s* and *p* sublevels (d) *f* sublevels
6. In moving down an A group on the periodic table, the number of electrons in the outermost energy level:
   (a) Increases regularly
   (b) Remains constant
   (c) Decreases regularly
   (d) Changes in an unpredictable manner
7. Which of the following would be an incorrect formula?
   (a) NaCl (b) K_2O (c) AlO (d) BaO
8. Elements of the noble gas family:
   (a) Form no compounds at all
   (b) Have full outer electron shells
   (c) Have an outer-shell electron structure of ns^2np^6 (helium excepted), where *n* is the period number
   (d) No correct answer given
9. The lanthanide and actinide series of elements are:
   (a) Representative elements
   (b) Transition elements
   (c) Filling in *d* level electrons
   (d) No correct answer given
10. The element having the structure $1s^22s^22p^63s^23p^2$ is in Group:
    (a) IIA (b) IIB (c) IVA (d) IVB
11. In Group VA, the element having the smallest atomic radius is:
    (a) Bi (b) P (c) As (d) N
12. In Group IVA, the most metallic element is:
    (a) C (b) Si (c) Ge (d) Sn
13. Which group in the periodic table contains the least reactive elements?
    (a) IA (b) IIA (c) IIIA (d) Noble gases
14. Which group in the periodic table contains the alkali metals?
    (a) IA (b) IIA (c) IIIA (d) IVA

15. An atom of fluorine is smaller than an atom of oxygen. One possible explanation is that, compared to oxygen, fluorine has:
(a) A larger mass number
(b) A smaller atomic number
(c) A greater nuclear charge
(d) More unpaired electrons

CHAPTER 7 CHEMICAL BONDS: THE FORMATION OF COMPOUNDS FROM ATOMS

True–False. *Answer the following as either true or false.*

1. The amount of energy required to remove one electron from an atom is known as the ionization energy.
2. It requires less energy to remove a second electron from an atom than to remove the first electron.
3. The first ionization energy of potassium will be greater than that for sodium.
4. The first ionization energy of sulfur will be greater than that for sodium.
5. The electrons in the outermost shell of an element are called the valence electrons.
6. When an atom loses an electron it becomes a negative ion.
7. A bromide ion is smaller than a bromine atom.
8. Crystals of sodium chloride consist of arrays of NaCl molecules.
9. The sharing of a pair of electrons between a positive ion and a negative ion is called an ionic bond.
10. When elements combine by ionic bonding, they normally form molecules.
11. When a single atom furnishes both electrons that are shared between two atoms, the bond is called a coordinate covalent bond.
12. The NH_4^+ ion contains at least one coordinate covalent bond.
13. Oxygen has a greater electronegativity than carbon.
14. Bromine has a greater electronegativity than chlorine.
15. A dipole is a molecule that is electrically unsymmetrical, causing it to be oppositely charged at two points.
16. The water molecule is a dipole.
17. The formula of the aluminum ion is Al^+.
18. A stable group of atoms that has either a positive or a negative charge and behaves as a single unit is called a polyatomic ion.
19. A phosphate ion, PO_4^{3-}, is a polyatomic ion.
20. A sodium ion is smaller than a sodium atom.
21. In crystals of sodium chloride, each sodium ion is surrounded by six chloride ions, and each chloride ion is surrounded by six sodium ions.
22. Boron has a greater electronegativity than barium.
23. Metals tend to lose their valence electrons, forming positively charged ions.
24. When metals with low electronegativity combine with nonmetals of high electronegativity, they tend to form ionic compounds.
25. A pair of electrons shared between two atoms constitutes a covalent bond.
26. The ionic bond is the electrostatic attraction existing between oppositely charged ions.
27. Two atoms with different electronegativities that form a polar covalent bond will have partial ionic charges.
28. The covalent bond between a B atom and an N atom will be nonpolar.
29. The covalent bond between two carbon atoms will be nonpolar.
30. A covalent bond formed by sharing one pair of electrons is called a single bond.

Multiple Choice. *Choose the correct answer to each of the following.*

1. Which of the following formulas is not correct?
(a) Na^+ (b) S^- (c) Al^{3+} (d) F^-
2. Which of the following does not have a polar covalent bond?
(a) CH_4 (b) H_2O (c) CH_3OH (d) Cl_2
3. Which of the following molecules is a dipole?
(a) HBr (b) CH_4 (c) H_2 (d) CO_2
4. Which of the following has bonding that is ionic?
(a) H_2 (b) MgF_2 (c) H_2O (d) CH_4

5. Which of the following is a correct Lewis structure?

(a) :Ö:C:Ö: (b) :C̈l:C̈:C̈l: with :C̈l: above and below C

(c) C̈l::C̈l (d) :N̈:N̈:

6. Which of the following is an incorrect Lewis structure?

(a) H:N̈:H with H below N (b) :Ö:H with H below O

(c) H:C̈:H with H above and below C (d) :N:::N:

7. The correct Lewis structure for SO_2 is:

(a) :Ö:S̈:Ö: (b) :Ö:S̈::Ö:

(c) :Ö::S::Ö: (d) :Ö:S̤:Ö:

8. Which element has seven valence electrons?
   (a) S (b) Ne (c) Br (d) Ag
9. Carbon dioxide, CO_2, is a nonpolar molecule because:
   (a) Oxygen is more electronegative than carbon.
   (b) The two oxygen atoms are bonded to the carbon atom.
   (c) The molecule has a linear structure with the carbon atom in the middle.
   (d) The carbon–oxygen bonds are polar covalent.
10. When a magnesium atom participates in a chemical reaction, it is most likely to:
    (a) Lose 1 electron (c) Lose 2 electrons
    (b) Gain 1 electron (d) Gain 2 electrons
11. If X represents an element of Group IIIA, what is the general formula for its oxide?
    (a) X_3O_4 (b) X_3O_2 (c) XO (d) X_2O_3
12. Which of the following has the same electron structure as an argon atom?
    (a) Ca^{2+} (b) Cl^0 (c) Na^+ (d) K^0
13. As the difference in electronegativity between two elements decreases, the tendency for the elements to form a covalent bond:
    (a) Increases
    (b) Decreases
    (c) Remains the same
    (d) Sometimes increases and sometimes decreases
14. Which compound forms a tetrahedral molecule?
    (a) NaCl (b) CO_2 (c) CH_4 (d) $MgCl_2$
15. Which compound has a bent (V-shaped) molecular structure?
    (a) NaCl (b) CO_2 (c) CH_4 (d) H_2O
16. Which compound has double bonds within its molecular structure?
    (a) NaCl (b) CO_2 (c) CH_4 (d) H_2O
17. The total number of valence electrons in a nitrate ion, NO_3^-, is:
    (a) 12 (b) 18 (c) 23 (d) 24
18. The number of electrons in a triple bond is:
    (a) 3 (b) 4 (c) 6 (d) 8
19. The number of lone (unbonded) pairs of electrons in H_2O is:
    (a) 0 (b) 1 (c) 2 (d) 4
20. Which of the following does not have a noble-gas electron structure?
    (a) Na (b) Sc^{3+} (c) Ar (d) O^{2-}

CHAPTER **8**

Nomenclature of Inorganic Compounds

After studying Chapter 8, you should be able to

1. State the rules for assigning oxidation numbers.
2. Assign oxidation numbers to the representative elements in binary compounds from their positions in the periodic table.
3. Write the formulas of compounds formed by combining the ions from Tables 8.2, 8.3, and 8.4 (or from the inside back cover of this book) in the correct ratios.
4. Assign the oxidation number to each element in a compound or ion.
5. Write the names or formulas for inorganic binary compounds in which the metal has only one common oxidation state.
6. Write the names or formulas for inorganic binary compounds that contain metals of variable oxidation state, using either the Stock System or classical nomenclature.
7. Write the names or formulas for inorganic binary compounds that contain two nonmetals.
8. Write the names or formulas for binary acids.
9. Write the names or formulas for ternary oxy-acids.
10. Write the names or formulas for ternary salts.
11. Given the formula of a salt, write the name and formula of the acid from which the salt may be derived.
12. Write the names or formulas for salts that contain more than one kind of positive ion.
13. Write the names or formulas for inorganic bases.

8.1 Oxidation Numbers of Atoms

oxidation number

oxidation state

We have seen that atoms can combine to form compounds by losing, gaining, or sharing electrons. The **oxidation number** or **oxidation state** of an element is a number having a positive, a negative, or a zero value that may be assigned to an atom of that element in a compound. These positive and negative numbers are directly related to the positive and negative charges that result from the transfer of electrons from one atom to another in ionic compounds or from an unequal sharing of electrons between atoms forming covalent bonds. Oxidation numbers are assigned by a somewhat arbitrary system of rules. They are useful for writing formulas, naming compounds, and balancing chemical equations.

In a compound having ionic bonds, the oxidation number of an atom or group of atoms existing as an ion is the same as the *charge of the ion.* Thus, in sodium chloride, NaCl, the oxidation number of sodium is +1 and that of chlorine is −1; in magnesium oxide, MgO, the oxidation number of magnesium is +2 and that of oxygen is −2; in calcium chloride, $CaCl_2$, the oxidation number of calcium is +2 and that of chlorine is −1. The sum of the oxidation numbers of all the atoms in a compound is numerically equal to zero, because a compound is electrically neutral.

For practical purposes it is also convenient to assign oxidation numbers to the individual atoms comprising molecules and polyatomic ions. Here the electrons have not been completely transferred from one atom to another, and the assignment cannot be done solely on the basis of ionic charges. However, oxidation numbers can be readily assigned to the atoms in either molecules or polyatomic ions by this general method: For each covalent bond, first assign the shared pair of electrons to the more electronegative atom. Then assign an oxidation number to each atom corresponding to its apparent net charge based on the number of electrons gained or lost and on the fact that the sum of the oxidation numbers must equal zero for a compound or must equal the charge on a polyatomic ion. Consider these substances: H_2, H_2O, CH_4, and CCl_4,

	H:H	H:Ö: H	H H:C̈:H H	:C̈l :C̈l:C̈:C̈l: :C̈l:
oxidation number	H, 0	H, +1 O, −2	H, +1 C, −4	C, +4 Cl, −1

In H_2 the pair of electrons is shared equally between the two atoms; therefore, each H is assigned an oxidation number of zero. In H_2O, the oxygen is the more electronegative atom and is assigned the two pairs of shared electrons. The oxygen atom now has two more electrons than the neutral atom and therefore is assigned an oxidation number of −2. Each hydrogen atom in H_2O has one less electron than the neutral atom and is assigned an oxidation number of +1. In CH_4 all four shared pairs of electrons are assigned to the more electronegative carbon atom. The carbon atom then has an additional four electrons and is assigned an oxidation number of −4. Each hydrogen atom has one less electron than the neutral atom and is assigned an oxidation number of +1. In CCl_4 one pair of electrons is assigned to each of the four more electronegative chlorine atoms. The carbon atom therefore has four less electrons than the neutral atom and is assigned an oxidation number of +4. Each chlorine atom has one additional electron and is assigned an oxidation number of −1.

The following rules govern the assignment of oxidation numbers:

1. The oxidation number of any free element is zero, even when the atoms are combined with themselves. (*Examples*: Na, Mg, H_2, O_2, Cl_2)
2. Metals generally have positive oxidation numbers in compounds.
3. The oxidation number of hydrogen in a compound or an ion is +1 except in metal hydrides, where H is −1. (For example, NaH: Na, +1; H, −1)
4. The oxidation number of oxygen in a compound or an ion is −2 except in peroxides, where it is −1 and in OF_2 where it is +2.
5. The oxidation number of a monatomic ion is the same as the charge on the ion.
6. The oxidation number of an atom in a covalent compound is equal to the net apparent charge on the atom after each pair of shared electrons is assigned to the more electronegative element sharing the pair of electrons.
7. The algebraic sum of the oxidation numbers for all the atoms in a compound must equal zero.
8. The algebraic sum of the oxidation numbers for all the atoms in a polyatomic ion must equal the charge on the ion.

The oxidation numbers of many elements are predictable from their position in the periodic table. This predictability is especially true of the Group A elements because the number of electrons in their outer shells corresponds to the group number. Remember that metals lose electrons and become positively charged ions. Nonmetals tend to gain electrons and become negatively charged ions. Nonmetals can also share electrons with other atoms to assume a positive or negative oxidation number. Hydrogen can have a +1 or −1 oxidation number, depending on the relative electronegativity of the element with which it is combined.

The predictable oxidation numbers of the Group A elements are given in the following table:

Group number	IA	IIA	IIIA	IVA	VA	VIA	VIIA
Oxidation number	+1	+2	+3	+4 to −4	−3 to +5	−2 to +6	−1 to +7

Table 8.1 illustrates the use of oxidation numbers to predict formulas of binary compounds from representative members of these groups.

Table 8.1 Selected binary hydrogen, oxygen, and chlorine compounds of Group A elements

	Group number						
	IA	IIA	IIIA	IVA	VA	VIA	VIIA
Hydrogen compound	NaH	CaH_2	AlH_3	CH_4	NH_3	H_2S	HCl
Oxygen compound	Na_2O	CaO	Al_2O_3	CO_2	N_2O_5	SO_3	Cl_2O
Chlorine compound	$NaCl$	$CaCl_2$	$AlCl_3$	CCl_4	NCl_3	SCl_2	Cl_2

8.2 Oxidation Number Tables

Writing formulas of compounds and chemical equations is facilitated by a knowledge of oxidation numbers and ionic charges. Table 8.2 lists the names and ionic charges of common monatomic ions. Monatomic ions of Groups IA and IIA elements are not given in Table 8.2 because the charges and oxidation numbers of these ions are readily determined from the periodic table. The charges and oxidation numbers of the Groups IA, IIA, and IIIA metal ions are positive and correspond to the group number (for example, Na^+, Ca^{2+}, Al^{3+}). The negative charges and oxidation numbers of Groups VA, VIA, and VIIA monatomic ions can be determined by subtracting eight from the group number.

Table 8.2 Names, formulas, and charges of selected monatomic ions

Name	Formula	Oxidation number or charge	Name	Formula	Oxidation number or charge
Aluminum	Al^{3+}	+3	Lead(II)	Pb^{2+}	+2
Arsenic(III)	As^{3+}	+3	Magnesium	Mg^{2+}	+2
Barium	Ba^{2+}	+2	Manganese(II)	Mn^{2+}	+2
Cadmium	Cd^{2+}	+2	Mercury(I)	Hg^+	+1
Calcium	Ca^{2+}	+2	Mercury(II)	Hg^{2+}	+2
Chromium(III)	Cr^{3+}	+3	Nickel(II)	Ni^{2+}	+2
Copper(I)	Cu^+	+1	Silver	Ag^+	+1
Copper(II)	Cu^{2+}	+2	Tin(II)	Sn^{2+}	+2
Iron(II)	Fe^{2+}	+2	Tin(IV)	Sn^{4+}	+4
Iron(III)	Fe^{3+}	+3	Zinc	Zn^{2+}	+2
Bromide	Br^-	−1	Nitride	N^{3-}	−3
Chloride	Cl^-	−1	Oxide	O^{2-}	−2
Fluoride	F^-	−1	Sulfide	S^{2-}	−2
Iodide	I^-	−1			

Table 8.3 Names, formulas, and charges of some common polyatomic ions

Name	Formula	Charge	Name	Formula	Charge
Acetate	$C_2H_3O_2^-$	−1	Cyanide	CN^-	−1
Ammonium	NH_4^+	+1	Dichromate	$Cr_2O_7^{2-}$	−2
Arsenate	AsO_4^{3-}	−3	Hydroxide	OH^-	−1
Bicarbonate	HCO_3^-	−1	Nitrate	NO_3^-	−1
Bisulfate	HSO_4^-	−1	Nitrite	NO_2^-	−1
Bromate	BrO_3^-	−1	Permanganate	MnO_4^-	−1
Carbonate	CO_3^{2-}	−2	Phosphate	PO_4^{3-}	−3
Chlorate	ClO_3^-	−1	Sulfate	SO_4^{2-}	−2
Chromate	CrO_4^{2-}	−2	Sulfite	SO_3^{2-}	−2

Table 8.4 Principal oxidation numbers of some common elements that have variable oxidation states

Element	Oxidation number	Element	Oxidation number
Cu	+1, +2	Cl	−1, +1, +3, +5, +7
Hg	+1, +2	Br	−1, +1, +3, +5, +7
Sn	+2, +4	I	−1, +1, +3, +5, +7
Pb	+2, +4	S	−2, +4, +6
Fe	+2, +3	N	−3, +1, +2, +3, +4, +5
Au	+1, +3	P	−3, +3, +5
Ni	+2, +3	C	−4, +4
Co	+2, +3		
As	+3, +5		
Bi	+3, +5		
Cr	+2, +3, +6		

Sulfur, for example, is in Group VIA, and 6 − 8 = −2. Therefore the oxidation number of the sulfide ion (S^{2-}) is −2. All the halogens (F, Cl, Br, and I) in binary compounds with metals or hydrogen have an oxidation number of −1. The charges for Zn^{2+} and Cd^{2+} ions (Group IIB) are always +2.

The names, formulas, and ionic charges of some common polyatomic ions are given in Table 8.3. A more comprehensive list of both monatomic and polyatomic ions is given on the inside back cover of this book. Table 8.4 lists the principal oxidation numbers of common elements that have variable oxidation states.

8.3 Formulas of Ionic Compounds

The sum of the oxidation numbers of all the atoms in a compo statement applies to all substances, regardless of whether they covalently bonded. For ionically bonded compounds the sum o all the ions in the compound must also be zero. Hence the fo substances can easily be determined and written by simply combi ions in the simplest proportion that makes the sum of the ionic charges add up to zero.

To illustrate: Sodium chloride consists of Na^+ and Cl^- ions. Since $(+1) + (-1) = 0$, these ions combine in a one-to-one ratio, and the formula is written NaCl. Calcium fluoride is made up of Ca^{2+} and F^- ions; one Ca^{2+} ion $(+2)$ and two F^- ions (-2) are needed to make zero, so the formula is CaF_2. Aluminum oxide is a bit more complicated, because it consists of Al^{3+} and O^{2-} ions. Since 6 is the lowest common multiple of 3 and 2, we have $2(+3) + 3(-2) = 0$; that is, two Al^{3+} ions $(+6)$ and three O^{2-} ions (-6) are needed; therefore, the formula is Al_2O_3.

The foregoing compounds all are made up of monatomic ions. The same procedure is used for polyatomic ions. Consider calcium hydroxide, which is made up of Ca^{2+} and OH^- ions. Since $(+2) + 2(-1) = 0$, one Ca^{2+} and two OH^- ions are needed, so the formula is $Ca(OH)_2$. The parentheses are used to enclose the OH^- so that two hydroxide ions can be shown. It is not correct to write CaO_2H_2 in place of $Ca(OH)_2$ because the identity of the compound would be lost by so doing. Note that the positive ion is written first in formulas. The following table provides examples of formula writing for ionic compounds.

Name of compound	Ions	Lowest common multiple	Sum of charges on ions	Formula
Sodium bromide	Na^+, Br^-	1	$(+1) + (-1) = 0$	NaBr
Potassium sulfide	K^+, S^{2-}	2	$2(+1) + (-2) = 0$	K_2S
Zinc sulfate	Zn^{2+}, SO_4^{2-}	2	$(+2) + (-2) = 0$	$ZnSO_4$
Ammonium phosphate	NH_4^+, PO_4^{3-}	3	$3(+1) + (-3) = 0$	$(NH_4)_3PO_4$
Aluminum chromate	Al^{3+}, CrO_4^{2-}	6	$2(+3) + 3(-2) = 0$	$Al_2(CrO_4)_3$

The sum of the charges on the ions of an ionically bonded compound must equal zero.

PROBLEM 8.1 Write formulas for (a) calcium chloride; (b) iron(III) sulfide; (c) aluminum sulfate. Refer to Tables 8.2, 8.3, and 8.4 as needed.

(a) **Step 1** From the name we know that calcium chloride is composed of calcium and chloride ions. First write down the formulas of these ions.

Ca^{2+} and Cl^-

Step 2 To write the formula of the compound, combine the smallest numbers of Ca^{2+} and Cl^- ions to give a charge sum equal to zero. In this case the lowest common multiple of the charges is 2:

$$(Ca^{2+}) + 2(Cl^-) = 0$$
$$(+2) + 2(-1) = 0$$

Therefore, the formula is $CaCl_2$.

(b) Use the same procedure for iron(III) sulfide.

Step 1 Write down the formulas for the iron(III) and sulfide ions.

Fe^{3+} and S^{2-}

Step 2 Use the smallest numbers of these ions required to give a charge sum equal to zero. The lowest common multiple of the charges is 6:

$$2(Fe^{3+}) + 3(S^{2-}) = 0$$
$$2(+3) + 3(-2) = 0$$

Therefore, the formula is Fe_2S_3.

(c) Use the same procedure for aluminum sulfate.

Step 1 Write down the formulas for the aluminum and sulfate ions.

Al^{3+} and SO_4^{2-}

Step 2 Use the smallest numbers of these ions required to give a charge sum equal to zero. The lowest common multiple of the charges is 6:

$$2(Al^{3+}) + 3(SO_4^{2-}) = 0$$
$$2(+3) + 3(-2) = 0$$

Therefore, the formula is $Al_2(SO_4)_3$. Note the use of parentheses around the SO_4^{2-} ion.

8.4 Determining Oxidation Numbers and Ionic Charges from a Formula

If the formula of a compound is known, the oxidation number of an element or the charge on a polyatomic ion in the formula can often be determined by algebraic difference. To begin you must know the oxidation numbers of a few elements. Excellent ones with which to work are hydrogen, H^+, always $+1$ except in hydrides (a hydride is a compound of hydrogen combined with a metal);

oxygen, O^{2-}, always -2 except in peroxides; and sodium, Na^+, always $+1$. Using the compound sulfuric acid, H_2SO_4, as an example, let us determine the charge of the sulfate ion and the oxidation number of the sulfur atom. The sulfate ion is combined with two hydrogen atoms, each with a $+1$ oxidation number. The sulfate ion must then have a -2 charge in order for the net charge in the compound to be zero.

H^+	$+1$
H^+	$+1$
SO_4^{2-}	-2
	0

To find the oxidation number of sulfur, we proceed as follows:

Step 1 Write the oxidation number of a single atom of hydrogen and a single atom of oxygen below the atoms in the formula.

Step 2 Below this write the sums of the oxidation numbers of all the H and O atoms: $2(+1) = +2$ and $4(-2) = -8$.

Step 3 Then add together the total oxidation numbers of all the atoms, including the sulfur atom, and set them equal to zero: $+2 + S + (-8) = 0$. Solving the equation for S, we determine that the oxidation number of sulfur is $+6$, the value needed to give the sum of zero.

$$H_2\ S\ O_4$$

Step 1	$+1$	-2
Step 2	$2(+1) = +2$	$4(-2) = -8$
Step 3	$+2 + S + (-8) = 0$	
	$S = +6$	

The oxidation number of sulfur in H_2SO_4 is $+6$.

What is the oxidation number of chromium in sodium dichromate, $Na_2Cr_2O_7$? Using the same method as for H_2SO_4, we have

$$Na_2\ Cr_2\ O_7$$

Step 1	$+1$	-2
Step 2	$2(+1) = +2$	$7(-2) = -14$
Step 3	$+2 + 2\,Cr + (-14) = 0$	
	$2\,Cr = +12$	
	$Cr = +6$	

The oxidation number of chromium in $Na_2Cr_2O_7$ is $+6$.

The formula of radium chloride is $RaCl_2$. What is the oxidation number of radium? If you remember that the oxidation number of chloride is -1, then the value for radium is $+2$, because one radium ion is combined with two Cl^- ions. If you do not remember the oxidation number of chloride, then you should try to recall the formula of another chloride. One that might come to mind is sodium chloride, NaCl, in which the chloride is -1 because of its combination with one sodium ion of $+1$. This recollection establishes the oxidation number of chloride, which then enables you to calculate the value for radium.

What is the oxidation number of phosphorus in the phosphate ion, PO_4^{3-}? First of all, note that this ion is polyatomic, with a charge of -3. The sum of the oxidation numbers of phosphorus and oxygen must equal -3 and not zero. Four oxygen atoms, each with a -2, give a total of -8. The oxidation number of the phosphorus atom must then be $+5$:

$$\mathbf{P\,O_4^{3-}}$$

$$\begin{array}{l} \quad\;\; -2 \\ P + 4(-2) = -3 \\ P = +5 \end{array}$$

The sum of the oxidation numbers of the atoms in a polyatomic ion must equal the charge of the polyatomic ion.

8.5 Common, or Trivial, Names of Substances

Chemical nomenclature is the system of names that chemists use to identify compounds. When a new substance is formulated, it must be named in order to distinguish it from all other substances. Before chemistry was systematized, a substance was given a name that generally associated it with one of its outstanding physical or chemical properties. For example, *quicksilver* is a common name for mercury translated from the Latin to mean *liquid silver.* Nitrous oxide, N_2O, used as an anesthetic in dentistry, has been called *laughing gas* because it induces laughter when inhaled. The name *nitrous oxide* is now giving way to the more systematic name *dinitrogen monoxide.* Nonsystematic names are called *common,* or *trivial,* names.

Common names for chemicals are widely used in many industries because the systematic name frequently is too long or too technical for everyday use. For example, CaO is called *lime*, not *calcium oxide*, by plasterers; photographers refer to $Na_2S_2O_3$ as *hypo*, rather than *sodium thiosulfate*; nutritionists refer to $C_{27}H_{44}O$ as *vitamin* D_3, not as 9,10-*secocholesta*-5,7,10(19)-*trien*-3β-*ol.* These examples, particularly that of vitamin D_3, show the practical need for short, common names. Table 8.5 lists the common names, formulas, and chemical names of some familiar substances.

Table 8.5 Common, or trivial, names, formulas, and chemical names of some familiar substances

Common name	Formula	Chemical name
Acetylene	C_2H_2	Ethyne
Lime	CaO	Calcium oxide
Slaked lime	$Ca(OH)_2$	Calcium hydroxide
Water	H_2O	Water
Galena	PbS	Lead(II) sulfide
Alumina	Al_2O_3	Aluminum oxide
Baking soda	$NaHCO_3$	Sodium hydrogen carbonate
Cane or beet sugar	$C_{12}H_{22}O_{11}$	Sucrose
Blue stone, blue vitriol	$CuSO_4 \cdot 5\ H_2O$	Copper(II) sulfate pentahydrate
Borax	$Na_2B_4O_7 \cdot 10\ H_2O$	Sodium tetraborate decahydrate
Brimstone	S	Sulfur
Calcite, marble, limestone	$CaCO_3$	Calcium carbonate
Cream of tartar	$KHC_4H_4O_6$	Potassium hydrogen tartrate
Epsom salts	$MgSO_4 \cdot 7\ H_2O$	Magnesium sulfate heptahydrate
Gypsum	$CaSO_4 \cdot 2\ H_2O$	Calcium sulfate dihydrate
Grain alcohol	C_2H_5OH	Ethanol, ethyl alcohol
Hypo	$Na_2S_2O_3$	Sodium thiosulfate
Laughing gas	N_2O	Dinitrogen monoxide
Litharge	PbO	Lead(II) oxide
Lye, caustic soda	$NaOH$	Sodium hydroxide
Milk of magnesia	$Mg(OH)_2$	Magnesium hydroxide
Muriatic acid	HCl	Hydrochloric acid
Oil of vitriol	H_2SO_4	Sulfuric acid
Plaster of paris	$CaSO_4 \cdot \frac{1}{2}\ H_2O$	Calcium sulfate hemihydrate
Potash	K_2CO_3	Potassium carbonate
Pyrite (fool's gold)	FeS_2	Iron disulfide
Quicksilver	Hg	Mercury
Sal ammoniac	NH_4Cl	Ammonium chloride
Saltpeter (chile)	$NaNO_3$	Sodium nitrate
Table salt	$NaCl$	Sodium chloride
Washing soda	$Na_2CO_3 \cdot 10\ H_2O$	Sodium carbonate decahydrate
Wood alcohol	CH_3OH	Methanol, methyl alcohol

8.6 Systematic Chemical Nomenclature

The trivial name is not entirely satisfactory to the chemist, who requires a name that will identify precisely the composition of each substance. Therefore, as the number of known compounds increased, it became necessary to develop a scientific, systematic method of identifying compounds by name. The systematic method of naming inorganic compounds considers the compound to be composed of two parts, one positive and one negative. The positive part (positive ion or least electronegative element) is named and written first. The negative part, generally nonmetallic, follows. The names of the elements are modified with suffixes and prefixes to identify the different types or classes of compounds. Thus the compound composed of sodium ions and chloride ions is named sodium chloride; the covalent compound composed of carbon atoms and oxygen atoms in a 1:2 ratio is named carbon dioxide; the compound composed of iron(II) ions and chloride ions is named iron(II) chloride (read as "iron-two chloride").

We shall consider the naming of acids, bases, salts, and oxides. Refer to Tables 8.2, 8.3, and 8.4 for the names, formulas, and oxidation numbers of ions. For handy, quick reference, the names and formulas of some common ions are given on the inside back cover of this book.

These two general rules are very helpful in naming the different types of compounds:

Rule 1 The usual oxidation numbers of Groups IA, IIA, and IIIA elements are $+1$, $+2$, and $+3$, respectively. That is, for these elements the oxidation number is the same as the group number.

Rule 2 The oxidation numbers of elements in the other periodic groups are variable. However, in practically all compounds, combined oxygen has an oxidation number of -2; and the halogens (Group VIIA) have an oxidation number of -1 in binary compounds with metals.

8.7 Binary Compounds

Binary compounds contain two different elements, and their names have two parts: the name of the more positive element followed by the name of the more negative element modified to end in *ide*. [Some nonbinary compounds have names ending in *ide*, but they are exceptions to the rule and are discussed in part (d) of this section.]

(a) Binary compounds in which the positive element has a fixed oxidation state Most of these compounds contain a metal and a nonmetal. The chemical name is composed of the name of the metal followed by the name of the nonmetal, which has been modified to an identifying stem plus the suffix *ide*. For example, sodium chloride, NaCl, is composed of one atom each of sodium and chlorine. The name of the metal, sodium, is written first and is not modified. The second part of the name is derived from the nonmetal, chlorine, by using the stem *chlor* and adding the ending *ide*; it is named *chloride*. The compound name is sodium chloride.

NaCl

Elements:	Sodium (metal)
	Chlorine(nonmetal)
	name modified to the stem *chlor* + *ide*
Name of compound:	Sodium chloride

Stems of the more common negative-ion forming elements are shown in the following table. Table 8.6 shows some compounds with names ending in *ide*.

Symbol	Element	Stem	Binary name ending
B	Boron	Bor	Boride
Br	Bromine	Brom	Bromide
Cl	Chlorine	Chlor	Chloride
F	Fluorine	Fluor	Fluoride
H	Hydrogen	Hydr	Hydride
I	Iodine	Iod	Iodide
N	Nitrogen	Nitr	Nitride
O	Oxygen	Ox	Oxide
P	Phosphorus	Phosph	Phosphide
S	Sulfur	Sulf	Sulfide

Compounds may contain more than one atom of the same element, but as long as they contain only two different elements and if only one compound of these two elements exists, the name follows the rule for binary compounds:

Examples: $CaBr_2$ Calcium bromide; Mg_3N_2 Magnesium nitride; Ag_2O Silver oxide

Table 8.6 Examples of compounds with names ending in *ide*

Formula	Name	Formula	Name
$AlCl_3$	Aluminum chloride	PbS	Lead(II) sulfide
Al_2O_3	Aluminum oxide	LiI	Lithium iodide
CaC_2	Calcium carbide	$MgBr_2$	Magnesium bromide
HCl	Hydrogen chloride	NaH	Sodium hydride
HI	Hydrogen iodide	Na_2O	Sodium oxide

(b) Binary compounds containing metals of variable oxidation numbers Two systems are commonly used for compounds in this category. The official system, designated by the International Union of Pure and Applied Chemistry (IUPAC), is known as the *Stock System*. In the Stock System, when a compound contains a metal that can have more than one oxidation number, the oxidation number of the metal is designated by a Roman numeral placed in parentheses immediately following the name of the metal. The negative element is treated in the usual manner for binary compounds.

Oxidation number	+1	+2	+3	+4	+5	+6
Roman numeral	(I)	(II)	(III)	(IV)	(V)	(VI)

Examples:			
	$FeCl_2$	Iron(II) chloride	Fe^{2+}
	$FeCl_3$	Iron(III) chloride	Fe^{3+}
	CuCl	Copper(I) chloride	Cu^{+}
	$CuCl_2$	Copper(II) chloride	Cu^{2+}

The fact that $FeCl_2$ has two chloride ions, each with a −1 charge, establishes that the oxidation number of Fe is +2. To distinguish between the two iron chlorides, $FeCl_2$ is named iron(II) chloride and $FeCl_3$ is named iron(III) chloride.

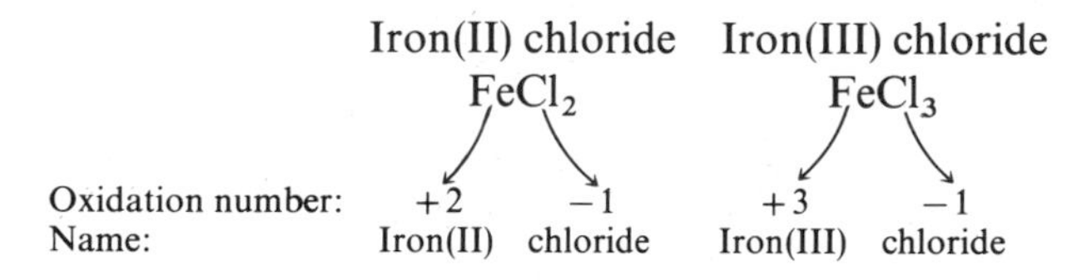

When a metal has only one possible oxidation state, we need not distinguish one oxidation state from another, so Roman numerals are not needed. Thus we do not say calcium(II) chloride for $CaCl_2$, but rather calcium chloride, since the oxidation number of calcium is understood to be +2.

In classical nomenclature, when the metallic ion has only two oxidation numbers, the name of the metal (usually the Latin name) is modified with the suffixes *ous* and *ic* to distinguish between the two. The lower oxidation state is given the *ous* ending, and the higher one, the *ic* ending.

Examples:				
	$FeCl_2$	Ferrous chloride	Fe^{2+}	(lower oxidation state)
	$FeCl_3$	Ferric chloride	Fe^{3+}	(higher oxidation state)
	CuCl	Cuprous chloride	Cu^{+}	(lower oxidation state)
	$CuCl_2$	Cupric chloride	Cu^{2+}	(higher oxidation state)

Table 8.7 lists some common metals that have more than one oxidation number.

Notice that the *ous–ic* naming system does not give the oxidation state of an element but merely indicates that at least two oxidation states exist. The Stock System avoids any possible uncertainty by clearly stating the oxidation number.

Table 8.7 Names and oxidation numbers of some common metal ions that have more than one oxidation number

Formula	Stock System name	Classical name
Cu^{1+}	Copper(I)	Cuprous
Cu^{2+}	Copper(II)	Cupric
Hg^{1+} $(Hg_2)^{2+}$	Mercury(I)	Mercurous
Hg^{2+}	Mercury(II)	Mercuric
Fe^{2+}	Iron(II)	Ferrous
Fe^{3+}	Iron(III)	Ferric
Sn^{2+}	Tin(II)	Stannous
Sn^{4+}	Tin(IV)	Stannic
Pb^{2+}	Lead(II)	Plumbous
Pb^{4+}	Lead(IV)	Plumbic
As^{3+}	Arsenic(III)	Arsenous
As^{5+}	Arsenic(V)	Arsenic
Ti^{3+}	Titanium(III)	Titanous
Ti^{4+}	Titanium(IV)	Titanic

(c) Binary compounds containing two nonmetals The chemical bond that exists between two nonmetals is predominantly covalent. In a covalent compound, positive and negative oxidation numbers are assigned to the elements according to their electronegativities. The least electronegative element is named first. The common nonmetals arranged in order of increasing electronegativity form this series:

Si, B, P, H, C, S, I, Br, N, Cl, O, F

In a compound between two of these elements, the element that occurs first in the series is written and named first. The name of the second element retains the modified binary ending. A Latin or Greek prefix (*mono*, *di*, *tri*, and so on) is attached to the name of each element to indicate the number of atoms of that element in the molecule. The prefix *mono* is generally omitted except when needed to distinguish between two or more compounds, such as carbon monoxide, CO, and carbon dioxide, CO_2. Some common prefixes and their numerical equivalences follow.

Mono = 1	*Hexa* = 6
Di = 2	*Hepta* = 7
Tri = 3	*Octa* = 8
Tetra = 4	*Nona* = 9
Penta = 5	*Deca* = 10

Here are some examples of compounds that illustrate this system:

CO	Carbon monoxide	N_2O	Dinitrogen monoxide
CO_2	Carbon dioxide	N_2O_4	Dinitrogen tetroxide
PCl_3	Phosphorus trichloride	NO	Nitrogen oxide
PCl_5	Phosphorus pentachloride	N_2O_3	Dinitrogen trioxide
P_2O_5	Diphosphorus pentoxide	S_2Cl_2	Disulfur dichloride
CCl_4	Carbon tetrachloride	S_2F_{10}	Disulfur decafluoride

N_2O_3

Dinitrogen — Indicates two nitrogen atoms

Trioxide — Indicates three oxygen atoms

Some special names such as ammonia and NH_3 are exceptions to the system.

(d) Exceptions that use *ide* endings Three notable exceptions that use the *ide* ending are hydroxides (OH^-), cyanides (CN^-), and ammonium (NH_4^+) compounds. These polyatomic ions, when combined with another element, take the ending *ide*, even though more than two elements are present in the compound.

NH_4I	Ammonium iodide
$Ca(OH)_2$	Calcium hydroxide
KCN	Potassium cyanide

(e) Acids derived from binary compounds Certain binary hydrogen compounds, when dissolved in water, form solutions that have *acid* properties. Because of this property, these compounds are given acid names in addition to their regular *ide* names. For example, HCl is a gas and is called *hydrogen chloride*, but its water solution is known as *hydrochloric acid*. Binary acids are composed of hydrogen and one other nonmetallic element. However, not all binary hydrogen compounds are acids. To express the formula of a binary acid it is customary to write the symbol of hydrogen first, followed by the symbol of the second element (for example, HCl, HBr, H_2S). When we see formulas such as CH_4 or NH_3, we understand that these compounds are not normally considered to be acids.

To name a binary acid, place the prefix *hydro* in front of, and the suffix *ic* after, the stem of the nonmetal name. Then add the word *acid*.

	HCl	H_2S
Examples:	*Hydro* chlor/*ic acid*	*Hydro* sulfur/*ic acid*
	(hydrochloric acid)	(hydrosulfuric acid)

Acids are hydrogen-containing substances that liberate hydrogen ions when dissolved in water. The same formula is often used to express binary hydrogen compounds such as HCl, regardless of whether they are dissolved in water. Table 8.8 shows several examples of binary acids.

(f) Salts of binary acids Salts are ionic compounds of cations and anions. When the hydrogen of an acid is replaced by a metal ion or an ammonium ion,

Table 8.8 Names and formulas of selected binary acids

Formula	Acid name
HF	Hydrofluoric acid
HCl	Hydrochloric acid
HBr	Hydrobromic acid
HI	Hydriodic acid
H_2S	Hydrosulfuric acid
H_2Se	Hydroselenic acid

the compound formed is a *salt*. Salts are named by combining the names of the cation and anion, the name of the cation being given first. Thus table salt (NaCl), made of sodium ions (Na^+) and chloride ions (Cl^-), is named sodium chloride. It is the sodium salt of hydrochloric acid.

For salts of binary acids, the name of the negative ion is taken from the name of the compound from which the acid was derived. To illustrate, the name of hydrobromic acid (HBr) is derived from hydrogen *bromide*; therefore, the series of salts obtained from this acid are *bromides*. The names and formulas of the sodium, calcium, and aluminum salts of hydrobromic acid are:

NaBr	$CaBr_2$	$AlBr_3$
Sodium bromide	Calcium bromide	Aluminum bromide

8.8 Ternary Compounds

Ternary compounds contain three elements and generally consist of an electropositive group, either hydrogen or a metal, combined with a polyatomic negative group or ion. Most ternary compounds are either ternary oxy-acids or salts of ternary oxy-acids. In general, when naming ternary compounds, the positive group is given first followed by the name of the negative ion. We shall consider the naming of compounds in which one of the three elements is oxygen.

(a) Ternary oxy-acids Inorganic ternary compounds containing hydrogen, oxygen, and one other element are called *oxy-acids*. The element other than hydrogen or oxygen in these acids is usually a nonmetal, but in some cases it can be a metal. The *ous–ic* system is used in naming ternary acids. The suffixes *ous* and *ic* are used to indicate different oxidation states of the element other than hydrogen and oxygen. The *ous* ending again indicates the lower oxidation state and the *ic* ending, the higher oxidation state.

To name these acids, we place the ending *ic* or *ous* after the stem of the element other than hydrogen and oxygen and add the word *acid*. If an element has only one usual oxidation state, the *ic* ending is used. Hydrogen in a ternary

Table 8.9 Oxy-acids and oxy-anions of chlorine

Acid formula	Acid name	Chlorine oxidation number	Anion formula	Anion name
$HClO$	*Hypo*chlor*ous* acid	+1	ClO^-	*Hypo*chlor*ite*
$HClO_2$	Chlor*ous* acid	+3	ClO_2^-	Chlor*ite*
$HClO_3$	Chlor*ic* acid	+5	ClO_3^-	Chlor*ate*
$HClO_4$	*Per*chlor*ic* acid	+7	ClO_4^-	*Per*chlor*ate*

Table 8.10 Formulas and names of selected ternary oxy-acids

Formula	Acid name	Formula	Acid name
H_2SO_3	Sulfurous acid	$HC_2H_3O_2$	Acetic acid
H_2SO_4	Sulfuric acid	$H_2C_2O_4$	Oxalic acid
HNO_2	Nitrous acid	H_2CO_3	Carbonic acid
HNO_3	Nitric acid	$HBrO_3$	Bromic acid
H_3PO_2	Hypophosphorous acid	HIO_3	Iodic acid
H_3PO_3	Phosphorous acid	H_3BO_3	Boric acid
H_3PO_4	Phosphoric acid		

oxy-acid is not specifically designated in the acid name, but its presence is implied by use of the word *acid*.

Examples: H_2SO_3 Sulfur/*ous acid* (S is +4)
H_2SO_4 Sulfur/*ic acid* (S is +6)

Once again, the acid name is associated with the water solution of the pure compound. In the pure state the usual ternary name may be used. Thus, H_2SO_4 is called both sulfuric acid and hydrogen sulfate.

In cases in which a series has more than two oxy-acids, the *ous–ic* names are further modified with the prefixes *per* and *hypo*. *Per* is placed before the stem of the element other than hydrogen and oxygen when the element has a higher oxidation number than in the *ic* acid. *Hypo* is used as a prefix before the stem when the element has a lower oxidation number than in the *ous* acid. The complete system for naming ternary oxy-acids is shown for the oxy-acids of chlorine in Table 8.9.

The Lewis structures of the oxy-acids of chlorine are

H:Ö:Cl: — Hypochlorous acid

H:Ö:Cl: with :Ö: on Cl — Chlorous acid

H:Ö:Cl:Ö: with :Ö: on Cl — Chloric acid

H:Ö:Cl:Ö: with :Ö: above and below Cl — Perchloric acid

Table 8.11 Comparison of acid, anion, and salt names for selected ternary oxy-compounds

Acid	Anion	Formula and names of representative salts	
H_2SO_4 Sulfuric acid	SO_4^{2-} Sulfate ion	$CaSO_4$ $Fe_2(SO_4)_3$	Calcium sulfate Iron(III) sulfate or ferric sulfate
H_2SO_3 Sulfurous acid	SO_3^{2-} Sulfite ion	Na_2SO_3 Ag_2SO_3	Sodium sulfite Silver sulfite
HNO_3 Nitric acid	NO_3^- Nitrate ion	KNO_3 $Hg(NO_3)_2$	Potassium nitrate Mercury(II) nitrate or mercuric nitrate
HNO_2 Nitrous acid	NO_2^- Nitrite ion	KNO_2 $Co(NO_2)_2$	Potassium nitrite Cobalt(II) nitrite or cobaltous nitrite
H_2CO_3 Carbonic acid	CO_3^{2-} Carbonate ion	Li_2CO_3 $BaCO_3$	Lithium carbonate Barium carbonate
H_3PO_4 Phosphoric acid	PO_4^{3-} Phosphate ion	$AlPO_4$ $Zn_3(PO_4)_2$	Aluminum phosphate Zinc phosphate
H_3PO_3 Phosphorous acid	PO_3^{3-} Phosphite ion	Na_3PO_3 $Zn_3(PO_3)_2$	Sodium phosphite Zinc phosphite
HIO_3 Iodic acid	IO_3^- Iodate ion	$AgIO_3$ $Cu(IO_3)_2$	Silver iodate Copper(II) iodate or cupric iodate
$HC_2H_3O_2$ Acetic acid	$C_2H_3O_2^-$ Acetate ion	$Pb(C_2H_3O_2)_2$ $NH_4C_2H_3O_2$	Lead(II) acetate Ammonium acetate
$H_2C_2O_4$ Oxalic acid	$C_2O_4^{2-}$ Oxalate ion	CaC_2O_4 $(NH_4)_2C_2O_4$	Calcium oxalate Ammonium oxalate

Check the oxidation number of chlorine in each of these oxy-acids using the method for assigning oxidation numbers.

Examples of other ternary oxy-acids and their names are shown in Table 8.10.

(b) Salts of ternary acids As with binary acids a salt is formed when the hydrogen of the acid is replaced by a metal or ammonium ion. These salts are named in much the same way as the salts of binary acids; that is, the name of the cation is given first, followed by the name of the polyatomic anion. The names of the anions are derived from the names of the corresponding ternary acids by changing the *ous* and *ic* endings to *ite* and *ate*, respectively. The stem portion of the acid name is not changed.

Ternary oxy-acid		*Ternary oxy-salt*
ous ending of acid	becomes	*ite* ending in salt
ic ending of acid	becomes	*ate* ending in salt

Thus the sulf*ite* ion (SO_3^{2-}) is derived from sulfur*ous* acid (H_2SO_3) and the sulf*ate* ion (SO_4^{2-}), from sulfur*ic* acid (H_2SO_4). The names and formulas of the sodium, calcium, and aluminum salts of sulfurous and sulfuric acids are:

Na_2SO_3	$CaSO_3$	$Al_2(SO_3)_3$
Sodium sulfite	Calcium sulfite	Aluminum sulfite
Na_2SO_4	$CaSO_4$	$Al_2(SO_4)_3$
Sodium sulfate	Calcium sulfate	Aluminum sulfate

The complete system for naming ternary acids and their anions, using the oxy-acids of chlorine as a model, is shown in Table 8.9. A comparison of the acid, anion, and salt names for a number of ternary oxy-compounds is presented in Table 8.11.

The endings *ous*, *ic*, *ite*, and *ate* are part of classical nomenclature; they are not used in the Stock System to indicate different oxidation states of the elements. These endings are still used, however, in naming many common compounds. The Stock name for H_2SO_4 is tetraoxosulfuric(VI) acid, and that for H_2SO_3 is trioxosulfuric(IV) acid. These Stock names are awkward and are not commonly used.

8.9 Salts with More than One Kind of Positive Ion

Salts can be formed from acids that contain two or more acid hydrogen atoms by replacing only one of the hydrogen atoms with a metal or by replacing both hydrogen atoms with different metals. Each positive group is named first, and then the appropriate salt ending is added.

Acid	Salt	Name of salt
H_2CO_3	$NaHCO_3$	Sodium hydrogen carbonate or sodium bicarbonate
H_2S	$NaHS$	Sodium hydrogen sulfide or sodium bisulfide
H_3PO_4	$MgNH_4PO_4$	Magnesium ammonium phosphate
H_2SO_4	$NaKSO_4$	Sodium potassium sulfate

Table 8.12 Names of selected salts that contain more than one kind of positive ion

Acid	Salt	Name of salt
H_2SO_4	$KHSO_4$	Potassium hydrogen sulfate or potassium bisulfate
H_2SO_3	$Ca(HSO_3)_2$	Calcium hydrogen sulfite or calcium bisulfite
H_2S	NH_4HS	Ammonium hydrogen sulfide or ammonium bisulfide
H_3PO_4	$MgNH_4PO_4$	Magnesium ammonium phosphate
H_3PO_4	NaH_2PO_4	Sodium dihydrogen phosphate
H_3PO_4	Na_2HPO_4	Disodium hydrogen phosphate
$H_2C_2O_4$	KHC_2O_4	Potassium hydrogen oxalate or potassium binoxalate
H_2SO_4	$KAl(SO_4)_2$	Potassium aluminum sulfate
H_2CO_3	$Al(HCO_3)_3$	Aluminum hydrogen carbonate or aluminum bicarbonate

Note the name *sodium bicarbonate* given in the table. The prefix *bi* is commonly used to indicate a compound in which one of two acid hydrogen atoms has been replaced by a metal. The HCO_3^- ion is known as the hydrogen carbonate ion or the bicarbonate ion. Another example is sodium bisulfate, which has the formula $NaHSO_4$. Table 8.12 shows examples of other salts that contain more than one kind of positive ion.

Note that prefixes are also used in chemical nomenclature to give special clarity or emphasis to certain compounds as well as to distinguish between two or more compounds.

Examples:

Na_3PO_4	Trisodium phosphate
Na_2HPO_4	Disodium hydrogen phosphate
NaH_2PO_4	Sodium dihydrogen phosphate

8.10 Bases

Inorganic bases contain the hydroxide ion, OH^-, in chemical combination with a metal ion. These compounds are called *hydroxides*. The OH^- group is named as a single ion and is given the ending *ide*. Several common bases are listed below:

NaOH	Sodium hydroxide
KOH	Potassium hydroxide

NH_4OH Ammonium hydroxide
$Ca(OH)_2$ Calcium hydroxide
$Ba(OH)_2$ Barium hydroxide

We have now looked at ways of naming a variety of inorganic compounds—binary compounds consisting of a metal and a nonmetal and of two nonmetals, and binary and ternary acids, salts, and bases. These compounds are just a small part of the classified chemical compounds. Most of the remaining classes are in the broad field of organic chemistry under such categories as hydrocarbons, alcohols, ethers, aldehydes, ketones, phenols, and carboxylic acids.

PROBLEM 8.2 Name the compound CaS.

Step 1 From the formula it is a two-element compound and follows the rules for binary compounds.
Step 2 The compound is composed of Ca, a Group IIA metal, and S, a nonmetal. Since Group IIA elements have only one oxidation state, we name the positive part of the compound *calcium*.
Step 3 Modify the name of the second element to the identifying stem *sulf* and add the binary ending *ide* to form the name of the negative part, *sulfide*.
Step 4 The name of the compound, therefore, is *calcium sulfide*.

PROBLEM 8.3 Name the compound FeS.

Step 1 This compound follows the rules for a binary compound and, like CaS, must be a sulfide.
Step 2 It is a compound of Fe, a metal, and S, a nonmetal. In the oxidation number tables we see that Fe has two oxidation numbers. In sulfides, the oxidation number of S is -2. Therefore, the oxidation number of Fe must be $+2$, and the name of the positive part of the compound is *iron(II)* or *ferrous*.
Step 3 We have already determined that the name of the negative part of the compound will be *sulfide*.
Step 4 The name of FeS is *iron(II) sulfide*, or *ferrous sulfide*.

PROBLEM 8.4 Name the compound PCl_5.

Step 1 Phosphorus and chlorine are nonmetals, so the rules for naming binary compounds containing two nonmetals apply. Phosphorus is named first because it is less electronegative than chlorine. Therefore the compound is a chloride.
Step 2 No prefix is needed for phosphorus because each molecule has only one atom of phosphorus. The prefix *penta* is used with chloride to indicate the five chlorine atoms. (PCl_3 is also a known compound.)
Step 3 The name for PCl_5 is *phosphorus pentachloride*.

PROBLEM 8.5 (a) Name the salt KNO_3 and (b) name the acid HNO_3 from which this salt can be derived.

(a) **Step 1** The compound contains three elements and follows the rules for ternary compounds.

Step 2 The salt is composed of a K^+ ion and a NO_3^- ion. The name of the positive part of the compound is *potassium*.

Step 3 Since it is a ternary salt, the name will end in *ite* or *ate*. In the oxidation number tables, we see that the name of the NO_3^- ion is *nitrate*.

Step 4 The name of the compound is *potassium nitrate*.

(b) The name of the acid follows the rules for ternary oxy-acids. Because the name of the salt KNO_3 ends in *ate*, the name of the corresponding acid will end in *ic acid*. Change the *ate* ending of nitrate to *ic*. Thus *nitrate* becomes *nitric*, and the name of the acid is *nitric acid*.

QUESTIONS

In naming compounds be careful to use correct spelling.

1. Using the principle employed in Table 8.1, write formulas for:
   (a) The hydrogen compounds of Li, C, I, N, Sr
   (b) The oxygen compounds of Ca, Si, Br, Al, P
   (c) The iodine compounds of Li, Al, C, Sr, S
2. Use the oxidation number tables and determine the formulas for compounds composed of the following ions:
   (a) Sodium and chlorate
   (b) Hydrogen and sulfate
   (c) Tin(II) and acetate
   (d) Copper(I) and oxide
   (e) Zinc and bicarbonate
   (f) Iron(III) and carbonate
3. Determine the oxidation number of the element underlined in each formula:
   (a) $\underline{Mn}CO_3$ (d) $K\underline{Mn}O_4$ (g) $\underline{W}Cl_5$
   (b) $\underline{Sn}F_4$ (e) $Ba\underline{C}O_3$ (h) $K_2\underline{Cr}_2O_7$
   (c) $K\underline{N}O_3$ (f) $\underline{P}Cl_3$
4. Determine the oxidation number of the element underlined in each formula:
   (a) $\underline{In}I_3$ (d) $\underline{C}_2H_5OH$ (g) $Fe_2(\underline{C}O_3)_3$
   (b) $K\underline{Cl}O_3$ (e) $Mg(\underline{N}O_3)_2$ (h) $Na\underline{Cl}O_4$
   (c) $Na_2\underline{S}O_4$ (f) $\underline{Sn}O_2$
5. Write the formula of the compound that would be formed between the given elements:
   (a) Na and I (c) Al and O (e) Cs and Cl
   (b) Ba and F (d) K and S (f) Sr and Br
6. Write the formula of the compound that would be formed between the given elements:
   (a) Ba and O (c) Ga and Cl (e) Li and Si
   (b) H and S (d) Be and Br (f) Mg and P
7. Does the fact that two elements combine in a one-to-one atomic ratio mean that their oxidation numbers are both 1? Explain.
8. What might be the formula of a compound formed between elements X and Z where:
   (a) X has 1 electron and Z has 6 electrons in the outer energy level.
   (b) X has 4 electrons and Z has 7 electrons in the outer energy level.
   (c) X has 1 electron and Z has 7 electrons in the outer energy level.
9. Write formulas for the following cations (do not forget to include the charges): sodium, magnesium, aluminum, copper(II), iron(II), ferric, lead(II), silver, cobalt(II), barium, hydrogen, mercury(II), tin(II), chromium(III), stannic, manganese(II) bismuth(III).
10. Write formulas for the following anions (do not forget to include the charges): chloride, bromide, fluoride, iodide, cyanide, oxide, hydroxide, sulfide, sulfate, bisulfate, bisulfite, chromate, carbonate, bicarbonate, acetate, chlorate, permanganate, oxalate.

11. Complete the table, filling in each box with the proper formula.

Cations \ Anions	Br^-	O^{2-}	NO_3^-	PO_4^{3-}	CO_3^{2-}
K^+	KBr				
Mg^{2+}					
Al^{3+}					
Zn^{2+}				$Zn_3(PO_4)_2$	
H^+					

12. Complete the table, filling in each box with the proper formula.

Cations \ Anions	SO_4^{2-}	Cl^-	AsO_4^{3-}	$C_2H_3O_2^-$	CrO_4^{2-}
NH_4^+			$(NH_4)_3AsO_4$		
Ca^{2+}					
Fe^{3+}	$Fe_2(SO_4)_3$				
Ag^+					
Cu^{2+}					

13. State how each of the following is used in naming inorganic compounds: *ide*, *ous*, *ic*, *hypo*, *per*, *ite*, *ate*, Roman numerals.

14. Write formulas for the following binary compounds, all of which are composed of nonmetals:
(a) Carbon monoxide
(b) Sulfur trioxide
(c) Carbon tetrabromide
(d) Phosphorus trichloride
(e) Nitrogen dioxide
(f) Dinitrogen pentoxide

15. Name the following binary compounds, all of which are composed of nonmetals:
(a) CO_2 (c) PCl_5 (e) SO_2
(b) N_2O (d) CCl_4 (f) N_2O_4
(g) P_2O_5 (i) NF_3
(h) OF_2 (j) CS_2

16. Name the following compounds:
(a) K_2O (d) $BaCO_3$ (g) $Zn(NO_3)_2$
(b) NH_4Br (e) Na_3PO_4 (h) Ag_2SO_4
(c) CaI_2 (f) Al_2O_3

17. Write formulas for the following compounds:
(a) Sodium nitrate
(b) Magnesium fluoride
(c) Barium hydroxide
(d) Ammonium sulfate
(e) Silver carbonate
(f) Calcium phosphate

18. Name each of the following compounds by both the Stock (IUPAC) System, and the *ous–ic* system:

(a) $CuCl_2$ (d) $FeCl_3$ (g) $As(C_2H_3O_2)_3$
(b) $CuBr$ (e) SnF_2 (h) TiI_3
(c) $Fe(NO_3)_2$ (f) $HgCO_3$

19. Write formulas for the following compounds:
(a) Tin(IV) bromide (d) Mercuric nitrite
(b) Copper(I) sulfate (e) Titanic sulfide
(c) Ferric carbonate (f) Iron(II) acetate

20. Write formulas for the following acids:
(a) Hydrochloric acid (d) Carbonic acid
(b) Chloric acid (e) Sulfurous acid
(c) Nitric acid (f) Phosphoric acid

21. Name the following acids:
(a) HNO_2 (d) HBr (g) HF
(b) H_2SO_4 (e) H_3PO_3 (h) $HBrO_3$
(c) $H_2C_2O_4$ (f) $HC_2H_3O_2$

22. Write formulas for the following acids:
(a) Acetic acid (d) Boric acid
(b) Hydrofluoric acid (e) Nitrous acid
(c) Hypochlorous acid (f) Hydrosulfuric acid

23. Name the following acids:
(a) H_3PO_4 (c) HIO_3 (e) $HClO$ (g) HI
(b) H_2CO_3 (d) HCl (f) HNO_3 (h) $HClO_4$

24. Name the following compounds:
(a) $Ba(NO_3)_2$ (d) $MgSO_4$ (g) NiS
(b) $NaC_2H_3O_2$ (e) $CdCrO_4$ (h) $Sn(NO_3)_4$
(c) PbI_2 (f) $BiCl_3$ (i) $Ca(OH)_2$

25. Write formulas for the following compounds:
(a) Silver sulfite
(b) Cobalt(II) bromide
(c) Tin(II) hydroxide
(d) Aluminum sulfate
(e) Manganese(II) fluoride
(f) Ammonium carbonate
(g) Chromium(III) oxide
(h) Cupric chloride
(i) Potassium permanganate
(j) Barium nitrite
(k) Sodium peroxide
(l) Ferrous sulfate
(m) Potassium dichromate
(n) Bismuth(III) chromate

26. Write formulas for the following compounds:
(a) Sodium chromate
(b) Magnesium hydride
(c) Nickel(II) acetate
(d) Calcium chlorate
(e) Lead(II) nitrate
(f) Potassium dihydrogen phosphate
(g) Manganese(II) hydroxide
(h) Cobalt(II) bicarbonate
(i) Sodium hypochlorite
(j) Arsenic(V) carbonate
(k) Chromium(III) sulfite
(l) Antimony(III) sulfate
(m) Sodium oxalate
(n) Potassium thiocyanate

27. Write the name of each salt and the formula and name of the acid from which the salt may be derived. [*Example*: NiC_2O_4, nickel(II) oxalate, $H_2C_2O_4$, oxalic acid.]
(a) $ZnSO_4$ (k) $Ca(HSO_4)_2$
(b) $HgCl_2$ (l) $As_2(SO_3)_3$
(c) $CuCO_3$ (m) $Sn(NO_2)_2$
(d) $Cd(NO_3)_2$ (n) $FeBr_3$
(e) $Al(C_2H_3O_2)_3$ (o) $KHCO_3$
(f) CoF_2 (p) $BiAsO_4$
(g) $Cr(ClO_3)_3$ (q) $Fe(BrO_3)_2$
(h) Ag_3PO_4 (r) $(NH_4)_2HPO_4$
(i) NiS (s) $NaClO$
(j) $BaCrO_4$ (t) $KMnO_4$

28. Write the chemical formula for each of the following substances:
(a) Baking soda (i) Fool's gold
(b) Lime (j) Saltpeter
(c) Epsom salts (k) Limestone
(d) Muriatic acid (l) Cane sugar
(e) Vinegar (m) Milk of magnesia
(f) Potash (n) Washing soda
(g) Lye (o) Grain alcohol
(h) Quicksilver

29. Give the name and formula of three salts with *ide* endings that are not binary compounds. Do not use the exact salts used as examples in the chapter.

30. Which of these statements are correct?
(a) An oxidation number can be positive, negative, or zero.
(b) The oxidation number of Group IA elements is -1.
(c) The sum of the oxidation numbers for all the atoms in a polyatomic ion is zero.
(d) The formula for calcium hydride is CaH_2.
(e) All the following compounds are acids: H_2SO_4, HCl, HNO_3, $NaC_2H_3O_2$.
(f) The ions of all the following metals have an oxidation number of $+2$: Ca, Ba, Sr, Cd, Zn.
(g) The formulas for nitrous and sulfurous acids are HNO_2 and H_2SO_3.

(h) The formula for the compound between Fe^{3+} and O^{2-} is Fe_3O_2.
(i) The oxidation number of Cr in K_2CrO_4 is +6.
(j) The oxidation number of Sn in $SnCl_4$ is +4.
(k) The oxidation number of Co in $CoCl_2$ is +4.
(l) The name for $NaNO_2$ is sodium nitrite.
(m) The name for $Ca(ClO_3)_2$ is calcium chlorate.
(n) The name for CuO is copper(I) oxide.
(o) The name for SO_4^{2-} is sulfate ion.
(p) The name for N_2O_4 is dinitrogen tetroxide.
(q) The name for Na_2O is disodium oxide.
(r) If the name of an anion ends with *ide*, the name of the corresponding acid will start with *hydro*.
(s) If the name of an anion ends with *ite*, the corresponding acid name will end with *ic*.
(t) If the name of an acid ends with *ous*, the corresponding salt name will end with *ate*.
(u) In FeI_2, the iron is iron(II) because it is combined with two I^- ions.
(v) In Cu_2SO_4, the copper is copper(II) because there are two copper ions.
(w) In Ru_2O_3, we can deduce the oxidation state of Ru as +3 because two ions are combined with three oxide ions.
(x) When two nonmetals combine, prefixes of *di*, *tri*, *tetra*, and so on are used to specify how many atoms of each element are in a molecule.
(y) N_2O_3 is called dinitrogen trioxide.
(z) $Sn(CrO_4)_2$ is called tin dichromate.
(aa) In the Stock System of nomenclature, when a compound contains a metal that can have more than one oxidation number, the oxidation number of the metal is designated by a Roman numeral written immediately after the name of the metal.

CHAPTER 9

Quantitative Composition of Compounds

After studying Chapter 9, you should be able to

1 Understand the new terms listed in Question A at the end of the chapter.
2 Explain the meaning of the mole.
3 Discuss the relationship between a mole and Avogadro's number.
4 Convert grams, atoms, molecules, molecular weights, and formula weights to moles, and vice versa.
5 Determine the formula weight or molecular weight of a compound from the formula.
6 Calculate the percentage composition by weight of a compound from its formula.
7 Calculate the percentage composition of a compound from experimental data on combining weights.
8 Explain the relationship between an empirical formula and a molecular formula.
9 Determine the empirical formula of a compound from its percentage composition.
10 Calculate the molecular formula of a compound from its percentage composition and molecular weight.

9.1 The Mole

In the laboratory we normally weigh substances on a balance in units of grams. But, when we run a chemical reaction, the reaction occurs between atoms and molecules. For example, in the reaction between magnesium and sulfur, one atom of sulfur reacts with one atom of magnesium.

$$Mg + S \longrightarrow MgS$$

However, when we measure the masses of these two elements that react, we find that 24.305 g of Mg are required to react with 32.06 g of S. Because magnesium and sulfur react in a 1:1 atom ratio, we can conclude from this experiment that 24.305 g of Mg contains the same number of atoms as 32.06 g of S. How many atoms are in 24.305 g of Mg or 32.06 g of S? These two amounts each contain one mole of atoms.

The mole (abbreviated mol) is one of the seven base units in the International System and is the unit for an amount of substance. The mole is a counting unit as in other things that we count, such as a dozen (12) eggs or a gross (144) of pencils. But a mole is a much larger number of things, namely 6.022×10^{23}. Thus one mole contains 6.022×10^{23} entities of anything. In reference to our reaction between magnesium and sulfur, 1 mol Mg (24.305 g) contains 6.022×10^{23} atoms of magnesium and 1 mol S (32.06 g) contains 6.022×10^{23} atoms of sulfur.

Avogadro's number

The number 6.022×10^{23} is known as **Avogadro's number** in honor of Amedeo Avogadro (1776–1856), an Italian physicist. Avogadro's number is an important constant in chemistry and physics and has been experimentally determined by several independent methods.

Avogadro's number = 6.022×10^{23}

It is difficult to imagine how large Avogadro's number really is, but perhaps the following analogy will help express it: If 10,000 people started to count Avogadro's number and each counted at the rate of 100 numbers per minute each minute of the day, it would take them over 1 trillion (10^{12}) years to count the total number. So, even the minutest amount of matter contains extremely large numbers of atoms. For example, 1 mg (0.001 g) of sulfur contains 2×10^{19} atoms of sulfur.

Avogadro's number is the basis for the amount of substance that is used to express a particular number of chemical species, such as atoms, molecules, formula units, ions, or electrons. This amount of substance is the mole. We define

mole

a **mole** as an amount of a substance containing the same number of formula units as there are atoms in exactly 12 g of carbon-12. (Recall that carbon-12 is the reference isotope for atomic weights.) Other definitions are used, but they all relate to a mole being Avogadro's number of formula units of a substance. A *formula unit* is the atom or molecule indicated by the formula of the substance under consideration—for example, Mg, MgS, H_2O, O_2, $^{75}_{33}As$

From the above definition we can say that the atomic weight in grams of any element contains one mole of atoms. For example,

$$1 \text{ mol H} = 1.0079 \text{ g H} = 6.022 \times 10^{23} \text{ atoms}$$
$$1 \text{ mol Mg} = 24.305 \text{ g Mg} = 6.022 \times 10^{23} \text{ atoms}$$
$$1 \text{ mol S} = 32.06 \text{ g S} = 6.022 \times 10^{23} \text{ atoms}$$
$$1 \text{ mol O} = 15.9994 \text{ g O} = 6.022 \times 10^{23} \text{ atoms}$$
$$1 \text{ mol Na} = 22.9898 \text{ g Na} = 6.022 \times 10^{23} \text{ atoms}$$

1 atomic weight (g) = 1 mole of atoms
= Avogadro's number (6.022×10^{23}) of atoms

The term *mole* is so commonplace in chemical jargon that chemists use it as freely as the words *atom* and *molecule*. The mole is used in conjunction with many different particles, such as atoms, molecules, ions, and electrons, to represent Avogadro's number of these particles. If we can speak of a mole of atoms, we can also speak of a mole of molecules, a mole of electrons, a mole of ions, understanding that in each case we mean 6.022×10^{23} formula units of these particles.

1 mole of atoms = 6.022×10^{23} atoms
1 mole of molecules = 6.022×10^{23} molecules
1 mole of ions = 6.022×10^{23} ions

We frequently encounter problems that require conversions involving quantities of mass, numbers, and moles of atoms of an element. Conversion factors that can be used for this purpose are

(a) Grams to atoms: $\dfrac{6.022 \times 10^{23} \text{ atoms of the element}}{1 \text{ at. wt of the element}}$

(b) Atoms to grams: $\dfrac{1 \text{ at. wt of the element}}{6.022 \times 10^{23} \text{ atoms of the element}}$

(c) Grams to moles: $\dfrac{1 \text{ mole of the element}}{1 \text{ at. wt of the element}}$
(Monatomic elements)

(d) Moles to grams: $\dfrac{1 \text{ at. wt of the element}}{1 \text{ mole of the element}}$
(Monatomic elements)

PROBLEM 9.1 How many moles of iron does 25.0 g of Fe represent?

The problem requires that we change grams of Fe to moles of Fe. We look up the atomic weight of Fe in the atomic weight table and find it to be 55.8. Then we use the proper conversion factor to obtain moles. The conversion factor is (c).

$$\text{grams Fe} \longrightarrow \text{moles Fe} \qquad \text{grams Fe} \times \frac{1 \text{ mole Fe}}{1 \text{ at. wt Fe}}$$

$$25.0 \cancel{\text{g Fe}} \times \frac{1 \text{ mol Fe}}{55.8 \cancel{\text{g Fe}}} = 0.448 \text{ mol Fe} \qquad \text{(Answer)}$$

PROBLEM 9.2 How many magnesium atoms are contained in 5.00 g of Mg?

The problem requires that we change grams of magnesium to atoms of magnesium.

$$\text{grams Mg} \longrightarrow \text{atoms Mg}$$

We find the atomic weight of magnesium to be 24.3 and set up the calculation using conversion factor (a):

$$\text{grams Mg} \times \frac{6.022 \times 10^{23} \text{ atoms Mg}}{1 \text{ at. wt Mg}}$$

$$5.00 \cancel{\text{g Mg}} \times \frac{6.022 \times 10^{23} \text{ atoms Mg}}{24.3 \cancel{\text{g Mg}}} = 1.24 \times 10^{23} \text{ atoms Mg} \qquad \text{(Answer)}$$

An alternative solution is first to convert grams of magnesium to moles of magnesium, which are then changed to atoms of magnesium.

$$\text{grams Mg} \longrightarrow \text{moles Mg} \longrightarrow \text{atoms Mg}$$

Use conversion factor (c) followed by (a). The calculation setup is

$$5.00 \cancel{\text{g Mg}} \times \frac{\cancel{1 \text{ mol Mg}}}{24.3 \cancel{\text{g Mg}}} \times \frac{6.022 \times 10^{23} \text{ atoms Mg}}{\cancel{1 \text{ mol Mg}}} = 1.24 \times 10^{23} \text{ atoms Mg}$$

Thus 1.24×10^{23} atoms of Mg are contained in 5.00 g of Mg.

PROBLEM 9.3 How many grams does one atom of carbon weigh?

From the table of atomic weights we see that the atomic weight of carbon is 12.0 g. The factor needed to convert atoms to grams is conversion factor (b).

$$\text{atoms C} \longrightarrow \text{grams C} \qquad \text{atoms C} \times \frac{1 \text{ at. wt C}}{6.022 \times 10^{23} \text{ atoms C}}$$

$$1 \cancel{\text{atom C}} \times \frac{12.0 \text{ g C}}{6.022 \times 10^{23} \cancel{\text{atoms C}}} = 1.99 \times 10^{-23} \text{ g C} \qquad \text{(Answer)}$$

PROBLEM 9.4

How many grams do 3.01×10^{23} atoms of sodium weigh?

The information needed to solve this problem is the atomic weight of Na (23.0 g) and conversion factor (b).

$$\text{atoms Na} \longrightarrow \text{grams Na} \qquad \text{atoms Na} \times \frac{1 \text{ at. wt Na}}{6.022 \times 10^{23} \text{ atoms Na}}$$

$$3.01 \times 10^{23} \cancel{\text{atoms Na}} \times \frac{23.0 \text{ g Na}}{6.022 \times 10^{23} \cancel{\text{atoms Na}}} = 11.5 \text{ g Na} \qquad \text{(Answer)}$$

PROBLEM 9.5

How many grams does 0.252 mol of Cu weigh?

The information needed to solve this problem is the atomic weight of Cu (63.5 g) and conversion factor (d).

$$\text{moles Cu} \longrightarrow \text{grams Cu} \qquad \text{moles Cu} \times \frac{1 \text{ at. wt Cu}}{1 \text{ mole Cu}}$$

$$0.252 \cancel{\text{mol Cu}} \times \frac{63.5 \text{ g Cu}}{1 \cancel{\text{mol Cu}}} = 16.0 \text{ g Cu} \qquad \text{(Answer)}$$

PROBLEM 9.6

How many oxygen atoms are present in 1.00 mol of oxygen molecules?

Oxygen is a diatomic molecule with the formula O_2. Therefore a molecule of oxygen contains two atoms of oxygen.

$$\frac{2 \text{ atoms O}}{1 \text{ molecule } O_2}$$

The sequence of conversions is

$$\text{moles } O_2 \longrightarrow \text{molecules } O_2 \longrightarrow \text{atoms O}$$

Two conversion factors are needed; they are

$$\frac{6.022 \times 10^{23} \text{ molecules } O_2}{1 \text{ mol } O_2} \quad \text{and} \quad \frac{2 \text{ atoms O}}{1 \text{ molecule } O_2}$$

The calculation is

$$1.00 \cancel{\text{mol } O_2} \times \frac{6.022 \times 10^{23} \cancel{\text{molecules } O_2}}{1 \cancel{\text{mol } O_2}} \times \frac{2 \text{ atoms O}}{1 \cancel{\text{molecule } O_2}} = 1.20 \times 10^{24} \text{ atoms O}$$

(Answer)

9.2 Formula Weight or Molecular Weight

formula weight

molecular weight

Because compounds are composed of atoms, their masses may be represented by the formula weight or the molecular weight. The **formula weight** (form. wt) of a substance is the total mass of all the atoms in the chemical formula of that substance. The **molecular weight** (mol. wt) of a substance is the total mass of all the atoms in a molecule of that substance. Formula weight and molecular weight are often used interchangeably. However, the term *formula weight* is more inclusive, since it includes both molecular and ionic substances.

If the formula of a substance is known, its formula weight or molecular weight may be determined by adding together the atomic weights of all the atoms in the formula. If more than one atom of any element is present, it must be added as many times as it is used in the formula. When atomic weights are in grams, the formula weight and molecular weight will also be in grams. We will continue to use atomic weights in grams in the future.

PROBLEM 9.7

The molecular formula for water is H_2O. What is its molecular weight?

Proceed by looking up the atomic weights of H (1.008) and O (15.999) and adding together the masses of all the atoms in the formula unit. Water contains two atoms of H and one atom of O. Thus,

$$\begin{aligned} 2\,\text{H} &= 2 \times 1.008 = 2.016\ \text{g} \\ 1\,\text{O} &= 1 \times 15.999 = \underline{15.999\ \text{g}} \\ & \qquad 18.015\ \text{g} = \text{molecular or formula weight} \end{aligned}$$

PROBLEM 9.8

Calculate the formula weight of calcium hydroxide, $Ca(OH)_2$.

The formula of this substance contains one atom of Ca and two atoms each of O and H. Proceed as in Problem 9.7. Thus

$$\begin{aligned} 1\,\text{Ca} &= 1 \times 40.08 = 40.08\ \text{g} \\ 2\,\text{O} &= 2 \times 15.999 = 31.998\ \text{g} \\ 2\,\text{H} &= 2 \times 1.008 = \underline{2.016\ \text{g}} \\ & \qquad 74.094\ \text{g} = \text{formula weight} \end{aligned}$$

The atomic weights of elements are often rounded off to one decimal place to simplify calculations. (However, this simplification cannot be made in the more exacting chemical work.) If we calculate the formula weight of $Ca(OH)_2$ on the basis of one decimal place, we find the value to be 74.1 g instead of 74.094 g. The formula weight of $Ca(OH)_2$ would then be calculated as follows:

$$\begin{aligned} 1\,\text{Ca} &= 1 \times 40.1 = 40.1\ \text{g} \\ 2\,\text{O} &= 2 \times 16.0 = 32.0\ \text{g} \\ 2\,\text{H} &= 2 \times 1.0 = \underline{2.0\ \text{g}} \\ & \qquad 74.1\ \text{g} = \text{formula weight} \end{aligned}$$

A formula weight or molecular weight of a compound is one mole of that compound and therefore contains Avogadro's number of formula units or molecules.

As an illustration consider the compound hydrogen chloride, HCl. One atom of H combines with one atom of Cl to form one molecule of HCl. When 1 atomic weight of H (1.0 g of H representing 1 mol or 6.022×10^{23} H atoms) combines with 1 atomic weight of Cl (35.5 g of Cl representing 1 mol or 6.022×10^{23} Cl atoms), 1 molecular weight of HCl (36.5 g of HCl representing 1 mol or 6.022×10^{23} HCl molecules) is produced. Since 36.5 g of HCl contains 6.022×10^{23} molecules, we may refer to this quantity as a molecular weight, as a formula weight, or as a mole of HCl. These relationships are summarized in the following table.

H	Cl	HCl
6.022×10^{23} H *atoms*	6.022×10^{23} Cl *atoms*	6.022×10^{23} HCl *molecules*
1 mol H *atoms*	1 mol Cl *atoms*	1 mol HCl *molecules*
1.0 g H	35.5 g Cl	36.5 g HCl
1 at. wt H	1 at. wt Cl	1 mol. wt HCl or 1 form. wt HCl

In dealing with diatomic elements (H_2, O_2, N_2, F_2, Cl_2, Br_2, and I_2), special care must be taken to distinguish between a mole of atoms (atomic weight) and a mole of molecules (molecular weight). For example consider *one* mole of oxygen molecules, which weighs 32.0 g. This quantity is equal to *two* atomic weights of the element oxygen and thus represents *two* moles of oxygen atoms. The key concept is that one mole represents Avogadro's number of the particular chemical entity—atoms, molecules, formula units, and so forth—that is under consideration.

$$
\begin{aligned}
1 \text{ mol } H_2O &= 18.0 \text{ g } H_2O &&= 6.022 \times 10^{23} \text{ molecules} \\
1 \text{ mol NaCl} &= 58.5 \text{ g NaCl} &&= 6.022 \times 10^{23} \text{ formula units} \\
1 \text{ mol } H_2 &= 2.0 \text{ g } H_2 &&= 6.022 \times 10^{23} \text{ molecules} \\
1 \text{ mol } HNO_3 &= 63.0 \text{ g } HNO_3 &&= 6.022 \times 10^{23} \text{ molecules} \\
1 \text{ mol } K_2SO_4 &= 174.3 \text{ g } K_2SO_4 &&= 6.022 \times 10^{23} \text{ formula units}
\end{aligned}
$$

1 mol = 6.022×10^{23} formula units or molecules
1 mol = 1 formula weight or 1 molecular weight of a compound

We often need to convert moles of a compound to grams, and grams of a compound to moles. The factors for these conversions are

$$\text{grams to moles: } \frac{1 \text{ mol of a substance}}{1 \text{ form. wt of the substance}}$$

$$\text{moles to grams: } \frac{\text{1 form. wt of a substance}}{\text{1 mol of the substance}}$$

PROBLEM 9.9

What is the weight of 1 mol of sulfuric acid, H_2SO_4?

One mole of H_2SO_4 is one formula weight of H_2SO_4. The problem, therefore, is solved in a similar manner to Problems 9.7 and 9.8. Look up the atomic weights of H, S, and O, and solve.

$$\begin{aligned} 2\text{ H} &= 2 \times 1.0 = 2.0\text{ g} \\ 1\text{ S} &= 1 \times 32.1 = 32.1\text{ g} \\ 4\text{ O} &= 4 \times 16.0 = \underline{64.0\text{ g}} \\ & \qquad\qquad\;\; 98.1\text{ g} = \text{weight of 1 mol (1 form. wt) of } H_2SO_4 \end{aligned}$$

PROBLEM 9.10

How many moles of NaOH are there in 1.00 kg of sodium hydroxide?

First we know that

$$1\text{ mol} = 1\text{ form. wt} = (23.0 + 16.0 + 1.0\text{ g}) \text{ or } 40.0\text{ g NaOH}$$

$$1\text{ kg} = 1000\text{ g}$$

To convert grams to moles we use the conversion factor

$$\frac{1\text{ mol}}{1\text{ form. wt}} \quad \text{or} \quad \frac{1\text{ mol NaOH}}{40.0\text{ g NaOH}}$$

Use this conversion sequence:

$$\text{kg NaOH} \longrightarrow \text{g NaOH} \longrightarrow \text{mol NaOH}$$

The calculation is

$$1.00\,\cancel{\text{kg NaOH}} \times \frac{1000\,\cancel{\text{g NaOH}}}{\cancel{\text{kg NaOH}}} \times \frac{1\text{ mol NaOH}}{40.0\,\cancel{\text{g NaOH}}} = 25.0\text{ mol NaOH}$$

$$1.00\text{ kg NaOH} = 25.0\text{ mol NaOH}$$

PROBLEM 9.11

How many grams does 5.00 moles of water weigh?

First we know that

$$1\text{ mol } H_2O = 18.0\text{ g} \qquad \text{(Problem 9.7)}$$

To convert moles to grams use the conversion factor

$$\frac{1\text{ mol. wt } H_2O}{1\text{ mol } H_2O} \quad \text{or} \quad \frac{18.0\text{ g } H_2O}{1\text{ mol } H_2O}$$

The conversion is

$$\text{mol } H_2O \longrightarrow \text{g } H_2O$$

The calculation is

$$5.00 \ \cancel{\text{mol } H_2O} \times \frac{18.0 \text{ g } H_2O}{1 \ \cancel{\text{mol } H_2O}} = 90.0 \text{ g } H_2O \qquad \text{(Answer)}$$

PROBLEM 9.12 How many molecules of HCl are there in 25.0 g of hydrogen chloride?

From the formula we find that the molecular weight of HCl is 36.5 g (1.0 + 35.5 g). The sequence of conversions is

$$\text{g HCl} \longrightarrow \text{mol HCl} \longrightarrow \text{molecules HCl}$$

using the conversion factors

$$\frac{1 \text{ mol HCl}}{36.5 \text{ g HCl}} \quad \text{and} \quad \frac{6.022 \times 10^{23} \text{ molecules HCl}}{1 \text{ mol HCl}}$$

$$25.0 \ \cancel{\text{g HCl}} \times \frac{1 \ \cancel{\text{mol HCl}}}{36.5 \ \cancel{\text{g HCl}}} \times \frac{6.022 \times 10^{23} \text{ molecules HCl}}{1 \ \cancel{\text{mol HCl}}} = 4.12 \times 10^{23} \text{ molecules HCl}$$

(Answer)

9.3 Percentage Composition of Compounds

percentage composition of a compound

Percent means parts per one hundred parts. Just as each piece of pie is a percentage of the whole pie, each element in a compound is a percentage of the whole compound. The **percentage composition of a compound** is the *weight-percent* of each element in the compound. The formula weight represents the total mass, or 100%, of the compound. Thus the percentage composition of water, H_2O, is 11.1% H and 88.9% O by weight. According to the Law of Definite Composition, the percentage composition must be the same no matter what size sample is taken.

The percentage composition of a compound can be determined if its formula is known or if the weights of two or more elements that have combined with each other are known or are experimentally determined.

If the formula is known, it is essentially a two-step process to determine the percentage composition.

Step 1 Calculate the formula weight as was done in Section 9.2.

Step 2 Divide the total weight of each element in the formula weight by the formula weight and multiply by 100%. This gives the percentage composition.

$$\frac{\text{total weight of the element}}{\text{formula weight}} \times 100\% = \text{percentage of the element}$$

It is necessary to multiply by 100% and not just 100; otherwise the answer is unitless.

PROBLEM 9.13 Calculate the percentage composition of sodium chloride, NaCl.

Step 1 Calculate the formula weight of NaCl:

Atomic weights: Na, 23.0; Cl, 35.5

1 Na = 23.0 g
1 Cl = 35.5 g
58.5 g (formula weight)

Step 2 Now calculate the percentage composition. We know there are 23.0 g Na and 35.5 g Cl in 58.5 g NaCl.

$$\text{Na: } \frac{23.0 \not{g} \text{ Na}}{58.5 \not{g} \text{ NaCl}} \times 100\% = 39.3\% \text{ Na}$$

$$\text{Cl: } \frac{35.5 \not{g} \text{ Cl}}{58.5 \not{g} \text{ NaCl}} \times 100\% = \underline{60.7\% \text{ Cl}}$$

$$100.0\% \text{ total}$$

In any two-component system, if one percentage is known, the other is automatically defined by difference; that is, if Na = 39.3%, then Cl = 100% − 39.3% = 60.7%. However, the calculation of the percentage of each component should be carried out, since this provides a check against possible error. The percentage composition data should add up to 100 ± 0.2%.

PROBLEM 9.14 Calculate the percentage composition of potassium chloride, KCl.

Step 1 Calculate the formula weight of KCl:

Atomic weights: K, 39.1; Cl, 35.5

1 K = 39.1 g
1 Cl = 35.5 g
74.6 g (formula weight)

Step 2 Now calculate the percentage composition. We know there are 39.1 g K and 35.5 g Cl in 74.6 g KCl.

$$\text{K: } \frac{39.1 \not{g} \text{ K}}{74.6 \not{g} \text{ KCl}} \times 100\% = 52.4\% \text{ K}$$

$$\text{Cl: } \frac{35.5 \not{g} \text{ Cl}}{74.6 \not{g} \text{ KCl}} \times 100\% = \underline{47.6\% \text{ Cl}}$$

$$100.0\% \text{ total}$$

Comparing the data calculated for NaCl and for KCl, we see that NaCl contains a higher percentage of Cl by weight, although each compound has a one-to-one atom ratio of Cl to Na and K. The reason for this weight percentage difference is that Na and K do not have the same atomic weights.

It is important to realize that, when we compare 1 mole of NaCl with 1 mole of KCl, each quantity contains the same number of Cl atoms—namely, 1 mole of Cl atoms. However, if we compare equal masses of NaCl and KCl, there will be more Cl atoms in the mass of NaCl since NaCl has a higher weight percent of Cl.

1 mole NaCl contains	100 g NaCl contains
1 mol Na 1 mol Cl 60.7% Cl	39.3 g Na 60.7 g Cl

1 mole KCl contains	100 g KCl contains
1 mol K 1 mol Cl 47.6% Cl	52.4 g K 47.6 g Cl

PROBLEM 9.15 Calculate the percentage composition of potassium sulfate, K_2SO_4.

Step 1 Calculate the formula weight of K_2SO_4:

Atomic weights: K, 39.1; S, 32.1; O, 16.0

$$
\begin{aligned}
2\,\text{K} &= 2 \times 39.1 = 78.2\text{ g} \\
1\,\text{S} &= 1 \times 32.1 = 32.1\text{ g} \\
4\,\text{O} &= 4 \times 16.0 = \underline{64.0\text{ g}} \\
& \qquad\qquad\quad 174.3\text{ g} \quad \text{(formula weight)}
\end{aligned}
$$

Step 2 Now calculate the percentage composition. We know there are 78.2 g of K, 32.1 g of S, and 64.0 g of O in 174.3 g of K_2SO_4.

$$\text{K:}\quad \frac{78.2\text{ g K}}{174.3\text{ g }K_2SO_4} \times 100\% = 44.9\%\text{ K}$$

$$\text{S:}\quad \frac{32.1\text{ g S}}{174.3\text{ g }K_2SO_4} \times 100\% = 18.4\%\text{ S}$$

$$\text{O:}\quad \frac{64.0\text{ g O}}{174.3\text{ g }K_2SO_4} \times 100\% = \underline{36.7\%\text{ O}}$$

$$100.0\%\text{ total}$$

The percentage composition can be determined from experimental data without knowing the formula of a compound. This determination is done by calculating the weight of each element in a compound as a percentage of the total weight of the compound formed (see Problem 9.16).

PROBLEM 9.16 When heated in the air, 1.63 g of zinc, Zn, combine with 0.40 g of oxygen, O_2, to form zinc oxide. Calculate the percentage composition of the compound formed.

First, calculate the total weight of the compound formed.

$$\begin{array}{ll} 1.63 \text{ g Zn} & \\ \underline{0.40 \text{ g } O_2} & \\ 2.03 \text{ g} & = \text{total weight of product} \end{array}$$

Then divide the weight of each element by the total weight (2.03 g) and multiply by 100%.

$$\frac{1.63 \text{ g}}{2.03 \text{ g}} \times 100\% = 80.3\% \text{ Zn}$$

$$\frac{0.40 \text{ g}}{2.03 \text{ g}} \times 100\% = \frac{19.7\% \text{ O}}{100.0\% \text{ total}}$$

The compound formed contains 80.3% Zn and 19.7% O.

9.4 Empirical Formula versus Molecular Formula

empirical formula

molecular formula

The **empirical formula**, or *simplest formula*, gives the smallest whole-number ratio of the atoms that are present in a compound. This formula gives the relative number of atoms of each element in the compound.

The **molecular formula** is the true formula, representing the total number of atoms of each element present in one molecule of a compound. It is entirely possible that two or more substances will have the same percentage composition, yet be distinctly different compounds. For example, acetylene, C_2H_2, is a common gas used in welding; benzene, C_6H_6, is an important solvent obtained from coal tar and is used in the synthesis of styrene and nylon. Both acetylene and benzene contain 92.3% C and 7.7% H. The smallest ratio of C and H corresponding to these percentages is CH (1:1). Therefore the empirical formula for both acetylene and benzene is CH, even though it is known that the molecular formulas are C_2H_2 and C_6H_6, respectively. It is not uncommon for the molecular formula to be the same as the empirical formula. If the molecular formula is not the same, it will be an integral (whole number) multiple of the empirical formula.

$$CH = \text{empirical formula}$$

$$(CH)_2 = C_2H_2 = \text{acetylene} \quad \text{(molecular formula)}$$

$$(CH)_6 = C_6H_6 = \text{benzene} \quad \text{(molecular formula)}$$

Table 9.1 summarizes the data concerning these CH formulas. Table 9.2 shows empirical and molecular formula relationships of other compounds.

Table 9.1 Molecular formulas of two compounds having an empirical formula with a 1:1 ratio of carbon and hydrogen atoms.

Formula	Composition % C	% H	Formula weight
CH (empirical)	92.3	7.7	13.0 (empirical)
C_2H_2 (acetylene)	92.3	7.7	26.0 (2 × 13.0)
C_6H_6 (benzene)	92.3	7.7	78.0 (6 × 13.0)

Table 9.2 Some empirical and molecular formulas

Compound	Empirical formula	Molecular formula	Compound	Empirical formula	Molecular formula
Acetylene	CH	C_2H_2	Diborane	BH_3	B_2H_6
Benzene	CH	C_6H_6	Hydrazine	NH_2	N_2H_4
Ethylene	CH_2	C_2H_4	Hydrogen	H	H_2
Formaldehyde	CH_2O	CH_2O	Chlorine	Cl	Cl_2
Acetic acid	CH_2O	$C_2H_4O_2$	Bromine	Br	Br_2
Glucose	CH_2O	$C_6H_{12}O_6$	Oxygen	O	O_2
Hydrogen chloride	HCl	HCl	Nitrogen	N	N_2
Carbon dioxide	CO_2	CO_2			

9.5 Calculation of Empirical Formula

It is possible to establish an empirical formula because (1) the individual atoms in a compound are combined in whole-number ratios and (2) each element has a specific atomic weight.

In order to calculate the empirical formula we need to know (1) the elements that are combined, (2) their atomic weights, and (3) the ratio by weight or percentage in which they are combined. If elements A and B form a compound, we may represent the empirical formula as A_xB_y, where x and y are small whole numbers that represent the number of atoms of A and B. To write the empirical formula we must determine x and y.

The solution to this problem requires three or four steps.

Step 1 Assume a definite starting quantity (usually 100 g) of the compound, if not given, and express the weight of each element in grams.

Step 2 Multiply the weight (in grams) of each element by the factor 1 mol/1 at. wt to convert grams to moles. This conversion gives the number of moles of atoms of each element in the quantity assumed. At this point these numbers will usually not be whole numbers.

Step 3 Divide each of the values obtained in Step 2 by the smallest of these values. If the numbers obtained by this procedure are whole numbers, use them as subscripts in writing the empirical formula. If the numbers obtained are not whole numbers, go on to Step 4.

Step 4 Multiply the values obtained in Step 3 by the smallest number that will convert them to whole numbers. Use these whole numbers as the subscripts in the empirical formula. For example, if the ratio of A to B is 1.0:1.5, multiply both numbers by 2 to obtain a ratio of 2:3. The empirical formula then is A_2B_3.

PROBLEM 9.17 Calculate the empirical formula of a compound containing 11.19% hydrogen, H, and 88.89% oxygen, O.

Step 1 Express each element in grams. If we assume that there are 100 g of material, then the percentage of each element is equal to the grams of each element in 100 g.

$$H = 11.19\% = \frac{11.19 \text{ g}}{100 \text{ g}}$$

$$O = 88.89\% = \frac{88.89 \text{ g}}{100 \text{ g}}$$

Step 2 Multiply the grams of each element by the proper mol/at. wt factor to obtain the relative number of moles of atoms:

$$\text{H:} \quad 11.19 \text{ g H} \times \frac{1 \text{ mol H atoms}}{1.01 \text{ g H}} = 11.1 \text{ mol H atoms}$$

$$\text{O:} \quad 88.89 \text{ g O} \times \frac{1 \text{ mol O atoms}}{16.0 \text{ g O}} = 5.55 \text{ mol O atoms}$$

The formula could be expressed as $H_{11.1}O_{5.55}$. However, it is customary to use the smallest whole-number ratio of atoms. This ratio is calculated in Step 3.

Step 3 Change these numbers to whole numbers by dividing each by the smaller.

$$H = \frac{11.1 \text{ mol}}{5.55 \text{ mol}} = 2 \qquad O = \frac{5.55 \text{ mol}}{5.55 \text{ mol}} = 1$$

In this step the ratio of atoms has not changed, because we divided the number of moles of each element by the same number.

The simplest ratio of H to O is 2:1.

Empirical formula = H_2O

PROBLEM 9.18 The analysis of a salt showed that it contained 56.58% potassium, K, 8.68% carbon, C, and 34.73% oxygen, O. Calculate the empirical formula for this substance.

Steps 1 and 2 After changing the percentage of each element to grams, find the relative number of moles of each element by multiplying by the proper mol/at. wt factor.

$$\text{K:}\quad 56.58\ \cancel{\text{g K}} \times \frac{1\ \text{mol K atoms}}{39.1\ \cancel{\text{g K}}} = 1.45\ \text{mol K atoms}$$

$$\text{C:}\quad 8.68\ \cancel{\text{g C}} \times \frac{1\ \text{mol C atoms}}{12.0\ \cancel{\text{g C}}} = 0.720\ \text{mol C atoms}$$

$$\text{O:}\quad 34.73\ \cancel{\text{g O}} \times \frac{1\ \text{mol O atoms}}{16.0\ \cancel{\text{g O}}} = 2.17\ \text{mol O atoms}$$

Step 3 Divide each number of moles by the smallest value.

$$\text{K} = \frac{1.45\ \text{mol}}{0.720\ \text{mol}} = 2.01$$

$$\text{C} = \frac{0.720\ \text{mol}}{0.720\ \text{mol}} = 1.00$$

$$\text{O} = \frac{2.17\ \text{mol}}{0.720\ \text{mol}} = 3.01$$

The simplest ratio of K:C:O is 2:1:3.

Empirical formula = K_2CO_3

PROBLEM 9.19 A sulfide of iron was formed by combining 2.233 g of iron, Fe, with 1.926 g of sulfur, S. What is the empirical formula of the compound?

Steps 1 and 2 The grams of each element are given, so we use them directly in our calculations. Calculate the relative number of moles of each element by multiplying grams of each element by the proper mol/at. wt factor.

$$\text{Fe:}\quad 2.233\ \cancel{\text{g Fe}} \times \frac{1\ \text{mol Fe atoms}}{55.8\ \cancel{\text{g Fe}}} = 0.0400\ \text{mol Fe atoms}$$

$$\text{S:}\quad 1.926\ \cancel{\text{g S}} \times \frac{1\ \text{mol S atoms}}{32.1\ \cancel{\text{g S}}} = 0.0600\ \text{mol S atoms}$$

Step 3 Divide each number of moles by the smaller of the two numbers.

$$\text{Fe} = \frac{0.0400\ \text{mol}}{0.0400\ \text{mol}} = 1.00$$

$$\text{S} = \frac{0.0600\ \text{mol}}{0.0400\ \text{mol}} = 1.50$$

Step 4 We still have not reached a ratio that will give a formula containing whole numbers of atoms, so we must double each value to obtain a ratio of 2.00 atoms of Fe to 3.00 atoms of S. Doubling both values does not change the ratio of Fe and S atoms.

Fe: $1.00 \times 2 = 2.00$

S: $1.50 \times 2 = 3.00$

Empirical formula = Fe_2S_3

In many of these calculations results may vary somewhat from an exact whole number, which can be due to experimental errors in obtaining the data or from rounding off numbers. Calculations that vary by no more than ± 0.1 from a whole number can usually be rounded off to the nearest whole number. Deviations greater than about 0.1 unit usually mean that the calculated ratios need to be multiplied by a factor to make them all whole numbers. For example, an atom ratio of 1:1.33 should be multiplied by 3 to make the ratio 3:4.

9.6 Calculation of the Molecular Formula from the Empirical Formula

The molecular formula can be calculated from the empirical formula if the molecular weight, in addition to data for calculating the empirical formula, is known. The molecular formula, as stated in Section 9.4, will be equal to or some multiple of the empirical formula. For example, if the empirical formula of a compound of hydrogen and fluorine is HF, the molecular formula can be expressed as $(HF)_n$, where $n = 1, 2, 3, 4, \ldots$ This n means that the molecular formula could be HF, H_2F_2, H_3F_3, H_4F_4, and so on. To determine the molecular formula, we must evaluate n.

$$n = \frac{\text{molecular weight}}{\text{empirical formula weight}} = \text{number of empirical formula units}$$

What we actually calculate is the number of units of the empirical formula that is contained in the molecular formula.

PROBLEM 9.20 A compound of nitrogen and oxygen with a molecular weight of 92.0 g/mol was found to have an empirical formula of NO_2. What is its molecular formula?

Step 1 Let n be the number of (NO_2) units in a molecule; then the molecular formula is $(NO_2)_n$.

Step 2 Each (NO_2) unit weighs $[14 + (2 \times 16)]$ or 46.0 g. The molecular weight of $(NO_2)_n$ is 92.0 g and the number of (46.0) units in 92.0 is 2.

$$n = \frac{92.0 \text{ g}}{46.0 \text{ g}} = 2 \quad \text{(empirical formula units)}$$

Step 3 The molecular formula is $(NO_2)_2$, or N_2O_4.

PROBLEM 9.21 The hydrocarbon propylene has a molecular weight of 42.0 g/mol and contains 14.3% H and 85.7% C. What is its molecular formula?

Step 1 First find the empirical formula:

$$\text{C:}\quad 85.7\ \cancel{\text{g C}} \times \frac{1\ \text{mol C atoms}}{12.0\ \cancel{\text{g C}}} = 7.14\ \text{mol C atoms}$$

$$\text{H:}\quad 14.3\ \cancel{\text{g H}} \times \frac{1\ \text{mol H atoms}}{1.0\ \cancel{\text{g H}}} = 14.3\ \text{mol H atoms}$$

Divide each value by the smaller number of moles.

$$\text{C} = \frac{7.14\ \text{mol}}{7.14\ \text{mol}} = 1.0$$

$$\text{H} = \frac{14.3\ \text{mol}}{7.14\ \text{mol}} = 2.0$$

Empirical formula $= CH_2$

Step 2 Determine the molecular formula from the empirical formula and molecular weight.

Molecular formula $= (CH_2)_n$

Molecular weight $= 42.0$

Each CH_2 unit weighs $(12.0 + 2.0)$ or 14.0. The number of CH_2 units in 42.0 is 3.

$$n = \frac{42.0}{14.0} = 3 \quad \text{(empirical formula units)}$$

The molecular formula is $(CH_2)_3$, or C_3H_6.

QUESTIONS

A. *Review the meanings of the new terms introduced in this chapter.*

1. Avogadro's number
2. Mole
3. Formula weight
4. Molecular weight
5. Percentage composition of a compound
6. Empirical formula
7. Molecular formula

B. *Review questions*

1. What is a mole?
2. Which would weigh more: a mole of potassium atoms, or a mole of gold atoms?
3. Which would contain more atoms: a mole of potassium atoms, or a mole of gold atoms?
4. Which would contain more electrons: a mole of potassium atoms, or a mole of gold atoms?

5. If the atomic weight scale had been defined differently, with an atom of $^{12}_{6}C$ being defined as weighing 50 amu, would this have any effect on the value of Avogadro's number? Explain.
6. What is the numerical value of Avogadro's number?
7. What is the relationship between Avogadro's number and the mole?
8. Complete the following statements, supplying the proper quantity:
   (a) A mole of oxygen atoms (O) contains ________ atoms.
   (b) A mole of oxygen molecules (O_2) contains ________ molecules.
   (c) A mole of oxygen molecules (O_2) contains ________ atoms.
   (d) A mole of oxygen atoms (O) weighs ________ grams.
   (e) A mole of oxygen molecules (O_2) weighs ________ grams.
9. Which of the following statements are correct?
   (a) One atomic weight of any element contains 6.022×10^{23} atoms.
   (b) The mass of one atom of chlorine is $\frac{35.5 \text{ g}}{6.022 \times 10^{23} \text{ atoms}}$.
   (c) A mole of magnesium atoms (24.3 g) contains the same number of atoms as a mole of sodium atoms (23.0 g).
   (d) A mole of bromine atoms contains 6.022×10^{23} atoms of bromine.
   (e) A mole of chlorine molecules (Cl_2) contains 6.022×10^{23} atoms of chlorine.
   (f) A mole of aluminum atoms weighs the same as a mole of tin atoms.
   (g) A mole of H_2O contains 6.022×10^{23} atoms.
   (h) A mole of hydrogen molecules (H_2) contains 1.204×10^{24} electrons.
10. How are formula weight and molecular weight related to each other? In what respects are they different?
11. How many molecules are present in 1 mol. wt of sulfuric acid, H_2SO_4? How many atoms are present?
12. What is the relationship between the following?
    (a) Mole and molecular weight
    (b) Mole and formula weight
13. Why is it correct to refer to the weight of 1 mole of sodium chloride, but incorrect to refer to a molecular weight of sodium chloride?
14. In calculating the empirical formula of a compound from its percentage composition, why do we choose to start with 100 g of the compound?
15. Which of the following statements are correct?
    (a) A mole of sodium and a mole of sodium chloride contain the same number of sodium atoms.
    (b) A compound such as NaCl has a formula weight but no true molecular weight.
    (c) One mole of nitrogen gas (N_2) weighs 14.0 g.
    (d) The percentage of oxygen is higher in K_2CrO_4 than it is in Na_2CrO_4.
    (e) The number of Cr atoms is the same in a mole of K_2CrO_4 as it is in a mole of Na_2CrO_4.
    (f) Both K_2CrO_4 and Na_2CrO_4 contain the same percentage by weight of Cr.
    (g) A molecular weight of sucrose, $C_{12}H_{22}O_{11}$, contains 1 mole of sucrose molecules.
    (h) Two moles of nitric acid, HNO_3, contain 6 moles of oxygen atoms.
    (i) The empirical formula of sucrose, $C_{12}H_{22}O_{11}$, is CH_2O.
    (j) A hydrocarbon that has a molecular weight of 280 and an empirical formula of CH_2 has a molecular formula of $C_{22}H_{44}$.
    (k) The empirical formula is often called the simplest formula.
    (l) The empirical formula of a compound gives the smallest whole-number ratio of the atoms that are present in a compound.
    (m) If the molecular formula and the empirical formula of a compound are not the same, the empirical formula will be an integral multiple of the molecular formula.
    (n) The empirical formula of benzene, C_6H_6, is CH.
    (o) A compound having an empirical formula of CH_2O, and a molecular weight of 60, has a molecular formula of $C_3H_6O_3$.

C. *Problems*

Formula weight, molecular weight

1. Determine the molecular or formula weight of the following compounds:
   (a) KBr
   (b) Na_2SO_4
   (c) $Pb(NO_3)_2$
   (d) C_2H_5OH
   (e) $HC_2H_3O_2$
   (f) Fe_3O_4
   (g) $C_{12}H_{22}O_{11}$
   (h) $Al_2(SO_4)_3$
   (i) $(NH_4)_2HPO_4$

2. Determine the formula weight of the following compounds:
   (a) NaOH (f) C_6H_5COOH
   (b) Ag_2CO_3 (g) $C_6H_{12}O_6$
   (c) Cr_2O_3 (h) $K_4Fe(CN)_6$
   (d) $(NH_4)_2CO_3$ (i) $BaCl_2 \cdot 2\ H_2O$
   (e) $Mg(HCO_3)_2$

Moles and Avogadro's number

3. How many moles of atoms are contained in the following?
   (a) 22.5 g Zn
   (b) 0.688 g Mg
   (c) 4.5×10^{22} atoms Cu
   (d) 382 g Co
   (e) 0.055 g Sn
   (f) 8.5×10^{24} molecules N_2
4. How many moles are contained in the following?
   (a) 25.0 g NaOH (d) 14.8 g CH_3OH
   (b) 44.0 g Br_2 (e) 2.88 g Na_2SO_4
   (c) 0.684 g $MgCl_2$ (f) 4.20 lb ZnI_2
5. Calculate the number of grams in each of the following:
   (a) 0.550 mol Au
   (b) 15.8 mol H_2O
   (c) 12.5 mol Cl_2
   (d) 3.15 mol NH_4NO_3
   (e) 4.25×10^{-4} mol H_2SO_4
   (f) 4.5×10^{22} molecules CCl_4
   (g) 0.00255 mol Ti
   (h) 1.5×10^{16} atoms S
6. How many molecules are contained in each of the following:
   (a) 1.26 mol O_2 (c) 16.0 g CH_4
   (b) 0.56 mol C_6H_6 (d) 1000. g HCl
7. Calculate the weight in grams of each of the following:
   (a) 1 atom Pb
   (b) 1 atom Ag
   (c) 1 molecule H_2O
   (d) 1 molecule $C_3H_5(NO_3)_3$
8. Make the following conversions:
   (a) 8.66 mol Cu to grams Cu
   (b) 125 mol Au to kilograms Au
   (c) 10 atoms C to moles C
   (d) 5000 molecules CO_2 to moles CO_2
   (e) 28.4 g S to moles S
   (f) 2.50 kg NaCl to moles NaCl
   (g) 42.4 g Mg to atoms Mg
   (h) 485 mL Br_2 (d = 3.12 g/mL) to moles Br_2
9. One mole of carbon disulfide (CS_2) contains
   (a) How many carbon disulfide molecules?
   (b) How many carbon atoms?
   (c) How many sulfur atoms?
   (d) How many total atoms of all kinds?
10. White phosphorus is one of several forms of phosphorus and exists as a waxy solid consisting of P_4 molecules. How many atoms are present in 0.350 mol of P_4?
11. How many grams of sodium contain the same number of atoms as 10.0 g of potassium?
12. One atom of an unknown element is found to have a mass of 1.79×10^{-23} g. What is the atomic weight of this element?

*13. If a stack of 500 sheets of paper is 4.60 cm high, what will be the height, in meters, of a stack of Avogadro's number of sheets of paper?

14. There are about 4.2 billion (4.2×10^9) people on earth. If 1 mole of dollars were distributed equally among these people, how many dollars would each person receive?

*15. If 20 drops of water equal 1.0 mL (1.0 cm^3),
   (a) How many drops of water are there in a cubic mile of water?
   (b) What would be the volume in cubic miles of a mole of drops of water?

*16. Silver has a density of 10.5 g/cm^3. If 1.00 mol of silver were shaped into a cube,
   (a) What would be the volume of the cube?
   (b) What would be the length of one side of the cube?

17. How many atoms of oxygen are contained in each of the following?
   (a) 16.0 g O_2
   (b) 0.622 mol MgO
   (c) 6.00×10^{22} molecules $C_6H_{12}O_6$
   (d) 5.0 mol MnO_2
   (e) 250 g $MgCO_3$
   (f) 5.0×10^{18} molecules H_2O
18. Calculate the number of:
   (a) Grams of silver in 25.0 g AgBr
   (b) Grams of chlorine in 5.00 g $PbCl_2$
   (c) Grams of nitrogen in 6.34 mol $(NH_4)_3PO_4$
   (d) Grams of oxygen in 8.45×10^{22} molecules SO_3
   (e) Grams of hydrogen in 45.0 g C_3H_8O

*19. A sulfuric acid solution contains 65.0% H_2SO_4 by weight and has a density of 1.55 g/mL. How many moles of the acid are present in 1.00 L of the solution?

*20. A nitric acid solution containing 72.0% HNO_3 by weight has a density of 1.42 g/mL. How many moles of HNO_3 are present in 100 mL of the solution?

21. Given 1.00 g samples of each of the following compounds: CO_2, O_2, H_2O, and CH_3OH.
(a) Which sample will contain the largest number of molecules?
(b) Which sample will contain the largest number of atoms?
Show proof for your answers.

22. How many grams of Fe_2S_3 will contain a total number of atoms equal to Avogadro's number?

Percentage composition

23. Calculate the percentage composition by weight of the following compounds:
(a) NaBr (c) $FeCl_3$ (e) $Al_2(SO_4)_3$
(b) $KHCO_3$ (d) $SiCl_4$ (f) $AgNO_3$

24. Calculate the percentage composition by weight of the following compounds:
(a) $ZnCl_2$ (d) $(NH_4)_2SO_4$
(b) $NH_4C_2H_3O_2$ (e) $Fe(NO_3)_3$
(c) MgP_2O_7 (f) ICl_3

25. Calculate the percentage of iron, Fe, in the following compounds:
(a) FeO (c) Fe_3O_4
(b) Fe_2O_3 (d) $K_4Fe(CN)_6$

26. Which of the following chlorides has the highest and which has the lowest percentage of chlorine, Cl, by weight, in its formula?
(a) KCl (b) $BaCl_2$ (c) $SiCl_4$ (d) LiCl

27. A 6.20 g sample of phosphorus was reacted with oxygen to form an oxide weighing 14.20 g. Calculate the percentage composition of the compound.

28. A sample of ethylene chloride was analyzed to contain 6.00 g of C, 1.00 g of H, and 17.75 g of Cl. Calculate the percentage composition of ethylene chloride.

29. How many grams of lithium will combine with 20.0 grams of sulfur to form the compound Li_2S?

30. Calculate the percentage of
(a) Mercury in $HgCO_3$
(b) Oxygen in $Ca(ClO_3)_2$
(c) Nitrogen in $C_{10}H_{14}N_2$ (nicotine)
(d) Mg in $C_{55}H_{72}MgN_4O_5$ (chlorophyll)

31. Answer the following by examination of the formulas. Check your answers by calculations if you wish. Which compound has the:
(a) Higher percent by weight of hydrogen, H_2O or H_2O_2?
(b) Lower percent by weight of nitrogen, NO or N_2O_3?
(c) Higher percent by weight of oxygen, NO_2 or N_2O_4?
(d) Lower percent by weight of chlorine, $NaClO_3$ or $KClO_3$?
(e) Higher percent by weight of sulfur, $KHSO_4$ or K_2SO_4?
(f) Lower percent by weight of chromium, Na_2CrO_4 or $Na_2Cr_2O_7$?

Empirical and molecular formulas

32. Calculate the empirical formula of each compound from the percentage compositions given.
(a) 63.6% N, 36.4% O
(b) 46.7% N, 53.3% O
(c) 25.9% N, 74.1% O
(d) 43.4% Na, 11.3% C, 45.3% O
(e) 18.8% Na, 29.0% Cl, 52.3% O
(f) 72.02% Mn, 27.98% O

33. Calculate the empirical formula of each compound from the percentage compositions given.
(a) 64.1% Cu, 35.9% Cl
(b) 47.2% Cu, 52.8% Cl
(c) 51.9% Cr, 48.1% S
(d) 55.3% K, 14.6% P, 30.1% O
(e) 38.9% Ba, 29.4% Cr, 31.7% O
(f) 3.99% P, 82.3% Br, 13.7% Cl

34. A sample of tin (Sn) weighing 3.996 g was oxidized and found to have combined with 1.077 g of oxygen. Calculate the empirical formula of this oxide of tin.

35. A 3.054 g sample of vanadium (V) combined with oxygen to form 5.454 g of product. Calculate the empirical formula for this compound.

*36. Zinc and sulfur react to form zinc sulfide, ZnS. If we mix 19.5 g of zinc and 9.40 g of sulfur, have we added sufficient sulfur to fully react all the zinc? Show evidence for your answer.

37. Hydroquinone is an organic compound commonly used as a photographic developer. It has a molecular weight of 110 g/mol and a composition of 65.45% C, 5.45% H, and 29.09% O. Calculate the molecular formula of hydroquinone.

38. Fructose is a very sweet natural sugar that is present in honey, fruits, and fruit juices. It has a molecular weight of 180 g/mol and a com-

position of 40.0% C, 6.7% H, and 53.3% O. Calculate the molecular formula of fructose.

39. Aspirin is well known as a pain reliever (analgesic) and as a fever reducer (anti-pyretic). It has a molecular weight of 180 and a composition of 60.0% C, 4.48% H, and 35.5% O. Calculate the molecular formula of aspirin.

40. How many grams of oxygen are contained in 8.50 g $Al_2(SO_4)_3$?

41. Gallium arsenide is one of the newer materials used to make semiconductor chips for use in supercomputers. Its composition is 48.2% Ga and 51.8% As. What is the empirical formula?

42. Listed below are the compositions of four different compounds of carbon and chlorine. Determine both the empirical formula and the molecular formula for each.

	Percentage C	Percentage Cl	Molecular weight
(a)	7.79	92.21	154
(b)	10.13	89.87	237
(c)	25.26	74.74	285
(d)	11.25	88.75	320

CHAPTER

10 Chemical Equations

After studying Chapter 10, you should be able to

1. Understand the terms listed in Question A at the end of the chapter.
2. Know the format used in setting up chemical equations.
3. Recognize the various symbols commonly used in writing chemical equations.
4. Balance simple chemical equations.
5. Interpret a balanced equation in terms of the relative numbers or amounts of molecules, atoms, grams, or moles of each substance represented.
6. Classify equations as representing combination, decomposition, single-displacement, or double-displacement reactions.
7. Use the activity series to predict whether a single-displacement reaction will occur.
8. Complete and balance equations for simple combination, decomposition, single-displacement, and double-displacement reactions when given the reactants.
9. Distinguish between exothermic and endothermic reactions, and relate the quantity of heat to the amounts of substances involved in the reaction.
10. Identify the major sources of chemical energy and their uses.

10.1 The Chemical Equation

Chemical reactions and equations were introduced in Chapter 4. We saw that in a chemical reaction the substances entering the reaction are called reactants and the substances formed are called the products. In a chemical reaction atoms, molecules, or ions interact and rearrange themselves to form the products. During this process chemical bonds are broken and new bonds are formed. The reactants and products may be in the solid, liquid, or gaseous state, or in solution.

chemical equation

word equation

A **chemical equation** is a shorthand expression for a chemical change or reaction. It shows, among other things, the rearrangement of the atoms that are involved in the reaction. A **word equation** states in words, in equation form, the substances involved in a chemical reaction. For example, when mercury(II) oxide is heated, it decomposes to form mercury and oxygen. The word equation for this decomposition is

$$\text{mercury(II) oxide} + \text{heat} \longrightarrow \text{mercury} + \text{oxygen}$$

From the chemist's point of view this method of describing a chemical reaction is inadequate. It is bulky and cumbersome to use and does not give quantitative information. The chemical equation, using symbols and formulas, is a far better way to describe the decomposition of mercury(II) oxide:

$$2\ HgO \xrightarrow{\Delta} 2\ Hg + O_2\uparrow$$

This equation gives all the information from the word equation plus formulas, composition, reactive amounts of all the substances involved in the reaction, and much additional information (see Section 10.4). Even though a chemical equation provides much quantitative information, it is still not a complete description; it does not tell us how much energy is needed to cause decomposition, what we observe during the reaction, or anything about the rate of reaction. This information must be obtained from other sources or from experimentation.

10.2 Format for Writing Chemical Equations

A chemical equation uses the chemical symbols and formulas of the reactants and products and other symbolic terms to represent a chemical reaction. Equations are written according to this general format:

1 The reactants are separated from the products by an arrow (→) that indicates the direction of the reaction. A double arrow (⇄) indicates that the reaction goes in both directions and establishes an equilibrium between the reactants and the products.
2 The reactants are placed to the left and the products to the right of the arrow. A plus sign (+) is placed between reactants and between products when needed.

3 Conditions required to carry out the reaction may, if desired, be placed above or below the arrow or equality sign. For example, a delta sign placed over the arrow ($\xrightarrow{\Delta}$) indicates that heat is supplied to the reaction.
4 Coefficients (small integral numbers) are placed in front of substances (for example, 2 H_2O) to balance the equation and to indicate the number of formula units (atoms, molecules, moles, ions) of each substance reacting or being produced. When no number is shown, it is understood that one formula unit of the substance is indicated.
5 The physical state of a substance is indicated by the following symbols: (*s*) for solid state; (*l*) for liquid state; (*g*) for gaseous state; and (*aq*) for substances in aqueous solution.

Symbols commonly used in equations are given in Table 10.1.

Table 10.1 Symbols commonly used in chemical equations

Symbol	Meaning
→	Yields; produces (points to products)
⇄	Reversible reaction; equilibrium between reactants and products
↑	Gas evolved (written after a substance)
↓	Solid or precipitate formed (written after a substance)
(*s*)	Solid state (written after a substance)
(*l*)	Liquid state (written after a substance)
(*g*)	Gaseous state (written after a substance)
(*aq*)	Aqueous solution (substance dissolved in water)
Δ	Heat
+	Plus or added to (placed between substances)

10.3 Writing and Balancing Equations

balanced equation

To represent the quantitative relationships of a reaction, the chemical equation must be balanced. A **balanced equation** contains the same number of each kind of atom on each side of the equation. The balanced equation, therefore, obeys the Law of Conservation of Mass.

The ability to balance equations must be acquired by every chemistry student. Simple equations are easy to balance, but some care and attention to detail are required. The way to balance an equation is to adjust the number of atoms of each element so that it is the same on each side of the equation. But we must not change a correct formula in order to achieve a balanced equation. Each equation must be treated on its own merits; we have no simple "plug in" formula for balancing equations. The following outline gives a general procedure for balancing equations. Study this outline and refer to it as needed when working examples. There is no substitute for practice in learning to write and balance chemical equations.

1 Identify the reaction for which the equation is to be written. Formulate a description or word equation for the reaction if needed (for example, mercury(II) oxide decomposes yielding mercury and oxygen).
2 Write the unbalanced, or skeleton, equation. Make sure that the formula for each substance is correct and that the reactants are written to the left and the products to the right of the arrow (for example, $HgO \rightarrow Hg + O_2$). The correct formulas must be known or ascertained from the periodic table, oxidation numbers, lists of ions, or experimental data.
3 Balance the equation. Use the following steps as necessary:
   (a) Count and compare the number of atoms of each element on each side of the equation and determine those that must be balanced.
   (b) Balance each element, one at a time, by placing small whole numbers (coefficients) in front of the formulas containing the unbalanced element. It is usually best to balance metals first, then nonmetals, then hydrogen and oxygen. Select the smallest coefficients that will give the same number of atoms of the element on each side. A coefficient placed before a formula multiplies every atom in the formula by that number (for example, 2 H_2SO_4 means two molecules of sulfuric acid and also means four H atoms, two S atoms, and eight O atoms.)
   (c) Check all other elements after each individual element is balanced to see whether, in balancing one, other elements have become unbalanced. Make adjustments as needed.
   (d) Balance polyatomic ions such as SO_4^{2-}, which remain unchanged from one side of the equation to the other, in the same way as individual atoms.
   (e) Do a final check, making sure that each element and/or polyatomic ion is balanced and that the smallest possible set of whole-number coefficients has been used.

$$4\ HgO \longrightarrow 4\ Hg + 2\ O_2 \quad \text{(incorrect form)}$$
$$2\ HgO \longrightarrow 2\ Hg + O_2 \quad \text{(correct form)}$$

Not all chemical equations can be balanced by the simple method of inspection just described. The balancing of more complex equations is described in Chapter 17.

The following examples show stepwise sequences leading to balanced equations. Study each example carefully.

EXAMPLE 10.1 Write the balanced equation for the reaction that takes place when magnesium metal is burned in air to produce magnesium oxide.

1 *Word equation*

magnesium + oxygen $\longrightarrow$ magnesium oxide

2 *Skeleton equation*

$Mg + O_2 \longrightarrow MgO$ (unbalanced)

3 *Balance*

(a) Oxygen is not balanced. Two O atoms appear on the left side and one on the right side.

(b) Place the coefficient 2 before MgO.

$$Mg + O_2 \longrightarrow 2\ MgO \quad \text{(unbalanced)}$$

(c) Now Mg is not balanced. One Mg atom appears on the left side and two on the right side. Place a 2 before Mg.

$$2\ Mg + O_2 \longrightarrow 2\ MgO \quad \text{(balanced)}$$

(d) *Check:* Each side has two Mg and two O atoms.

EXAMPLE 10.2 When methane, CH_4, undergoes complete combustion, it reacts with oxygen to produce carbon dioxide and water. Write the balanced equation for this reaction.

1 *Word equation*

$$\text{methane} + \text{oxygen} \longrightarrow \text{carbon dioxide} + \text{water}$$

2 *Skeleton equation*

$$CH_4 + O_2 \longrightarrow CO_2 + H_2O \quad \text{(unbalanced)}$$

3 *Balance*

(a) Carbon is balanced. Hydrogen and oxygen are not balanced.

(b) Balance H atoms by placing a 2 before H_2O.

$$CH_4 + O_2 \longrightarrow CO_2 + 2\ H_2O \quad \text{(unbalanced)}$$

Each side of the equation has four H atoms; oxygen is still not balanced. Place a 2 before O_2 to balance the oxygen atoms.

$$CH_4 + 2\ O_2 \longrightarrow CO_2 + 2\ H_2O \quad \text{(balanced)}$$

(c) *Check:* The equation is correctly balanced; it has one C, four O, and four H atoms on each side.

EXAMPLE 10.3 Oxygen and potassium chloride are formed by heating potassium chlorate. Write a balanced equation for this reaction.

1 *Word equation*

$$\text{potassium chlorate} \xrightarrow{\Delta} \text{potassium chloride} + \text{oxygen}$$

2 *Skeleton equation*

$$KClO_3 \xrightarrow{\Delta} KCl + O_2 \quad \text{(unbalanced)}$$

3 *Balance*

(a) Oxygen is unbalanced (three O atoms on the left and two on the right side).

(b) How many oxygen atoms are needed? The subscripts of oxygen (3 and 2) in $KClO_3$ and O_2 have a common denominator of 6. Therefore coefficients for $KClO_3$ and O_2 are needed to give six oxygen atoms on each side. Place a 2 before $KClO_3$ and a 3 before O_2 to give six O atoms on each side.

$$2\ KClO_3 \xrightarrow{\Delta} KCl + 3\ O_2 \quad \text{(unbalanced)}$$

Now K and Cl are not balanced. Place a 2 before KCl, which balances both K and Cl at the same time.

$$2\ KClO_3 \xrightarrow{\Delta} 2\ KCl + 3\ O_2 \quad \text{(balanced)}$$

(c) *Check*: Each side contains two K, two Cl, and six O atoms.

EXAMPLE 10.4 Balance by starting with the word equation given.

1 *Word equation*

silver nitrate + hydrogen sulfide ⟶ silver sulfide + nitric acid

2 *Skeleton equation*

$$AgNO_3 + H_2S \longrightarrow Ag_2S + HNO_3 \quad \text{(unbalanced)}$$

3 *Balance*

(a) Ag and H are unbalanced.

(b) Place a 2 in front of $AgNO_3$ to balance Ag.

$$2\ AgNO_3 + H_2S \longrightarrow Ag_2S + HNO_3 \quad \text{(unbalanced)}$$

(c) H and NO_3^- are still unbalanced. Balance by placing a 2 in front of HNO_3.

$$2\ AgNO_3 + H_2S \longrightarrow Ag_2S + 2\ HNO_3 \quad \text{(balanced)}$$

(d) In this example N and O atoms are balanced by balancing the NO_3^- ion as a unit.

(e) *Check*: Each side has two Ag, two H, and one S atom. Also, each side has two NO_3^- ions.

EXAMPLE 10.5 Balance by starting with the word equation given.

1 *Word equation*

aluminum hydroxide + sulfuric acid ⟶ aluminum sulfate + water

2 *Skeleton equation*

$$Al(OH)_3 + H_2SO_4 \longrightarrow Al_2(SO_4)_3 + H_2O \quad \text{(unbalanced)}$$

3 *Balance*

(a) All elements are unbalanced.

(b) Balance Al by placing a 2 in front of $Al(OH)_3$. Treat the unbalanced SO_4^{2-} ion as a unit and balance by placing a 3 before H_2SO_4. Note that Step 3(d) may sometimes be combined with Step 3(b).

$$2\ Al(OH)_3 + 3\ H_2SO_4 \longrightarrow Al_2(SO_4)_3 + H_2O \quad \text{(unbalanced)}$$

Balance the unbalanced H and O by placing a 6 in front of H_2O.

$$2\ Al(OH)_3 + 3\ H_2SO_4 \longrightarrow Al_2(SO_4)_3 + 6\ H_2O \quad \text{(balanced)}$$

(c) *Check*: Each side has two Al, twelve H, three S, and eighteen O atoms.

EXAMPLE 10.6 When the fuel in a butane gas stove undergoes complete combustion, it reacts with oxygen to form carbon dioxide and water. Write the balanced equation for this reaction.

1 *Word equation*

butane + oxygen ⟶ carbon dioxide + water

2 *Skeleton equation*

$$C_4H_{10} + O_2 \longrightarrow CO_2 + H_2O \quad \text{(unbalanced)}$$

3 *Balance*

(a) All elements are unbalanced.

(b) Balance C by placing a 4 in front of CO_2.

$$C_4H_{10} + O_2 \longrightarrow 4\ CO_2 + H_2O \quad \text{(unbalanced)}$$

Balance H by placing a 5 in front of H_2O.

$$C_4H_{10} + O_2 \longrightarrow 4\ CO_2 + 5\ H_2O \quad \text{(unbalanced)}$$

Oxygen remains unbalanced. The oxygen atoms on the right side are fixed, because 4 CO_2 and 5 H_2O are derived from the single C_4H_{10} molecule on the left. When we try to balance the O atoms, we find that there is no integer (whole number) that can be placed in front of O_2 to bring about a balance. The equation can be balanced if we use $6\frac{1}{2}\ O_2$ and then double the coefficients of each substance, including the $6\frac{1}{2}\ O_2$, to obtain the balanced equation.

$$C_4H_{10} + 6\tfrac{1}{2}\ O_2 \longrightarrow 4\ CO_2 + 5\ H_2O \quad \text{(balanced—incorrect form)}$$

$$2\ C_4H_{10} + 13\ O_2 \longrightarrow 8\ CO_2 + 10\ H_2O \quad \text{(balanced)}$$

(c) *Check*: Each side now has eight C, twenty H, and twenty-six O atoms.

10.4 What Information Does an Equation Tell Us?

Depending on the particular context in which it is used, a formula can have different meanings. The meanings refer either to an individual chemical entity (atom, ion, molecule, or formula unit) or to a mole of that chemical entity. For example, the formula H_2O can be used to indicate either

$$\text{(a)}\ H_2O = \begin{cases} \text{1 molecule of water} \\ \text{2 hydrogen atoms + 1 oxygen atom} \\ \text{18 amu of water} \end{cases}$$

$$\text{(b)}\ H_2O = \begin{cases} \text{1 mole of water} \\ 6.022 \times 10^{23} \text{ molecules of water} \\ \text{18.0 grams of water} \end{cases}$$

Formulas used in equations can be expressed in units of individual chemical entities or as moles, the latter being more commonly used. For example, in the reaction of hydrogen and oxygen to form water,

$$2\,H_2 + O_2 \longrightarrow 2\,H_2O$$

the 2 H_2 can represent 2 molecules or 2 moles of hydrogen; the O_2, 1 molecule or 1 mole of oxygen; and the 2 H_2O, 2 molecules or 2 moles of water. In terms of moles, this equation is stated: 2 moles of H_2 react with 1 mole of O_2 to give 2 moles of H_2O.

As indicated earlier, a chemical equation is a shorthand description of a chemical reaction. Interpretation of the balanced equation gives us the following information:

1 What the reactants are and what the products are
2 The formulas of the reactants and products
3 The number of molecules or formula units of reactants and products in the reaction
4 The number of atoms of each element involved in the reaction
5 The number of molecular weights or formula weights of each substance used or produced
6 The number of moles of each substance
7 The number of grams of each substance used or produced

Consider the equation

$$H_2(g) + Cl_2(g) \longrightarrow 2\,HCl(g)$$

This equation states that hydrogen gas reacts with chlorine gas to produce hydrogen chloride, also a gas. Let us summarize all the information relating to the equation. The information that can be stated about the relative amount of each substance, with respect to all other substances in the balanced equation, is written below its formula in the following equation:

$$H_2(g) \quad + \quad Cl_2(g) \quad \longrightarrow \quad 2\ HCl(g)$$

Hydrogen	Chlorine	Hydrogen chloride
1 molecule	1 molecule	2 molecules
2 atoms	2 atoms	2 atoms H + 2 atoms Cl
1 mol. wt	1 mol. wt	2 mol. wt
1 mole	1 mole	2 moles
2.0 g	71.0 g	2 × 36.5 g or 73.0 g

These data are very useful in calculating quantitative relationships that exist among substances in a chemical reaction. For example, if we react 2 moles of hydrogen (twice as much as is indicated by the equation) with 2 moles of chlorine, we can expect to obtain 4 moles, or 146 g, of hydrogen chloride as a product. We will study this phase of using equations in more detail in the next chapter.

Let us try another equation. When propane gas is burned in air, the products are carbon dioxide, CO_2, and water, H_2O. The equation and its interpretation are as follows:

$$C_3H_8(g) \quad + \quad 5\ O_2(g) \quad \longrightarrow \quad 3\ CO_2(g) \quad + \quad 4\ H_2O(g)$$

Propane	Oxygen	Carbon dioxide	Water
1 molecule	5 molecules	3 molecules	4 molecules
3 atoms C	10 atoms O	3 atoms C	8 atoms H
8 atoms H		6 atoms O	4 atoms O
1 mol. wt	5 mol. wt	3 mol. wt	4 mol. wt
1 mole	5 moles	3 moles	4 moles
44.0 g	5 × 32.0 g (160.0 g)	3 × 44.0 g (132.0 g)	4 × 18.0 g (72.0 g)

10.5 Types of Chemical Equations

Chemical equations represent chemical changes or reactions. To be of any significance an equation must represent an actual or possible reaction. Part of the problem of writing equations is determining the products formed. We have no sure method of predicting products, nor do we have time to carry out experimentally all the reactions we may wish to consider. Therefore we must use data reported in the writings of other workers, certain rules to aid in our predictions, and the atomic structure and combining capacities of the elements to help us predict the formulas of the products of a chemical reaction. The final proof of the existence of any reaction, of course, is in the actual observation of the reaction in the laboratory (or elsewhere).

Reactions are classified into types to assist in writing equations and to aid in predicting other reactions. Many chemical reactions fit one or another of the four principal reaction types that are discussed in the following paragraphs. Reactions are also classified as oxidation–reduction. Special methods are used to balance complex oxidation–reduction equations (see Chapter 17).

combination or synthesis reaction

1. Combination or synthesis reaction In a **combination reaction**, two reactants combine to give one product. The general form of the equation is

$$A + B \longrightarrow AB$$

in which A and B are either elements or compounds and AB is a compound. The formula of the compound in many cases can be determined from a knowledge of the oxidation numbers of the reactants in their combined states. Some reactions that fall into this category are the following:

(a) metal + oxygen $\longrightarrow$ metal oxide

$$2\ Mg + O_2 \xrightarrow{\Delta} 2\ MgO$$

$$4\ Al + 3\ O_2 \xrightarrow{\Delta} 2\ Al_2O_3$$

(b) nonmetal + oxygen $\longrightarrow$ nonmetal oxide

$$S(s) + O_2(g) \xrightarrow{\Delta} SO_2(g)$$

$$N_2(g) + O_2(g) \xrightarrow{\Delta} 2\ NO(g)$$

(c) metal + nonmetal $\longrightarrow$ salt

$$2\ Na(s) + Cl_2(g) \longrightarrow 2\ NaCl(s)$$

$$2\ Al(s) + 3\ Br_2(l) \longrightarrow 2\ AlBr_3(s)$$

(d) metal oxide + water $\longrightarrow$ base (metal hydroxide)

$$Na_2O(s) + H_2O(l) \longrightarrow 2\ NaOH(aq)$$

$$CaO(s) + H_2O(l) \longrightarrow Ca(OH)_2(aq)$$

(e) nonmetal oxide + water $\longrightarrow$ oxy-acid

$$SO_3(g) + H_2O(l) \longrightarrow H_2SO_4(aq)$$

$$N_2O_5(s) + H_2O(l) \longrightarrow 2\ HNO_3(aq)$$

decomposition reaction

2. Decomposition reaction In a **decomposition reaction** a single substance is decomposed or broken down to give two or more different substances. The reaction may be considered the reverse of combination. The starting material must be a compound, and the products may be elements or compounds. The general form of the equation is

$$AB \longrightarrow A + B$$

Predicting the products of a decomposition reaction can be difficult and requires an understanding of each individual reaction. Heating oxygen-containing compounds often results in decomposition. Some reactions that fall into this category are the following:

(a) Metal oxides. Some metal oxides decompose to yield the free metal plus oxygen, others give a lower oxide, and some are very stable, resisting decomposition by heating.

$$2\ HgO(s) \xrightarrow{\Delta} 2\ Hg(l) + O_2(g)$$

$$2\ PbO_2(s) \xrightarrow{\Delta} 2\ PbO(s) + O_2(g)$$

(b) Carbonates and bicarbonates decompose to yield CO_2 when heated.

$$CaCO_3(s) \xrightarrow{\Delta} CaO(s) + CO_2(g)$$

$$2\ NaHCO_3(s) \xrightarrow{\Delta} Na_2CO_3(s) + H_2O(g) + CO_2(g)$$

(c) Miscellaneous

$$2\ KClO_3(s) \xrightarrow{\Delta} 2\ KCl(s) + 3\ O_2(g)$$

$$2\ NaNO_3(s) \xrightarrow{\Delta} 2\ NaNO_2(s) + O_2(g)$$

$$2\ H_2O_2(l) \xrightarrow{\Delta} 2\ H_2O(l) + O_2(g)$$

single-displacement reaction

3. Single-displacement reaction In a **single-displacement reaction** one element reacts with a compound to take the place of one of the elements of that compound. A different element and a different compound are formed. The general form of the equation is

$$\underset{\text{metal}}{A} + BC \longrightarrow B + AC \qquad \text{or} \qquad \underset{\text{halogen}}{A} + BC \longrightarrow C + BA$$

If A is a metal, A will replace B to form AC, providing A is a more reactive metal than B. If A is a halogen, it will replace C to form BA, providing A is a more reactive halogen than C.

Brief Activity Series of selected metals (and hydrogen) and halogens follow. The series are listed in descending order of chemical reactivity, with the most active metals and halogens at the left. From such series it is possible to predict many chemical reactions. Any metal on the list will replace the ions of those metals that appear anywhere to the right of it on the list. For example, zinc metal will replace hydrogen from a hydrochloric acid solution. But copper metal, which is to the right of hydrogen on the list and thus less reactive than hydrogen, will not replace hydrogen from a hydrochloric acid solution.

Activity Series

metals: K, Ca, Na, Mg, Al, Zn, Fe, Ni, Sn, Pb, H, Cu, Ag, Hg, Au
halogens: F_2, Cl_2, Br_2, I_2

Some reactions that fall into this category follow.

(a) metal + acid $\longrightarrow$ hydrogen + salt

$$Zn(s) + 2\ HCl(aq) \longrightarrow H_2(g) + ZnCl_2(aq)$$
$$2\ Al(s) + 3\ H_2SO_4(aq) \longrightarrow 3\ H_2(g) + Al_2(SO_4)_3(aq)$$

(b) metal + water $\longrightarrow$ hydrogen + metal hydroxide or metal oxide

$$2\ Na(s) + 2\ H_2O \longrightarrow H_2(g) + 2\ NaOH(aq)$$
$$Ca(s) + 2\ H_2O \longrightarrow H_2(g) + Ca(OH)_2(aq)$$
$$3\ Fe(s) + \underset{\text{Steam}}{4\ H_2O(g)} \longrightarrow 4\ H_2(g) + Fe_3O_4(s)$$

(c) metal + salt $\longrightarrow$ metal + salt

$$Fe(s) + CuSO_4(aq) \longrightarrow Cu(s) + FeSO_4(aq)$$
$$Cu(s) + 2\ AgNO_3(aq) \qquad 2\ Ag(s) + Cu(NO_3)_2(aq)$$

(d) halogen + halogen salt $\longrightarrow$ halogen + halogen salt

$$Cl_2(g) + 2\ NaBr(aq) \longrightarrow Br_2(l) + 2\ NaCl(aq)$$
$$Cl_2(g) + 2\ KI(aq) \longrightarrow I_2(s) + 2\ KCl(aq)$$

A common chemical reaction is the displacement of hydrogen from water or acids. This reaction illustrates well the relative reactivity of the metals and the use of the activity series.

K, Ca, and Na displace hydrogen from cold water, steam, and acids.
Mg, Al, Zn, and Fe displace hydrogen from steam and acids.
Ni, Sn, and Pb displace hydrogen only from acids.
Cu, Ag, Hg, and Au do not displace hydrogen.

EXAMPLE 10.7 Will a reaction occur between (a) nickel metal and hydrochloric acid and (b) tin metal and a solution of aluminum chloride. Write balanced equations for the reactions.

(a) Nickel is more reactive than hydrogen so it will displace hydrogen from hydrochloric acid. The products are hydrogen gas and a salt of Ni^{2+} and Cl^- ions.

$$Ni(s) + HCl(aq) \longrightarrow H_2(g) + NiCl_2(aq)$$

(b) According to the activity series, tin is less reactive than aluminum so no reaction will occur.

$$Sn(s) + AlCl_3(aq) \longrightarrow \text{no reaction}$$

---

double-displacement or metathesis reaction

4. Double-displacement or metathesis reaction In a **double-displacement reaction**, two compounds exchange partners with each other to produce two different compounds. The general form of the equation is

$$AB + CD \longrightarrow AD + CB$$

This reaction may be thought of as an exchange of positive and negative groups, in which A combines with D and C combines with B. In writing the formulas of the products, we must take into account the oxidation numbers or charges of the combining groups. Some reactions that fall into this category follow.

(a) Neutralization of an acid and a base

$$\text{acid} + \text{base} \longrightarrow \text{salt} + \text{water}$$
$$HCl(aq) + NaOH(aq) \longrightarrow NaCl(aq) + H_2O$$
$$H_2SO_4(aq) + Ba(OH)_2(aq) \longrightarrow BaSO_4\downarrow + 2\ H_2O$$

(b) Formation of an insoluble precipitate

$$BaCl_2(aq) + 2\ AgNO_3(aq) \longrightarrow 2\ AgCl\downarrow + Ba(NO_3)_2(aq)$$
$$FeCl_3(aq) + 3\ NH_4OH(aq) \longrightarrow Fe(OH)_3\downarrow + 3\ NH_4Cl(aq)$$

(c) metal oxide + acid $\longrightarrow$ salt + water

$$CuO(s) + 2\ HNO_3(aq) \longrightarrow Cu(NO_3)_2(aq) + H_2O$$
$$CaO(s) + 2\ HCl(aq) \longrightarrow CaCl_2(aq) + H_2O$$

(d) Formation of a gas

$$H_2SO_4(l) + NaCl(s) \longrightarrow NaHSO_4(s) + HCl\uparrow$$
$$2\ HCl(aq) + ZnS(s) \longrightarrow ZnCl_2(aq) + H_2S\uparrow$$
$$2\ HCl(aq) + Na_2CO_3(s) \longrightarrow 2\ NaCl(aq) + H_2O(l) + CO_2\uparrow$$

Some reactions we attempt may fail because the substances are not reactive, or the proper conditions for reaction may not be present. For example, mercury(II) oxide does not decompose until it is heated; magnesium does not burn in air or oxygen until the temperature is raised to the point at which it begins to react. When silver is placed in a solution of copper(II) sulfate, no reaction takes place; however, when a strip of copper is placed in a solution of silver nitrate, the single-displacement reaction as shown in 3(c) takes place because copper is a more reactive metal than silver. The successful prediction of the products of a reaction is not always easy. The ability to predict products correctly comes with knowledge and experience. Although you may not be able to predict many reactions at this point, as you continue you will find that reactions can be categorized, and that prediction of the products thereby becomes easier, if not always certain.

EXAMPLE 10.8 Write the equation for the reaction between aqueous solutions of hydrobromic acid and potassium hydroxide.

First write the formulas for the reactants. They are HBr and KOH. Then classify the type of reaction that would occur between them. Because the reactants are compounds and one is an acid and the other is a base, the reaction will be of the neutralization type:

$$\text{acid} + \text{base} \longrightarrow \text{salt} + \text{water}$$

Now rewrite the equation by putting down the formulas for the known substances:

$$HBr(aq) + KOH(aq) \longrightarrow \text{salt} + H_2O$$

In this reaction, which is a double-displacement type, the H^+ from the acid combines with the OH^- from the base to form water. The salt must be composed of the other two ions, K^+ and Br^-. We determine the formula of the salt to be KBr from the fact that K is a +1 cation and Br is a −1 anion. The final balanced equation is

$$HBr(aq) + KOH(aq) \longrightarrow KBr(aq) + H_2O(l)$$

EXAMPLE 10.9 Complete and balance the equation for the reaction between aqueous solutions of barium chloride and sodium sulfate.

First determine the formulas for the reactants. They are $BaCl_2$ and Na_2SO_4. Then classify these substances as acids, bases, or salts. Both substances are salts. Since both substances are compounds, the reaction looks as though it will be of the double-displacement type. Start writing the equation with the reactants:

$$BaCl_2(aq) + Na_2SO_4(aq) \longrightarrow$$

If the reaction is double-displacement, Ba^{2+} will be written combined with SO_4^{2-} and Na^+ with Cl^- as the products. The balanced equation is

$$BaCl_2(aq) + Na_2SO_4(aq) \longrightarrow BaSO_4 + 2\ NaCl$$

The final step is to determine the nature of the products, which controls whether or not the reaction will take place. If both products are soluble, we may merely have a mixture of all the possible products in solution. But if an insoluble precipitate is formed, the reaction will definitely occur. We know from experience that NaCl is fairly soluble in water, but what about $BaSO_4$? The Solubility Table in Appendix IV can give us this information. From this table we see that $BaSO_4$ is insoluble in water, so it will be a precipitate in the reaction. Thus the reaction will occur, a white precipitate will form, and the equation is

$$BaCl_2(aq) + Na_2SO_4(aq) \longrightarrow BaSO_4\downarrow + 2\ NaCl(aq)$$

We have a great deal yet to learn about which substances react with each other, how they react, and what conditions are necessary to bring about their reaction. It is possible to make accurate predictions concerning the occurrence of proposed reactions. Such predictions require, in addition to appropriate data, a good knowledge of thermodynamics, a subject usually reserved for advanced courses in chemistry and physics. But even without the formal use of thermodynamics your knowledge of such generalities as the four reaction types just cited, the periodic table, atomic structure, oxidation numbers, and so on, can be put to good use in predicting reactions and in writing equations. Indeed, such applications serve to make chemistry an interesting and fascinating study.

10.6 Heat in Chemical Reactions

Energy changes always accompany chemical reactions. One reason why reactions occur is that the products attain a lower, more stable energy state than the reactants. For the products to attain this more stable state, energy must be liberated and given off to the surroundings as heat (or as heat and work). When a solution of a base is neutralized by the addition of an acid, the liberation of heat energy is signaled by an immediate rise in the temperature of the solution. When an automobile engine burns gasoline, heat is certainly liberated; at the same time, part of the liberated energy does the work of moving the automobile.

exothermic reaction

endothermic reaction

Reactions are either exothermic or endothermic. **Exothermic reactions** liberate heat; **endothermic reactions** absorb heat. In an exothermic reaction heat is a product and may be written on the right side of the equation for the reaction. In an endothermic reaction heat can be regarded as a reactant and is written on the left side of the equation. Examples indicating heat in an exothermic and an endothermic reaction follow.

$$H_2(g) + Cl_2(g) \longrightarrow 2\ HCl(g) + 185\ \text{kJ}\ (44.2\ \text{kcal}) \quad \text{(exothermic)}$$
$$N_2(g) + O_2(g) + 181\ \text{kJ}\ (43.2\ \text{kcal}) \longrightarrow 2\ NO(g) \quad \text{(endothermic)}$$

heat of reaction

The quantity of heat produced by a reaction is known as the **heat of reaction.** The units used can be kilojoules or kilocalories. Consider the reaction represented by this equation

$$C(s) + O_2(g) \longrightarrow CO_2(g) + 393\ \text{kJ}\ (94.0\ \text{kcal})$$

When the heat liberated is expressed as part of the equation, the substances are expressed in units of moles. Thus, when 1 mole (12.0 g) of C combines with 1 mole (32.0 g) of O_2, 1 mole (44.0 g) of CO_2 is formed and 393 kJ (94.0 kcal) of heat are liberated. In this reaction, as in many others, the heat energy is more useful than the chemical products.

Aside from relatively small amounts of energy from nuclear processes, the sun is the major provider of energy for life on earth. The sun maintains the temperature necessary for life and also supplies light energy for the endothermic photosynthetic reactions carried on by green plants. In photosynthesis carbon dioxide and water are converted to free oxygen and glucose.

$$6\ CO_2 + 6\ H_2O + 2519\ \text{kJ}\ (673\ \text{kcal}) \longrightarrow \underset{\text{Glucose}}{C_6H_{12}O_6} + 6\ O_2$$

Nearly all of the chemical energy used by living organisms is obtained from glucose or compounds derived from glucose. Modern technology depends for energy on fossil fuels— coal, petroleum, and natural gas. The energy is obtained from the combustion (burning) of these fuels, which are converted to carbon dioxide and water. **Combustion** is the term for a chemical reaction in which heat and light are given off.

combustion

Fossil fuels constitute a huge energy reservoir. Some coal is about 90% carbon. Since 393 kJ are obtained from the combustion of 1 mol (12.0 g) of carbon, the combustion of a single ton of coal yields about 2.68×10^{10} J (6.40×10^9 cal) of energy. At 4.184 J per gram per degree, this energy is enough to heat about 21,000 gallons of water from room temperature to the boiling point (20° to 100°C).

Natural gas is primarily methane, CH_4. Petroleum is a mixture of hydrocarbons (compounds of carbon and hydrogen). Liquefied petroleum gas (LPG) is a mixture of propane (C_3H_8) and butane (C_4H_{10}). The combustion of these fuels is a major source of energy.

$$CH_4(g) + 2\ O_2(g) \longrightarrow CO_2(g) + 2\ H_2O(g) + 890\ \text{kJ}\ (213\ \text{kcal})$$

$$C_3H_8(g) + 5\ O_2(g) \longrightarrow 3\ CO_2(g) + 4\ H_2O(g) + 2200\ \text{kJ}\ (526\ \text{kcal})$$

$$\underset{\text{octane in gasoline}}{2\ C_8H_{18}(l)} + 25\ O_2(g) \longrightarrow 16\ CO_2(g) + 18\ H_2O(g) + 10{,}900\ \text{kJ}\ (2606\ \text{kcal})$$

Be careful not to confuse an exothermic reaction that merely requires heat (activation energy) to get it started with a truly endothermic process. The combustion of magnesium is highly exothermic, yet magnesium must be heated to a fairly high temperature in air before combustion begins. Once started, however, the combustion reaction goes very vigorously until either the magnesium or the available supply of oxygen is exhausted. The electrolytic decomposition of water to hydrogen and oxygen is highly endothermic. If the electric current is shut off when this process is going on, the reaction stops instantly. The relative energy levels of reactants and products in exothermic and in endothermic processes are presented graphically in Figure 10.1.

In reaction (a) of Figure 10.1, the products are at a lower energy level than the reactants. Energy (heat) is given off, and the reaction is exothermic. In reaction (b) the products are at a higher energy level than the reactants. Energy has therefore been absorbed, and the reaction is endothermic.

Examples of endothermic and exothermic processes that can be demonstrated easily in the laboratory are shown in Figure 10.2. When dissolving ammonium chloride, NH_4Cl, in water, we observe an endothermic process. The temperature changes from 24.5°C to 18.1°C when 10 g of NH_4Cl are added to

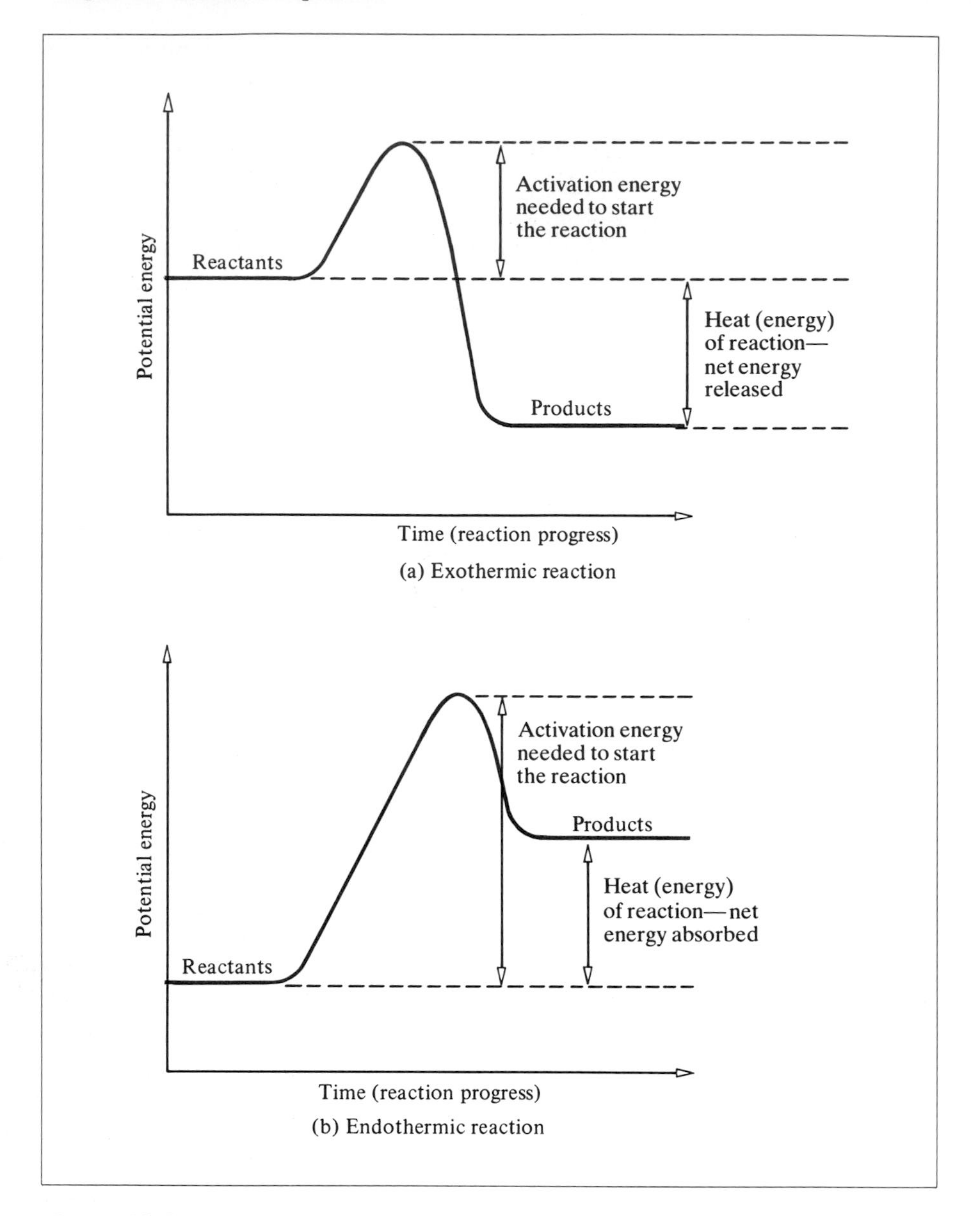

Figure 10.1

Energy levels in exothermic and endothermic reactions

100 mL of water. This energy, in the form of heat, is taken from the immediate surroundings, the water, causing the salt solution to become colder. In the second example, we observe a temperature change from 24.8°C to 47.0°C when 10 mL of concentrated sulfuric acid, H_2SO_4, are dissolved in 100 mL of water. This reaction is an exothermic process. In both examples the temperature changes are large enough to be detected by touching the containers.

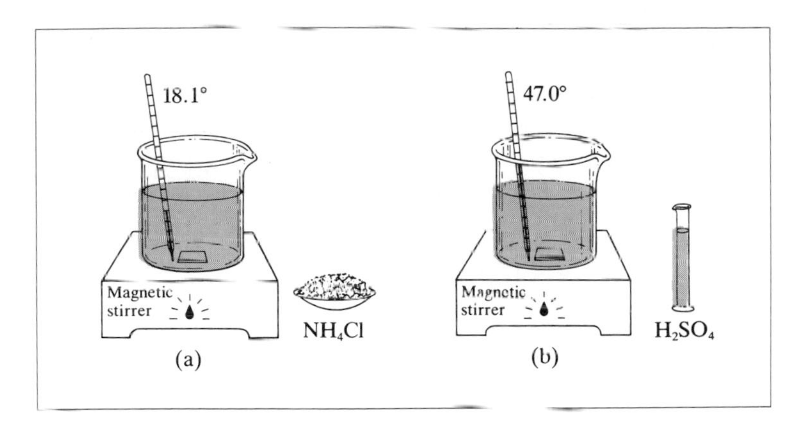

Figure 10.2
In the endothermic process (a) dissolving 10 g of NH_4Cl in 100 mL of H_2O caused the temperature of the solution to decrease from 24.5°C to 18.1°C. In the exothermic process (b) dissolving 10 mL of concentrated H_2SO_4 in 100 mL of H_2O caused the temperature of the solution to increase from 24.8°C to 47.0°C.

QUESTIONS

A. *Review the meanings of the new terms introduced in this chapter.*

1. Chemical equation
2. Word equation
3. Balanced equation
4. Combination or synthesis reaction
5. Decomposition reaction
6. Single-displacement reaction
7. Double-displacement or metathesis reaction
8. Exothermic reaction
9. Endothermic reaction
10. Heat of reaction
11. Combustion

B. *Study Table 10.1 so that you will be familiar with the most common symbols used in equations.*

C. *Review questions*

1. Balance the following equations:
   (a) $H_2 + O_2 \longrightarrow H_2O$
   (b) $H_2 + Br_2 \longrightarrow HBr$
   (c) $C + Fe_2O_3 \longrightarrow Fe + CO$
   (d) $H_2O_2 \longrightarrow H_2O + O_2$
   (e) $Ba(ClO_3)_2 \xrightarrow{\Delta} BaCl_2 + O_2$
   (f) $H_2SO_4 + NaOH \longrightarrow H_2O + Na_2SO_4$
   (g) $NH_4I + Cl_2 \longrightarrow NH_4Cl + I_2$
   (h) $CrCl_3 + AgNO_3 \longrightarrow Cr(NO_3)_3 + AgCl$
   (i) $Al_2(CO_3)_3 \xrightarrow{\Delta} Al_2O_3 + CO_2$
   (j) $Al + C \xrightarrow{\Delta} Al_4C_3$
2. Classify the reactions in Question C.1. as combination, decomposition, single displacement, or double displacement.
3. Balance the following equations:
   (a) $SO_2 + O_2 \longrightarrow SO_3$
   (b) $Al + MnO_2 \xrightarrow{\Delta} Mn + Al_2O_3$
   (c) $Na + H_2O \longrightarrow NaOH + H_2$
   (d) $AgNO_3 + Ni \longrightarrow Ni(NO_3)_2 + Ag$
   (e) $Bi_2S_3 + HCl \longrightarrow BiCl_3 + H_2S$
   (f) $PbO_2 \xrightarrow{\Delta} PbO + O_2$
   (g) $LiAlH_4 \xrightarrow{\Delta} LiH + Al + H_2$
   (h) $KI + Br_2 \longrightarrow KBr + I_2$
   (i) $K_3PO_4 + BaCl_2 \longrightarrow KCl + Ba_3(PO_4)_2$

4. Balance the following equations.
   (a) $MnO_2 + CO \longrightarrow Mn_2O_3 + CO_2$
   (b) $Mg_3N_2 + H_2O \longrightarrow Mg(OH)_2 + NH_3$
   (c) $C_3H_5(NO_3)_3 \longrightarrow CO_2 + H_2O + N_2 + O_2$
   (d) $FeS_2 + O_2 \longrightarrow Fe_2O_3 + SO_2$
   (e) $Cu(NO_3)_2 \longrightarrow CuO + NO_2 + O_2$
   (f) $NO_2 + H_2O \longrightarrow HNO_3 + NO$
   (g) $Al + H_2SO_4 \longrightarrow Al_2(SO_4)_3 + H_2$
   (h) $HCN + O_2 \longrightarrow N_2 + CO_2 + H_2O$
   (i) $B_5H_9 + O_2 \longrightarrow B_2O_3 + H_2O$
   (j) $NH_3 + O_2 \longrightarrow NO + H_2O$
5. Change the following word equations into formula equations and balance them:
   (a) Copper + Sulfur $\xrightarrow{\Delta}$ Copper(I) sulfide
   (b) Phosphoric acid + Calcium hydroxide $\longrightarrow$ Calcium phosphate + Water
   (c) Silver oxide $\xrightarrow{\Delta}$ Silver + Oxygen
   (d) Iron(III) chloride + Sodium hydroxide $\longrightarrow$ Iron(III) hydroxide + Sodium chloride
   (e) Nickel(II) phosphate + Sulfuric acid $\longrightarrow$ Nickel(II) sulfate + Phosphoric acid
   (f) Zinc carbonate + Hydrochloric acid $\longrightarrow$ Zinc chloride + Water + Carbon dioxide
   (g) Silver nitrate + Aluminum chloride $\longrightarrow$ Silver chloride + Aluminum nitrate
6. Change the following word equations into formula equations and balance them:
   (a) Water $\longrightarrow$ Hydrogen + Oxygen
   (b) Acetic acid + Potassium hydroxide $\longrightarrow$ Potassium acetate + Water
   (c) Phosphorus + Iodine $\longrightarrow$ Phosphorus triiodide
   (d) Aluminum + Copper(II) sulfate $\longrightarrow$ Copper + Aluminum sulfate
   (e) Ammonium sulfate + Barium chloride $\longrightarrow$ Ammonium chloride + Barium sulfate
   (f) Sulfur tetrafluoride + Water $\longrightarrow$ Sulfur dioxide + Hydrogen fluoride
   (g) Chromium(III) carbonate $\xrightarrow{\Delta}$ Chromium(III) oxide + Carbon dioxide
7. Complete and balance the equations for these combination reactions:
   (a) $K + O_2 \longrightarrow$ (c) $CO_2 + H_2O \longrightarrow$
   (b) $Al + Cl_2 \longrightarrow$ (d) $CaO + H_2O \longrightarrow$
8. Complete and balance the equations for these decomposition reactions:
   (a) $HgO \xrightarrow{\Delta}$ (c) $MgCO_3 \xrightarrow{\Delta}$
   (b) $NaClO_3 \xrightarrow{\Delta}$ (d) $PbO_2 \xrightarrow{\Delta} PbO +$
9. Complete and balance the equations for these single-displacement reactions:
   (a) $Zn + H_2SO_4 \longrightarrow$
   (b) $AlI_3 + Cl_2 \longrightarrow$
   (c) $Mg + AgNO_3 \longrightarrow$
   (d) $Al + CoSO_4 \longrightarrow$
10. Use the activity series to predict which of the following reactions will occur. Complete and balance the equations. Where no reaction will occur, write "no reaction" as the product.
    (a) $Ag + H_2SO_4(aq) \longrightarrow$
    (b) $Cl_2 + NaBr(aq) \longrightarrow$
    (c) $Mg + ZnCl_2(aq) \longrightarrow$
    (d) $Pb + AgNO_3(aq) \longrightarrow$
    (e) $Cu + FeCl_3(aq) \longrightarrow$
    (f) $H_2 + Al_2O_3(s) \xrightarrow{\Delta}$
    (g) $Al + HBr(aq) \longrightarrow$
    (h) $I_2 + HCl(aq) \longrightarrow$
11. Complete and balance the equations for these double-displacement reactions:
    (a) $ZnCl_2 + KOH \longrightarrow$
    (b) $CuSO_4 + H_2S \longrightarrow$
    (c) $Ca(OH)_2 + H_3PO_4 \longrightarrow$
    (d) $(NH_4)_3PO_4 + Ni(NO_3)_2 \longrightarrow$
    (e) $Ba(OH)_2 + HNO_3 \longrightarrow$
    (f) $(NH_4)_2S + HCl \longrightarrow$
12. Complete and balance the equations for these reactions:
    (a) $H_2 + I_2 \longrightarrow$
    (b) $CaCO_3 \xrightarrow{\Delta}$
    (c) $Mg + H_2SO_4 \longrightarrow$
    (d) $FeCl_2 + NaOH \longrightarrow$
    (e) $SO_2 + H_2O \longrightarrow$
    (f) $SO_3 + H_2O \longrightarrow$
    (g) $Ca + H_2O \longrightarrow$
    (h) $Bi(NO_3)_3 + H_2S \longrightarrow$
13. Complete and balance the equations for the following reactions:
    (a) $Ba + O_2 \longrightarrow$
    (b) $NaHCO_3 \xrightarrow{\Delta} Na_2CO_3 +$
    (c) $Ni + CuSO_4 \longrightarrow$
    (d) $MgO + HCl \longrightarrow$
    (e) $H_3PO_4 + KOH \longrightarrow$
    (f) $C + O_2 \longrightarrow$
    (g) $Al(ClO_3)_3 \xrightarrow{\Delta} O_2 +$
    (h) $CuBr_2 + Cl_2 \longrightarrow$
    (i) $SbCl_3 + (NH_4)_2S \longrightarrow$
    (j) $NaNO_3 \xrightarrow{\Delta} NaNO_2 +$

14. What is the purpose of balancing equations?
15. What is represented by the numbers (coefficients) that are placed in front of the formulas in a balanced equation?
16. Interpret the following chemical reactions in terms of the number of moles of each reactant and product:
    (a) $MgBr_2 + 2\ AgNO_3 \longrightarrow Mg(NO_3)_2 + 2\ AgBr$
    (b) $N_2 + 3\ H_2 \longrightarrow 2\ NH_3$
    (c) $2\ C_3H_7OH + 9\ O_2 \longrightarrow 6\ CO_2 + 8\ H_2O$
17. Interpret each of the following equations in terms of the relative number of moles of each substance involved and indicate whether the reaction is exothermic or endothermic:
    (a) $2\ Na + Cl_2 \longrightarrow 2\ NaCl + 822$ kJ (196.4 kcal)
    (b) $PCl_5 + 92.9$ kJ (22.2 kcal) $\longrightarrow PCl_3 + Cl_2$
18. Write balanced equations for each of these reactions, including the heat term:
    (a) Lime, CaO, is converted to slaked lime, $Ca(OH)_2$, by reaction with water. The reaction liberates 65.3 kJ (15.6 kcal) of heat for each mole of lime reacted.
    (b) The industrial production of aluminum metal from aluminum oxide is an endothermic electrolytic process requiring 1630 kJ per mole of Al_2O_3. Oxygen is also a product.
19. Write balanced equations for the combustion of the following hydrocarbons.
    (a) ethane, C_2H_6 (c) heptane, C_7H_{16}
    (b) benzene, C_6H_6
20. Which of the following statements are correct?
    (a) The coefficients in front of the formulas in a balanced chemical equation give the relative number of moles of the reactants and products in the reaction.
    (b) A balanced chemical equation is one that has the same number of moles on each side of the equation.
    (c) In a chemical equation, the symbol $\xrightarrow{\Delta}$ indicates that the reaction is exothermic.
    (d) A chemical change that absorbs heat energy is said to be endothermic.
    (e) In the reaction $H_2 + Cl_2 \longrightarrow 2\ HCl$, 100 molecules of HCl are produced for every 50 molecules of H_2 reacted.
    (f) The symbol (*aq*) after a substance in an equation means that the substance is in a water solution.
    (g) The equation $H_2O \longrightarrow H_2 + O_2$ can be balanced by placing a 2 in front of H_2O.
    (h) In the equation $3\ H_2 + N_2 \longrightarrow 2\ NH_3$ there are fewer moles of product than there are moles of reactants.
    (i) The total number of moles of reactants and products represented by this equation is 5 moles:
    $Mg + 2\ HCl \longrightarrow MgCl_2 + H_2$
    (j) One mole of glucose, $C_6H_{12}O_6$, contains 6 moles of carbon atoms.
    (k) The reactants are the substances produced by the chemical reaction.
    (l) In a balanced equation each side of the equation contains the same number of atoms of each element.
    (m) When a precipitate is formed in a chemical reaction, it can be indicated in the equation with the symbol $\downarrow$ or (*s*) immediately before the formula of the substance precipitated.
    (n) When a gas is evolved in a chemical reaction, it can be indicated in the equation with the symbol $\uparrow$ or (*g*) immediately following the formula of the gas.
    (o) According to the equation $3\ H_2 + N_2 \longrightarrow 2\ NH_3$, 4 mol of NH_3 will be formed when 6 mol of H_2 and 2 mol of N_2 react.
    (p) The products of an exothermic reaction are at a lower energy state than the reactants.
    (q) The combustion of hydrocarbons produces carbon dioxide and water as products.

Review Exercises for Chapters 8–10

CHAPTER 8 NOMENCLATURE OF INORGANIC COMPOUNDS

True–False. *Answer the following as either true or false.*

1. The oxidation number of an element can have a positive, negative, or zero value.
2. The sum of the oxidation numbers of all the elements in a compound equals zero.
3. The oxidation number of an ion in an ionic compound is equal to the charge of the ion.
4. The oxidation numbers of the elements in Group IIIA are 0 or +3.
5. The oxidation number of oxygen in a compound is usually −1.
6. In $KMnO_4$ the oxidation number of Mn is +6.
7. Binary compounds have names ending in *ide*.
8. In binary compounds with metals, the halogens have a −2 oxidation number.
9. The compound formed from Ga^{3+} and O^{2-} is Ga_3O_2.
10. The compound formed from NH_4^+ and SO_4^{2-} is $(NH_4)_2SO_4$.
11. The nitrite ion has three oxygen atoms and the nitrate ion has four oxygen atoms.
12. The prefixes *tetra* and *penta* mean four and five, respectively.
13. If the name of an acid ends in *ous*, the corresponding salt name will end in *ate*.
14. The lower and higher oxidation states of iron (Fe) are called *ferrous* and *ferric*, respectively.
15. The common name for sulfur is brimstone.
16. The formula for muriatic acid is HNO_3.
17. The formula for cane or beet sugar is $C_6H_{12}O_6$.
18. The name for Cl_2O_3 is dichloroheptoxide.
19. The formula for copper(II) oxide is Cu_2O.
20. The formula for barium hydroxide is BaOH.

Names and Formulas. *In which of the following is the formula correct for the name given?*

1. Copper(II) sulfate, $CuSO_4$
2. Ammonium hydroxide, NH_4OH
3. Mercury(I) carbonate, $HgCO_3$
4. Phosphorus triiodide, PI_3
5. Calcium acetate, $Ca(C_2H_3O_2)_2$
6. Hypochlorous acid, HClO
7. Dichlorine heptoxide, Cl_2O_7
8. Magnesium iodide, MgI
9. Sulfurous acid, H_2SO_3
10. Potassium manganate, $KMnO_4$
11. Lead(II) chromate, $PbCrO_4$
12. Ammonium bicarbonate, NH_4HCO_3
13. Iron(II) phosphate, $FePO_4$
14. Calcium hydrogen sulfate, $CaHSO_4$
15. Mercury(II) sulfate, $HgSO_4$
16. Dinitrogen pentoxide, N_2O_5
17. Sodium hypochlorite, NaClO
18. Sodium dichromate, $Na_2Cr_2O_7$
19. Cadmium cyanide, $Cd(CN)_2$
20. Bismuth(III) oxide, Bi_3O_2
21. Carbonic acid, H_2CO_3
22. Silver oxide, Ag_2O
23. Ferric iodide, FeI_2
24. Tin(II) fluoride, TiF_2
25. Carbon monoxide, CO
26. Phosphoric acid, H_3PO_3
27. Sodium bromate, Na_2BrO_3
28. Hydrosulfuric acid, H_2S
29. Potassium hydroxide, POH
30. Sodium carbonate, Na_2CO_3
31. Zinc sulfate, $ZnSO_3$
32. Sulfur trioxide, SO_3
33. Tin(IV) nitrate, $Sn(NO_3)_4$
34. Ferrous sulfate, $FeSO_4$
35. Chloric acid, HCl
36. Aluminum sulfide, Al_2S_3
37. Cobalt(II) chloride, $CoCl_2$
38. Acetic acid, $HC_2H_3O_2$
39. Zinc oxide, ZnO_2
40. Stannous fluoride, SnF_2

CHAPTER 9 QUANTITATIVE COMPOSITION OF COMPOUNDS

True–False. *Answer the following as either true or false.*

1. A mole contains Avogadro's number of atoms, molecules, or formula units.
2. A mole of Ag (107.9 g) contains the same number of atoms as a mole of Na (23.0 g).
3. One mole of chlorine molecules contains two moles of chlorine atoms.
4. One mole of glucose, $C_6H_{12}O_6$, contains 24 moles of atoms.
5. Two moles of hydrogen molecules weigh 4.0 g.
6. One gram of sulfur contains 6.022×10^{23} atoms.
7. A mole of NaCl weighs less than a mole of KCl.
8. The formula weight is the sum of the atomic weights of all the atoms in a formula unit.
9. $CaCl_2$ has a higher percentage of chlorine than $MgCl_2$.
10. A compound has an empirical formula of C_2H_2O and a molecular weight of 168.0. The molecular formula is $C_6H_6O_3$.
11. The percentage composition of a compound is the weight percent of each element in the compound.
12. A compound such as NaCl has a formula weight but no true molecular weight.
13. One mole of $HC_2H_3O_2$ weighs 60.0 g.
14. If the molecular formula and empirical formula of a compound are not the same, the empirical formula will be an integral multiple of the molecular formula.
15. The empirical formula of a compound gives the smallest ratio of the atoms that are present in a compound.
16. A mole of magnesium and a mole of magnesium oxide, MgO, contain the same number of magnesium atoms.
17. The number of sulfur atoms is the same in 1 mole of Na_2SO_4 as in 1 mole of K_2SO_4.
18. The number of sulfur atoms is the same in 1 g of Na_2SO_4 as in 1 g of K_2SO_4.
19. There are 14 moles of chlorine atoms in 3.5 moles of CCl_4.
20. A compound that has a carbon to hydrogen ratio of 1:2 can have a molecular weight of 48.0.

Multiple Choice. *Choose the correct answer to each of the following.*

1. 4.0 g of oxygen contains:
   (a) 1.5×10^{23} atoms of oxygen
   (b) 4.0 atomic weights of oxygen
   (c) 0.50 mol of oxygen
   (d) 6.022×10^{23} atoms of oxygen
2. One mole of hydrogen atoms contains:
   (a) 2.0 g of hydrogen
   (b) 6.022×10^{23} atoms of hydrogen
   (c) 1 atom of hydrogen
   (d) 12 g of carbon-12
3. One atom of magnesium weighs:
   (a) 24.3 g (c) 12.0 g
   (b) 54.9 g (d) 4.035×10^{-23} g
4. Avogadro's number of magnesium atoms:
   (a) Weigh 1.0 g
   (b) Weigh the same as Avogadro's number of sulfur atoms
   (c) Weigh 12.0 g
   (d) Are 1 mol of magnesium atoms
5. Which of the following contains the largest number of moles?
   (a) 1.0 g Li (c) 1.0 g Al
   (b) 1.0 g Na (d) 1.0 g Ag
6. The number of moles in 112 g of acetylsalicylic acid (aspirin), $C_9H_8O_4$, is:
   (a) 1.61 (b) 0.622 (c) 112 (d) 0.161
7. How many moles of aluminum hydroxide are in one antacid tablet containing 400 mg of $Al(OH)_3$?
   (a) 5.13×10^{-3} (c) 5.13
   (b) 0.400 (d) 9.09×10^{-3}
8. How many grams of Au_2S can be obtained from 1.17 mol of Au?
   (a) 182 g (b) 249 g (c) 364 g (d) 499 g
9. The formula weight of $Ba(NO_3)_2$ is:
   (a) 199.3 (b) 261.3 (c) 247.3 (d) 167.3
10. A 16 g sample of O_2:
    (a) Is 1 mol of O_2
    (b) Contains 6.022×10^{23} molecules of O_2
    (c) Is 0.50 molecule of O_2
    (d) Is 0.50 mol. wt of O_2

11. 2.00 mol of CO_2:
    (a) Weigh 2.00 g
    (b) Contain 1.20×10^{24} molecules
    (c) Weigh 56.0 g
    (d) Contain 6.00 molecular weights of CO_2
12. In Ag_2CO_3, the percentage by weight of:
    (a) Carbon is 43.5% (c) Oxygen is 17.4%
    (b) Silver is 64.2% (d) Oxygen is 21.9%
13. The empirical formula of the compound containing 31.0% Ti and 69.0% Cl is:
    (a) TiCl (c) $TiCl_3$
    (b) $TiCl_2$ (d) $TiCl_4$
14. A compound contains 54.3% C, 5.6% H, and 40.1% Cl. The empirical formula is:
    (a) CH_3Cl (c) $C_2H_4Cl_2$
    (b) C_2H_5Cl (d) C_4H_5Cl
15. A compound contains 40.0% C, 6.7% H, and 53.3% O. The molecular weight is 60.0. The molecular formula is:
    (a) $C_2H_3O_2$ (c) C_2HCl
    (b) C_3H_8O (d) $C_2H_4O_2$
16. How many chlorine atoms are in 4.0 mol of PCl_3?
    (a) 3 (c) 12.0
    (b) 7.2×10^{24} (d) 2.4×10^{24}
17. What is the weight of 4.53 mol of Na_2SO_4?
    (a) 142.1 g (c) 31.4 g
    (b) 644 g (d) 3.19×10^{-2} g
18. The percentage composition of Mg_3N_2 is:
    (a) 72.2% Mg, 27.8% N
    (b) 63.4% Mg, 36.6% N
    (c) 83.9% Mg, 16.1% N
    (d) No correct answer given
19. How many grams of oxygen are contained in 0.500 mol of Na_2SO_4?
    (a) 16.0 g (c) 64.0 g
    (b) 32.0 g (d) No correct answer given
20. The empirical formula of a compound is CH. If the molecular weight of this compound is 78.0, then the molecular formula is:
    (a) C_2H_2 (c) C_6H_6
    (b) C_5H_{18} (d) No correct answer given

CHAPTER 10 CHEMICAL EQUATIONS

True–False. *Answer the following as either true or false.*

1. In balancing an equation, we change the formulas of compounds to make the number of atoms on each side of the equation balance.
2. The equation $N_2 + 3\ H_2 \rightleftharpoons 2\ NH_3$ can be interpreted as saying that 1 mole of N_2 reacts with 3 moles of H_2 to form 2 moles of NH_3.
3. The equation $N_2 + 3\ H_2 \rightleftharpoons 2\ NH_3$ can be interpreted as saying that 1 g of N_2 reacts with 3 g of H_2 to form 2 g of NH_3.
4. The substances on the right side of a chemical equation are called the products.
5. When a gas is evolved in a chemical reaction, it can be indicated in the equation with the symbol ↑ or (*g*) immediately following the formula of the gas.
6. A balanced chemical equation is one that has the same number of moles on each side of the equation.
7. The coefficients in front of the formulas in a balanced chemical equation give the relative number of moles of the reactants and products in the reaction.
8. Water is formed in a neutralization reaction.
9. When carbonates or bicarbonates react with acids, carbon monoxide is formed.
10. In the reaction

$$Cu(s) + 2\ AgNO_3(aq) \longrightarrow Cu(NO_3)_2(aq) + 2\ Ag(s)$$

    Cu replaces Ag^+ because copper is a more reactive element than silver.
11. The combustion of hydrocarbons is an exothermic reaction.
12. The addition of sulfuric acid to water is an endothermic process.

Multiple Choice. *Choose the correct answer to each of the following.*

1. The reaction $BaCl_2 + (NH_4)_2CO_3 \longrightarrow BaCO_3 + 2\ NH_4Cl$ is an example of:
    (a) combination
    (b) decomposition
    (c) single displacement
    (d) double displacement

2. The reaction $2\ Al + 3\ Br_2 \longrightarrow 2\ AlBr_3$ is an example of:
   (a) combination
   (b) single displacement
   (c) decomposition
   (d) double displacement
3. When the equation $PbO_2 \xrightarrow{\Delta} PbO + O_2$ is balanced, one of the terms in the balanced equation is:
   (a) PbO_2 (b) $3\ O_2$ (c) $3\ PbO$ (d) O_2
4. When the equation $Cr_2S_3 + HCl \longrightarrow CrCl_3 + H_2S$ is balanced, one of the terms in the balanced equation is:
   (a) $3\ HCl$ (b) $CrCl_3$ (c) $3\ H_2S$ (d) $2\ Cr_2S_3$
5. When the equation $F_2 + H_2O \longrightarrow HF + O_2$ is balanced, a term in the balanced equation is:
   (a) $2\ HF$ (b) $3\ O_2$ (c) $4\ HF$ (d) $4\ H_2O$
6. When the equation $NH_4OH + H_2SO_4 \longrightarrow$ is completed and balanced, one of the terms in the balanced equation is:
   (a) NH_4SO_4 (c) H_2OH
   (b) $2\ H_2O$ (d) $2\ (NH_4)_2SO_4$
7. When the equation $H_2 + V_2O_5 \longrightarrow V +$ is completed and balanced, a term in the balanced equation is:
   (a) $2\ V_2O_5$ (b) $3\ H_2O$ (c) $2\ V$ (d) $8\ H_2$
8. When the equation $Fe_2(SO_4)_3 + Ba(OH)_2 \longrightarrow$ is completed and balanced, a term in the balanced equation is:
   (a) $Ba_2(SO_4)_3$ (c) $2\ Fe_2(SO_4)_3$
   (b) $2\ Fe(OH)$ (d) $2\ Fe(OH)_3$
9. For the reaction $2\ H_2 + O_2 \longrightarrow 2\ H_2O + 572.4\ kJ$, which of the following is not true?
   (a) The reaction is exothermic.
   (b) 572.4 kJ of heat are liberated for each mole of water formed.
   (c) Two moles of hydrogen react with 1 mole of oxygen.
   (d) 572.4 kJ of heat are liberated for each 2 moles of hydrogen reacted.
10. When a nonmetal oxide reacts with water,
    (a) A base is formed
    (b) An acid is formed
    (c) A salt is formed
    (d) A nonmetal oxide is formed

CHAPTER 11

Calculations from Chemical Equations

After studying Chapter 11, you should be able to

1 Understand the new terms listed in Question A at the end of the chapter.

2 Write mole ratios for any two substances involved in a chemical reaction.

3 Outline the mole or mole-ratio method for making stoichiometric calculations.

4 Calculate the number of moles of a desired substance obtainable from a given number of moles of a starting substance in a chemical reaction (mole to mole calculations).

5 Calculate the mass of a desired substance obtainable from a given number of moles of a starting substance in a chemical reaction and vice versa (mole to mass and mass to mole calculations).

6 Calculate the mass of a desired substance involved in a chemical reaction from a given mass of a starting substance (mass to mass calculations).

7 Deduce the limiting reactant or reagent when given the amounts of starting substances, and then calculate the moles or mass of desired substance obtainable from a given chemical reaction (limiting reactant calculations).

8 Apply theoretical yield or actual yield to any of the foregoing types of problems, or calculate theoretical and actual yields of a chemical reaction.

This chapter shows the quantitative relationship between reactants and products in chemical reactions and also reviews and correlates such concepts as molecular weight, the molecule, the mole, and balancing equations.

11.1 A Short Review

(a) Molecular Weight or Formula Weight The molecular weight is the sum of the atomic weights of all the atoms in a molecule. The formula weight applies to the mass of any formula unit—atoms, molecules, or ions; it is the atomic weight of an atom, or the sum of the atomic weights in a molecule or an ion.

(b) Relationship between Molecule and Mole A molecule is the smallest unit of a molecular substance (for example, Cl_2), and a mole is Avogadro's number, 6.022×10^{23}, of molecules of that substance. A mole of chlorine (Cl_2) has the same number of molecules as a mole of carbon dioxide, a mole of water, or a mole of any other molecular substance. When we relate molecules to molecular weight, 1 mol. wt = 1 mol, or 6.022×10^{23} molecules.

In addition to referring to molecular substances, the term *mole* may refer to any chemical species. It represents a quantity in grams equal to the formula weight (form. wt) and may be applied to atoms, ions, electrons, and formula units of nonmolecular substances.

$$1 \text{ mole} = \begin{cases} 1 \text{ mol. wt} = 6.022 \times 10^{23} \text{ molecules} \\ 1 \text{ form. wt} = 6.022 \times 10^{23} \text{ formula units} \\ 1 \text{ atomic wt} = 6.022 \times 10^{23} \text{ atoms} \\ 1 \text{ ionic wt} = 6.022 \times 10^{23} \text{ ions} \end{cases}$$

Other useful mole relationships are

$$\text{number of moles} = \frac{\text{grams of a substance}}{\text{molecular weight of the substance}}$$

$$\text{number of moles} = \frac{\text{grams of a monatomic element}}{\text{atomic weight of the element}}$$

$$\text{number of moles} = \frac{\text{number of molecules}}{6.022 \times 10^{23} \text{ molecules/mole}}$$

Two other useful equalities can be derived algebraically from each of these mole relationships. What are they?

(c) Balanced Equations When using chemical equations for calculations of mole–mass–volume relationships between reactants and products, the equations must be balanced. Remember that the number in front of a formula in a balanced chemical equation can represent the number of moles of that substance in the chemical reaction.

11.2 Calculations from Chemical Equations: The Mole Method

stoichiometry

It is often necessary to calculate the amount of a substance that is produced from, or needed to react with, a given quantity of another substance. The area of chemistry that deals with the quantitative relationships among reactants and products is known as **stoichiometry** (*stoy-key-ah-meh-tree*). Although several methods are known, we firmly believe that the *mole* or *mole-ratio* method is generally best for solving problems in stoichiometry. This method is straightforward and, in our opinion, makes it easy to see and understand the relationships of the reacting species.

mole ratio

A **mole ratio** is a ratio between the number of moles of any two species involved in a chemical reaction. For example, in the reaction

$$\underset{2\text{ mol}}{2\,H_2} + \underset{1\text{ mol}}{O_2} \longrightarrow \underset{2\text{ mol}}{2\,H_2O}$$

six mole ratios apply only to this reaction:

$$\frac{2\text{ mol }H_2}{1\text{ mol }O_2} \qquad \frac{2\text{ mol }H_2}{2\text{ mol }H_2O} \qquad \frac{1\text{ mol }O_2}{2\text{ mol }H_2}$$

$$\frac{1\text{ mol }O_2}{2\text{ mol }H_2O} \qquad \frac{2\text{ mol }H_2O}{2\text{ mol }H_2} \qquad \frac{2\text{ mol }H_2O}{1\text{ mol }O_2}$$

The mole ratio is a conversion factor used to convert the number of moles of one substance to the corresponding number of moles of another substance in a chemical reaction. For example, if we want to calculate the number of moles of H_2O that can be obtained from 4.0 mol of O_2, we use the mole ratio 2 mol H_2O/1 mol O_2.

$$4.0\,\cancel{\text{mol }O_2} \times \frac{2\text{ mol }H_2O}{1\,\cancel{\text{mol }O_2}} = 8.0\text{ mol }H_2O$$

Since stoichiometric problems are encountered throughout the entire field of chemistry, it is profitable to master this general method for their solution. The mole method makes use of three simple basic operations.

A Convert the quantity of starting substance to moles (if it is not given in moles).

B Convert the moles of starting substance to moles of desired substance.

C Convert the moles of desired substance to the units specified in the problem.

Like learning to balance chemical equations, learning to make stoichiometric calculations requires practice. A detailed step-by-step description of the general method, together with a variety of worked examples, is given in the following paragraphs. Study this material and apply the method to the problems at the end of this chapter.

Step 1 Use a balanced equation. Write a balanced equation for the chemical reaction in question or check to see that the equation given is balanced.

Step 2 Determine the number of moles of starting substance. Identify the starting substance from the data given in the statement of the problem. When the starting substance is given in moles, use it in that form; if it is not in moles, convert the quantity of the starting substance to moles.

Step 3 Determine the mole ratio of the desired substance to the starting substance. The number of moles of each substance in the balanced equation is indicated by the coefficient in front of each substance. Use these coefficients to set up the mole ratio:

$$\text{mole ratio} = \frac{\text{moles of desired substance in the equation}}{\text{moles of starting substance in the equation}}$$

Step 4 Calculate the number of moles of the desired substance. Multiply the number of moles of starting substance (from Step 2) by the mole ratio (from Step 3) to obtain the number of moles of desired substance:

$$\begin{pmatrix}\text{moles of desired}\\ \text{substance}\end{pmatrix} = \underbrace{\begin{pmatrix}\text{moles of starting}\\ \text{substance}\end{pmatrix}}_{\text{From Step 2}} \times \underbrace{\frac{\begin{pmatrix}\text{moles of desired substance}\\ \text{in the equation}\end{pmatrix}}{\begin{pmatrix}\text{moles of starting substance}\\ \text{in the equation}\end{pmatrix}}}_{\text{Mole ratio from Step 3}}$$

Note that the units of moles of starting substance cancel out in the numerator and the denominator.

Step 5 Calculate the desired substance in the units specified in the problem. If the answer is to be in moles, the problem is finished in Step 4. If units other than moles are wanted, multiply the moles of the desired substance (from Step 4) by the appropriate factor to convert moles to the units required.

For example, if grams of the desired substance are wanted,

$$\begin{pmatrix}\text{grams of desired}\\ \text{substance}\end{pmatrix} = \underbrace{\begin{pmatrix}\text{moles of desired}\\ \text{substance}\end{pmatrix}}_{\text{From Step 4}} \times \frac{\begin{pmatrix}\text{formula weight of}\\ \text{desired substance}\end{pmatrix}}{\begin{pmatrix}\text{one mole of}\\ \text{desired substance}\end{pmatrix}}$$

Use the conversion factors 6.022×10^{23} atoms/mol or 6.022×10^{23} molecules/mol when the problem asks for the answer in atoms or molecules.

The steps for converting the mass of starting substance A to either mass, atoms, or molecules of desired substance B are summarized below:

grams of A $\xrightarrow{\text{Step 2}}$ moles of A $\xrightarrow{\text{Steps 3 and 4}}$ moles B $\xrightarrow{\text{Step 5}}$ grams of B

moles B $\xrightarrow{\text{Step 5}}$ atoms or molecules of B

11.3 Mole–Mole Calculations

The object of this type of problem is to calculate the moles of one substance that react with, or are produced from, a given number of moles of another substance. Illustrative problems follow.

PROBLEM 11.1 How many moles of carbon dioxide will be produced by the complete oxidation of 2.5 mol of glucose, $C_6H_{12}O_6$, according to the following reaction?

$$\underset{1\text{ mol}}{C_6H_{12}O_6} + \underset{6\text{ mol}}{6\,O_2} \longrightarrow \underset{6\text{ mol}}{6\,CO_2} + \underset{6\text{ mol}}{6\,H_2O}$$

The balanced equation states that 6 mol of CO_2 will be produced from 1 mol of $C_6H_{12}O_6$. Even though we can readily see that 15 mol of CO_2 will be formed from 2.5 mol of $C_6H_{12}O_6$, the mole method of solving the problem is shown here. This method, using mole ratios, will be very helpful in solving later problems.

Step 1 The equation given is balanced.

Step 2 The number of moles of starting substance is 2.5 mol $C_6H_{12}O_6$. The conversion needed is

$$\text{moles } C_6H_{12}O_6 \longrightarrow \text{moles } CO_2$$

Step 3 From the balanced equation, set up the mole ratio between the two substances in question, placing the moles of the substance being sought in the numerator and the moles of the starting substance in the denominator. The number of moles, in each case, is the same as the coefficient in front of the substance in the balanced equation.

$$\text{mole ratio} = \frac{6 \text{ mol } CO_2}{1 \text{ mol } C_6H_{12}O_6} \quad \text{(from equation)}$$

Step 4 Multiply 2.5 mol of glucose (given in the problem) by this mole ratio.

$$2.5\ \text{mol}\ C_6H_{12}O_6 \times \frac{6\ \text{mol}\ CO_2}{1\ \text{mol}\ C_6H_{12}O_6} = 15\ \text{mol}\ CO_2 \quad \text{(Answer)}$$

Again note the use of units. The moles of $C_6H_{12}O_6$ cancel, leaving the answer in units of moles of CO_2.

PROBLEM 11.2 How many moles of ammonia can be produced from 8.00 mol of hydrogen reacting with nitrogen?

Step 1 First we need the balanced equation

$$3\ H_2 + N_2 \longrightarrow 2\ NH_3$$

Step 2 The moles of starting substance are 8.00 mol of hydrogen. The conversion needed is

$$\text{moles}\ H_2 \longrightarrow \text{moles}\ NH_3$$

Step 3 The balanced equation states that we get 2 mol of NH_3 for every 3 mol of H_2 that react. Set up the mole ratio of desired substance (NH_3) to starting substance (H_2):

$$\text{mole ratio} = \frac{2\ \text{mol}\ NH_3}{3\ \text{mol}\ H_2}$$

Step 4 Multiplying the 8.00 mol of starting H_2 by this mole ratio, we get

$$8.00\ \text{mol}\ H_2 \times \frac{2\ \text{mol}\ NH_3}{3\ \text{mol}\ H_2} = 5.33\ \text{mol}\ NH_3 \quad \text{(Answer)}$$

PROBLEM 11.3 Given the balanced equation

$$\underset{1\ \text{mol}}{K_2Cr_2O_7} + \underset{6\ \text{mol}}{6\ KI} + 7\ H_2SO_4 \longrightarrow Cr_2(SO_4)_3 + 4\ K_2SO_4 + \underset{3\ \text{mol}}{3\ I_2} + 7\ H_2O$$

calculate (a) the number of moles of potassium dichromate ($K_2Cr_2O_7$) that will react with 2.0 mol of potassium iodide (KI); (b) the number of moles of iodine (I_2) that will be produced from 2.0 mol of potassium iodide.

After the equation is balanced, we are concerned only with $K_2Cr_2O_7$, KI, and I_2, and we can ignore all the other substances. The equation states that 1 mol of $K_2Cr_2O_7$ will react with 6 mol of KI to produce 3 mol of I_2.

(a) Calculate the number of moles of $K_2Cr_2O_7$.

Step 1 The equation given is balanced.
Step 2 The moles of starting substance are 2.0 mol of KI. The conversion needed is

$$\text{moles KI} \longrightarrow \text{moles}\ K_2Cr_2O_7$$

Step 3 Set up the mole ratio of desired substance to starting substance:

$$\text{mole ratio} = \frac{1\ \text{mol}\ K_2Cr_2O_7}{6\ \text{mol KI}} \quad \text{(from equation)}$$

Step 4 Multiply the moles of starting material by this ratio to obtain the answer.

$$2.0\ \cancel{\text{mol KI}} \times \frac{1\ \text{mol}\ K_2Cr_2O_7}{6\ \cancel{\text{mol KI}}} = 0.33\ \text{mol}\ K_2Cr_2O_7 \quad \text{(Answer)}$$

(b) Calculate the number of moles of I_2.

Steps 1 and 2 The equation given is balanced and the moles of starting substance are 2.0 mol KI as in part (a). The conversion needed is

$$\text{moles KI} \longrightarrow \text{moles}\ I_2$$

Step 3 Set up the mole ratio of desired substance to starting substance:

$$\text{mole ratio} = \frac{3\ \text{mol}\ I_2}{6\ \text{mol KI}} \quad \text{(from equation)}$$

Step 4 Multiply the moles of starting material by this ratio to obtain the answer.

$$2.0\ \cancel{\text{mol KI}} \times \frac{3\ \text{mol}\ I_2}{6\ \cancel{\text{mol KI}}} = 1.0\ \text{mol}\ I_2 \quad \text{(Answer)}$$

Thus 2.0 mol KI will react with 0.33 mol $K_2Cr_2O_7$ to produce 1.0 mol I_2.

PROBLEM 11.4 How many molecules of water can be produced by reacting 0.010 mol of oxygen with hydrogen?

The sequence of conversions needed in the calculation is

$$\text{moles}\ O_2 \longrightarrow \text{moles}\ H_2O \longrightarrow \text{molecules}\ H_2O$$

Step 1 First we write the balanced equation

$$\underset{}{2\,H_2} + \underset{1\ \text{mol}}{O_2} \longrightarrow \underset{2\ \text{mol}}{2\,H_2O}$$

Step 2 The moles of starting substance is 0.010 mol O_2.
Step 3 Set up the mole ratio of desired substance to starting substance:

$$\text{mole ratio} = \frac{2\ \text{mol}\ H_2O}{1\ \text{mol}\ O_2} \quad \text{(from equation)}$$

Step 4 Multiplying the 0.010 mol of oxygen by this ratio, we obtain

$$0.010\ \cancel{\text{mol}\ O_2} \times \frac{2\ \text{mol}\ H_2O}{1\ \cancel{\text{mol}\ O_2}} = 0.020\ \text{mol}\ H_2O$$

Step 5 Since the problem asks for molecules instead of moles of H_2O, we must convert moles to molecules. Use the conversion factor (6.022×10^{23} molecules)/mole.

$$0.020\ \cancel{\text{mol}}\ H_2O \times \frac{6.022 \times 10^{23}\ \text{molecules}}{1\ \cancel{\text{mol}}} = 1.2 \times 10^{22}\ \text{molecules}\ H_2O \quad \text{(Answer)}$$

Note that 0.020 mol is still quite a large number of water molecules.

11.4 Mole–Mass and Mass–Mass Calculations

The object of these types of problems is to calculate the mass of one substance that reacts with, or is produced from, a given number of moles or a given mass of another substance in a chemical reaction. The mole ratio is used to convert from moles of starting substance to moles of desired substance.

PROBLEM 11.5 What mass of hydrogen can be produced by reacting 6.0 mol of aluminum with hydrochloric acid?

First calculate the moles of hydrogen produced, using the mole-ratio method, and then calculate the mass of hydrogen by multiplying the moles of hydrogen by its mass per mole. The sequence of conversions in the calculation is

$$\text{moles Al} \longrightarrow \text{moles } H_2 \longrightarrow \text{grams } H_2$$

Step 1 The balanced equation is

$$\underset{2\text{ mol}}{2\,Al(s)} + 6\,HCl(aq) \longrightarrow 2\,AlCl_3(aq) + \underset{3\text{ mol}}{3\,H_2(g)}$$

Step 2 The moles of starting substance are 6.0 mol of aluminum.
Steps 3 and 4 Calculate moles of H_2 by the mole-ratio method.

$$6.0\,\cancel{\text{mol Al}} \times \frac{3\text{ mol }H_2}{2\,\cancel{\text{mol Al}}} = 9.0\text{ mol }H_2$$

Step 5 Convert moles of H_2 to grams [$g = mol \times (g/mol)$]:

$$9.0\,\cancel{\text{mol }H_2} \times \frac{2.0\text{ g }H_2}{1\,\cancel{\text{mol }H_2}} = 18\text{ g }H_2$$

We see that 18 g of H_2 can be produced by reacting 6.0 mol of Al with HCl. The following setup combines all the above steps into one continuous calculation:

$$6.0\,\cancel{\text{mol Al}} \times \frac{3\,\cancel{\text{mol }H_2}}{2\,\cancel{\text{mol Al}}} \times \frac{2.0\text{ g }H_2}{1\,\cancel{\text{mol }H_2}} = 18\text{ g }H_2$$

PROBLEM 11.6 What mass of carbon dioxide is produced by the complete combustion of 100 g of the hydrocarbon pentane, C_5H_{12}?

The sequence of conversions in the calculation is

$$\text{grams } C_5H_{12} \longrightarrow \text{moles } C_5H_{12} \longrightarrow \text{moles } CO_2 \longrightarrow \text{grams } CO_2$$

Formula weights: C_5H_{12}, 72.0; CO_2, 44.0

Step 1 The balanced equation is

$$\underset{1\text{ mol}}{C_5H_{12}} + 8\,O_2 \longrightarrow \underset{5\text{ mol}}{5\,CO_2} + 6\,H_2O$$

Step 2 The starting substance is 100 g of C_5H_{12}. Convert 100 g of C_5H_{12} to moles:

$$100\ \cancel{\text{g C}_5\text{H}_{12}} \times \frac{1\ \text{mol C}_5\text{H}_{12}}{72.0\ \cancel{\text{g C}_5\text{H}_{12}}} = 1.39\ \text{mol C}_5\text{H}_{12}$$

Steps 3 and 4 Calculate the moles of CO_2 by the mole-ratio method:

$$1.39\ \cancel{\text{mol C}_5\text{H}_{12}} \times \frac{5\ \text{moles CO}_2}{1\ \cancel{\text{mol C}_5\text{H}_{12}}} = 6.95\ \text{mol CO}_2$$

Step 5 Convert moles of CO_2 to grams:

$$\text{mol CO}_2 \times \frac{\text{form. wt CO}_2}{1\ \text{mol CO}_2} = \text{grams CO}_2$$

$$6.95\ \cancel{\text{mol CO}_2} \times \frac{44.0\ \text{g CO}_2}{1\ \cancel{\text{mol CO}_2}} = 306\ \text{g CO}_2 \quad \text{(Answer)}$$

We see that 306 g of CO_2 are produced from the complete combustion of 100 g of C_5H_{12}. The calculation in a continuous setup is

$$\boxed{100\ \cancel{\text{g C}_5\text{H}_{12}}} \times \boxed{\frac{1\ \cancel{\text{mol C}_5\text{H}_{12}}}{72.0\ \cancel{\text{g C}_5\text{H}_{12}}}} \times \boxed{\frac{5\ \cancel{\text{mol CO}_2}}{1\ \cancel{\text{mol C}_5\text{H}_{12}}}} \times \boxed{\frac{44.0\ \text{g CO}_2}{1\ \cancel{\text{mol CO}_2}}} = 306\ \text{g CO}_2$$

$$\text{grams C}_5\text{H}_{12} \longrightarrow \text{moles C}_5\text{H}_{12} \longrightarrow \text{moles CO}_2 \longrightarrow \text{grams CO}_2$$

PROBLEM 11.7 How many grams of nitric acid, HNO_3, are required to produce 8.75 g of dinitrogen monoxide, N_2O, according to the following equation?

$$\underset{10\ \text{mol}}{4\ \text{Zn}(s) + 10\ \text{HNO}_3(aq)} \longrightarrow 4\ \text{Zn(NO}_3)_2(aq) + \underset{1\ \text{mol}}{\text{N}_2\text{O}(g)} + 5\ \text{H}_2\text{O}(l)$$

In this problem, the amount of product is given, and we are asked to calculate the amount of reactant required to produce that product. The calculation is no different than in Problem 11.6. The starting substance for the calculation is 8.75 g of N_2O. The grams of HNO_3 need to be calculated. We shall do this problem in a continuous calculation setup.

Step 1 The equation for the reaction is balanced.
Step 2 The starting substance is 8.75 g of N_2O. The sequence of conversions needed is

$$\text{grams N}_2\text{O} \longrightarrow \text{moles N}_2\text{O} \longrightarrow \text{moles HNO}_3 \longrightarrow \text{grams HNO}_3$$

Formula weights: N_2O, 44.0; HNO_3, 63.0

Steps 3–5 The calculation is

$$8.75\ \cancel{\text{g N}_2\text{O}} \times \frac{1\ \cancel{\text{mol N}_2\text{O}}}{44.0\ \cancel{\text{g N}_2\text{O}}} \times \frac{10\ \cancel{\text{mol HNO}_3}}{1\ \cancel{\text{mol N}_2\text{O}}} \times \frac{63.0\ \text{g HNO}_3}{1\ \cancel{\text{mol HNO}_3}} = 125\ \text{g HNO}_3$$

$$\text{grams N}_2\text{O} \longrightarrow \text{moles N}_2\text{O} \longrightarrow \text{moles HNO}_3 \longrightarrow \text{grams HNO}_3$$

Thus, 125 g of HNO_3 are required to produce 8.75 g of N_2O in this reaction.

11.5 Limiting Reactant and Yield Calculations

In many chemical processes the quantities of the reactants used are such that the amount of one reactant is in excess of the amount of a second reactant in the reaction. The amount of the product(s) formed in such a case will depend on the reactant that is not in excess. Thus the reactant that is not in excess is known as the **limiting reactant** (sometimes called the limiting reagent), because it limits the amount of product that can be formed.

limiting reactant

As an example, consider the case in which solutions containing 1.0 mol of sodium hydroxide and 1.5 mol of hydrochloric acid are mixed:

$$\underset{1\text{ mol}}{NaOH} + \underset{1\text{ mol}}{HCl} \longrightarrow \underset{1\text{ mol}}{NaCl} + \underset{1\text{ mol}}{H_2O}$$

According to the equation it is possible to obtain 1.0 mol of NaCl from 1.0 mol of NaOH and 1.5 mol of NaCl from 1.5 mol of HCl. However, we cannot have two different yields of NaCl from the reaction. When 1.0 mol of NaOH and 1.5 mol of HCl are mixed, there is insufficient NaOH to react with all of the HCl. Therefore, HCl is the reactant in excess and NaOH is the limiting reactant. Since the amount of NaCl formed is dependent on the limiting reactant, only 1.0 mol of NaCl will be formed. Because 1.0 mol of NaOH reacts with 1.0 mol of HCl, 0.5 mol of HCl remains unreacted.

$$\left.\begin{matrix}1.0\text{ mol NaOH}\\1.5\text{ mol HCl}\end{matrix}\right\}\longrightarrow\begin{matrix}1.0\text{ mol NaCl}\\1.0\text{ mol }H_2O\end{matrix}+0.5\text{ mol HCl unreacted}$$

Problems giving the amounts of two reactants are generally of the limiting-reactant type. Several methods can be used to identify the limiting reactant in a chemical reaction. In the most direct method two steps are needed to determine the limiting reactant, the amount of product, and the amount of reactant in excess or unreacted.

1 Calculate the amount of product (moles or grams, as needed) that can be formed from each reactant. The reactant that gives the least amount of product is the limiting reactant. The other reactant is in excess, and some of it will remain at the end of the reaction. The least amount of product is the amount that will be formed by the reaction.
2 Calculate the amount of the reactant in excess that will react with the limiting reactant. Subtract the amount that reacts from the starting quantity of the reactant in excess. This result is the amount of that substance that remains unreacted.

PROBLEM 11.8 How many moles of Fe_3O_4 can be obtained by reacting 16.8 g Fe with 10.0 g H_2O? Which substance is the limiting reactant? Which substance is in excess?

$$3\ Fe(s) + 4\ H_2O(g) \xrightarrow{\Delta} Fe_3O_4(s) + 4\ H_2(g)$$

This is a typical problem in which one of the starting substances will control or limit the yield of product.

Calculate the moles of Fe_3O_4 that can be formed from each reactant.

g reactant $\longrightarrow$ mol reactant $\longrightarrow$ mol Fe_3O_4

Formula weights: Fe, 55.8; H_2O, 18.0

$$16.8 \text{ g Fe} \times \frac{1 \text{ mol Fe}}{55.8 \text{ g Fe}} \times \frac{1 \text{ mol } Fe_3O_4}{3 \text{ mol Fe}} = 0.100 \text{ mol } Fe_3O_4$$

$$10.0 \text{ g } H_2O \times \frac{1 \text{ mol } H_2O}{18.0 \text{ g } H_2O} \times \frac{1 \text{ mol } Fe_3O_4}{4 \text{ mol } H_2O} = 0.139 \text{ mol } Fe_3O_4$$

The limiting reactant is Fe because it produces less Fe_3O_4. H_2O is in excess. The yield of product is 0.100 mol of Fe_3O_4.

PROBLEM 11.9 How many grams of silver bromide, AgBr, can be formed when solutions containing 50.0 g of $MgBr_2$ and 100 g of $AgNO_3$ are mixed together? How many grams of the excess reactant remain unreacted?

First identify the limiting reactant and the reactant in excess.

$$MgBr_2(aq) + 2\ AgNO_3(aq) \longrightarrow 2\ AgBr\downarrow + Mg(NO_3)_2(aq)$$

Formula weights: $MgBr_2$, 184.1; $AgNO_3$, 169.9; AgBr, 187.8

Step 1 Calculate the grams of AgBr that can be formed from each reactant.

g reactant $\longrightarrow$ mol reactant $\longrightarrow$ mol AgBr $\longrightarrow$ g AgBr

$$50.0 \text{ g } MgBr_2 \times \frac{1 \text{ mol } MgBr_2}{184.1 \text{ g } MgBr_2} \times \frac{2 \text{ mol AgBr}}{1 \text{ mol } MgBr_2} \times \frac{187.8 \text{ g AgBr}}{1 \text{ mol AgBr}} = 102 \text{ g AgBr}$$

$$100 \text{ g } AgNO_3 \times \frac{1 \text{ mol } AgNO_3}{169.9 \text{ g } AgNO_3} \times \frac{2 \text{ mol AgBr}}{2 \text{ mol } AgNO_3} \times \frac{187.8 \text{ g AgBr}}{1 \text{ mol AgBr}} = 111 \text{ g AgBr}$$

The limiting reactant is $MgBr_2$ because it gives less AgBr. $AgNO_3$ is in excess. The yield is 102 g AgBr. The final mixture will contain 102 g AgBr, $Mg(NO_3)_2$, and some unreacted $AgNO_3$.

Step 2 Calculation of the grams of unreacted $AgNO_3$. Calculate the grams of $AgNO_3$ that will react with 50.0 g of $MgBr_2$.

g $MgBr_2$ $\longrightarrow$ mol $MgBr_2$ $\longrightarrow$ mol $AgNO_3$ $\longrightarrow$ g $AgNO_3$

$$50.0 \text{ g } MgBr_2 \times \frac{1 \text{ mol } MgBr_2}{184.1 \text{ g } MgBr_2} \times \frac{2 \text{ mol } AgNO_3}{1 \text{ mol } MgBr_2} \times \frac{169.9 \text{ g } AgNO_3}{1 \text{ mol } AgNO_3} = 92.3 \text{ g } AgNO_3$$

Thus 92.3 g of $AgNO_3$ react with 50.0 g of $MgBr_2$. The amount of $AgNO_3$ that remains unreacted is

$$100 \text{ g } AgNO_3 - 92.3 \text{ g } AgNO_3 = 7.7 \text{ g } AgNO_3 \text{ unreacted}$$

The quantities of the products that we have been calculating from equations represent the maximum yield (100%) of product according to the reaction represented by the equation. Many reactions, especially those involving organic substances, fail to give a 100% yield of product. The main reasons for this failure are the side reactions that give products other than the main product and the fact that many reactions are reversible. In addition, some product may be lost in handling and transferring from one vessel to another. The **theoretical yield** of a reaction is the calculated amount of product that can be obtained from a given amount of reactant, according to the chemical equation. The **actual yield** is the amount of product that we finally obtain.

theoretical yield

actual yield

percentage yield

The **percentage yield** is the ratio of the actual yield to the theoretical yield multiplied by 100%. Both yields must have the same units.

$$\frac{\text{actual yield}}{\text{theoretical yield}} \times 100\% = \text{percentage yield}$$

For example, if the theoretical yield calculated for a reaction is 14.8 g, and the amount of product obtained is 9.25 g, the percentage yield is

$$\frac{9.25\text{ g}}{14.8\text{ g}} \times 100\% = 62.5\%$$

PROBLEM 11.10 Carbon tetrachloride was prepared by reacting 100 g of carbon disulfide and 100 g of chlorine. Calculate the percentage yield if 65.0 g of CCl_4 was obtained from the reaction.

$$CS_2 + 3\,Cl_2 \longrightarrow CCl_4 + S_2Cl_2$$

Formula weights: CS_2, 76.2; Cl_2, 71.0; CCl_4, 154

In this problem we need to determine the limiting reactant in order to calculate the quantity of CCl_4 (theoretical yield) that can be formed. Then we can compare this amount with the 65.0 g of CCl_4 actual yield to calculate the percentage yield.

Step 1 Calculate the grams of CCl_4 that can be formed from each reactant.

$$\text{g reactant} \longrightarrow \text{mol reactant} \longrightarrow \text{mol } CCl_4 \longrightarrow \text{g } CCl_4$$

$$100\text{ g }\cancel{CS_2} \times \frac{1\ \cancel{\text{mol }CS_2}}{76.2\text{ g }\cancel{CS_2}} \times \frac{1\ \cancel{\text{mol }CCl_4}}{1\ \cancel{\text{mol }CS_2}} \times \frac{154\text{ g }CCl_4}{1\ \cancel{\text{mol }CCl_4}} = 202\text{ g }CCl_4$$

$$100\text{ g }\cancel{Cl_2} \times \frac{1\ \cancel{\text{mol }Cl_2}}{71.0\text{ g }\cancel{Cl_2}} \times \frac{1\ \cancel{\text{mol }CCl_4}}{3\ \cancel{\text{mol }Cl_2}} \times \frac{154\text{ g }CCl_4}{1\ \cancel{\text{mol }CCl_4}} = 72.3\text{ g }CCl_4$$

The limiting reactant is Cl_2. CS_2 is in excess. The theoretical yield is 72.3 g CCl_4.

Step 2 Calculate the percentage yield. According to the equation, 72.3 g of CCl_4 is the maximum amount or theoretical yield of CCl_4 possible from 100 g of Cl_2. Actual yield is 65.0 g of CCl_4.

$$\text{percentage yield} = \frac{65.0\text{ g}}{72.3\text{ g}} \times 100\% = 89.9\%$$

When solving problems, you will achieve better results if at first you do not try to take shortcuts. Write the data and numbers in a logical, orderly manner. Make certain that the equations are balanced and that the computations are accurate and expressed to the correct number of significant figures. Remember that units are very important; a number without units has little meaning. Finally, an electronic calculator can save you many hours of tedious computations.

QUESTIONS

A. *Review the meanings of the new terms introduced in this chapter.*

1. Stoichiometry
2. Mole ratio
3. Limiting reactant
4. Theoretical yield
5. Actual yield
6. Percentage yield

B. *Review problems*
In some of the following problems, equations have not been balanced.

Mole Review Problems

1. Calculate the number of moles in each of the following quantities:
   (a) 25.0 g KNO_3
   (b) 10.8 g $Ca(NO_3)_2$
   (c) 5.4×10^2 g $(NH_4)_2C_2O_4$
   (d) 2.10 kg $NaHCO_3$
   (e) 525 mg $ZnCl_2$
   (f) 56 millimol NaOH
   (g) 9.8×10^{24} molecules CO_2
   (h) 250 mL ethyl alcohol, C_2H_5OH ($d = 0.789$ g/mL)
   *(i) 16.8 mL H_2SO_4 solution ($d = 1.727$ g/mL, 80.0% H_2SO_4 by weight)
2. Calculate the number of grams in each of the following quantities:
   (a) 2.55 mol $Fe(OH)_3$
   (b) 0.00844 mol $NiSO_4$
   (c) 125 kg $CaCO_3$
   (d) 0.0600 mol $HC_2H_3O_2$
   (e) 10.5 mol NH_3
   (f) 0.725 mol Bi_2S_3
   (g) 72 millimol HCl
   (h) 4.50×10^{21} molecules glucose, $C_6H_{12}O_6$
   (i) 500 mL of liquid Br_2 ($d = 3.119$ g/mL)
   *(j) 75 mL K_2CrO_4 solution ($d = 1.175$ g/mL, 20.0% K_2CrO_4 by weight)
3. Which contains the larger number of molecules?
   (a) 10.0 g H_2O or 10.0 g H_2O_2
   (b) 25.0 g HCl or 85.0 g $C_6H_{12}O_6$

Mole–Ratio Problems

4. Given the equation for the combustion of isopropyl alcohol:

$$2\,C_3H_7OH + 9\,O_2 \longrightarrow 6\,CO_2 + 8\,H_2O$$

   what is the mole ratio of:
   (a) CO_2 to C_3H_7OH
   (b) C_3H_7OH to O_2
   (c) O_2 to CO_2
   (d) H_2O to C_3H_7OH
   (e) CO_2 to H_2O
   (f) H_2O to O_2
5. For the reaction

$$3\,CaCl_2 + 2\,H_3PO_4 \longrightarrow Ca_3(PO_4)_2 + 6\,HCl$$

   set up the mole ratio of
   (a) $CaCl_2$ to $Ca_3(PO_4)_2$
   (b) HCl to H_3PO_4
   (c) $CaCl_2$ to H_3PO_4
   (d) $Ca_3(PO_4)_2$ to H_3PO_4
   (e) HCl to $Ca_3(PO_4)_2$
   (f) H_3PO_4 to HCl

Mole–Mass Problems

6. How many moles of Cl_2 can be produced from 5.60 mol HCl?

$$4\,HCl + O_2 \longrightarrow 2\,Cl_2 + 2\,H_2O$$

7. How many grams of sodium hydroxide can be produced from 500 g of calcium hydroxide according to this equation?

$Ca(OH)_2 + Na_2CO_3 \longrightarrow 2\ NaOH + CaCO_3$

8. Given the equation

$$Al_4C_3 + 12\ H_2O \longrightarrow 4\ Al(OH)_3 + 3\ CH_4$$

(a) How many moles of water are needed to react with 100 g of Al_4C_3?
(b) How many moles of $Al(OH)_3$ will be produced when 0.600 mol of CH_4 is formed?

9. How many grams of zinc phosphate, $Zn_3(PO_4)_2$, are formed when 10.0 g of Zn are reacted with phosphoric acid?

10. Given the equation

$$4\ FeS_2 + 11\ O_2 \longrightarrow 2\ Fe_2O_3 + 8\ SO_2$$

(a) How many moles of Fe_2O_3 can be made from 1.00 mol of FeS_2?
(b) How many moles of O_2 are required to react with 4.50 mol of FeS_2?
(c) If the reaction produces 1.55 mol of Fe_2O_3, how many moles of SO_2 are produced?
(d) How many grams of SO_2 can be formed from 0.512 mol of FeS_2?
(e) If the reaction produces 40.6 g of SO_2, how many moles of O_2 were reacted?
(f) How many grams of FeS_2 are needed to produce 221 g of Fe_2O_3?

11. An early method of producing chlorine was by the reaction of pyrolusite, MnO_2, and hydrochloric acid. How many moles of HCl will react with 1.05 mol of MnO_2? (Balance the equation first.)

$$MnO_2(s) + HCl(aq) \longrightarrow Cl_2(g) + MnCl_2(aq) + H_2O$$

12. Given the reaction

$$Zn + HCl \longrightarrow ZnCl_2 + H_2$$

180 g of zinc were dropped into a beaker of hydrochloric acid. After the reaction ceased, 35 g of unreacted zinc remained in the beaker.
(a) How many moles of hydrogen gas were produced?
(b) How many grams of HCl were reacted?

13. In a blast furnace, iron(III) oxide reacts with coke (carbon) to produce molten iron and carbon monoxide:

$$Fe_2O_3 + 3\ C \longrightarrow 2\ Fe + 3\ CO$$

How many kilograms of iron would be formed from 125 kg of Fe_2O_3?

14. How many grams of steam and iron must react to produce 375 g of magnetic iron oxide, Fe_3O_4?

$$3\ Fe(s) + 4\ H_2O(g) \longrightarrow Fe_3O_4(s) + 4\ H_2(g)$$

15. Ethane gas, C_2H_6, burns in air (i.e., reacts with the oxygen in air) to form carbon dioxide and water:

$$2\ C_2H_6 + 7\ O_2 \longrightarrow 4\ CO_2 + 6\ H_2O$$

(a) How many moles of O_2 are needed for the complete combustion of 15.0 mol of ethane?
(b) How many grams of CO_2 are produced for each 8.00 g of H_2O produced?
(c) How many grams of CO_2 will be produced by the combustion of 75.0 g of C_2H_6?

Limiting-Reactant and Percentage-Yield Problems

16. In the following equations, determine which reactant is the limiting reactant and which reactant is in excess. The amounts mixed together are shown below each reactant. Show evidence for your answers.

(a) $KOH + HNO_3 \longrightarrow KNO_3 + H_2O$
16.0 g 12.0 g

(b) $2\,NaOH + H_2SO_4 \longrightarrow Na_2SO_4 + 2\,H_2O$
10.0 g 10.0 g

(c) $2\,Bi(NO_3)_3 + 3\,H_2S \longrightarrow Bi_2S_3 + 6\,HNO_3$
50.0 g 6.00 g

(d) $3\ Fe + 4\ H_2O \longrightarrow Fe_3O_4 + 4\ H_2$
40.0 g 16.0 g

17. Given the equation

$$Fe(s) + CuSO_4(aq) \longrightarrow Cu(s) + FeSO_4(aq)$$

(a) When 2.0 mol of Fe and 3.0 mol of $CuSO_4$ are reacted, what substances will be present when the reaction is over? How many moles of each substance are present?
(b) When 20.0 g of Fe and 40.0 g of $CuSO_4$ are reacted, what substances will be present when the reaction is over? How many grams of each substance are present?

18. The reaction for the combustion of propane, C_3H_8, is:

$$C_3H_8 + 5\ O_2 \longrightarrow 3\ CO_2 + 4\ H_2O$$

(a) If 5.0 mol of C_3H_8 and 5.0 mol of O_2 are reacted, how many moles of CO_2 can be produced?
(b) If 3.0 mol of C_3H_8 and 20.0 mol of O_2 are reacted, how many moles of CO_2 can be produced?
(c) If 20.0 mol of C_3H_8 and 3.0 mol of O_2 are reacted, how many moles of CO_2 can be produced?
*(d) If 2.0 mol of C_3H_8 and 14.0 mol of O_2 are placed in a closed container, and they react to completion (until one reactant is completely used up), what compounds are present in the container after the reaction, and how many moles of each compound are present?
(e) If 20.0 g of C_3H_8 and 20.0 g of O_2 are reacted, how many grams of CO_2 can be produced?
(f) If 20.0 g of C_3H_8 and 80.0 g of O_2 are reacted, how many grams of CO_2 can be produced?
(g) If 20.0 g of C_8H_8 and 200 g of O_2 are reacted, how many grams of CO_2 can be produced?

19. Aluminum reacts with bromine to form aluminum bromide:

$$2\ Al + 3\ Br_2 \longrightarrow 2\ AlBr_3$$

If 25.0 g of Al and 100 g of Br_2 are reacted, and 64.2 g of $AlBr_3$ product are recovered, what is the percentage yield for the reaction?

20. Methyl alcohol, CH_3OH, is made by reacting carbon monoxide and hydrogen in the presence of certain metal oxide catalysts. How much alcohol can be obtained by reacting 40.0 g of CO and 10.0 g of H_2? How many grams of excess reactant remain unreacted?

$$CO(g) + 2\ H_2(g) \longrightarrow CH_3OH(l)$$

21. Iron was reacted with a solution containing 400 g of copper(II) sulfate. The reaction was stopped after 1 hour, and 151 g of copper were obtained. Calculate the percentage yield of copper obtained.

$$Fe(s) + CuSO_4(aq) \longrightarrow Cu(s) + FeSO_4(aq)$$

***22.** Ethyl alcohol, C_2H_5OH, also called grain alcohol, can be made by the fermentation of sugar, which often comes from starch in grain:

$$\underset{\text{Glucose}}{C_6H_{12}O_6} \longrightarrow 2\ \underset{\text{Ethyl alcohol}}{C_2H_5OH} + 2\ CO_2$$

If an 84.6% yield of ethyl alcohol is obtained,
(a) What weight of ethyl alcohol will be produced from 750 g of glucose?
(b) What weight of glucose should be used to produce 475 g of C_2H_5OH?

***23.** Carbon disulfide, CS_2, can be made from coke, C, and sulfur dioxide, SO_2:

$$3\ C + 2\ SO_2 \longrightarrow CS_2 + 2\ CO_2$$

If the actual yield of CS_2 is 86.0% of the theoretical yield, what weight of coke is needed to produce 950 g of CS_2?

***24.** Acetylene, C_2H_2, can be manufactured by the reaction of water and calcium carbide, CaC_2:

$$CaC_2 + 2\ H_2O \longrightarrow C_2H_2(g) + Ca(OH)_2$$

When 44.5 g of commercial grade (impure) calcium carbide were reacted, 0.540 mol of C_2H_2 was produced. Assuming that all of the CaC_2 was reacted to C_2H_2, what is the percent of CaC_2 in the commercial grade material?

Additional Problems

***25.** Both $CaCl_2$ and $MgCl_2$ react with $AgNO_3$ to precipitate AgCl. When solutions containing equal weights of $CaCl_2$ and $MgCl_2$ are reacted, which salt will produce the larger amount of AgCl? Show proof.

***26.** An astronaut excretes about 2500 g of water a day. If lithium oxide, Li_2O, is used in the spaceship to absorb this water, how many kilograms of Li_2O must be carried for a 30-day space trip for three astronauts?

$$Li_2O + H_2O \longrightarrow 2\ LiOH$$

***27.** Much commercial hydrochloric acid is prepared by the reaction of concentrated sulfuric acid with sodium chloride:

$$H_2SO_4 + 2\ NaCl \longrightarrow Na_2SO_4 + 2\ HCl$$

How many kilograms of concentrated H_2SO_4, 96% H_2SO_4 by weight, are required to produce 20.0 L of concentrated hydrochloric acid (d = 1.20 g/mL, 42.0% HCl by weight)?

*** 28.** Gastric juice contains about 3.0 g of HCl per liter. If a person produces about 2.5 L of gastric juice per day, how many antacid tablets, each containing 400 mg of $Al(OH)_3$, are needed to neutralize all the HCl produced in one day?

$$Al(OH)_3(s) + 3\,HCl(aq) \longrightarrow AlCl_3(aq) + 6\,H_2O(l)$$

*** 29.** 12.82 g of a mixture of $KClO_3$ and NaCl are heated strongly. The $KClO_3$ reacts according to the following equation:

$$2\,KClO_3(s) \longrightarrow 2\,KCl(s) + 3\,O_2(g)$$

The NaCl does not undergo any reaction. After the heating, the residue (KCl and NaCl) weighs 9.45 g. Assuming that all the loss of weight represents loss of oxygen gas, calculate the percentage of $KClO_3$ in the original mixture.

30. Phosphine, PH_3, can be prepared by the hydrolysis of calcium phosphide, Ca_3P_2:

$$Ca_3P_2 + 6\,H_2O \longrightarrow 3\,Ca(OH)_2 + 2\,PH_3$$

Based on this equation, which of the following statements are correct?

(a) One mole of Ca_3P_2 produces 2 mol of PH_3.
(b) One gram of Ca_3P_2 produces 2 g of PH_3.
(c) Three moles of $Ca(OH)_2$ are produced for each 2 mol of PH_3 produced.
(d) The mole ratio between phosphine and calcium phosphide is

$$\frac{2\text{ mol }PH_3}{1\text{ mol }Ca_3P_2}$$

(e) When 2.0 mol of Ca_3P_2 and 3.0 mol of H_2O react, 4.0 mol of PH_3 can be formed.
(f) When 2.0 mol of Ca_3P_2 and 15.0 mol of H_2O react, 6.0 mol of $Ca(OH)_2$ can be formed.
(g) When 200 g of Ca_3P_2 and 100 g of H_2O react, Ca_3P_2 is the limiting reactant.
(h) When 200 g of Ca_3P_2 and 100 g of H_2O react, the theoretical yield of PH_3 is 57.4 g.

31. The equation representing the reaction used for the commercial preparation of hydrogen cyanide is

$$2\,CH_4 + 3\,O_2 + 2\,NH_3 \longrightarrow 2\,HCN + 6\,H_2O$$

Based on this equation, which of the statements below are correct?

(a) Three moles of O_2 are required for 2 moles of NH_3.
(b) Twelve moles of HCN are produced for every 16 moles of O_2 that react.
(c) The mole ratio between H_2O and CH_4 is

$$\frac{6\text{ mol }H_2O}{2\text{ mol }CH_4}$$

(d) When 12 moles of HCN are produced, 4 moles of H_2O will be formed.
(e) When 10 moles of CH_4, 10 moles of O_2, and 10 moles of NH_3 are mixed and reacted, O_2 is the limiting reactant.
(f) When 3 moles each of CH_4, O_2, and NH_3 are mixed and reacted, 3 moles of HCN will be produced.

CHAPTER

12 The Gaseous State of Matter

After studying Chapter 12, you should be able to

1 Understand the terms listed in Question A at the end of the chapter.
2 State the principal assumptions of the Kinetic-Molecular Theory (KMT).
3 Estimate the relative rates of effusion of two gases of known molecular weight.
4 Sketch and explain the operation of a mercury barometer.
5 Tell what two factors determine gas pressure in a vessel of fixed volume.
6 Work problems involving (a) Boyle's and (b) Charles' gas laws.
7 State what is meant by standard temperature and pressure (STP).
8 Give the equation for the combined gas law that deals with the pressure, volume, and temperature relationships expressed in Boyle's, Charles', and Gay-Lussac's gas laws.
9 Use Dalton's Law of Partial Pressures and the combined gas laws to calculate the dry STP volume of a gas collected over water.
10 State Avogadro's law.
11 Understand the mole–weight–volume relationship of gases.
12 Determine the density of any gas at STP.
13 Determine the molecular weight of a gas from its density at a known temperature and pressure.
14 Solve problems involving the ideal gas equation.
15 Make mole-to-volume, weight-to-volume, and volume-to-volume stoichiometric calculations from balanced chemical equations.
16 State two valid reasons why real gases may deviate from the behavior predicted for an ideal gas.

12.1 General Properties of Gases

In Chapter 3, solids, liquids, and gases are described in a brief outline. In this chapter we shall consider the behavior of gases in greater detail.

Gases are the least compact and most mobile of the three states of matter. A solid has a rigid structure, and its particles remain in essentially fixed positions. When a solid absorbs sufficient heat, it melts and changes into a liquid. Melting occurs because the molecules (or ions) have absorbed enough energy to break out of the rigid crystal lattice structure of the solid. The molecules or ions in the liquid are more energetic than they were in the solid, as shown by their increased mobility. Molecules in the liquid state are *coherent*; that is, they cling to one another. When the liquid absorbs additional heat, the more energetic molecules break away from the liquid surface and go into the gaseous state. Gases represent the most mobile state of matter. Gas molecules move with very high velocities and have high kinetic energy (KE). The average velocity of hydrogen molecules at 0°C is over 1600 meters (1 mile) per second. Because of the high velocities of their molecules, mixtures of gases are uniformly distributed within the container in which they are confined.

A quantity of a substance occupies a much greater volume as a gas than it does as a liquid or a solid. For example, 1 mole of water (18 g) has a volume of 18 mL at 4°C. This same amount of water would occupy about 22,400 mL in the gaseous state—more than a 1200-fold increase in volume. We may assume from this difference in volume that (1) gas molecules are relatively far apart, (2) gases can be greatly compressed, and (3) the volume occupied by a gas is mostly empty space.

12.2 The Kinetic-Molecular Theory

Careful scientific studies of the behavior and properties of gases were begun in the 17th century by Robert Boyle (1627–1691). This work was carried forward by many investigators after Boyle. The accumulated data were used in the second half of the 19th century to formulate a general theory to explain the behavior and properties of gases. This theory is called the **Kinetic-Molecular Theory (KMT)**. The KMT has since been extended to cover, in part, the behavior of liquids and solids. It ranks today with the atomic theory as one of the greatest generalizations of modern science.

Kinetic-Molecular Theory (KMT)

The KMT is based on the motion of particles, particularly gas molecules. A gas that behaves exactly as outlined by the theory is known as an **ideal**, or **perfect**, **gas**. Actually no ideal gases exist, but under certain conditions of temperature and pressure gases approach ideal behavior, or at least show only small deviations from it. Under extreme conditions, such as very high pressure and low temperature, real gases may deviate greatly from ideal behavior. For example, at low temperature and high pressure many gases become liquids.

ideal, or perfect, gas

The principal assumptions of the Kinetic-Molecular Theory are

1. Gases consist of tiny (submicroscopic) molecules.
2. The distance between molecules is large compared with the size of the molecules themselves. The volume occupied by a gas consists mostly of empty space.
3. Gas molecules have no attraction for one another.
4. Gas molecules move in straight lines in all directions, colliding frequently with one another and with the walls of the container.
5. No energy is lost by the collision of a gas molecule with another gas molecule or with the walls of the container. All collisions are perfectly elastic.
6. The average kinetic energy for molecules is the same for all gases at the same temperature, and its value is directly proportional to the Kelvin temperature.

The kinetic energy (KE) of a molecule is one-half its mass times its velocity squared. It is expressed by the equation

$$KE = \frac{1}{2}mv^2$$

where m is the mass and v is the velocity of the molecule.

All gases have the same kinetic energy at the same temperature. Therefore, from the kinetic energy equation we can see that, if we compare the velocities of the molecules of two gases, the lighter molecules will have a greater velocity than the heavier ones. For example, calculations show that the velocity of a hydrogen molecule is four times the velocity of an oxygen molecule.

diffusion

Due to their molecular motion gases have the property of **diffusion**, the ability of two or more gases to mix spontaneously until they form a uniform mixture. The diffusion of gases may be illustrated by use of the apparatus shown in Figure 12.1. Two large flasks, one containing reddish-brown bromine vapors and the other dry air, are connected by a side tube. When the stopcock between the flasks is opened, the bromine and air will diffuse into each other. After standing awhile, both flasks will contain bromine and air.

effusion

If we put a pinhole in a balloon, the gas inside will effuse or flow out of the balloon. **Effusion** is a process by which gas molecules pass through a very small orifice (opening) from a container at higher pressure to one at lower pressure.

Graham's law of effusion

Thomas Graham (1805–1869), a Scottish chemist, observed that the rate of effusion was dependent on the density of a gas. This observation led to **Graham's law of effusion**: The rates of effusion of two gases at the same temperature and pressure are inversely proportional to the square roots of their densities or molecular weights,

$$\frac{\text{rate of effusion of gas A}}{\text{rate of effusion of gas B}} = \sqrt{\frac{d\text{B}}{d\text{A}}} = \sqrt{\frac{\text{mol. wt B}}{\text{mol. wt A}}}$$

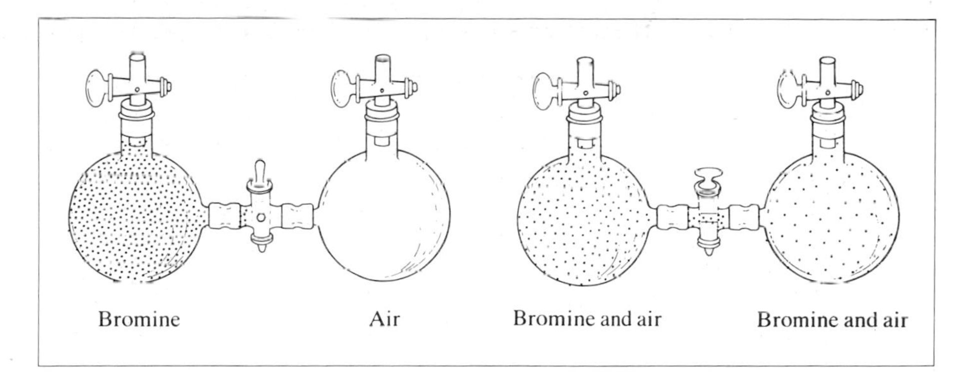

Figure 12.1
Diffusion of gases. When the stopcock between the two flasks is opened, colored bromine molecules can be seen diffusing into the flask containing air.

A major application of Graham's law occurred during World War II with the separation of the isotopes of uranium-235 (U-235) and uranium-238 (U-238). Naturally occurring uranium consists of 0.7% U-235, 99.3% U-238, and a trace of U-234. However, only U-235 is useful as fuel for nuclear reactors and atomic bombs, so the concentration of U-235 in the mixture of isotopes had to be increased.

Uranium was first changed to uranium hexafluoride, UF_6, a white solid that readily goes into the gaseous state. The gaseous mixture of $^{235}UF_6$ and $^{238}UF_6$ was then allowed to effuse through porous walls. Although the effusion rate of the lighter gas is only slightly faster than that of the heavier one,

$$\frac{\text{effusion rate } ^{235}UF_6}{\text{effusion rate } ^{238}UF_6} = \sqrt{\frac{\text{mol. wt } ^{238}UF_6}{\text{mol. wt } ^{235}UF_6}} = \sqrt{\frac{352}{349}} = 1.0043$$

the separation and enrichment of U-235 was accomplished by subjecting the gaseous mixture to several thousand stages of effusion.

The properties of an ideal gas are independent of the molecular constitution of the gas. Mixtures of gases also obey the Kinetic-Molecular Theory if the gases in the mixture do not enter into a chemical reaction with one another.

12.3 Measurement of Pressure of Gases

pressure

Pressure is defined as force per unit area. Do gases exert pressure? Yes. When a rubber balloon is inflated with air, it stretches and maintains its larger size because the pressure on the inside is greater than that on the outside. Pressure results from the collisions of gas molecules with the walls of the balloon (see

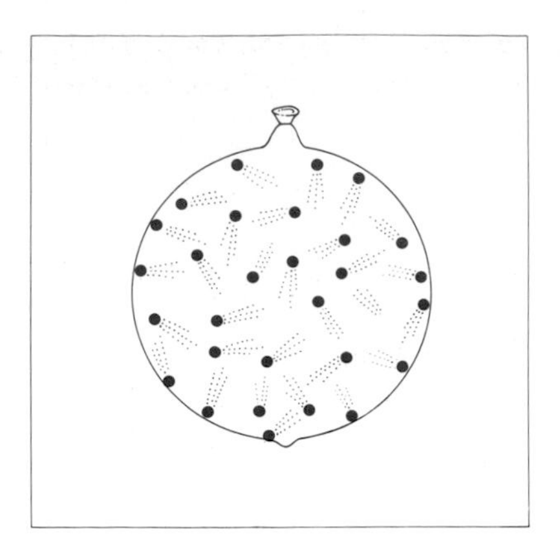

Figure 12.2
Pressure resulting from the collisions of gas molecules with the walls of the balloon keeps the balloon inflated.

Figure 12.2). When the gas is released, the force or pressure of the air escaping from the small neck propels the balloon in a rapid, irregular flight. If the balloon is inflated until it bursts, the gas escaping all at once causes an explosive noise. This pressure that gases display can be measured; it can also be transformed into useful work. Steam under pressure, used in the steam locomotive, played an important role in the early development of the United States. Compressed steam is used today to generate at least part of the electricity for many cities. Compressed air is used to operate many different kinds of mechanical equipment.

The mass of air surrounding the earth is called the *atmosphere*. It is composed of about 78% nitrogen, 21% oxygen, and 1% argon, and other minor constituents by volume (see Table 12.1). The outer boundary of the atmosphere is not known precisely, but more than 99% of the atmosphere is below an altitude of 20 miles (32 km). Thus, the concentration of gas molecules in the atmosphere decreases with altitude, and at about 4 miles the amount of oxygen is insufficient to sustain human life. The gases in the atmosphere exert a pressure known as **atmospheric pressure**. The pressure exerted by a gas depends on the number of molecules of gas present, the temperature, and the volume in which the gas is confined. Gravitational forces hold the atmosphere relatively close to the earth and prevent air molecules from flying off into outer space. Thus, the atmospheric pressure at any point is due to the mass of the atmosphere pressing downward at that point.

atmospheric pressure

The pressure of a gas can be measured with a pressure gauge, a manometer, or a **barometer**. A mercury barometer is commonly used in the laboratory to measure atmospheric pressure. A simple barometer of this type may be prepared by completely filling a long tube with pure, dry mercury and inverting the open end into an open dish of mercury. If the tube is longer than 760 mm, the mercury level will drop to a point at which the column of mercury in the tube is just supported by the pressure of the atmosphere. If the tube is properly prepared, a vacuum will exist above the mercury column. The mass of mercury, per unit area, is equal to the pressure of the atmosphere. The column of mercury is supported by the pressure of the atmosphere, and the height of the column is a measure of this pressure (see Figure 12.3). The mercury barometer was invented in 1643 by

barometer

Table 12.1 Average composition of normal dry air

Gas	Percent by volume	Gas	Percent by volume
N_2	78.08	He	0.0005
O_2	20.95	CH_4	0.0002
Ar	0.93	Kr	0.0001
CO_2	0.033	Xe, H_2, and N_2O	Trace
Ne	0.0018		

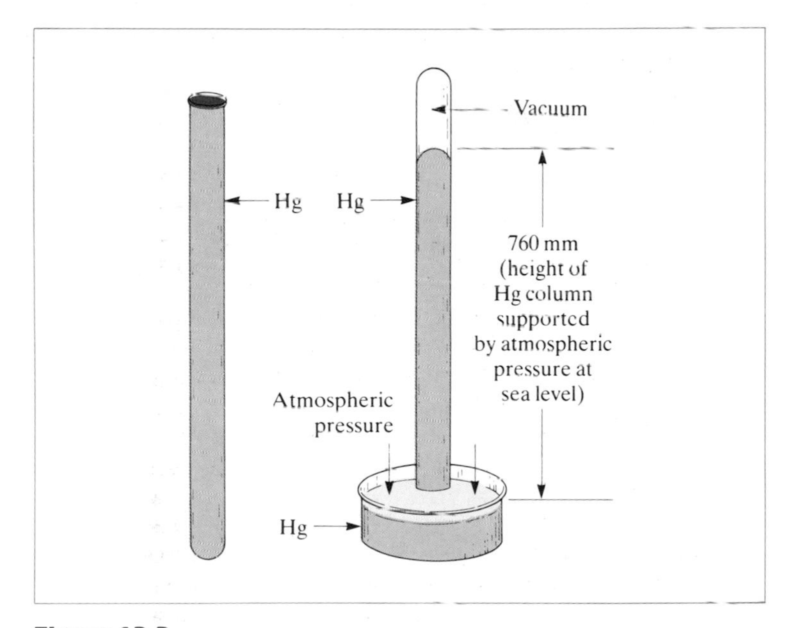

Figure 12.3

Preparation of a mercury barometer. The full tube of mercury at the left is inverted and placed in a dish of mercury.

the Italian physicist E. Torricelli (1608–1647), for whom the unit of pressure *torr* was named.

1 atmosphere

Air pressure is measured and expressed in many units. The standard atmospheric pressure, or simply **1 atmosphere**, is the pressure exerted by a column of mercury 760 mm high at a temperature of 0°C. The abbreviation for atmosphere is atm. The normal pressure of the atmosphere at sea level is 1 atm or 760 torr or 760 mm Hg. The SI unit for pressure is the pascal (Pa), where 1 atm = 101,325 Pa or 101.3 kPa. Other units for expressing pressure are inches of mercury, centimeters of mercury, the millibar (mbar), and pounds per square inch (lb/in.^{2} or psi). The meteorologist uses inches of mercury in reporting atmospheric pressure. The values of these units equivalent to 1 atm are summarized in Table 12.2 (1 atm $\equiv$ 760 torr $\equiv$ 760 mm Hg $\equiv$ 76 cm Hg $\equiv$ 101,325 Pa $\equiv$ 1013 mbar $\equiv$ 29.9 in. Hg $\equiv$ 14.7 lb/in.^{2}). (The symbol $\equiv$ means *identical with*.)

Table 12.2 Pressure units equivalent to 1 atmosphere

1 atm
760 torr
760 mm Hg
76 cm Hg
101,325 Pa
1013 mbar
29.9 in. Hg
14.7 lb/in.^{2}

Atmospheric pressure varies with altitude. The average pressure at Denver, Colorado, 1.61 km (1 mile) above sea level, is 630 torr (0.83 atm). Atmospheric pressure is 0.5 atm at about 5.5 km (3.4 miles) altitude.

Other liquids besides mercury may be employed for barometers, but they are not as useful as mercury because of the difficulty of maintaining a vacuum above the liquid and because of impractical heights of the liquid column. For example, a pressure of 1 atm will support a column of water about 10,336 mm (33.9 ft) high.

Pressure is often measured by reading the heights of mercury columns in millimeters on barometers and manometers. Thus pressure may be recorded as mm Hg. But in many applications the torr is superseding mm Hg as a unit of pressure. In problems dealing with gases it is necessary to make interconversions among the various pressure units. Since atm, torr, and mm Hg are common pressure units, illustrative problems involving all three of these units are used in this text.

$$1 \text{ atm} = 760 \text{ torr} = 760 \text{ mm Hg}$$

PROBLEM 12.1 The average atmospheric pressure at Walnut, California, is 740 mm Hg. Calculate this pressure in (a) torr and (b) atmospheres.

This problem can be solved using conversion factors relating one unit of pressure to another.

(a) To convert mm Hg to torr, use the conversion factor 760 torr/760 mm Hg (1 torr/1 mm Hg):

$$740 \cancel{\text{mm Hg}} \times \frac{1 \text{ torr}}{1 \cancel{\text{mm Hg}}} = 740 \text{ torr}$$

(b) To convert mm Hg to atm, use the conversion factor 1 atm/760 mm Hg:

$$740 \cancel{\text{mm Hg}} \times \frac{1 \text{ atm}}{760 \cancel{\text{mm Hg}}} = 0.934 \text{ atm}$$

12.4 Dependence of Pressure on Number of Molecules and Temperature

Pressure is produced by gas molecules colliding with the walls of a container. At a specific temperature and volume, the number of collisions depends on the number of gas molecules present. The number of collisions can be increased by increasing the number of gas molecules present. If we double the number of molecules, the frequency of collisions and the pressure should double. We find, for an ideal gas, that this doubling is actually what happens: When the temperature and volume are kept constant, the pressure is directly proportional to the number of moles or molecules of gas present. Figure 12.4 illustrates this concept.

A good example of this molecule–pressure relationship may be observed on an ordinary cylinder of compressed gas that is equipped with a pressure gauge. When the valve is opened, gas escapes from the cylinder. The volume of the cylinder is constant, and the decrease in quantity (moles) of gas is registered by a drop in pressure indicated on the gauge.

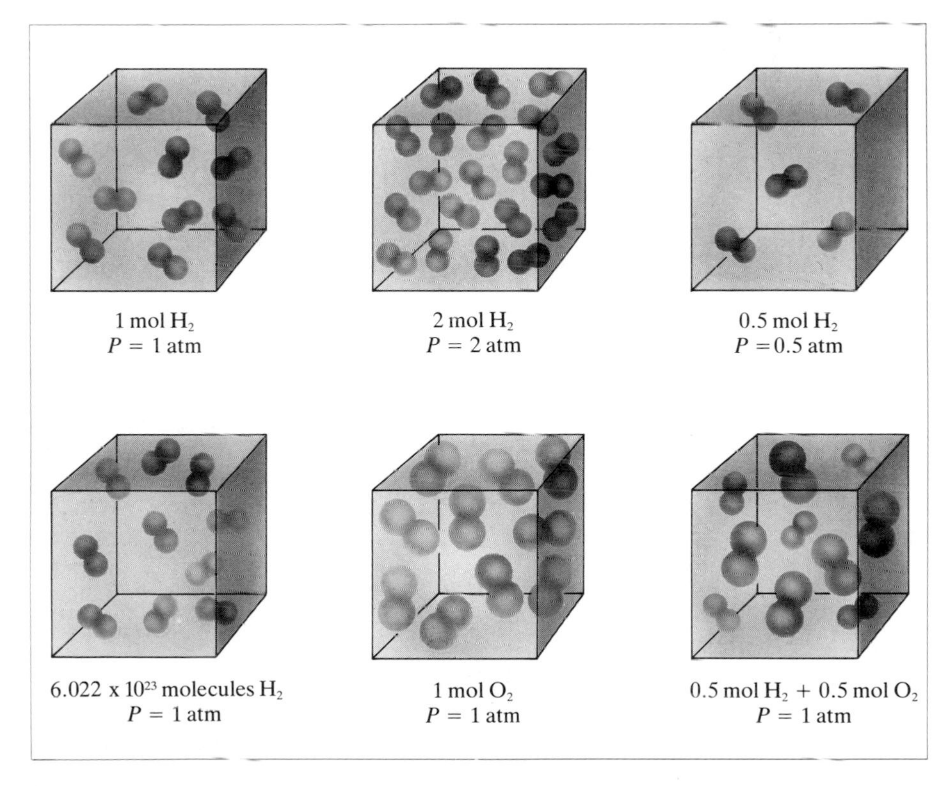

Figure 12.4

The pressure exerted by a gas is directly proportional to the number of molecules present. In each case shown, the volume is 22.4 L and the temperature is 0°C.

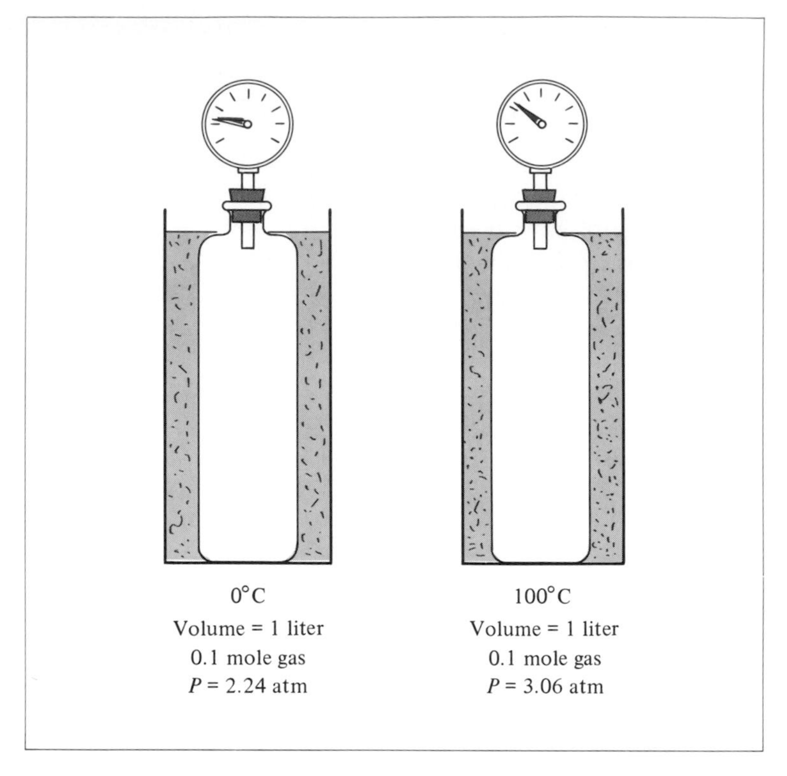

Figure 12.5

The pressure of a gas in a fixed volume increases with increasing temperature. The increased pressure is due to more frequent collisions of the gas molecules with the walls of the container at the higher temperature.

The pressure of a gas in a fixed volume also varies with temperature. When the temperature is increased, the kinetic energy of the molecules increases, causing more frequent collisions of the molecules with the walls of the container. This increase in collision frequency results in a pressure increase (see Figure 12.5).

12.5 Boyle's Law: The Relationship of the Volume and Pressure of a Gas

Boyle's Law

Robert Boyle demonstrated experimentally that, at constant temperature (T), the volume (V) of a fixed mass of a gas is inversely proportional to the pressure (P). This relationship of P and V is known as **Boyle's law**. Mathematically, Boyle's law may be expressed

$$V \propto \frac{1}{P} \quad \text{(mass and temperature are constant)}$$

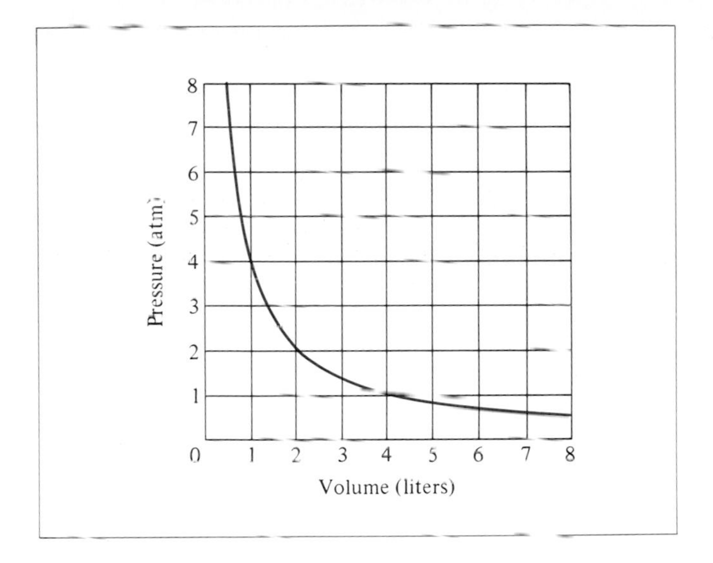

Figure 12.6

Graph of pressure versus volume showing the inverse PV relationship of an ideal gas

This equation says that the volume varies ($\propto$) inversely with the pressure, at constant mass and temperature. When the pressure on a gas is increased, its volume will decrease, and vice versa. The inverse relationship of pressure and volume is shown graphically in Figure 12.6.

Boyle demonstrated that, when he doubled the pressure on a specific quantity of a gas, keeping the temperature constant, the volume was reduced to one-half the original volume; when he tripled the pressure on the system, the new volume was one-third the original volume; and so on. His demonstration showed that the product of volume and pressure is constant if the temperature is not changed:

$$\text{PV} = \text{constant} \qquad \text{or} \qquad PV = k \quad \text{(mass and temperature are constant)}$$

Let us demonstrate this law by using a cylinder with a movable piston so that the volume of gas inside the cylinder may be varied by changing the external pressure (see Figure 12.7). We assume that the temperature and the number of gas molecules do not change. Let us start with a volume of 1000 mL and a pressure of 1 atm. When we change the pressure to 2 atm, the gas molecules are crowded closer together, and the volume is reduced to 500 mL. When we increase the pressure to 4 atm, the volume becomes 250 mL.

Note that the product of the pressure times the volume in each case is the same number, substantiating Boyle's law. We may then say that

$$P_1V_1 = P_2V_2$$

where P_1V_1 is the pressure–volume product at one set of conditions, and P_2V_2 is the product at another set of conditions. In each case the new volume may be calculated by multiplying the starting volume by a ratio of the two pressures

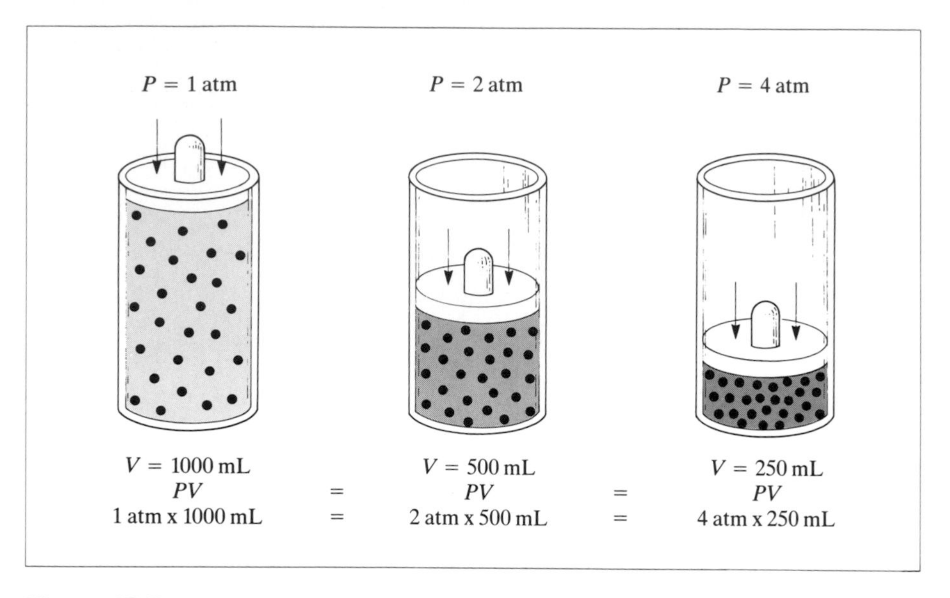

Figure 12.7

The effect of pressure on the volume of a gas

involved. Of course, the ratio of pressures used must reflect the direction in which the volume should change. When the pressure is changed from 1 atm to 2 atm, the ratio to be used is 1 atm/2 atm. Now we can verify the results given in Figure 12.7:

(a) Starting volume, 1000 mL; pressure change, 1 atm $\longrightarrow$ 2 atm

$$1000\text{ mL} \times \frac{1\ \cancel{\text{atm}}}{2\ \cancel{\text{atm}}} = 500\text{ mL}$$

(b) Starting volume, 1000 mL; pressure change, 1 atm $\longrightarrow$ 4 atm

$$1000\text{ mL} \times \frac{1\ \cancel{\text{atm}}}{4\ \cancel{\text{atm}}} = 250\text{ mL}$$

(c) Starting volume, 500 mL; pressure change, 2 atm $\longrightarrow$ 4 atm

$$500\text{ mL} \times \frac{2\ \cancel{\text{atm}}}{4\ \cancel{\text{atm}}} = 250\text{ mL}$$

In summary, a change in the volume of a gas due to a change in pressure can be calculated by multiplying the original volume by a ratio of the two pressures. If the pressure is increased, the ratio should have the smaller pressure in the numerator and the larger pressure in the denominator. If the pressure is decreased, the larger pressure should be in the numerator and the smaller pressure in the denominator.

new volume = original volume × ratio of pressures

Examples of problems based on Boyle's law follow. If no mention is made of temperature, assume that it remains constant.

PROBLEM 12.2

What volume will 2.50 L of a gas occupy if the pressure is changed from 760 mm Hg to 630 mm Hg?

First we must determine whether the pressure is being increased or decreased. In this case it is being decreased. This decrease in pressure should result in an increase in volume. Therefore, we need to multiply 2.50 L by a ratio of the pressures that will give us an increase in volume. This ratio is 760 mm Hg/630 mm Hg. The calculation is

$$V = 2.50 \text{ L} \times \frac{760 \cancel{\text{mm Hg}}}{630 \cancel{\text{mm Hg}}} = 3.02 \text{ L} \quad \text{(new volume)}$$

Alternatively an algebraic approach may be used, solving $P_1V_1 = P_2V_2$ for V_2:

$$V_2 = V_1 \times \frac{P_1}{P_2} = 2.50 \text{ L} \times \frac{760 \cancel{\text{mm Hg}}}{630 \cancel{\text{mm Hg}}} = 3.02 \text{ L}$$

where $V_1 = 2.50$ L, $P_1 = 760$ mm Hg, and $P_2 = 630$ mm Hg.

PROBLEM 12.3

A given mass of hydrogen occupies 40.0 L at 700 torr pressure. What volume will it occupy at 5.00 atm pressure?

Since the units of the two pressures are not the same they must be made the same; otherwise, the units will not cancel in the final calculation. Because the pressure is increased, the volume should decrease. Therefore, we need to multiply 40.0 L by a ratio of the pressures that will give us a decrease in volume.

Step 1 Convert 700 torr to atmospheres by multiplying by the conversion factor 1 atm/760 torr:

$$700 \cancel{\text{torr}} \times \frac{1 \text{ atm}}{760 \cancel{\text{torr}}} = 0.921 \text{ atm}$$

Step 2 Multiply the volume (40.0 L) by a pressure ratio that will give a volume decrease.

$$V = 40.0 \text{ L} \times \frac{0.921 \cancel{\text{atm}}}{5.00 \cancel{\text{atm}}} = 7.37 \text{ L} \quad \text{(Answer)}$$

PROBLEM 12.4

A gas occupies a volume of 200 mL at 400 torr pressure. To what pressure must the gas be subjected in order to change the volume to 75.0 mL?

In order to reduce the volume from 200 mL to 75.0 mL, it will be necessary to increase the pressure. In the same way that we calculated volume change affected by a change in pressure, we must multiply the original pressure by a ratio of the two volumes. The volume ratio in this case is 200 mL/75.0 mL. The calculation is new pressure = original pressure × ratio of volumes

$$P = 400 \text{ torr} \times \frac{200 \text{ mL}}{75.0 \text{ mL}} = 1067 \text{ torr} \quad \text{or} \quad 1.07 \times 10^3 \text{ torr} \quad \text{(new pressure)}$$

Algebraically, $P_1V_1 = P_2V_2$ may be solved for P_2:

$$P_2 = P_1 \times \frac{V_1}{V_2} = 400 \text{ torr} \times \frac{200 \text{ mL}}{75.0 \text{ mL}} = 1.07 \times 10^3 \text{ torr}$$

where $P_1 = 400$ torr, $V_1 = 200$ mL, and $V_2 = 75.0$ mL.

In problems of this type it is good practice to check the answers to see if they are consistent with the given facts. For example, if the data indicate that the pressure is increased, the final volume should be smaller than the initial volume.

12.6 Charles' Law: The Effect of Temperature on the Volume of a Gas

The effect of temperature on the volume of a gas was observed in about 1787 by the French physicist J. A. C. Charles (1746–1823). Charles found that various gases expanded by the same fractional amount when heated through the same temperature interval. Later it was found that if a given volume of any gas initially at 0°C was cooled by 1°C, the volume decreased by $\frac{1}{273}$; if cooled by 2°C, by $\frac{2}{273}$; if cooled by 20°C, by $\frac{20}{273}$; and so on. Since each degree of cooling reduced the volume by $\frac{1}{273}$, it was apparent that any quantity of any gas would have zero volume, if it could be cooled to −273°C. Of course no real gas can be cooled to −273°C for the simple reason that it liquefies before that temperature is reached. However, −273°C (more precisely −273.16°C) is referred to as *absolute zero*; this temperature is the zero point on the Kelvin (absolute) temperature scale. It is the temperature at which the volume of an ideal, or perfect, gas would become zero.

The volume–temperature relationship for gases is shown graphically in Figure 12.8. Experimental data show the graph to be a straight line that, when extrapolated, crosses the temperature axis at −273.16°C, or absolute zero.

Charles' law

In modern form, **Charles' law** states that at constant pressure the volume of a fixed mass of any gas is directly proportional to the absolute temperature. Mathematically, Charles' law may be expressed as

$$V \propto T \quad (P \text{ is constant})$$

which means that the volume of a gas varies directly with the absolute temperature when the pressure remains constant. In equation form Charles' law may be written as

$$V = kT \quad \text{or} \quad \frac{V}{T} = k \quad \text{(at constant pressure)}$$

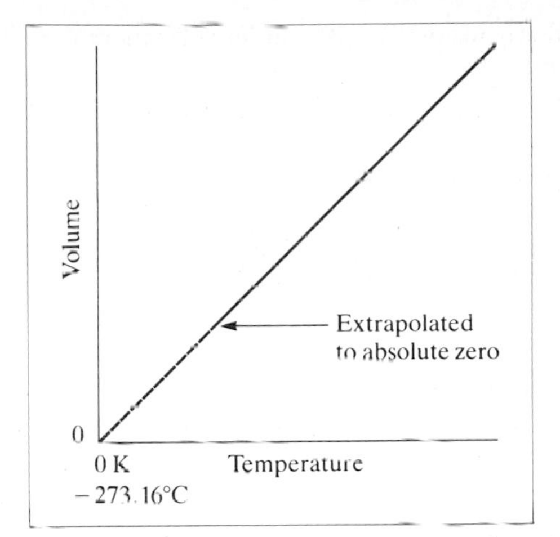

Figure 12.8

Volume–temperature relationship of gases. Extrapolated portion of the graph is shown by the broken line.

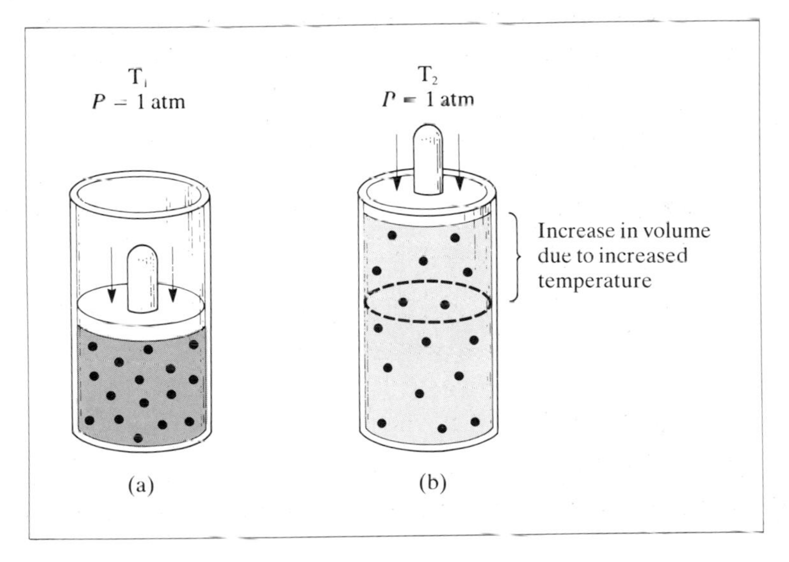

Figure 12.9

The effect of temperature on the volume of a gas. The gas in cylinder (a) is heated from T_1 to T_2. With the external pressure constant at 1 atm, the free-floating piston rises, resulting in an increased volume, as shown in cylinder (b).

where k is a constant for a fixed mass of the gas. If the absolute temperature of a gas is doubled, the volume will double. (A capital T is usually used for absolute temperature, K, and a small t for °C.)

To illustrate, let us return to the gas cylinder with the movable or free-floating piston, (see Figure 12.9). Assume that the cylinder labeled (a) contains a quantity of gas and the pressure on it is 1 atm. When the gas is heated, the molecules move faster, and their kinetic energy increases. This action should increase the number of collisions per unit of time and therefore increase the pressure. However, the increased internal pressure will cause the piston to rise to a level at which the internal and external pressures again equal 1 atm, as we see in cylinder (b). The net result is an increase in volume due to an increase in temperature.

Another equation relating the volume of a gas at two different temperatures is

$$\frac{V_1}{T_1} = \frac{V_2}{T_2} \quad \text{(constant } P\text{)}$$

where V_1 and T_1 are one set of conditions and V_2 and T_2 are another set of conditions.

A simple experiment showing the variation of the volume of a gas with temperature is illustrated in Figure 12.10. A flask to which a balloon is attached is immersed in either ice water or hot water. In ice water the volume is reduced, as shown by the collapse of the balloon; in hot water the gas expands and the balloon increases in size.

The calculation of changes in volume due to changes in temperature involves two basic steps: (1) changing the temperatures to K and (2) multiplying the original volume by a ratio of the initial and final temperatures. If the temperature is increased, the higher temperature is placed in the numerator of the ratio and the lower temperature in the denominator. If the temperature is decreased, the lower temperature is placed in the numerator of the ratio and the higher temperature in the denominator.

$$\text{new volume} = \text{original volume} \times \text{ratio of temperatures (K)}$$

Problems based on Charles' law follow.

PROBLEM 12.5 Three liters of hydrogen at $-20°C$ are allowed to warm to a room temperature of 27°C. What is the volume at room temperature if the pressure remains constant?

Step 1 Change °C to K:

$$°C + 273 = K$$
$$-20°C + 273 = 253 \text{ K}$$
$$27°C + 273 = 300 \text{ K}$$

Step 2 Since the temperature is increased, the volume should increase. The original volume should be multiplied by the temperature ratio of 300 K/253 K. The calculation is

$$V = 3.00 \text{ L} \times \frac{300 \not{K}}{253 \not{K}} = 3.56 \text{ L} \quad \text{(new volume)}$$

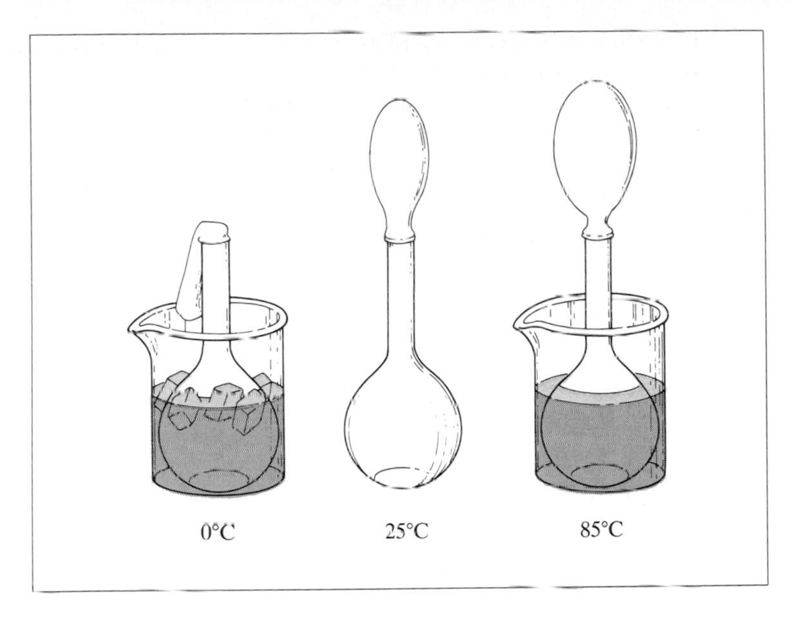

Figure 12.10
The effect of temperature on the volume of a gas. A volume decrease occurs when a flask to which a balloon is attached is immersed in ice water; the volume increases when the flask is immersed in hot water.

To obtain the answer by algebra, solve $V_1/T_1 = V_2/T_2$ for V_2:

$$V_2 = V_1 \times \frac{T_2}{T_1} = 3.00\text{ L} \times \frac{300\,\cancel{\text{K}}}{253\,\cancel{\text{K}}} = 3.56\text{ L}$$

where $V_1 = 3.00$ L, $T_1 = 253$ K, and $T_2 = 300$ K.

PROBLEM 12.6 If 20.0 L of nitrogen are cooled from 100°C to 0°C, what is the new volume?
Since no mention is made of pressure, assume that pressure does not change.

Step 1 Change °C to K:

$$100°\text{C} + 273 = 373\text{ K}$$
$$0°\text{C} + 273 = 273\text{ K}$$

Step 2 The ratio of temperature to be used is 273 K/373 K, because the final volume should be smaller than the original volume. The calculation is

$$V = 20.0\text{ L} \times \frac{273\,\cancel{\text{K}}}{373\,\cancel{\text{K}}} = 14.6\text{ L} \quad \text{(new volume)}$$

Three variables—pressure, P; volume, V; and temperature, T—are needed to describe a fixed amount of a gas. Boyle's law, $PV = k$, relates pressure and volume at constant temperature; Charles' law, $V = kT$, relates volume and temperature at constant pressure. A third relationship involving pressure and temperature at constant volume is stated: The pressure of a fixed mass of a gas, at constant volume, is directly proportional to the Kelvin temperature. In equation form the relationship is

$$P = kT \quad \text{(at constant volume)} \qquad \frac{P_1}{T_1} = \frac{P_2}{T_2}$$

This relationship is a modification of Charles' law and is sometimes called Gay-Lussac's law.

We may summarize the effects of changes in pressure, temperature, and quantity of a gas as follows:

1 In the case of a fixed or constant volume,
   (a) when the temperature is increased, the pressure increases.
   (b) when the quantity of a gas is increased, the pressure increases (T remaining constant).
2 In the case of a variable volume,
   (a) when the external pressure is increased, the volume decreases (T remaining constant).
   (b) when the temperature of a gas is increased, the volume increases (P remaining constant).
   (c) when the quantity of a gas is increased, the volume increases (P and T remaining constant).

12.7 Standard Temperature and Pressure

standard conditions

standard temperature and pressure (STP)

In order to compare volumes of gases, common reference points of temperature and pressure were selected and called **standard conditions** or **standard temperature and pressure** (abbreviated **STP**). Standard temperature is 273 K (0°C), and standard pressure is 1 atm or 760 torr or 760 mm Hg. For purposes of comparison volumes of gases are usually changed to STP conditions.

standard temperature = 273 K or 0°C
standard pressure = 1 atm or 760 torr or 760 mm Hg

12.8 Combined Gas Laws: Simultaneous Changes in Pressure, Volume, and Temperature

When temperature and pressure change at the same time, the new volume may be calculated by multiplying the initial volume by the correct ratios of both pressure and temperature, as follows:

$$\text{final volume} = \text{initial volume} \times \begin{pmatrix}\text{ratio of}\\ \text{pressures}\end{pmatrix} \times \begin{pmatrix}\text{ratio of}\\ \text{temperatures}\end{pmatrix}$$

This equation combines Boyle's and Charles' laws, and the same considerations for the pressure and the temperature ratios should be used in the calculation. The four possible variations are:

1. Both T and P cause an increase in volume.
2. Both T and P cause a decrease in volume.
3. T causes an increase and P causes a decrease in volume.
4. T causes a decrease and P causes an increase in volume.

The P, V, and T relationships for a given mass of any gas, in fact, may be expressed as a single equation, $PV/T = k$. For problem solving this equation is usually written

$$\frac{P_1V_1}{T_1} = \frac{P_2V_2}{T_2}$$

where P_1, V_1, and T_1 are the initial conditions and P_2, V_2, and T_2 are the final conditions.

This equation can be solved for any one of the six variables and is useful in dealing with the pressure–volume–temperature relationships of gases. Note that when T is constant ($T_1 = T_2$), Boyle's law is represented; when P is constant ($P_1 = P_2$), Charles' law is represented; and when V is constant ($V_1 = V_2$), the modified Charles' or Gay-Lussac's law is represented.

PROBLEM 12.7 Given 20.0 L of ammonia gas at 5°C and 730 torr pressure, calculate the volume at 50°C and 800 torr.

In order to get a better look at the data, tabulate the initial and final conditions:

	Initial	Final
V	20.0 L	V_2
T	5°C	50°C
P	730 torr	800 torr

Step 1 Change °C to K:

$$5°C + 273 = 278 \text{ K}$$
$$50°C + 273 = 323 \text{ K}$$

Step 2 Set up ratios of T and P:

$$T \text{ ratio} = \frac{323 \text{ K}}{278 \text{ K}} \quad \text{(increase in } T \text{ should increase } V\text{)}$$

$$P \text{ ratio} = \frac{730 \text{ torr}}{800 \text{ torr}} \quad \text{(increase in } P \text{ should decrease } V\text{)}$$

Step 3 The calculation is

$$V_2 = 20.0 \text{ L} \times \frac{730 \cancel{\text{torr}}}{800 \cancel{\text{torr}}} \times \frac{323 \cancel{\text{K}}}{278 \cancel{\text{K}}} = 21.2 \text{ L}$$

The algebraic solution is:

Solve $\dfrac{P_1 V_1}{T_1} = \dfrac{P_2 V_2}{T_2}$ for V_2 by multiplying both sides of the equation by T_2/P_2 and rearranging to obtain

$$V_2 = \frac{V_1 \times P_1 \times T_2}{P_2 \times T_1}$$

Tabulate the known values:

$V_1 = 20.0$ L $\qquad V_2 = ?$

$T_1 = 5°C + 273 = 278$ K $\qquad T_2 = 50°C + 273 = 323$ K

$P_1 = 730$ torr $\qquad P_2 = 800$ torr

Substitute these values in the equation and calculate the value of V_2:

$$V_2 = \frac{20.0 \text{ L} \times 730 \cancel{\text{torr}} \times 323 \cancel{\text{K}}}{800 \cancel{\text{torr}} \times 278 \cancel{\text{K}}} = 21.2 \text{ L}$$

PROBLEM 12.8 To what temperature (°C) must 10.0 L of nitrogen at 25°C and 700 torr be heated in order to have a volume of 15.0 L and a pressure of 760 torr?

This problem is conveniently handled by an algebraic solution.

Solve $\dfrac{P_1 V_1}{T_1} = \dfrac{P_2 V_2}{T_2}$ for T_2 to obtain $\quad T_2 = \dfrac{T_1 \times P_2 \times V_2}{P_1 \times V_1}$

Tabulate the known values:

$P_1 = 700$ torr $\qquad P_2 = 760$ torr

$V_1 = 10.0$ L $\qquad V_2 = 15.0$ L

$T_1 = 25°C + 273 = 298$ K $\qquad T_2 = ?$

Substitute these known values in the equation and calculate T_2:

$$T_2 = \frac{298 \text{ K} \times 760 \cancel{\text{torr}} \times 15.0 \cancel{\text{L}}}{700 \cancel{\text{torr}} \times 10.0 \cancel{\text{L}}} = 485 \text{ K}$$

Since the problem asks for °C, we must subtract 273 from the K answer:

485 K − 273 = 212°C (Answer)

PROBLEM 12.9 The volume of a gas-filled balloon is 50.0 L at 20°C and 742 torr. What volume will it occupy at standard temperature and pressure (STP)?

Tabulate the data.

	Initial	Final
V	50.0 L	V_2
T	20°C	273 K
P	742 torr	760 torr

Step 1 STP conditions are 273 K and 760 torr. First change °C to K:

20°C + 273 = 293 K

Step 2 Set up ratios of T and P:

T ratio $= \dfrac{273 \text{ K}}{293 \text{ K}}$ [decrease in T (293 K to 273 K) should decrease V]

P ratio $= \dfrac{742 \text{ torr}}{760 \text{ torr}}$ [increase in P (742 torr to 760 torr) should decrease V]

Step 3 The calculation is

$$V_2 = 50.0 \text{ L} \times \frac{273 \cancel{\text{K}}}{293 \cancel{\text{K}}} \times \frac{742 \cancel{\text{torr}}}{760 \cancel{\text{torr}}} = 45.5 \text{ L}$$

The problem can also be done by solving $\dfrac{P_1V_1}{T_1} = \dfrac{P_2V_2}{T_2}$ for V_2 and substituting PVT values in the new equation.

12.9 Dalton's Law of Partial Pressures

If gases behave according to the Kinetic-Molecular Theory, there should be no difference in the pressure–volume–temperature relationships whether the gas molecules are all the same or different. This similarity in the behavior of gases is

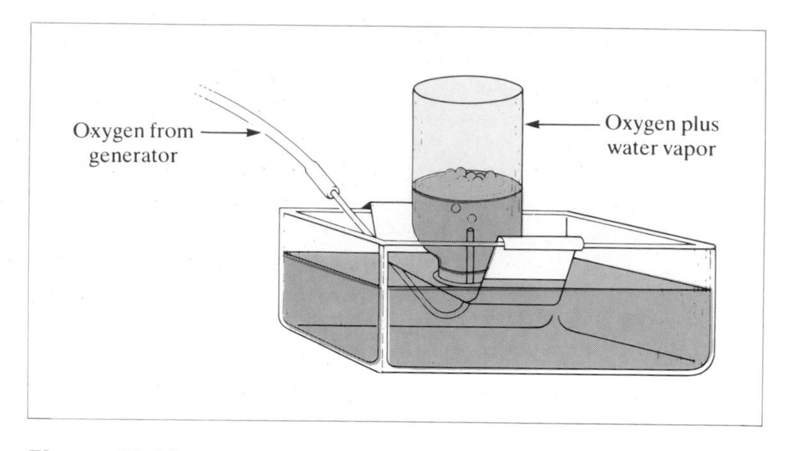

Figure 12.11
Oxygen collected over water

Dalton's Law of Partial Pressures

the basis for an understanding of **Dalton's Law of Partial Pressures**, which states that the total pressure of a mixture of gases is the sum of the partial pressures exerted by each of the gases in the mixture. Each gas in the mixture exerts a pressure that is independent of the other gases present. These pressures are called *partial pressures.* Thus, if we have a mixture of three gases, A, B, and C, exerting partial pressures of 50 torr, 150 torr, and 400 torr, respectively, the total pressure will be 600 torr.

$$P_{\text{Total}} = p_{\text{A}} + p_{\text{B}} + p_{\text{C}}$$
$$P_{\text{Total}} = 50 \text{ torr} + 150 \text{ torr} + 400 \text{ torr} = 600 \text{ torr}$$

We can see an application of Dalton's law in the collection of insoluble gases over water. When prepared in the laboratory, oxygen is commonly collected by the downward displacement of water. Thus the oxygen is not pure but is mixed with water vapor (see Figure 12.11). When the water levels are adjusted to the same height inside and outside the bottle, the pressure of the oxygen plus water vapor inside the bottle is equal to the atmospheric pressure:

$$P_{\text{atm}} = p_{O_2} + p_{H_2O}$$
$$p_{O_2} = P_{\text{atm}} - p_{H_2O}$$

To determine the amount of O_2 or any other gas collected over water, we must subtract the pressure of the water vapor from the total pressure of the gas. The vapor pressure of water at various temperatures is tabulated in Appendix II.

An illustrative problem follows.

PROBLEM 12.10 A 500 mL sample of oxygen, O_2, was collected over water at 23°C and 760 torr pressure. What volume will the dry O_2 occupy at 23°C and 760 torr? The vapor pressure of water at 23°C is 21.2 torr.

To solve this problem we must first find the pressure of the O_2 alone by subtracting the pressure of the water vapor present.

$$P_{\text{Total}} = 760 \text{ torr} = p_{O_2} + p_{H_2O}$$
$$p_{O_2} = 760 \text{ torr} - 21.2 \text{ torr} = 739 \text{ torr}$$

Thus the pressure of dry O_2 is 739 torr.

The problem is now of the Boyle's law type. It is treated as if we had 500 mL of dry O_2 at 739 torr pressure, which is then changed to 760 torr pressure with the temperature remaining constant. The calculation is

$$V = 500 \text{ mL} \times \frac{739 \text{ torr}}{760 \text{ torr}} = 486 \text{ mL dry } O_2$$

This means that 486 mL of the 500 mL mixture of O_2 and water vapor is pure O_2.

12.10 Avogadro's Law

Gay-Lussac's Law of Combining Volumes of Gases

Early in the 19th century J. L. Gay-Lussac (1778–1850) of France studied the volume relationships of reacting gases. His results, published in 1809, were summarized in a statement known as **Gay-Lussac's Law of Combining Volumes of Gases**: *When measured at the same temperature and pressure, the ratios of the volumes of reacting gases are small whole numbers.* Thus, H_2 and O_2 combine to form water vapor in a volume ratio of 2 to 1 (Figure 12.12); H_2 and Cl_2 react to form HCl in a volume ratio of 1 to 1; and H_2 and N_2 react to form NH_3 in a volume ratio of 3 to 1.

Avogadro's Law

Two years later, in 1811, Amedeo Avogadro used the Law of Combining Volumes of Gases to make a simple but significant and far-reaching generalization concerning gases. **Avogadro's law** states:

Equal volumes of different gases at the same temperature and pressure contain the same number of molecules.

This law was a real breakthrough in understanding the nature of gases. (1) It offered a rational explanation of Gay-Lussac's Law of Combining Volumes of Gases and indicated the diatomic nature of such elemental gases as hydrogen, chlorine, and oxygen; (2) it provided a method for determining the molecular weights of gases and for comparing the densities of gases of known molecular weight (see Sections 12.11 and 12.12); and (3) it afforded a firm foundation for the development of the Kinetic-Molecular Theory.

On a volume basis hydrogen and chlorine react thus:

hydrogen +	chlorine ⟶	hydrogen chloride
1 volume	1 volume	2 volumes

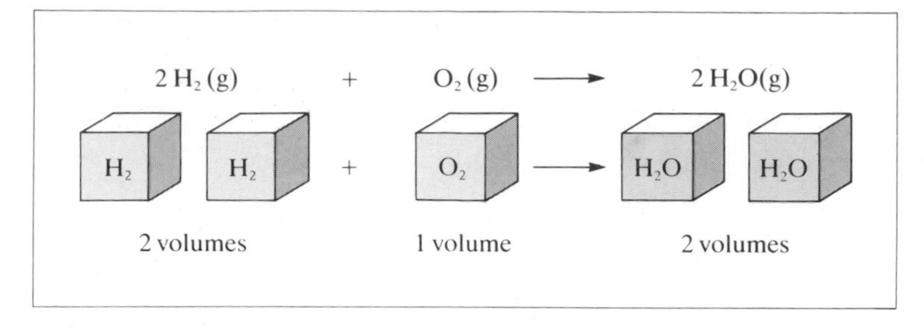

Figure 12.12

Gay-Lussac's Law of Combining Volumes of Gases applied to the reaction of hydrogen and oxygen. When measured at the same temperature and pressure, hydrogen and oxygen react in a volume ratio of 2 to 1.

By Avogadro's law, equal volumes of hydrogen and chlorine contain the same number of molecules. Therefore, hydrogen molecules react with chlorine molecules in a 1:1 ratio. Since two volumes of hydrogen chloride are produced, one molecule of hydrogen and one molecule of chlorine must produce two molecules of hydrogen chloride. Therefore, each hydrogen molecule and each chlorine molecule must be made up of two atoms. The coefficients of the balanced equation for the reaction give the correct ratios for volumes, molecules, and moles of reactants and products:

$$H_2 + Cl_2 \longrightarrow 2\,HCl$$

H_2	Cl_2	$2\,HCl$
1 volume	1 volume	2 volumes
1 molecule	1 molecule	2 molecules
1 mol	1 mol	2 mol

By like reasoning oxygen molecules also must contain at least two atoms because one volume of oxygen reacts with two volumes of hydrogen to produce two volumes of water vapor.

The volume of a gas depends on the temperature, the pressure, and the number of gas molecules. Different gases at the same temperature have the same average kinetic energy. Hence, if two different gases are at the same temperature, occupy equal volumes, and exhibit equal pressures, each gas must contain the same number of molecules. This statement is true because systems with identical *PVT* properties can be produced only by equal numbers of molecules having the same average kinetic energy.

12.11 Mole–Mass–Volume Relationship of Gases

Because a mole contains 6.022×10^{23} molecules (Avogadro's number), a mole of any gas will have the same volume as a mole of any other gas at the same temperature and pressure. It has been experimentally determined that the volume

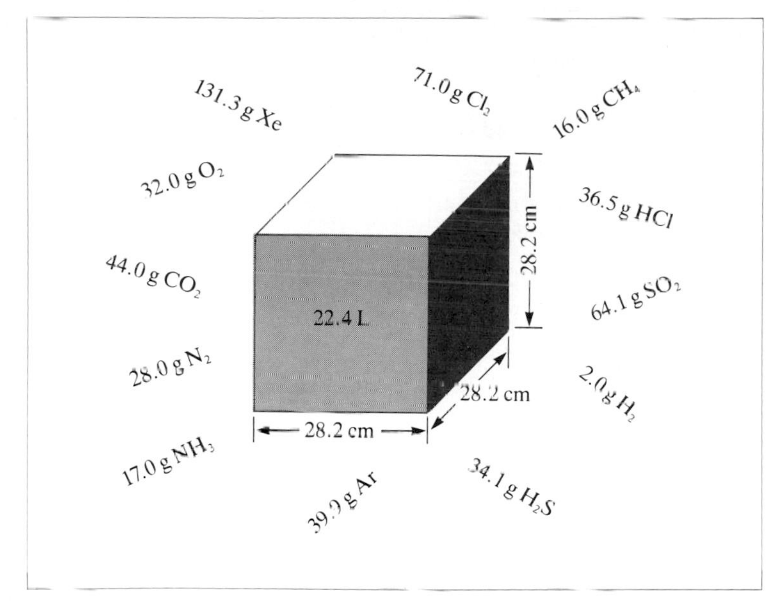

Figure 12.13
One mole of a gas occupies 22.4 liters at STP. The weight given for each gas is the weight of one mole.

molar volume of a gas

occupied by a mole of any gas is 22.4 liters at standard temperature and pressure. This volume, 22.4 liters, is known as the **molar volume of a gas**. The molar volume is a cube about 28.2 cm (11.1 in.) on a side (see Figure 12.13). The molecular weights of several gases, each occupying 22.4 L at STP, are also shown in Figure 12.13.

One mole of a gas occupies 22.4 liters at STP.

The molar volume is useful for determining the molecular weight of a gas or of substances that can be easily vaporized. If the mass and the volume of a gas at STP are known, we can calculate its molecular weight. For example, 1 liter of pure oxygen at STP weighs 1.429 g. The molecular weight of oxygen may be calculated by multiplying the weight of 1 liter by 22.4 L/mol.

$$\frac{1.429\text{ g}}{1\text{ L}} \times \frac{22.4\text{ L}}{1\text{ mol}} = 32.0\text{ g/mol} \quad \text{(mol. wt)}$$

If the mass and volume are at other than standard conditions, we change the volume to STP and then calculate the molecular weight. Note that we do not correct the mass to standard conditions—only the volume.

The molar volume, 22.4 L/mol, is used as a conversion factor to convert grams per liter to grams per mole (mol. wt) and also to convert liters to moles. The two conversion factors are

$$\frac{22.4\text{ L}}{1\text{ mol}} \quad \text{and} \quad \frac{1\text{ mol}}{22.4\text{ L}}$$

These conversions must be done at STP except under certain special circumstances. Examples follow.

PROBLEM 12.11 If 2.00 L of a gas measured at STP weigh 3.23 g, what is the molecular weight of the gas?

The unit of molecular weight is g/mol; the conversion is from

$$\frac{\text{g}}{\text{L}} \longrightarrow \frac{\text{g}}{\text{mol}}$$

The starting amount is $\frac{3.23\text{ g}}{2.00\text{ L}}$. The conversion factor is $\frac{22.4\text{ L}}{1\text{ mol}}$. The calculation is

$$\frac{3.23\text{ g}}{2.00\not{\text{L}}} \times \frac{22.4\not{\text{L}}}{1\text{ mol}} = 36.2\text{ g/mol} \quad \text{(mol. wt)}$$

PROBLEM 12.12 Measured at 40°C and 630 torr, 691 mL of ethyl ether weigh 1.65 g. Calculate the molecular weight of ethyl ether.

In order to use 22.4 L/mol we must first calculate the volume at STP:

$$630\text{ torr} \longrightarrow 760\text{ torr} \quad \text{(standard pressure)}$$

$$40°\text{C (313 K)} \longrightarrow 273\text{ K} \quad \text{(standard temperature)}$$

$$V = 691\text{ mL} \times \frac{273\not{\text{K}}}{313\not{\text{K}}} \times \frac{630\not{\text{torr}}}{760\not{\text{torr}}} = 500\text{ mL} = 0.500\text{ L} \quad \text{(at STP)}$$

The mass of the gas has not been altered by changing the volume to STP, so 0.500 L at STP weighs 1.65 g. The conversion is from g/L to g/mol.

$$\frac{1.65\text{ g}}{0.500\not{\text{L}}} \times \frac{22.4\not{\text{L}}}{\text{mol}} = 73.9\text{ g/mol} \quad \text{(mol. wt)}$$

12.12 Density and Specific Gravity of Gases

The density, d, of a gas is its mass per unit volume, which is generally expressed in grams per liter (g/L) as follows:

$$d = \frac{\text{mass}}{\text{volume}} = \frac{\text{g}}{\text{L}}$$

Because the volume of a gas depends on temperature and pressure, both should be given when stating the density of a gas. The volume of a solid or liquid is hardly affected by changes in pressure and is changed only slightly when the temperature is varied. Increasing the temperature from 0°C to 50°C will reduce the density of a gas by about 18% if the gas is allowed to expand, whereas a 50°C rise in the temperature of water (0°C ⟶ 50°C) will change its density by less than 0.2%.

The density of a gas at any temperature and pressure can be determined by calculating the weight of gas present in 1 liter. At STP, in particular, the density can be calculated by multiplying the molecular weight of the gas by 1 mol/22.4 L.

$$d\text{ (at STP)} = \text{mol. wt} \times \frac{1\text{ mol}}{22.4\text{ L}} \qquad \text{mol. wt} = d\text{ (at STP)} \times \frac{22.4\text{ L}}{1\text{ mol}}$$

PROBLEM 12.13 Calculate the density of Cl_2 at STP.

First calculate the molecular weight of Cl_2. It is 71.0 g/mol. Since $d = \text{g/L}$, the conversion is

$$\frac{\text{g}}{\text{mol}} \longrightarrow \frac{\text{g}}{\text{L}}$$

The conversion factor is $\dfrac{1\text{ mol}}{22.4\text{ L}}$.

$$d = \frac{71.0\text{ g}}{1\text{ mol}} \times \frac{1\text{ mol}}{22.4\text{ L}} = 3.17\text{ g/L}$$

The specific gravity (sp gr) of a gas is the ratio of the mass of any volume of the gas to the mass of an equal volume of some reference gas. Specific gravities of gases are commonly quoted in reference to air. The mass of air at STP is 1.29 g/L, which is the density of air.

The specific gravity of a gas can be calculated by dividing its density by the density of air. Both gases must be at the same temperature and pressure.

$$\text{sp gr} = \frac{\text{density of a gas}}{\text{density of air}}$$

The specific gravity of Cl_2, for example, is

$$\text{sp gr of } Cl_2 = \frac{\text{density of } Cl_2}{\text{density of air}} = \frac{3.17\text{ g/L}}{1.29\text{ g/L}} = 2.46$$

This indicates that Cl_2 is 2.46 times as heavy as air. Table 12.3 lists the densities and specific gravities of some common gases.

Table 12.3 Density and specific gravity of common gases at STP

Gas	Molecular weight	Density (g/L at STP)	Specific gravity (unitless)
H_2	2.0	0.090	0.070
CH_4	16.0	0.714	0.553
NH_3	17.0	0.760	0.589
C_2H_2	26.0	1.16	0.899
HCN	27.0	1.21	0.938
CO	28.0	1.25	0.969
N_2	28.0	1.25	0.969
O_2	32.0	1.43	1.11
H_2S	34.1	1.52	1.18
HCl	36.5	1.63	1.26
F_2	38.0	1.70	1.32
CO_2	44.0	1.96	1.52
C_3H_8	44.0	1.96	1.52
O_3	48.0	2.14	1.66
SO_2	64.1	2.86	2.22
Cl_2	71.0	3.17	2.46

12.13 Ideal Gas Equation

We have used four variables in calculations involving gases: the volume, V; the pressure, P; the absolute temperature, T; and the number of molecules or moles, which is abbreviated n. Combining these variables into a single expression, we obtain

$$V \propto \frac{nT}{P} \quad \text{or} \quad V = \frac{nRT}{P}$$

where R is a proportionality constant known as the *ideal gas constant*. The equation is commonly written as

$$PV = nRT$$

ideal gas equation

and is known as the **ideal gas equation**. This equation states in a single expression what we have considered in our earlier discussions: The volume of a gas varies directly with the number of gas molecules and the absolute temperature and varies inversely with the pressure. The value and units of R depend on the units of P, V, and T. We can calculate one value of R by taking 1 mol of a gas at STP conditions. Solve the equation for R:

$$R = \frac{PV}{nT} = \frac{1 \text{ atm} \times 22.4 \text{ L}}{1 \text{ mol} \times 273 \text{ K}} = 0.0821 \frac{\text{L-atm}}{\text{mol-K}}$$

The units of R in this case are liter-atmospheres (L-atm) per mol-K.

The ideal gas equation can be used to calculate any one of the four variables if the other three are known.

When the value of $R = 0.0821$ L-atm/mol-K, P is in atmospheres, n is in moles, V is in liters, and T is in Kelvin.

PROBLEM 12.14 What pressure will be exerted by 0.400 mol of a gas in a 5.00 L container at 17.0°C?

First solve the ideal gas equation for P:

$$PV = nRT \qquad \text{or} \qquad P = \frac{nRT}{V}$$

Then substitute the data in the problem into the equation and solve (change °C to K):

$$P = \frac{0.400 \text{ mol} \times 0.0821 \text{ L-atm} \times 290 \text{ K}}{5.00 \text{ L} \times \text{mol-K}} = 1.90 \text{ atm} \quad \text{(Answer)}$$

PROBLEM 12.15 How many moles of oxygen gas are in a 50.0 L tank at 22.0°C if the pressure gauge reads 2000 lb/in.^{2}?

First change the pressure to atmospheres and °C to K. Then solve the ideal gas equation for n (moles), and substitute the data into the equation to complete the calculation.

Step 1 Change pressure to atmospheres and °C to K:

$$\frac{2000 \text{ lb}}{\text{in.}^2} \times \frac{1 \text{ atm}}{14.7 \text{ lb/in.}^2} = 136 \text{ atm}$$

$$22.0^\circ\text{C} + 273 = 295 \text{ K}$$

Step 2 Solve for moles using the ideal gas equation:

$$PV = nRT \qquad \text{or} \qquad n = \frac{PV}{RT}$$

$$n = \frac{136 \text{ atm} \times 50.0 \text{ L}}{(0.0821 \text{ L-atm/mol-K}) \times 295 \text{ K}} = 281 \text{ mol } O_2 \quad \text{(Answer)}$$

The molecular weight of a gaseous substance can be determined using the ideal gas equation. Since mol. wt = g/mol, then mol = g/mol. wt. Using M for molecular weight we can substitute g/M for n (moles) in the ideal gas equation to get

$$PV = \frac{g}{M}RT \qquad \text{or} \qquad M = \frac{gRT}{PV}$$

which will allow us to calculate the molecular weight, M, for any substance in the gaseous state.

PROBLEM 12.16 Calculate the molecular weight of butane gas, if 3.69 g occupy 1.53 L at 20°C and 1 atm pressure.

Change 20°C to 293 K and substitute the data into the equation:

$$M = \frac{gRT}{PV} = \frac{3.69 \text{ g} \times 0.0821 \cancel{\text{L-atm}} \times 293 \cancel{\text{K}}}{1 \text{ atm} \times 1.53 \cancel{\text{L}} \times \text{mol-}\cancel{\text{K}}} = 58.0 \text{ g/mol} \quad \text{(Answer)}$$

12.14 Calculations from Chemical Equations Involving Gases

1. Mole–Volume (Gas) and Mass–Volume (Gas) Calculations

Stoichiometric problems involving gas volumes can be solved by the general mole-ratio method outlined in Chapter 11. The factors 1 mol/22.4 L and 22.4 L/1 mol are used for converting volume to moles and moles to volume, respectively. These conversion factors are used under the assumption that the gases are at STP and behave as ideal gases. In practice, gases are measured at other than STP conditions, and the volumes are converted to STP for stoichiometric calculations.

In a balanced equation, the number preceding the formula of a gaseous substance represents the number of moles or molar volumes (22.4 L at STP) of that substance.

The following are examples of typical problems involving gases and chemical equations.

PROBLEM 12.17 What volume of oxygen (at STP) can be formed from 0.500 mol of potassium chlorate?

Step 1 Write the balanced equation:

$$\underset{2 \text{ mol}}{2\,KClO_3} \longrightarrow 2\,KCl + \underset{3 \text{ mol}}{3\,O_2\uparrow}$$

Step 2 The starting amount is 0.500 mol $KClO_3$. The conversion is from

$$\text{moles } KClO_3 \longrightarrow \text{moles } O_2 \longrightarrow \text{liters } O_2$$

Step 3 Calculate the moles of O_2, using the mole-ratio method:

$$0.500 \cancel{\text{mol } KClO_3} \times \frac{3 \text{ mol } O_2}{2 \cancel{\text{mol } KClO_3}} = 0.750 \text{ mol } O_2$$

Step 4 Convert moles of O_2 to liters of O_2. The moles of a gas at STP are converted to liters by multiplying by the molar volume, 22.4 L/mole:

$$0.750 \cancel{\text{mol}}\, O_2 \times \frac{22.4 \text{ L}}{1 \cancel{\text{mol}}} = 16.8 \text{ L } O_2 \quad \text{(Answer)}$$

Setting up a continuous calculation, we obtain

$$0.500\ \text{mol KClO}_3 \times \frac{3\ \text{mol O}_2}{2\ \text{mol KClO}_3} \times \frac{22.4\ \text{L}}{1\ \text{mol}} = 16.8\ \text{L O}_2$$

PROBLEM 12.18 How many grams of aluminum must react with sulfuric acid to produce 1.25 L of hydrogen gas at STP?

Step 1 The balanced equation is

$$\underset{2\ \text{mol}}{2\ \text{Al}(s)} + 3\ \text{H}_2\text{SO}_4(aq) \longrightarrow \text{Al}_2(\text{SO}_4)_3(aq) + \underset{3\ \text{mol}}{3\ \text{H}_2(g)}$$

Step 2 We first convert liters of H_2 to moles of H_2. Then the familiar stoichiometric calculation from the equation is used. The conversion is:

$$\text{L H}_2 \longrightarrow \text{mol H}_2 \longrightarrow \text{mol Al} \longrightarrow \text{g Al}$$

$$1.25\ \text{L H}_2 \times \frac{1\ \text{mol}}{22.4\ \text{L}} \times \frac{2\ \text{mol Al}}{3\ \text{mol H}_2} \times \frac{27.0\ \text{g Al}}{1\ \text{mol Al}} = 1.00\ \text{g Al} \quad \text{(Answer)}$$

PROBLEM 12.19 What volume of hydrogen, collected at 30°C and 700 torr pressure, will be formed by reacting 50.0 g of aluminum with hydrochloric acid?

$$\underset{2\ \text{mol}}{2\ \text{Al}(s)} + 6\ \text{HCl}(aq) \longrightarrow 2\ \text{AlCl}_3(aq) + \underset{3\ \text{mol}}{3\ \text{H}_2(g)}$$

In this problem the conditions are not at STP, so we cannot use the method shown in Problem 12.17. Either we need to calculate the volume at STP from the equation and then convert this volume to the conditions given in the problem, or we can use the ideal gas equation. Let's use the ideal gas equation.

First calculate the moles of H_2 obtained from 50.0 g of Al. Then, using the ideal gas equation, calculate the volume of H_2 at the conditions given in the problem.

Step 1 Moles of H_2: The conversion is

$$\text{grams Al} \longrightarrow \text{moles Al} \longrightarrow \text{moles H}_2$$

$$50.0\ \text{g Al} \times \frac{1\ \text{mol Al}}{27.0\ \text{g Al}} \times \frac{3\ \text{mol H}_2}{2\ \text{mol Al}} = 2.78\ \text{mol H}_2$$

Step 2 Liters of H_2: Solve $PV = nRT$ for V and substitute the data into the equation.
Convert °C to K: 30°C + 273 = 303 K.
Convert torr to atm: 700 torr × 1 atm/760 torr = 0.921 atm.

$$V = \frac{nRT}{P} = \frac{2.78\ \text{mol H}_2 \times 0.0821\ \text{L-atm} \times 303\ \text{K}}{0.921\ \text{atm} \times \text{mol-K}} = 75.1\ \text{L H}_2 \quad \text{(Answer)}$$

Note: The volume at STP is 62.3 L H_2.

2. Volume–Volume Calculations When all substances in a reaction are in the gaseous state, simplifications in the calculation can be made that are based on Avogadro's law that gases under identical conditions of temperature and pressure contain the same number of molecules and occupy the same volume. Using this same law, we can also state that, under the same conditions of temperature and pressure, the volumes of gases reacting are proportional to the numbers of moles of the gases in the balanced equation. Consider the reaction

$H_2(g)$ +	$Cl_2(g)$ $\longrightarrow$	$2\ HCl(g)$
1 mol	1 mol	2 mol
22.4 L	22.4 L	2 × 22.4 L
1 volume	1 volume	2 volumes
Y volume	Y volume	2 Y volumes

In this reaction 22.4 L of hydrogen will react with 22.4 L of chlorine to give 2 × 22.4 = 44.8 L of hydrogen chloride gas. This statement is true because these volumes are equivalent to the number of reacting moles in the equation. Therefore, Y volume of H_2 will combine with Y volume of Cl_2 to give 2 Y volumes of HCl. For example, 100 L of H_2 react with 100 L of Cl_2 to give 200 L of HCl; if the 100 L of H_2 and of Cl_2 are at 50°C, they will give 200 L of HCl at 50°C. When the temperature and pressure before and after a reaction are the same, volumes can be calculated without changing the volumes to STP.

For reacting gases: Volume–volume relationships are the same as mole–mole relationships.

PROBLEM 12.20 What volume of oxygen will react with 150 L of hydrogen to form water vapor? What volume of water vapor will be formed?

Assume that both reactants and products are measured at the same conditions. Calculation by reacting volumes:

$2\ H_2(g)$ +	$O_2(g)$ $\longrightarrow$	$2\ H_2O(g)$
2 mol	1 mol	2 mol
2 × 22.4 L	22.4 L	2 × 22.4 L
2 volumes	1 volume	2 volumes
150 L	75 L	150 L

Thus, 150 L of H_2 will react with 75 L of O_2 to produce 150 L of $H_2O(g)$. For every two volumes of H_2 that react, one volume of O_2 and two volumes of $H_2O(g)$ are produced.

PROBLEM 12.21 The equation for the preparation of ammonia is

$$3\ H_2 + N_2 \xrightarrow{400^\circ C} 2\ NH_3$$

Assuming that the reaction goes to completion,

(a) What volume of H_2 will react with 50 L of N_2?
(b) What volume of NH_3 will be formed from 50 L of N_2?
(c) What volume of N_2 will react with 100 mL of H_2?
(d) What volume of NH_3 will be produced from 100 mL of H_2?
(e) If 600 mL of H_2 and 400 mL of N_2 are sealed in a flask and allowed to react, what amounts of H_2, N_2, and NH_3 are in the flask at the end of the reaction?

The answers to parts (a)–(d) are shown in the boxes and can be determined from the equation by inspection, using the principle of reacting volumes.

$$\underset{\text{3 volumes}}{3\,H_2} + \underset{\text{1 volume}}{N_2} \longrightarrow \underset{\text{2 volumes}}{2\,NH_3}$$

	$3\,H_2$	N_2	$2\,NH_3$
(a)	$\boxed{150\text{ L}}$	50 L	
(b)		50 L	$\boxed{100\text{ L}}$
(c)	100 mL	$\boxed{33.3\text{ mL}}$	
(d)	100 mL		$\boxed{66.7\text{ mL}}$

(e) Volume ratio from the equation $= \dfrac{3\text{ volumes } H_2}{1\text{ volume } N_2}$

$$\text{Volume ratio used} = \frac{600\text{ mL } H_2}{400\text{ mL } N_2} = \frac{3\text{ volumes } H_2}{2\text{ volumes } N_2}$$

Comparing these two ratios, we see that an excess of N_2 is present in the gas mixture. Therefore, the reactant limiting the amount of NH_3 that can be formed is H_2:

$$\underset{\text{600 mL}}{3\,H_2} + \underset{\text{200 mL}}{N_2} \longrightarrow \underset{\text{400 mL}}{2\,NH_3}$$

In order to have a 3:1 ratio of volumes reacting, 600 mL of H_2 will react with 200 mL of N_2 to produce 400 mL of NH_3, leaving 200 mL of N_2 unreacted. At the end of the reaction the flask will contain 400 mL of NH_3 and 200 mL of N_2.

All the gas laws are based on the behavior of an ideal gas—that is, a gas with a behavior that is described exactly by the gas laws for all possible values of *P*, *V*, and *T*. Most real gases actually do behave very nearly as predicted by the gas laws over a fairly wide range of temperatures and pressures. However, when conditions are such that the gas molecules are crowded closely together (high pressure and/or low temperature), they show marked deviations from ideal behavior. Deviations occur because molecules have finite volumes and also have intermolecular attractions, which results in less compressibility at high pressures and greater compressibility at low temperatures than predicted by the gas laws. Many gases become liquids at high pressure and low temperature.

QUESTIONS

A. *Review the meanings of the new terms introduced in this chapter.*

1. Kinetic-Molecular Theory (KMT)
2. Ideal, or perfect, gas
3. Diffusion
4. Effusion
5. Graham's law of effusion
6. Pressure
7. Atmospheric pressure
8. Barometer
9. 1 atmosphere
10. Boyle's law
11. Charles' law
12. Standard conditions
13. Standard temperature and pressure (STP)
14. Dalton's Law of Partial Pressures
15. Gay-Lussac's Law of Combining Volumes of Gases
16. Avogadro's law
17. Molar volume of a gas
18. Ideal gas equation

B. *Information useful in answering the following questions will be found in the tables and figures.*

1. What evidence is used to show diffusion in Figure 12.1? If hydrogen, H_2, and oxygen, O_2, were in the two flasks, how could we prove that diffusion had taken place?
2. How does the air pressure inside the balloon shown in Figure 12.2 compare with the air pressure outside the balloon? Explain.
3. According to Table 12.1, what two gases are the major constituents of dry air?
4. How does the pressure represented by 1 torr compare in magnitude to the pressure represented by 1 mm Hg? See Table 12.2.
5. In which container illustrated in Figure 12.5 are the molecules of gas moving faster? Assume both gases to be hydrogen.
6. In Figure 12.6, what gas pressure corresponds to a volume of 4 L?
7. How do the data illustrated in Figure 12.6 substantiate Boyle's law?
8. What effect would you observe in Figure 12.9 if T_2 were lower than T_1?
9. In the diagram shown in Figure 12.11, is the pressure of the oxygen plus water vapor inside the bottle equal to, greater than, or less than the atmospheric pressure outside the bottle? Explain.
10. List five gases in Table 12.3 that are denser than air. Explain the basis for your selections.

C. *Review questions*

1. What are the basic assumptions of the Kinetic-Molecular Theory?
2. Arrange the following gases, all at standard temperature, in order of increasing relative molecular velocities: H_2, CH_4, Rn, N_2, F_2, He. What is your basis for determining the order?
3. List, in descending order, the average kinetic energies of the molecules in Question 2.
4. What are the four parameters used to describe the behavior of a gas?
5. What are the characteristics of an ideal gas?
6. Under what condition of temperature, high or low, is a gas least likely to exhibit ideal behavior? Explain.
7. Under what conditions of pressure, high or low, is a gas least likely to exhibit ideal behavior? Explain.
8. Compare, at the same temperature and pressure, equal volumes of H_2 and O_2 as to:
   (a) Number of molecules
   (b) Mass
   (c) Number of moles
   (d) Average kinetic energy of the molecules
   (e) Rate of effusion
   (f) Density
9. How does the Kinetic-Molecular Theory account for the behavior of gases as described by:
   (a) Boyle's law
   (b) Charles' law
   (c) Dalton's Law of Partial Pressures
10. Explain how the reaction $N_2(g) + O_2(g) \xrightarrow{\Delta} 2\,NO(g)$ proves that nitrogen and oxygen are diatomic molecules.
11. What is the reason for referring gases to STP?
12. When constant pressure is maintained, what effect does heating a mole of N_2 gas have on the following?

(a) Its density
(b) Its specific gravity
(c) Its mass
(d) The average kinetic energy of its molecules
(e) The average velocity of its molecules
(f) The number of N_2 molecules in the sample

13. Assuming ideal gas behavior, which of the following statements are correct. (Try to answer without referring to your text.)

(a) The pressure exerted by a gas at constant volume is independent of the temperature of the gas.
(b) At constant temperature, increasing the pressure exerted on a gas sample will cause a decrease in the volume of the gas sample.
(c) At constant pressure, the volume of a gas is inversely proportional to the absolute temperature.
(d) At constant temperature, doubling the pressure on a gas sample will cause the volume of the gas sample to decrease to one-half its original volume.
(e) Compressing a gas at constant temperature will cause its density and mass to increase.
(f) Equal volumes of CO_2 and CH_4 gases at the same temperature and pressure contain
(1) The same number of molecules
(2) The same mass
(3) The same densities
(4) The same number of moles
(5) The same number of atoms
(g) At constant temperature, the average kinetic energy of O_2 molecules at 200 atm pressure is greater than the average kinetic energy of O_2 molecules at 100 atm pressure.
(h) According to Charles' law, the volume of a gas becomes zero at $-273°C$.
(i) One liter of O_2 gas at STP has the same mass as 1 L of O_2 gas at 273°C and 2 atm pressure.
(j) The volume occupied by a gas depends only on its temperature and pressure.
(k) In a mixture containing O_2 molecules and N_2 molecules, the O_2 molecules, on the average, are moving faster than the N_2 molecules.
(l) $PV = k$ is a statement of Charles' law.
(m) If the temperature of a sample of gas is increased from 25°C to 50°C, the volume of the gas will increase by 100%.
(n) One mole of chlorine, Cl_2, at 20°C and 600 torr pressure contains 6.022×10^{23} molecules
(o) One mole of H_2 plus 1 mole of O_2 in an 11.2 L container exert a pressure of 4 atm at 0°C.
(p) When the pressure on a sample of gas is halved, with the temperature remaining constant, the density of the gas is also halved.
(q) When the temperature of a sample of gas is increased at constant pressure, the density of the gas will decrease
(r) According to the equation

$2\,KClO_3(s) \xrightarrow{\Delta} 2\,KCl(s) + 3\,O_2(g)$, 1 mol of $KClO_3$ will produce 67.2 L of O_2 at STP.

(s) $PV = nRT$ is a statement of Avogadro's law.
(t) STP conditions are 1 atm and 0°C

D. *Review problems*

Pressure Units

1. The barometer reads 715 mm Hg. Calculate the corresponding pressure in
(a) atmospheres (d) torrs
(b) inches of Hg (e) millibars
(c) lb/in.^{2} (f) kilopascals

2. Express the following pressures in atmospheres:
(a) 28 mm Hg (c) 795 torr
(b) 6000 cm Hg (d) 5.00 kPa

Boyle's, Charles', and Gay-Lussac's Laws

3. A gas occupies a volume of 400 mL at 500 mm Hg pressure. What will be its volume, at constant temperature, if the pressure is changed to (a) 760 mm Hg; (b) 250 torr; (c) 2.00 atm?

4. A 500 mL sample of a gas is at a pressure of 640 mm Hg. What must be the pressure, at constant temperature, if the volume is changed to (a) 855 mL and (b) 450 mL?

5. At constant temperature, what pressure would be required to compress 2500 L of hydrogen gas at 1.0 atm pressure into a 25 L tank?

6. Given 6.00 L of N_2 gas at $-25°C$, what volume will the nitrogen occupy at (a) 0.0°C; (b) 0.0°F; (c) 100 K; (d) 345 K? (Assume constant pressure.)

7. Given a sample of a gas at 27°C, at what temperature would the volume of the gas sample be doubled, the pressure remaining constant?

*8. A gas sample at at 22°C and 740 torr pressure is heated until its volume is doubled. What pressure would restore the sample to its original volume?

9. A gas occupies 250 mL at 700 torr and 22°C. When the pressure is changed to 500 torr, what temperature (°C) is needed to maintain the same volume?

10. Hydrogen stored in a metal cylinder has a pressure of 252 atm at 25°C. What will be the pressure in the cylinder when the cylinder is lowered into liquid nitrogen at −196°C?

11. The tires on an automobile were filled with air to 30 psi at 71.0°F. When driving at high speeds, the tires become hot. If the tires have a bursting pressure of 44 psi, at what temperature (°F) will the tires "blow out"?

Combined Gas Laws

12. A gas occupies a volume of 410 mL at 27°C and 740 mm Hg pressure. Calculate the volume the gas would occupy at (a) STP; (b) 250°C and 680 mm Hg pressure.

13. What volume would 5.30 L of H_2 gas at STP occupy at 70°C and 830 torr pressure?

14. What pressure will 800 mL of a gas at STP exert when its volume is 250 mL at 30°C?

15. An expandable balloon contains 1400 L of He at 0.950 atm pressure and 18°C. At an altitude of 22 miles (temperature 2.0°C and pressure 4.0 torr), what will be the volume of the balloon?

16. A gas occupies 22.4 L at 2.50 atm and 27°C. What will be its volume at 1.50 atm and −5.00°C?

17. How many gas molecules are present in 600 mL of N_2O at 40°C and 400 torr pressure? How many atoms are present? What would be the volume of the sample at STP?

Dalton's Law of Partial Pressures

18. What would be the partial pressure of O_2 gas collected over water at 20°C and 720 torr pressure? (Check Appendix II for the vapor pressure of water.)

19. An equilibrium mixture contains H_2 at 600 torr pressure, N_2 at 200 torr pressure, and O_2 at 300 torr pressure. What is the total pressure of the gases in the system?

20. A sample of methane gas, CH_4, was collected over water at 25.0°C and 720 torr. The volume of the wet gas is 2.50 L. What will be the volume of the methane after the gas is dried?

21. 5.00 L of CO_2 at 500 torr and 3.00 L of CH_4 at 400 torr are put into a 10.0 L container. What is the pressure exerted by the gases in the container?

Mole–Mass–Volume Relationships

22. What volume will 2.5 mol of Cl_2 occupy at STP?

23. A steel cylinder contains 60 mol of H_2 at a pressure of 1500 lb/in.^{2}. (a) How many moles of H_2 are in the cylinder when the pressure reads 850 lb/in.^{2}? (b) How many grams of H_2 were initially in the cylinder?

24. How many grams of CO_2 are present in 2500 mL of CO_2 at STP?

25. At STP, 560 mL of a gas have a mass of 1.08 g. What is the molecular weight of the gas?

26. What volume will each of the following occupy at STP?
    (a) 1.0 mol of NO_2
    (b) 17.05 g of NO_2
    (c) 1.20×10^{24} molecules of NO_2

27. How many molecules of NH_3 gas are present in a 1.00 L flask at STP?

28. How many moles of Cl_2 are in one cubic meter (1.00 m^3) at STP?

Density of Gases

29. A gas has a density at STP of 1.78 g/L. What is its molecular weight?

30. Calculate the density of the following gases at STP:
    (a) Kr (b) He (c) SO_3 (d) C_4H_8

31. Calculate the density of:
    (a) F_2 gas at STP
    (b) F_2 gas at 27°C and 1 atm pressure

*32. At what temperature (°C) will the density of methane, CH_4, be 1.0 g/L at 1.0 atm pressure?

Ideal Gas Equation and Stoichiometry

33. Using the ideal gas equation, $PV = nRT$, calculate:
    (a) The volume of 0.510 mole of H_2 at 47°C and 1.6 atm pressure

(b) The number of grams in 16.0 L of CH_4 at 27°C and 600 torr pressure.

*(c) The density of CO_2 at 4.00 atm pressure and −20°C

*(d) The molecular weight of a gas having a density of 2.58 g/L at 27°C and 1 atm pressure.

Hints for (c) and (d): n = moles = g/mol. wt, and $d = g/V$.

34. What is the molecular weight of a gas if 1.15 g occupy 0.215 L at 0.813 atm and 30°C?

35. At 27°C and 750 torr pressure, what will be the volume of 2.3 mol of Ne?

36. What volume will a mixture of 5.00 mol of H_2 and 0.500 mol of CO_2 occupy at STP?

37. 4.50 mol of a gas occupy 0.250 L at 4.15 atm. What is the Kelvin temperature of the system?

38. How many moles of N_2 gas occupy 5.20 L at 250 K and 0.500 atm?

39. What volume of hydrogen at STP can be produced by reacting 8.30 mol of Al with sulfuric acid? The equation is

$$2\,Al(s) + 3\,H_2SO_4(aq) \longrightarrow Al_2(SO_4)_3(aq) + 3\,H_2(g)$$

40. Given the equation

$$4\,NH_3(g) + 5\,O_2(g) \longrightarrow 4\,NO(g) + 6\,H_2O(g)$$

(a) How many moles of NH_3 are required to produce 5.5 mol of NO?

(b) How many moles of NH_3 will react with 7.0 mol of O_2?

(c) How many liters of NO can be made from 12 L of O_2 and 10 L of NH_3 at STP?

(d) At constant temperature and pressure how many liters of NO can be made by the reaction of 800 mL of O_2?

(e) At constant temperature and pressure, what is the maximum volume, in liters, of NO that can be made from 3.0 L of NH_3 and 3.0 L of O_2?

(f) How many grams of O_2 must react to produce 60 L of NO measured at STP?

*(g) How many grams of NH_3 must react to produce a total of 32 L of products, NO plus H_2O, measured at STP?

41. Given the equation

$$4\,FeS_2(s) + 11\,O_2(g) \xrightarrow{\Delta} 2\,Fe_2O_3(s) + 8\,SO_2(g)$$

(a) How many liters of O_2, measured at STP, will react with 0.600 kg of FeS_2?

(b) How many liters of SO_2, measured at STP, will be produced from 0.600 kg of FeS_2?

***42.** Acetylene, C_2H_2, and hydrogen fluoride react to give difluoroethane.

$$C_2H_2(g) + 2\,HF(g) \longrightarrow C_2H_4F_2(g)$$

When 1.0 mol of C_2H_2 and 5.0 mol of HF are reacted in a 10.0 L flask, what will be the pressure in the flask at 0°C when the reaction is complete?

Additional Problems

43. What are the relative rates of effusion of N_2 and He?

***44.** (a) What are the relative rates of effusion of methane, CH_4, and helium, He?

(b) If these two gases are simultaneously introduced into opposite ends of a 100 cm tube and allowed to diffuse toward each other, at what distance from the helium end will molecules of the two gases meet?

***45.** A gas has a percent composition by weight of 85.7% carbon and 14.3% hydrogen. At STP the density of the gas is 2.50 g/L. What is the molecular formula of the gas?

***46.** Assume that the reaction $2\,CO(g) + O_2(g) \longrightarrow 2\,CO_2(g)$ goes to completion. When 10 mol of CO and 8.0 mol of O_2 react in a closed 10 L vessel,

(a) How many moles of CO, O_2, and CO_2 are present at the end of the reaction?

(b) What will be the total pressure in the flask at 0°C?

***47.** 250 mL of O_2, measured at STP, were obtained by the decomposition of the $KClO_3$ in a 1.20 g mixture of KCl and $KClO_3$:

$$2\,KClO_3(s) \xrightarrow{\Delta} 2\,KCl(s) + 3\,O_2(g)$$

What is the percentage by weight of $KClO_3$ in the mixture?

LAUREN
824 9857

Review Exercises for Chapters 11 and 12

CHAPTER 11 CALCULATIONS FROM CHEMICAL EQUATIONS

True–False. *Answer the following as either true or false.*

1. Stoichiometry is the section of chemistry involving calculations based on mass and mole relationships of substances in chemical reactions.
2. In a limiting-reactant problem, you determine which reactant has the fewest moles available.
3. The maximum amount of product that can be produced according to the equation, from the amounts of reactants supplied, is the theoretical yield.
4. A mole ratio is a ratio of the moles of products to the moles of reactants.
5. A mole ratio is used to convert the moles of starting substance to the moles of desired substance.
6. The limiting reactant of a chemical reaction limits the amount of product that can be formed.
7. The equation for a reaction should be balanced before doing stoichiometric calculations.
8. The mole ratio for converting moles of CO_2 to moles of O_2 in the reaction $2\,C_2H_6 + 7\,O_2 \longrightarrow 4\,CO_2 + 6\,H_2O$ is 4 mol CO_2/7 mol O_2.

Multiple Choice. *Choose the correct answer to each of the following.*

1. 20.0 g of Na_2CO_3 is how many moles?
   (a) 1.89 mol (c) 212 mol
   (b) 2.12×10^3 mol (d) 0.189 mol
2. What is the weight in grams of 0.30 mol of $BaSO_4$?
   (a) 7.0×10^3 g (c) 70 g
   (b) 0.13 g (d) 700.20 g
3. How many molecules are in 5.8 g of acetone, C_3H_6O?
   (a) 0.10 molecules
   (b) 6.0×10^{22} molecules
   (c) 3.5×10^{24} molecules
   (d) 6.022×10^{23} molecules

Problems 4 *through* 10 *refer to the reaction*

$$2\,C_2H_4 + 6\,O_2 \longrightarrow 4\,CO_2 + 4H_2O$$

The following molecular weights may be needed for these problems: $C_2H_4 = 28.0$, $O_2 = 32.0$, $CO_2 = 44.0$, $H_2O = 18.0$

4. If 6.0 mol of CO_2 are produced, how many moles of O_2 were reacted?
   (a) 4.0 mol (c) 9.0 mol
   (b) 7.5 mol (d) 15.0 mol
5. How many moles of O_2 are required for the complete reaction of 45 g of C_2H_4?
   (a) 1.3×10^2 mol (c) 112.5 mol
   (b) 0.64 mol (d) 4.8 mol
6. If 18.0 g of CO_2 are produced, how many grams of H_2O are produced?
   (a) 7.36 g (b) 3.68 g (c) 9.0 g (d) 14.7 g
7. How many moles of CO_2 can be produced by the reaction of 5.0 mol of C_2H_4 and 12.0 mol of O_2?
   (a) 4.0 mol (c) 8.0 mol
   (b) 5.0 mol (d) 10.0 mol
8. How many moles of CO_2 can be produced by the reaction of 0.480 mol of C_2H_4 and 1.08 mol of O_2?
   (a) 0.240 mol (c) 0.720 mol
   (b) 0.960 mol (d) 0.864 mol
9. How many grams of CO_2 could be produced from 2.0 g of C_2H_4 and 5.0 g of O_2?
   (a) 5.5 g (b) 4.6 g (c) 7.6 g (d) 6.3 g
10. If 14.0 g of C_2H_4 is reacted, and the actual yield of H_2O is 7.84 g, the percentage yield in the reaction is:
    (a) 0.56% (b) 43.6% (c) 87.1% (d) 56.0%

CHAPTER 12 THE GASEOUS STATE OF MATTER

True–False. *Answer the following as either true or false.*

1. The average kinetic energy of molecules of a gas increases with increased temperature.
2. Pressure is defined as force per unit area.
3. One torr is equal to 760 mm Hg.
4. One mole of hydrogen gas and 1 mole of oxygen gas, each in a box of equal volume, and each at the same temperature, will exert the same pressure.
5. Boyle's law states that the volume of a gas is inversely proportional to the pressure.
6. Charles' law states that the pressure of a gas is directly proportional to the temperature.
7. One mole of any gas always occupies 22.4 liters.
8. The specific gravity of a gas is the ratio of the mass of any volume of the gas to the mass of an equal volume of some reference gas at the same temperature and pressure.
9. Avogadro's law states that equal volumes of different gases at the same temperature and pressure contain the same weights of gas.
10. An ideal gas is one whose behavior is described exactly by the gas laws for all possible values of P, V, and T.
11. According to Avogadro, there are 6.022×10^{23} molecules of gas in a liter at STP.
12. When the temperature of a gas is decreased at fixed volume, the pressure decreases.
13. At constant temperature and pressure, N_2 will effuse more rapidly than O_2.
14. A barometer is an instrument used to measure the weight of a certain quantity of mercury.
15. The volume of a gas is dependent on the number of gas molecules, the pressure, the absolute temperature, and the gas constant, R.

Multiple Choice. *Choose the correct answer to each of the following.*

1. Which of the following is not one of the principal assumptions of the Kinetic-Molecular Theory for an ideal gas?
   (a) All collisions of gaseous molecules are perfectly elastic.
   (b) A mole of any gas occupies 22.4 L at STP.
   (c) Gas molecules have no attraction for one another.
   (d) The average kinetic energy for molecules is the same for all gases at the same temperature.
2. Which of the following is not equal to 1.00 atm pressure?
   (a) 760 cm Hg (c) 760 mm Hg
   (b) 29.9 in. Hg (d) 760 torr
3. If the pressure on 45 mL of gas is changed from 600 torr to 800 torr, the new volume will be:
   (a) 60 mL (c) 0.045 L
   (b) 34 mL (d) 22.4 L
4. The volume of a gas is 300 mL at 740 torr and 25°C. If the pressure remains constant and the temperature is raised to 100°C, the new volume will be:
   (a) 240 mL (c) 376 mL
   (b) 1.20 L (d) 75.0 mL
5. The volume of a dry gas is 4.00 L at 15°C and 745 torr. What volume will the gas occupy at 40°C and 700 torr?
   (a) 4.63 L (b) 3.46 L (c) 3.92 L (d) 4.08 L
6. A sample of Cl_2 occupies 8.50 L at 80°C and 740 mm Hg. What volume will the Cl_2 occupy at STP?
   (a) 10.7 L (b) 6.75 L (c) 11.3 L (d) 6.40 L
7. What volume will 8.00 g of O_2 occupy at 45°C and 2.00 atm?
   (a) 0.462 L (b) 104 L (c) 9.62 L (d) 3.26 L
8. The density of NH_3 gas at STP is:
   (a) 0.759 g/mL (c) 1.32 g/mL
   (b) 0.759 g/L (d) 1.32 g/L
9. The ratio of the relative rate of effusion of methane, CH_4, to sulfur dioxide, SO_2, is:
   (a) $\frac{64}{16}$ (b) $\frac{16}{64}$ (c) $\frac{1}{4}$ (d) $\frac{2}{1}$
10. Measured at 65°C and 500 torr, 3.21 L of a gas weighed 3.5 g. The molecular weight of this gas is:
    (a) 21 g/mole (c) 24 g/mole
    (b) 46 g/mole (d) 130 g/mole
11. Box A contains O_2 (mol. wt = 32.0) at a pressure of 200 torr. Box B, which is identical to Box A in volume, contains twice as many molecules of CH_4 (mol. wt = 16.0) as the molecules of O_2 in Box A. The temperatures of the gases are identical. The pressure in Box B is:
    (a) 100 torr (c) 400 torr
    (b) 200 torr (d) 800 torr

12. A 300 mL sample of oxygen, O_2, was collected over water at 23°C and 725 torr pressure. If the vapor pressure of water at 23°C is 21.0 torr, the volume of dry O_2 at STP would be:
(a) 256 mL (c) 341 mL
(b) 351 mL (d) 264 mL

13. A tank containing 0.01 mol of neon and 0.04 mol of helium shows a pressure of 1 atm. What is the partial pressure of neon in the tank?
(a) 0.8 atm (c) 0.2 atm
(b) 0.01 atm (d) 0.5 atm

14. How many liters of NO_2 (at STP) can be produced from 25.0 g of Cu reacting with concentrated nitric acid?

$$Cu(s) + 4\ HNO_3(aq) \longrightarrow Cu(NO_3)_2(aq) + 2\ H_2O(l) + 2\ NO_2\uparrow$$

(a) 4.41 L (b) 8.82 L (c) 17.6 L (d) 44.8 L

15. How many liters of butane vapor are required to produce 2.0 L of CO_2 at STP?

$$\underset{\text{Butane}}{2\ C_4H_{10}(g)} + 13\ O_2(g) \longrightarrow 8\ CO_2(g) + 10\ H_2O(g)$$

(a) 2.0 L (b) 4.0 L (c) 0.80 L (d) 0.50 L

16. What volume of CO_2 (at STP) can be produced when 15.0 g of C_2H_6 and 50.0 g of O_2 are reacted?

$$2\ C_2H_6(g) + 7\ O_2(g) \longrightarrow 4\ CO_2(g) + 6\ H_2O(g)$$

(a) 20.0 L (b) 22.4 L (c) 35.0 L (d) 5.6 L

17. Which of these gases has the highest density at STP?
(a) N_2O (b) NO_2 (c) Cl_2 (d) SO_2

18. What is the density of CO_2 at 25°C and 0.954 atm pressure?
(a) 1.72 g/L (c) 0.985 g/L
(b) 2.04 g/L (d) 1.52 g/L

CHAPTER 13

Water and the Properties of Liquids

After studying Chapter 13, you should be able to

1 Understand the terms listed in Question A at the end of the chapter.
2 Describe a water molecule with respect to Lewis structure, bond angle, and polarity.
3 Make sketches showing hydrogen bonding (a) between water molecules and (b) between hydrogen fluoride molecules.
4 Explain the effect of hydrogen bonding on the physical properties of water.
5 Determine whether a compound will or will not form hydrogen bonds.
6 Explain the process of evaporation from the standpoint of kinetic energy.
7 Relate vapor pressure data or vapor pressure curves of different substances to their relative rates of evaporation and to their relative boiling points.
8 Explain what is happening in the different segments of the time–temperature heating curve of water.
9 Complete and balance equations showing the formation of water (a) from hydrogen and oxygen, (b) by neutralization, and (c) by combustion of hydrogen-containing compounds.
10 Complete and balance equations for (a) the reaction of water with Na, K, and Ca; (b) the reaction of steam with Zn, Al, Fe, and C; and (c) the reaction of water with halogens.
11 Identify metal oxides as basic anhydrides and write balanced equations for their reactions with water.

12 Identify nonmetal oxides as acid anhydrides and write balanced equations for their reactions with water.
13 Deduce the formula of the acid anhydride or the basic anhydride when given the formula of the corresponding acid or base.
14 Identify, name, and write equations for the complete dehydration of hydrates.
15 Distinguish clearly between peroxides and ordinary oxides.
16 Discuss the occurrence of ozone and its effects on humans.
17 Outline the processes needed to prepare a potable water supply from a contaminated river source.
18 Describe how water may be softened by distillation, chemical precipitation, ion exchange, and demineralization. Write chemical equations for these processes where appropriate.

13.1 Occurrence of Water

Water is our most common natural resource; it covers about 70% of the earth's surface. Not only is it found in the oceans and seas, in lakes, rivers, streams, and in glacial ice deposits, it is also always present in the atmosphere and in cloud formations.

About 97% of the earth's water is in the oceans. This *saline* water contains vast amounts of dissolved minerals. More than 70 elements have been detected in the mineral content of seawater. Only four of these—chlorine, sodium, magnesium, and bromine—are now commercially obtained from the sea. The world's *fresh* water comprises the other 3%, of which about two-thirds is locked up in polar ice caps and glaciers. The remaining fresh water is found in ground water, lakes, and the atmosphere.

Water is an essential constituent of all living matter. It is the most abundant compound in the human body, making up about 70% of total body weight. About 92% of blood plasma is water; about 80% of muscle tissue is water; and about 60% of a red blood cell is water. Water is more important than food in the sense that a person can survive much longer without food than without water.

13.2 Physical Properties of Water

Water is a colorless, odorless, tasteless liquid with a melting point of 0°C and a boiling point of 100°C at 1 atm pressure. Two additional physical properties of matter are introduced with the study of water: heat of fusion and heat of vaporization. **Heat of fusion** is the amount of heat required to change 1 g of a solid into a liquid at its melting point. The heat of fusion of water is 335 J/g

heat of fusion

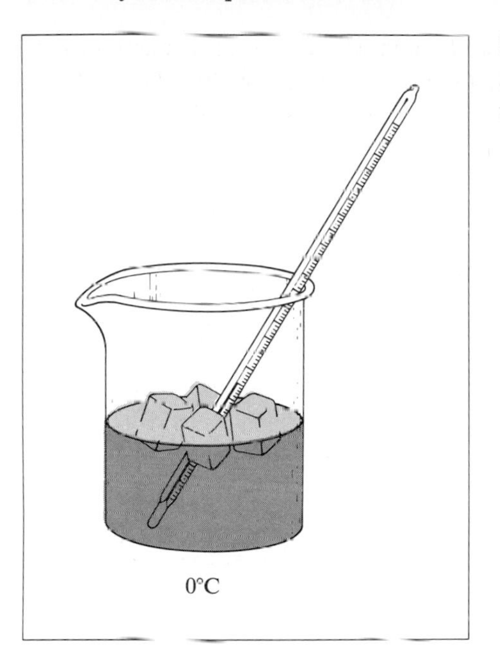

Figure 13.1

Ice and water in equilibrium at 0°C

(80 cal/g). The temperature of the solid–liquid system does not change during the absorption of this heat. The heat of fusion is the energy needed to break down the crystalline lattice of ice from a solid to a liquid. **Heat of vaporization** is the amount of heat required to change 1 g of liquid to a vapor at its normal boiling point. The value for water is 2.26 kJ/g (540 cal/g). Once again, the temperature does not change during the absorption of this heat. The heat of vaporization is the energy needed to overcome the attractive forces between molecules in changing a substance from the liquid to the gaseous state at its normal boiling point. The values for water for both the heat of fusion and the heat of vaporization are high compared with those for other substances; these high values indicate that strong attractive forces are acting between the molecules.

heat of vaporization

Ice and water exist together in equilibrium at 0°C, as shown in Figure 13.1. When ice at 0°C melts, it absorbs 335 J/g (80 cal/g) in changing into a liquid; the temperature remains at 0°C. In order to refreeze the water we have to remove 335 J/g (80 cal/g) from the liquid at 0°C.

In Figure 13.2 both boiling water and steam are shown to have a temperature of 100°C. It takes 418 J (100 cal) to heat 1 g of water from 0°C to 100°C, but water at its boiling point absorbs 2.26 kJ/g (540 cal/g) in changing to steam. Although boiling water and steam are both at the same temperature, steam contains considerably more heat per gram and can cause more severe burns than hot water. In Table 13.1 the physical properties of water are tabulated and compared with those of other hydrogen compounds of Group VIA elements.

The maximum density of water is 1.000 g/mL at 4°C. Water has the unusual property of contracting in volume as it is cooled to 4°C and then expanding when

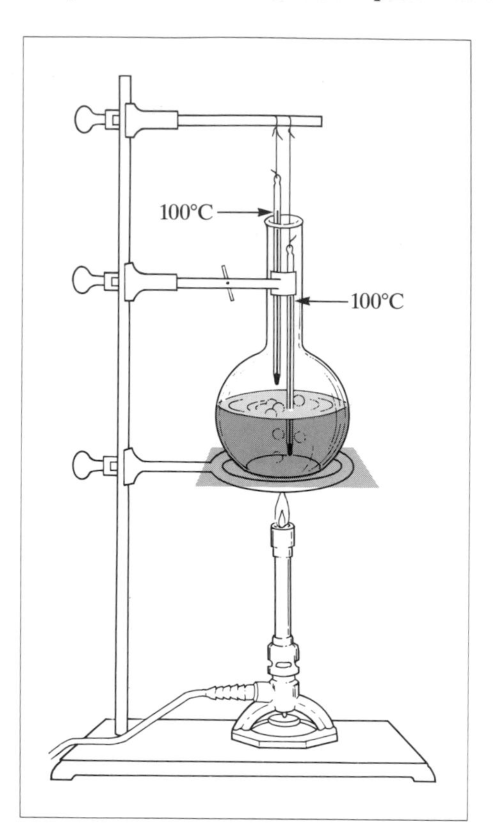

Figure 13.2
Boiling water and steam in equilibrium at 100°C

Table 13.1 Physical properties of water and other hydrogen compounds of Group VIA elements

Formula	Color	Molecular weight	Melting point (°C)	Boiling point, 1 atm (°C)	Heat of fusion J/g (cal/g)	Heat of vaporization J/g (cal/g)
H_2O	Colorless	18.0	0.00	100.0	335 (80.0)	2.26×10^3 (540)
H_2S	Colorless	34.1	−85.5	−60.3	69.9 (16.7)	548 (131)
H_2Se	Colorless	81.0	−65.7	−41.3	31 (7.4)	238 (57.0)
H_2Te	Colorless	129.6	−51	−2.3	—	179 (42.8)

cooled from 4°C to 0°C. Therefore 1 g of water occupies a volume greater than 1 mL at all temperatures except 4°C. Although most liquids contract in volume all the way down to the point at which they solidify, a large increase (about 9%) in volume occurs when water changes from a liquid at 0°C to a solid (ice) at 0°C. The density of ice at 0°C is 0.917 g/mL, which means that ice, being less dense than water, will float in water.

PROBLEM 13.1 How many joules of energy are needed to change 10.0 g of ice at 0°C to water at 20°C?

Ice will absorb 335 J/g (heat of fusion) in going from a solid at 0°C to a liquid at 0°C. Then an additional 4.184 J/g are needed to raise the temperature for each 1°C.

Joules needed to melt the ice:

$$10.0\ \text{g} \times \frac{335\ \text{J}}{1\ \text{g}} = 3.35 \times 10^3\ \text{J}\ (800\ \text{cal})$$

Joules needed to heat the water from 0°C to 20°C:

$$10.0\ \text{g} \times \frac{4.184\ \text{J}}{1\ \text{g}\ ^\circ\text{C}} \times 20^\circ\text{C} = 837\ \text{J}\ (200\ \text{cal})$$

Thus, 3350 J + 837 J = 4.18 × 10³ J (1000 cal) are needed.

PROBLEM 13.2 How many kilojoules of energy are needed to change 20.0 g of water at 20°C to steam at 100°C?

Kilojoules needed to heat the water from 20°C to 100°C:

$$20.0\ \text{g} \times \frac{4.184\ \text{J}}{1\ \text{g}\ ^\circ\text{C}} \times \frac{1\ \text{kJ}}{1000\ \text{J}} \times 80^\circ\text{C} = 6.69 \times 10^3\ \text{kJ}\ (1.60 \times 10^3\ \text{kcal})$$

Kilojoules needed to change water at 100°C to steam at 100°C:

$$20.0\ \text{g} \times \frac{2.26\ \text{kJ}}{1\ \text{g}} = 4.52 \times 10^4\ \text{kJ}\ (1.08 \times 10^4\ \text{kcal})$$

Thus, 6.69×10^3 kJ + 4.52×10^4 kJ = 5.19×10^4 kJ (1.24×10^3 kcal) are needed.

13.3 Structure of the Water Molecule

A single water molecule consists of two hydrogen atoms and one oxygen atom. Each H atom is attached to the O atom by a single covalent bond. This bond is formed by the overlap of the 1*s* orbital of hydrogen with an unpaired 2*p* orbital of oxygen. The average distance between the two nuclei is known as the *bond length*. The O—H bond length in water is 0.096 nm. The water molecule is nonlinear and has a V-shaped structure with an angle of about 105° between the two bonds (see Figure 13.3 on page 294).

Oxygen is the second most electronegative element. As a result, the two covalent OH bonds in water are polar. If the three atoms in a water molecule were aligned in a linear structure such as H ⟶ O ⟵ H, the two polar bonds would be acting in equal and opposite directions and the molecule would be nonpolar. However, water is a highly polar molecule. Therefore, it does not have a linear structure. When atoms are bonded together in a nonlinear fashion, the angle

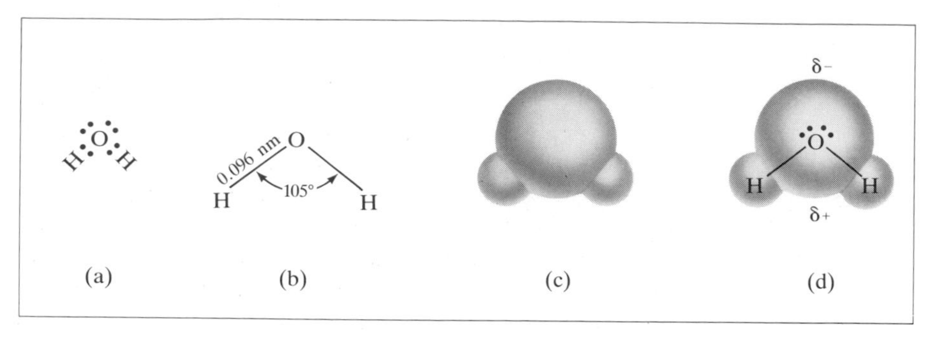

Figure 13.3

Diagrams of a water molecule: (a) electron distribution, (b) bond angle and O—H bond length, (c) molecular orbital structure, and (d) dipole representation

formed by the bonds is called the *bond angle*. In water the HOH bond angle is 105°. The two polar covalent bonds and the bent structure result in a partial negative charge on the oxygen atom and a partial positive charge on each hydrogen atom. The polar nature of water is responsible for many of its properties, including its behavior as a solvent.

13.4 The Hydrogen Bond

Table 13.1 compares the physical properties of H_2O, H_2S, H_2Se, and H_2Te. From this comparison it is apparent that four physical properties of water—melting point, boiling point, heat of fusion, and heat of vaporization—are extremely high and do not fit the trend relative to the molecular weights of the four compounds. If the properties of water followed the progression shown by the other three compounds, we would expect the melting point of water to be below −85°C and the boiling point to be below −60°C.

hydrogen bond

Why does water have these anomalous physical properties? The answer is that liquid water molecules are linked together by hydrogen bonds. A **hydrogen bond** is a chemical bond that is formed between polar molecules that contain hydrogen covalently bonded to a small, highly electronegative atom such as fluorine, oxygen, or nitrogen (F—H, O—H, N—H). The bond is actually the dipole–dipole attraction of polar molecules containing these three types of polar bonds.

Elements that have significant hydrogen bonding ability are F, O, and N.

What is a hydrogen bond, or H-bond? Because a hydrogen atom has only one electron, it can form only one covalent bond. When it is attached to a strong

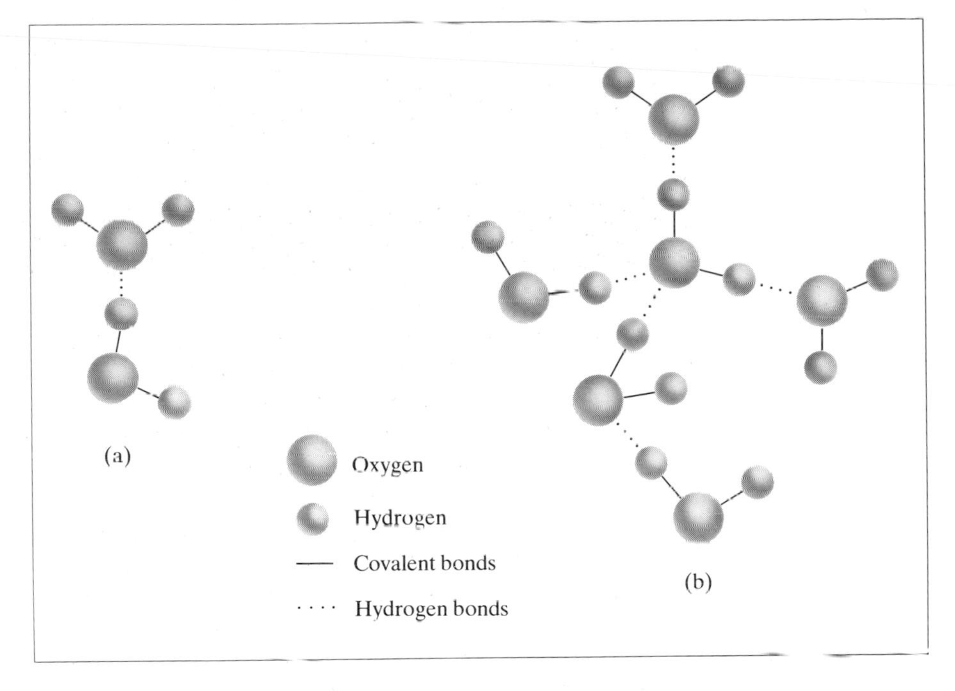

Figure 13.4

Hydrogen bonding. Water in the liquid and solid states exists as aggregates in which the water molecules are linked together by hydrogen bonds.

electronegative atom such as oxygen, a hydrogen atom will also be attracted to an oxygen atom of another molecule, forming a bond (or bridge) between the two molecules. Water has two types of bonds: covalent bonds that exist between hydrogen and oxygen atoms within a molecule and hydrogen bonds that exist between hydrogen and oxygen atoms in different water molecules.

Hydrogen bonds are *intermolecular* bonds; that is, they are formed between atoms in different molecules. They are somewhat ionic in character because they are formed by electrostatic attraction. Hydrogen bonds are much weaker than the ionic or covalent bonds that unite atoms to atoms to form compounds. Despite their weakness, hydrogen bonds are of great chemical importance.

The oxygen atom in water can form two hydrogen bonds—one through each of the unbonded pairs of electrons. Figure 13.4 shows (a) two water molecules linked by a hydrogen bond and (b) six water molecules linked by hydrogen bonds. A dash (—) is used for the covalent bond and a dotted line (····) for the hydrogen bond. In water each molecule is linked to others through hydrogen bonds to form a three-dimensional aggregate of water molecules. This intermolecular hydrogen bonding effectively gives water the properties of a much larger, heavier molecule, explaining in part its relatively high melting point, boiling point, heat of fusion, and heat of vaporization. As water is heated and energy is absorbed, hydrogen bonds are continually being broken until at 100°C, with the absorption of an additional 2.26 kJ/g (540 cal/g), water separates into individual molecules, going into the gaseous state. Sulfur, selenium, and tellurium are not sufficiently electronegative for their hydrogen compounds to behave like water. As a result, H-bonding in H_2S, H_2Se, and H_2Te is only of small consequence (if any) to their physical properties. For example, the lack of hydrogen bonding is one reason why H_2S is a gas and not a liquid at room temperature.

Fluorine, the most electronegative element, forms the strongest hydrogen bonds. This bonding is strong enough to link hydrogen fluoride molecules together as *dimers*, H_2F_2, or as larger, $(HF)_n$, molecular units. The dimer structure may be represented in this way:

```
                F
               /
  H       ...H
    \  ...
     F    ↖ H—bond
```

The existence of salts, such as KHF_2 and NH_4HF_2, verifies the hydrogen fluoride (bifluoride) structure, HF_2^- $(F—H \cdots\cdot F)^-$, where one H atom is bonded to two F atoms through one covalent bond and one hydrogen bond.

Hydrogen bonding can occur between two different atoms that are capable of forming H-bonds. Thus we may have an O···· H—N or O—H····N linkage in which the hydrogen atom forming the H-bond is between an oxygen and a nitrogen atom. This form of the H-bond exists in certain types of protein molecules and many biologically active substances.

PROBLEM 13.3 Would you expect hydrogen bonding to occur between molecules of the following substances?

```
        H   H                      H       H
        |   |                      |       |
(a) H—C—C—O—H              (b) H—C—O—C—H
        |   |                      |       |
        H   H                      H       H

    Ethyl alcohol                Dimethyl ether
```

(a) Hydrogen bonding should occur in ethyl alcohol because one hydrogen atom is bonded to an oxygen atom.

```
     H   H          H   H   H
     |   |          |   |   |
  H—C—C—O—H···O—C—C—H
     |   |          ↑   |   |
     H   H      H-bond  H   H
```

(b) There is no hydrogen bonding in dimethyl ether because all the hydrogen atoms are bonded only to carbon atoms.

Both ethyl alcohol and dimethyl ether have the same molecular weight (46.0). Although both compounds have the same molecular formula, C_2H_6O, ethyl alcohol has a much higher boiling point (78.4°C) than dimethyl ether (−23.7°C), because of hydrogen bonding between the alcohol molecules.

13.5 Evaporation

When beakers of water, ethyl ether, and ethyl alcohol are allowed to stand uncovered in an open room, the volumes of these liquids gradually decrease. The process by which this change takes place is called *evaporation.*

Attractive forces exist between molecules in the liquid state. Not all of these molecules, however, have the same kinetic energy. Molecules that have greater than average kinetic energy can overcome the attractive forces and break away from the surface of the liquid to become a gas. **Evaporation** or **vaporization** is the escape of molecules from the liquid state to the gas or vapor state.

evaporation

vaporization

In evaporation, molecules of higher than average kinetic energy escape from a liquid, leaving it cooler than it was before they escaped. For this reason, evaporation of perspiration is one way the human body cools itself and keeps its temperature constant. When volatile liquids such as ethyl chloride (C_2H_5Cl) are sprayed on the skin, they evaporate rapidly, cooling the area by removing heat. The numbing effect of the low temperature produced by evaporation of ethyl chloride allows it to be used as a local anesthetic for minor surgery.

Solids such as iodine, camphor, naphthalene (moth balls), and, to a small extent, even ice will go directly from the solid to the gaseous state, bypassing the liquid state. This change is a form of evaporation and is called **sublimation**:

sublimation

$$\text{liquid} \xrightarrow{\text{Evaporation}} \text{vapor}$$

$$\text{solid} \xrightarrow{\text{Sublimation}} \text{vapor}$$

13.6 Vapor Pressure

When a liquid vaporizes in a closed system as shown in Figure 13.5, part (b), some of the molecules in the vapor or gaseous state strike the surface and return to the liquid state by the process of *condensation.* The rate of condensation increases until it is equal to the rate of vaporization. At this point, the space above the liquid is said to be saturated with vapor, and an equilibrium, or steady state, exists between the liquid and the vapor. The equilibrium equation is

$$\text{liquid} \underset{\text{Condensation}}{\overset{\text{Vaporization}}{\rightleftarrows}} \text{vapor}$$

This equilibrium is dynamic; both processes—vaporization and condensation—are taking place, even though one cannot see or measure a change. The number of molecules leaving the liquid in a given time interval is equal to the number of molecules returning to the liquid.

At equilibrium the molecules in the vapor exert a pressure like any other gas. The pressure exerted by a vapor in equilibrium with its liquid is known as the **vapor pressure** of the liquid. The vapor pressure may be thought of as an internal pressure, a measure of the "escaping" tendency of molecules to go from the liquid

vapor pressure

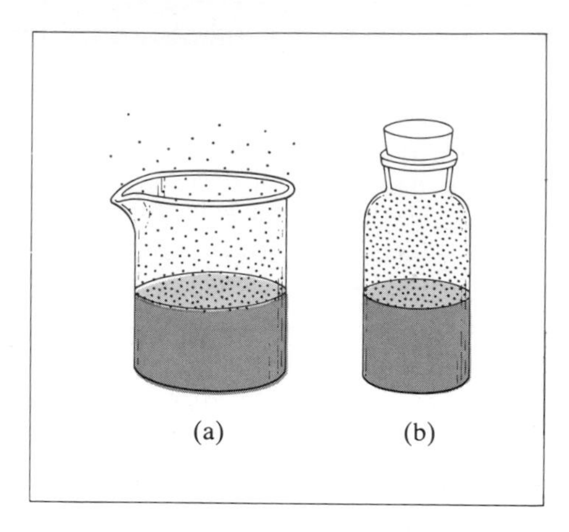

Figure 13.5
(a) Molecules in an open beaker evaporate from the liquid and disperse into the atmosphere. Under this condition, evaporation will continue until all the liquid is gone. (b) Molecules leaving the liquid are confined to a limited space. With time, the concentration in the vapor phase will increase to a point at which an equilibrium between liquid and vapor is established.

Table 13.2 The vapor pressure of water, ethyl alcohol, and ethyl ether at various temperatures

	Vapor pressure (torr)		
Temperature (°C)	Water	Ethyl alcohol	Ethyl ether[a]
0	4.6	12.2	185.3
10	9.2	23.6	291.7
20	17.5	43.9	442.2
30	31.8	78.8	647.3
40	55.3	135.3	921.3
50	92.5	222.2	1276.8
60	152.9	352.7	1729.0
70	233.7	542.5	2296.0
80	355.1	812.6	2993.6
90	525.8	1187.1	3841.0
100	760.0	1693.3	4859.4
110	1074.6	2361.3	6070.1

[a] Note that the vapor pressure of ethyl ether at temperatures of 40°C and higher exceeds standard pressure, 760 torr, which indicates that the substance has a low boiling point and therefore should be stored in a cool place in a tightly sealed container.

to the vapor state. The vapor pressure of a liquid is independent of the amount of liquid and vapor present, but it increases as the temperature rises (see Table 13.2). Figure 13.6 illustrates a liquid–vapor equilibrium and the measurement of vapor pressure.

When equal volumes of water, ethyl ether, and ethyl alcohol are placed in separate beakers and allowed to evaporate at the same temperature, we observe that the ether evaporates faster than the alcohol, which evaporates faster than the

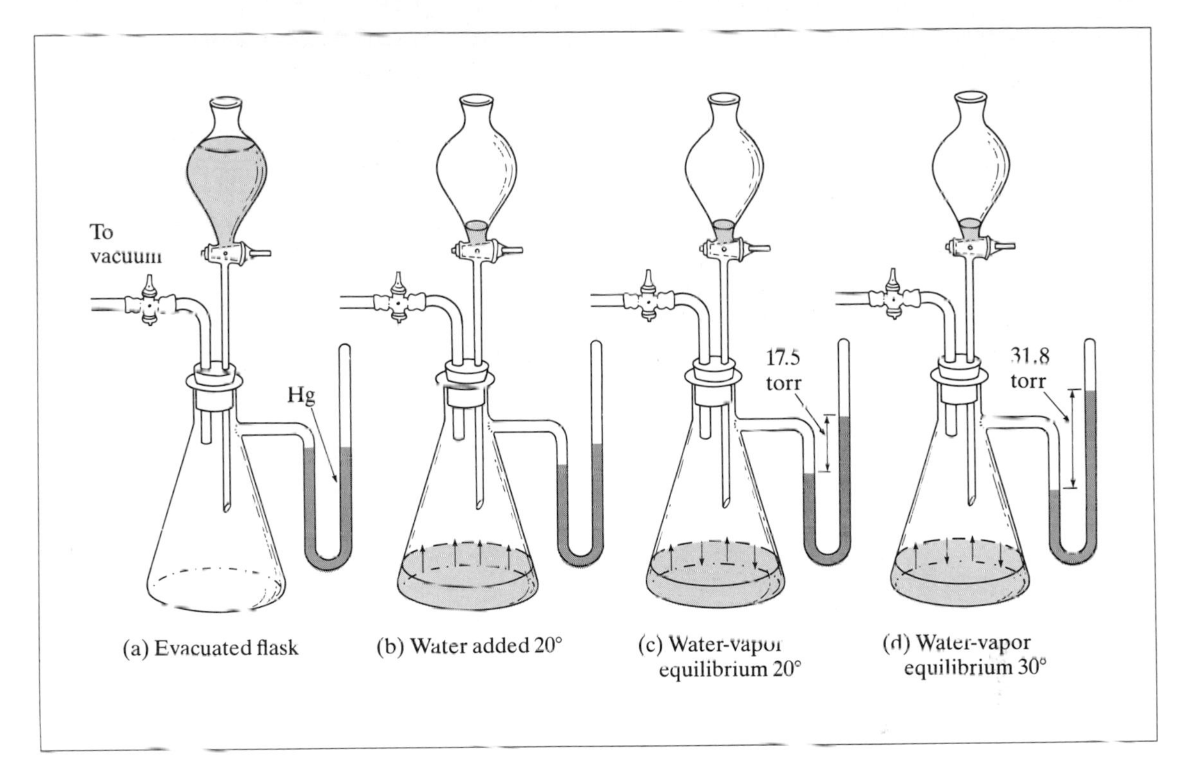

Figure 13.6
Measurement of the vapor pressure of water at 20°C and 30°C. In flask (a) the system is evacuated. The mercury manometer attached to the flask shows equal pressure in both legs. In (b) water has been added to the flask and begins to evaporate, exerting pressure as indicated by the manometer. In (c), when equilibrium is established, the pressure inside the flask remains constant at 17.5 torr. In (d) the temperature is changed to 30°C, and equilibrium is reestablished with the vapor pressure at 31.8 torr.

water. This order of evaporation is consistent with the fact that ether has a higher vapor pressure at any particular temperature than ethyl alcohol or water. One reason for this higher vapor pressure is that the attraction is less between ether molecules than between alcohol molecules or between water molecules. The vapor pressures of these three compounds at various temperatures are compared in Table 13.2.

volatile

Substances that evaporate readily are said to be **volatile**. A volatile liquid has a relatively high vapor pressure at room temperature. Ethyl ether is a very volatile liquid; water is not too volatile; mercury, which has a vapor pressure of 0.0012 torr at 20°C, is essentially a nonvolatile liquid. Most substances that are normally solids are nonvolatile (solids that sublime are exceptions).

13.7 Boiling Point

The boiling temperature of a liquid is associated with its vapor pressure. We have seen that the vapor pressure increases as the temperature increases. When the internal or vapor pressure of a liquid becomes equal to the external pressure, the liquid boils. (By external pressure we mean the pressure of the atmosphere above the liquid.) The boiling temperature of a pure liquid remains constant as long as the external pressure does not vary.

The boiling point (bp) of water is 100°C. Table 13.2 shows that the vapor pressure of water at 100°C is 760 torr, a figure we have seen many times before. The significant fact here is that the boiling point is the temperature at which the vapor pressure of the water or other liquid is equal to standard, or atmospheric, pressure at sea level. These relationships lead to the following definition: The **boiling point** is the temperature at which the vapor pressure of a liquid is equal to the external pressure above the liquid.

boiling point

We can readily see that a liquid has an infinite number of boiling points. When we give the boiling point of a liquid, we should also state the pressure. When we express the boiling point without stating the pressure, we mean it to be the **standard** or **normal boiling point** at standard pressure (760 torr). Using Table 13.2 again, we see that the normal boiling point of ethyl ether is between 30°C and 40°C, and for ethyl alcohol it is between 70°C and 80°C, because, for each compound, 760 torr pressure lies within these stated temperature ranges. At the normal boiling point, 1 g of a liquid changing to a vapor (gas) absorbs an amount of energy equal to its heat of vaporization (see Table 13.3).

standard or normal boiling point

The boiling point at various pressures may be evaluated by plotting the data of Table 13.2 on the graph in Figure 13.7, where temperature is plotted horizontally along the x axis and vapor pressure is plotted vertically along the y axis. The resulting curves are known as **vapor pressure curves**. Any point on these curves represents a vapor–liquid equilibrium at a particular temperature and pressure. We may find the boiling point at any pressure by tracing a horizontal

vapor pressure curves

Table 13.3 Physical properties of ethyl chloride, ethyl ether, ethyl alcohol, and water

	Boiling point (°C)	Melting point (°C)	Heat of vaporization J/g (cal/g)	Heat of fusion J/g (cal/g)
Ethyl chloride	13	−139	387 (92.5)	—
Ethyl ether	34.6	−116	351 (83.9)	—
Ethyl alcohol	78.4	−112	855 (204.3)	104 (24.9)
Water	100.0	0	2295 (540)	335 (80)

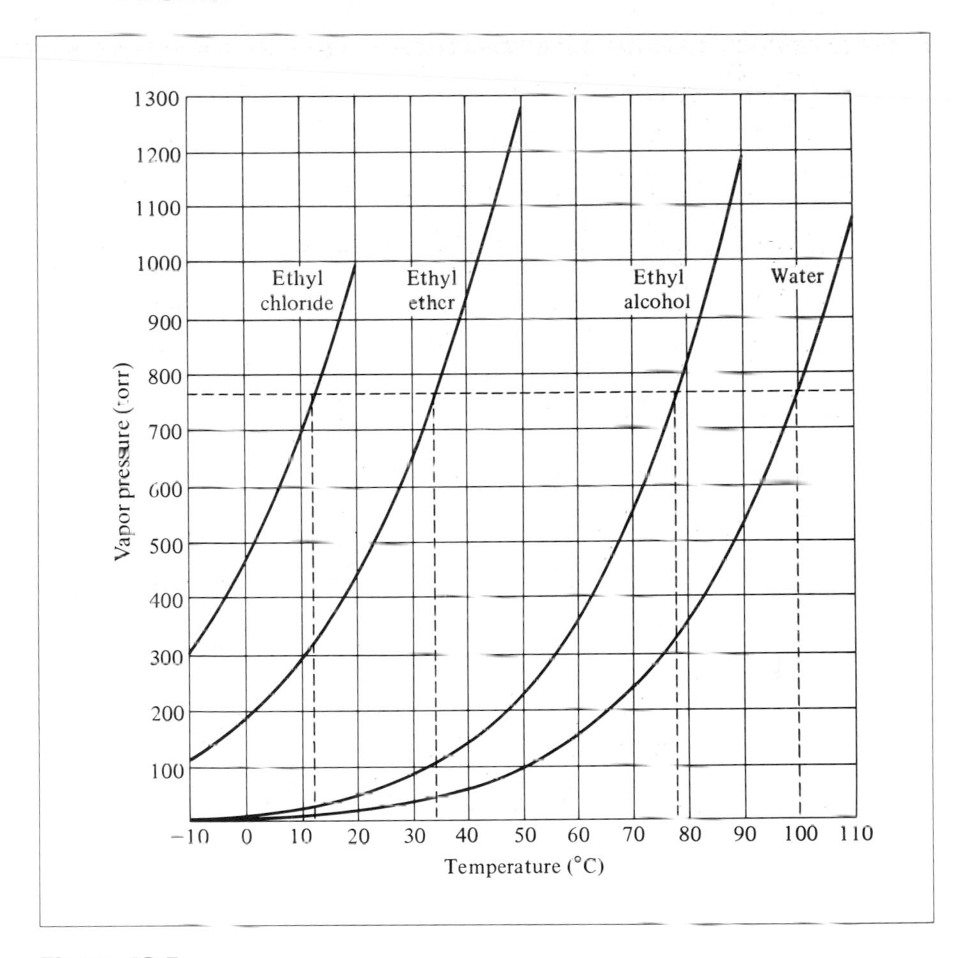

Figure 13.7

Vapor pressure–temperature curves for ethyl chloride, ethyl ether, ethyl alcohol, and water

line from the designated pressure to a point on the vapor pressure curve. From this point we draw a vertical line to obtain the boiling point on the temperature axis. Four such points are shown in Figure 13.7; they represent the normal boiling points of the four compounds at 760 torr pressure.

See if you can verify from the graph that the boiling points of ethyl chloride, ethyl ether, ethyl alcohol, and water at 600 torr pressure are 8.5°C, 28°C, 73°C, and 93°C, respectively. By reversing this process, you can ascertain at what pressure a substance will boil at a specific temperature. The boiling point is one of the most commonly used physical properties for characterizing and identifying substances.

13.8 Freezing Point or Melting Point

As heat is removed from a liquid, the liquid becomes colder and colder, until a temperature is reached at which it begins to solidify. A liquid that is changing into a solid is said to be *freezing*, or *solidifying*. When a solid is heated continuously, a temperature is reached at which the solid begins to liquefy. A solid that is changing into a liquid is said to be *melting*. The temperature at which the solid phase of a substance is in equilibrium with its liquid phase is known as the **freezing point** or **melting point** of that substance. The equilibrium equation is

freezing or melting point

$$\text{solid} \underset{\text{Freezing}}{\overset{\text{Melting}}{\rightleftarrows}} \text{liquid}$$

When a solid is slowly and carefully heated so that a solid–liquid equilibrium is achieved and then maintained, the temperature will remain constant as long as

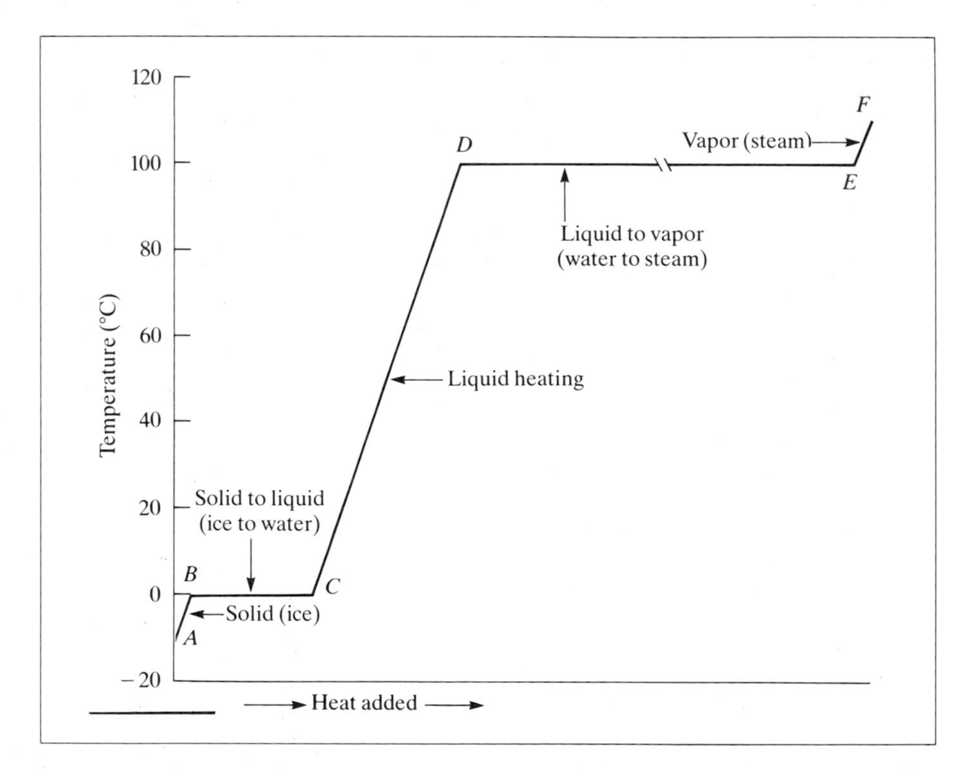

Figure 13.8
Heating curve. The absorption of heat by a substance from the solid state to the vapor state. Using water as an example, the interval *AB* represents the ice phase; *BC* interval, the melting of ice to water; *CD* interval, the elevation of the temperature of water from 0°C to 100°C; *DE* interval, the boiling of water to steam; and *EF* interval, the heating of steam.

both phases are present. One gram of a solid in changing into a liquid absorbs an amount of energy equal to its *heat of fusion* (see Table 13.3). The melting point is another physical property that is commonly used for characterizing substances.

The most common example of a solid–liquid equilibrium is ice and water (see Figure 13.1). In a well-stirred system of ice and water, the temperature remains at 0°C as long as both phases are present. The melting point changes with pressure but only slightly unless the pressure change is very large.

It has been known for a long time that dissolved substances markedly decrease the freezing point of a liquid. For example, salt–water–ice equilibrium mixtures can be obtained at temperatures as low as −20°C, 20 degrees below the usual freezing point of water.

If a liquid is heated continuously after all the solid has been melted, the temperature will rise until the liquid boils. The temperature will remain constant at the boiling point until all the liquid has boiled away. One gram of a liquid in changing into a gas at the normal boiling point absorbs an amount of energy equal to its *heat of vaporization* (see Table 13.3). The whole process of heating a substance (water) is illustrated graphically in Figure 13.8, which is called a heating curve. In the diagram, line *AB* represents the solid being heated, and line *BC* represents the time during which the solid is melting and is in equilibrium with the liquid. Along line *CD* the liquid absorbs heat; finally, at point *D*, it boils and continues to boil at a constant temperature (line *DE*). In the interval *EF* all the water exists as steam and is being further heated or superheated.

13.9 Formation of Water and Chemical Properties of Water

Water is very stable to heat; it decomposes to the extent of only about 1% at temperatures up to 2000°C. Pure water is a nonconductor of electricity. But when a small amount of sulfuric acid or sodium hydroxide is added, the solution is readily decomposed into hydrogen and oxygen by an electric current. Two volumes of hydrogen are produced for each volume of oxygen.

$$2\ H_2O(l) \xrightarrow[H_2SO_4 \text{ or NaOH}]{\text{Electrical energy}} 2\ H_2(g) + O_2(g)$$

Formation Water is formed when hydrogen burns in air. Pure hydrogen burns very smoothly in air, but mixtures of hydrogen and air or oxygen explode when ignited. The reaction is strongly exothermic.

$$2\ H_2(g) + O_2(g) \longrightarrow 2\ H_2O(g) + 484 \text{ kJ } (115.6 \text{ kcal})$$

Water is produced by a variety of other reactions, especially by (1) acid–base neutralizations, (2) combustion of hydrogen-containing materials, and (3) metabolic oxidation in living cells:

1. $HCl(aq) + NaOH(aq) \longrightarrow NaCl(aq) + H_2O(l)$
2. $2\,C_2H_2(g) + 5\,O_2(g) \longrightarrow 4\,CO_2(g) + 2\,H_2O(g) + 1212\text{ kJ } (289.6\text{ kcal})$
   Acetylene

   $CH_4 + 2\,O_2 \longrightarrow CO_2 + 2\,H_2O + 803\text{ kJ } (192\text{ kcal})$
   Methane
3. $C_6H_{12}O_6 + 6\,O_2 \xrightarrow{\text{Enzymes}} 6\,CO_2 + 6\,H_2O + 2519\text{ kJ } (673\text{ kcal})$
   Glucose

The combustion of acetylene shown in (2) is strongly exothermic and is capable of producing very high temperatures. It is used in oxygen–acetylene torches to cut and weld steel and other metals. Methane is known as natural gas and is commonly used as fuel for heating and cooking. The reaction of glucose with oxygen shown in (3) is the reverse of photosynthesis. It is the overall reaction by which living cells obtain needed energy by metabolizing glucose to carbon dioxide and water.

Reactions of Water with Metals and Nonmetals The reactions of metals with water at different temperatures show that these elements vary greatly in their reactivity. Metals such as sodium, potassium, and calcium react with cold water to produce hydrogen and a metal hydroxide. A small piece of sodium added to water melts from the heat produced by the reaction, forming a silvery metal ball, which rapidly flits back and forth on the surface of the water. Caution must be used when experimenting with this reaction, because the hydrogen produced is frequently ignited by the sparking of the sodium, and it will explode, spattering sodium. Potassium reacts even more vigorously than sodium. Calcium sinks in water and liberates a gentle stream of hydrogen. The equations for these reactions are

$$2\,Na(s) + 2\,H_2O(l) \longrightarrow H_2\uparrow + 2\,NaOH(aq)$$
$$2\,K(s) + 2\,H_2O(l) \longrightarrow H_2\uparrow + 2\,KOH(aq)$$
$$Ca(s) + 2\,H_2O(l) \longrightarrow H_2\uparrow + Ca(OH)_2(aq)$$

Zinc, aluminum, and iron do not react with cold water but will react with steam at high temperatures, forming hydrogen and a metallic oxide. The equations are

$$Zn(s) + H_2O(steam) \longrightarrow H_2\uparrow + ZnO(s)$$
$$2\,Al(s) + 3\,H_2O(steam) \longrightarrow 3\,H_2\uparrow + Al_2O_3(s)$$
$$3\,Fe(s) + 4\,H_2O(steam) \longrightarrow 4\,H_2\uparrow + Fe_3O_4(s)$$

Copper, silver, and mercury are examples of metals that do not react with cold water or steam to produce hydrogen. We conclude that sodium, potassium, and calcium are chemically more reactive than zinc, aluminum, and iron, which are more reactive than copper, silver, and mercury.

Certain nonmetals react with water under various conditions. For example, fluorine reacts violently with cold water, producing hydrogen fluoride and free oxygen. The reactions of chlorine and bromine are much milder, producing what is commonly known as "chlorine water" and "bromine water," respectively. Chlorine water contains HCl, HOCl, and dissolved Cl_2; the free chlorine gives it a yellow-green color. Bromine water contains HBr, HOBr, and dissolved Br_2; the free bromine gives it a reddish-brown color. Steam passed over hot coke (carbon) produces a mixture of carbon monoxide and hydrogen that is known as "water gas." Since water gas is combustible, it is useful as a fuel. It is also the starting material for the commercial production of several alcohols. The equations for these reactions are

$$2\ F_2(g) + 2\ H_2O(l) \longrightarrow 4\ HF(aq) + O_2(g)$$

$$Cl_2(g) + H_2O(l) \longrightarrow HCl(aq) + HOCl(aq)$$

$$Br_2(l) + H_2O(l) \longrightarrow HBr(aq) + HOBr(aq)$$

$$C(s) + H_2O(g) \xrightarrow{1000°C} CO(g) + H_2(g)$$

Reactions of Water with Metal and Nonmetal Oxides Metal oxides that react with water to form bases are known as **basic anhydrides**. Examples are

basic anhydride

$$CaO(s) + H_2O \longrightarrow Ca(OH)_2(aq)$$

Calcium hydroxide

$$Na_2O(s) + H_2O \longrightarrow 2\ NaOH(aq)$$

Sodium hydroxide

Certain metal oxides, such as CuO and Al_2O_3, do not form basic solutions because the oxides are insoluble in water.

acid anhydride

Nonmetal oxides that react with water to form acids are known as **acid anhydrides**. Examples are

$$CO_2(g) + H_2O(l) \rightleftharpoons H_2CO_3(aq)$$

Carbonic acid

$$SO_2(g) + H_2O(l) \rightleftharpoons H_2SO_3(aq)$$

Sulfurous acid

$$N_2O_5(s) + H_2O(l) \longrightarrow 2\ HNO_3(aq)$$

Nitric acid

The word *anhydrous* means "without water." An anhydride is a metal oxide or a nonmetal oxide derived from a base or an oxy-acid by the removal of water. To determine the formula of an anhydride, the elements of water, H_2O, are removed from an acid or base formula until all the hydrogen is removed. Sometimes more than one formula unit is needed to remove all the hydrogen as water. The formula of the anhydride then consists of the remaining metal or nonmetal and the remaining oxygen atoms. In calcium hydroxide, removal of water as indicated leaves CaO as the anhydride:

$$Ca\begin{matrix}O\boxed{H} \\ \boxed{OH}\end{matrix} \xrightarrow{\Delta} CaO + H_2O$$

In sodium hydroxide, H_2O cannot be removed from one formula unit, so two formula units of NaOH must be used, leaving Na_2O as the formula of the anhydride:

$$\begin{matrix}NaO\boxed{H} \\ Na\boxed{OH}\end{matrix} \xrightarrow{\Delta} Na_2O + H_2O$$

The removal of water from sulfuric acid, H_2SO_4, gives the acid anhydride SO_3:

$$H_2SO_4 \xrightarrow{\Delta} SO_3 + H_2O$$

The foregoing are examples of typical reactions of water but are by no means a complete list of the known reactions of water.

13.10 Hydrates

When certain salt solutions are allowed to evaporate, some water molecules remain as part of the crystalline salt that is left after evaporation is complete. Solids that contain water molecules as part of their crystalline structure are known as **hydrates**. Water in a hydrate is known as **water of hydration**, or **water of crystallization**.

hydrate

water of hydration

water of crystallization

Formulas for hydrates are expressed by first writing the usual anhydrous (without water) formula for the compound and then adding a dot followed by the number of water molecules present. An example is $BaCl_2 \cdot 2\ H_2O$. This formula tells us that each formula unit of this salt contains one barium ion, two chloride ions, and two water molecules. A crystal of the salt contains many of these units in its crystalline lattice.

In naming hydrates, we first name the compound exclusive of the water and then add the term *hydrate*, with the proper prefix representing the number of water molecules in the formula. For example, $BaCl_2 \cdot 2\ H_2O$ is called *barium chloride dihydrate*. Hydrates are true compounds and follow the Law of Definite Composition. The formula weight of $BaCl_2 \cdot 2\ H_2O$ is 244.3; it contains 56.20% barium, 29.06% chlorine, and 14.74% water.

Water molecules in hydrates are bonded by electrostatic forces between polar water molecules and the positive or negative ions of the compound. These forces are not as strong as covalent or ionic chemical bonds. As a result, water of crystallization can be removed by moderate heating of the compound. A partially dehydrated or completely anhydrous compound may result. When $BaCl_2 \cdot 2\ H_2O$ is heated, it loses its water at about 100°C:

$$BaCl_2 \cdot 2\ H_2O \xrightarrow{100^\circ C} BaCl_2 + 2\ H_2O\uparrow$$

When a solution of copper(II) sulfate ($CuSO_4$) is allowed to evaporate, beautiful blue crystals containing 5 moles of water per mole of $CuSO_4$ are formed. The formula for this hydrate is $CuSO_4 \cdot 5\ H_2O$; it is called copper(II) sulfate pentahydrate, or cupric sulfate pentahydrate. When $CuSO_4 \cdot 5\ H_2O$ is heated, water is lost, and a pale green-white powder, anhydrous $CuSO_4$, is formed.

$$CuSO_4 \cdot 5\ H_2O \xrightarrow{250^\circ C} CuSO_4 + 5\ H_2O\uparrow$$

When water is added to anhydrous copper(II) sulfate, the foregoing reaction is reversed, and the salt turns blue again. Because of this outstanding color change, anhydrous copper(II) sulfate has been used as an indicator to detect small amounts of water. The formation of the hydrate is noticeably exothermic.

The formula for plaster of paris is $(CaSO_4)_2 \cdot H_2O$. When mixed with the proper quantity of water, plaster of paris forms a dihydrate and sets to a hard mass. It is, therefore, useful for making patterns for the reproduction of art objects, molds, and surgical casts. The chemical reaction is

$$(CaSO_4)_2 \cdot H_2O(s) + 3\ H_2O(l) \longrightarrow 2\ CaSO_4 \cdot 2\ H_2O(s)$$

The occurrence of hydrates is commonplace in salts. Table 13.4 lists a number of common hydrates.

Table 13.4 Selected hydrates

Hydrate	Name
$CaCl_2 \cdot 2H_2O$	Calcium chloride dihydrate
$Ba(OH)_2 \cdot 8\ H_2O$	Barium hydroxide octahydrate
$MgSO_4 \cdot 7\ H_2O$	Magnesium sulfate heptahydrate
$SnCl_2 \cdot 2\ H_2O$	Tin(II) chloride dihydrate
$CoCl_2 \cdot 6\ H_2O$	Cobalt(II) chloride hexahydrate
$Na_2CO_3 \cdot 10\ H_2O$	Sodium carbonate decahydrate
$(NH_4)_2C_2O_4 \cdot H_2O$	Ammonium oxalate monohydrate
$NaC_2H_3O_2 \cdot 3\ H_2O$	Sodium acetate trihydrate
$Na_2B_4O_7 \cdot 10\ H_2O$	Sodium tetraborate decahydrate
$Na_2S_2O_3 \cdot 5\ H_2O$	Sodium thiosulfate pentahydrate

13.11 Hygroscopic Substances: Deliquescence; Efflorescence

hygroscopic substances

Many anhydrous salts and other substances readily absorb water from the atmosphere. Such substances are said to be **hygroscopic**. This property can be observed in the following simple experiment: Spread a weighed 10–20 g sample of anhydrous copper(II) sulfate on a watch glass and set it aside so that the salt is exposed to the air. Then weigh the sample periodically for 24 hours, noting the increase in weight and the change in color. Water is absorbed from the atmosphere, forming the blue pentahydrate $CuSO_4 \cdot 5\ H_2O$.

deliquescence

Some compounds continue to absorb water beyond the hydrate stage to form solutions. A substance that absorbs water from the air until it forms a solution is said to be **deliquescent**. A few granules of anhydrous calcium chloride or pellets of sodium hydroxide exposed to the air will appear moist in a few minutes, and within an hour will absorb enough water to form a puddle of solution. Phosphorus pentoxide (P_2O_5) picks up water so rapidly that it cannot be weighed accurately except in an anhydrous atmosphere.

Compounds that absorb water are useful as drying agents (desiccants). Refrigeration systems must be kept dry with such agents or the moisture will freeze and clog the tiny orifices in the mechanism. Bags of drying agents are often enclosed in packages containing iron or steel parts to absorb moisture and prevent rusting. Anhydrous calcium chloride, magnesium sulfate, sodium sulfate, calcium sulfate, silica gel, and phosphorus pentoxide are some of the compounds commonly used for drying liquids and gases that contain small amounts of moisture.

efflorescence

The process by which crystalline materials spontaneously lose water when exposed to the air is known as **efflorescence**. Glauber's salt ($Na_2SO_4 \cdot 10\ H_2O$), a transparent crystalline salt, loses water when exposed to the air. One can actually observe these well-defined, large crystals crumbling away as they lose water and form a white, noncrystalline-appearing powder. From our discussion of the decomposition of hydrates, we can predict that heat will increase the rate of efflorescence. The rate also depends on the concentration of moisture in the air. A dry atmosphere will allow the process to take place more rapidly.

13.12 Hydrogen Peroxide and Ozone

Although both water and hydrogen peroxide are compounds of hydrogen and oxygen, their properties are very different. A hydrogen peroxide molecule (H_2O_2) is composed of two H atoms and two O atoms. Its composition by weight is 94.1% oxygen and 5.9% hydrogen. Pure hydrogen peroxide is a pale blue liquid that has a melting point of $-0.41°C$, boils at 150°C, and has a density of 1.44 g/mL at 25°C. It is miscible with water in all proportions. Water solutions of hydrogen peroxide are slightly acidic. The structure of hydrogen peroxide may be represented as

$$:\ddot{O}:\ddot{O}: \text{ (with H above and below)} \quad \text{or} \quad H{-}\ddot{O}{-}\ddot{O}{-}H$$

Hydrogen peroxide is a common, useful source of oxygen, because it decomposes easily to give oxygen and water:

$$2\ H_2O_2(l) \longrightarrow 2\ H_2O(l) + O_2(g) + 192\ \text{kJ}\ (46.0\ \text{kcal})$$

This decomposition is accelerated by heat and light but can be minimized by storing peroxide solutions in brown bottles, keeping them cold, and adding stabilizers. The decomposition can also be accelerated by catalysts such as manganese dioxide.

The peroxide group, like an oxygen molecule (O_2), contains two oxygen atoms linked by a covalent bond. It has a -2 oxidation number and is written O_2^{2-} or $:\ddot{O}:\ddot{O}:^{2-}$; each O atom is considered to have an oxidation number of -1. Metal dioxides also contain two O atoms, but each is bonded individually to the metal ion. Thus, a metal dioxide contains two oxide ($:\ddot{O}:^{2-}$) ions.

Some discretion must be used when working with peroxide formulas. From their formulas, BaO_2 and TiO_2 appear to be similar compounds, but BaO_2 is a peroxide consisting of a $+2$ barium ion and a -2 peroxide ion, whereas titanium dioxide is an oxide consisting of a $+4$ titanium ion and two -2 oxide ions. Peroxides can be distinguished from dioxides chemically, because they generally yield H_2O_2 or O_2 when treated with acids or water. Two examples are the reactions of barium peroxide and sodium peroxide:

$$BaO_2(s) + H_2SO_4(aq) \longrightarrow H_2O_2(aq) + BaSO_4(s)$$
$$Na_2O_2(s) + 2\ H_2O(l) \longrightarrow H_2O_2(aq) + 2\ NaOH(aq)$$

A 3% solution of H_2O_2, which is commonly available at drugstores, is used as an antiseptic to cleanse open wounds. Somewhat stronger H_2O_2 solutions are widely used as bleaching agents for cotton, wood, and hair. For some oxidation processes the chemical industry uses a 30% solution. Concentrations of 85% and higher are used for oxidizing fuels in rocket propulsion. These highly concentrated solutions are extremely sensitive to decomposition and are a fire hazard if allowed to come into contact with organic material.

Ozone is another substance containing multiple oxygen linkages. One molecule, O_3, contains three atoms of oxygen:

$$:\ddot{O}::\ddot{O}: \qquad \ddot{O} \text{ bonded to two } O \text{ atoms}$$

Oxygen Ozone

Ozone can be prepared by passing air or oxygen through an electrical discharge:

$$3\,O_2(g) + 286\text{ kJ }(68.4\text{ kcal}) \xrightarrow[\text{discharge}]{\text{Electrical}} 2\,O_3(g)$$

The characteristic pungent odor of ozone is noticeable in the vicinity of electrical machines and power transmission lines. Ozone is formed in the atmosphere during electrical storms and by the photochemical action of ultraviolet radiation on a mixture of nitrogen dioxide and oxygen. Areas with high air pollution are subject to high atmospheric ozone concentrations.

Ozone is not a desirable low-altitude constituent of the atmosphere, because it is known to cause extensive plant damage, cracking of rubber, and the formation of eye-irritating substances. Concentrations of ozone greater than 0.1 part per million (ppm) of air cause coughing, choking, headache, fatigue, and reduced resistance to respiratory infection. Concentrations between 10 and 20 ppm are fatal to humans.

High-energy radiation from the sun also converts oxygen in the upper atmosphere (stratosphere) to ozone, forming a protective ozone layer in the stratosphere:

$$O_2 \xrightarrow{\text{Sunlight}} \underbrace{O + O}_{\text{Oxygen atoms}}$$

$$O_2 + O \longrightarrow O_3$$

Ultraviolet radiation from the sun is highly damaging to living tissues of plants and animals. The ozone layer, however, shields the earth by absorbing ultraviolet radiation and thus prevents most of this lethal radiation from reaching the earth's surface. The reaction that occurs is the reverse of the preceding one:

$$O_3 \xrightarrow[\text{radiation}]{\text{Ultraviolet}} O_2 + O + \text{heat}$$

Scientists have become concerned about a growing hazard to the ozone layer. Fluorocarbon propellants, such as the Freons, CCl_3F and CCl_2F_2, which were used in aerosol spray cans and are used in refrigeration and air conditioning units, are stable compounds and remain unchanged in the lower atmosphere. When these fluorocarbons are carried by convection currents to the stratosphere, they absorb ultraviolet radiation and produce chlorine atoms (chlorine free radicals) that in turn react with ozone. The following reaction sequence involving free radicals (see Section 16.15) has been proposed to explain the partial destruction of the ozone layer by fluorocarbons.

$$\underset{\text{Fluorocarbon molecule}}{CCl_3F} \xrightarrow[\text{radiation}]{\text{Ultraviolet}} \underset{\text{Fluorocarbon free radical}}{\cdot CCl_2F} + \underset{\text{Chlorine free radical (atom)}}{Cl\cdot}$$

$$Cl\cdot + O_3 \longrightarrow ClO\cdot + O_2$$

$$ClO\cdot + O \longrightarrow O_2 + Cl\cdot$$

Because a chlorine atom (radical) is regenerated for each ozone molecule that is destroyed, a single fluorocarbon molecule can be responsible for the destruction of many ozone molecules.

With the amount of fluorocarbons in the atmosphere already, scientists predict that up to 15% of the ozone layer could be destroyed by the year 2000. This destruction could result in damaging effects on life and major climate changes due to more heat passing into the lower atmosphere.

allotropy

Many elements exist in two or more molecular or crystalline forms. This phenomenon is known as **allotropy** (from the Greek *allotropia*, meaning "variety"). The individual forms of an element are known as allotropic forms, or allotropes. Oxygen (O_2) and ozone (O_3) are allotropic forms of the element oxygen. Two other common elements that exhibit allotropy are sulfur and carbon. Allotropic forms of sulfur are rhombic and monoclinic sulfur. Diamond and graphite are allotropic forms of carbon.

13.13 Natural Waters

Natural fresh waters are not pure, but contain dissolved minerals, suspended matter, and sometimes harmful bacteria. The water supplies of large cities are usually drawn from rivers or lakes. Such water is generally unsafe to drink without treatment. To make such water potable (that is, safe to drink), it is treated by some or all of the following processes:

1. **Screening.** Removal of relatively large objects, such as trash, fish, and so on.
2. **Flocculation and sedimentation.** Chemicals, usually lime and alum (aluminum sulfate), are added to form a flocculent jellylike precipitate of aluminum hydroxide. This precipitate traps most of the fine suspended matter in the water and carries it to the bottom of the sedimentation basin.
3. **Sand filtration.** Water is drawn from the top of the sedimentation basin and passed downward through fine sand filters. Nearly all the remaining suspended matter and bacteria are removed by the sand filters.
4. **Aeration.** Water is drawn from the bottom of the sand filters and is aerated by spraying. The purpose of this process is to remove objectionable odors and tastes.
5. **Disinfection.** In the final stage chlorine gas is injected into the water to kill harmful bacteria before the water is distributed to the public. Ozone is also used in some countries to disinfect water. In emergencies water may be disinfected by simply boiling for a few minutes.

If the drinking water of children contains an optimum amount of fluoride ion, their teeth will be more resistant to decay. Therefore, in many communities NaF or Na_2SiF_6 is added to the water supply to bring the fluoride ion concentration up to the optimum level of about 1.0 ppm. Excessively high concentrations of fluoride ion can cause mottling of the teeth.

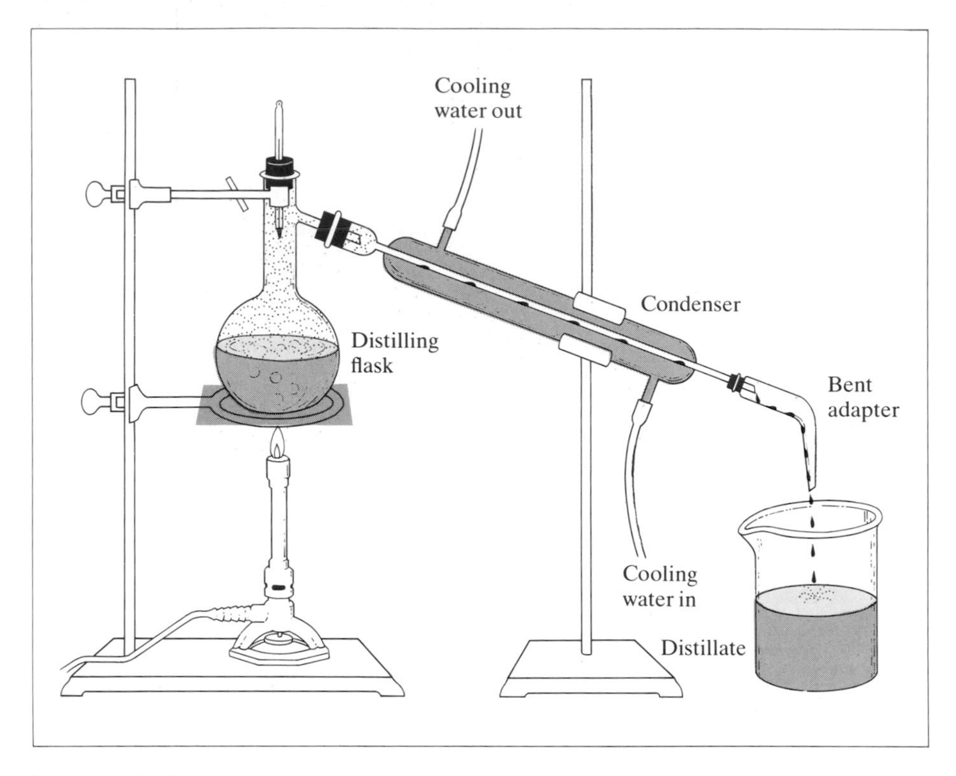

Figure 13.9

Simple laboratory setup for distillation of liquids.

Water that contains dissolved calcium and magnesium salts is called *hard water*. One drawback of hard water is that ordinary soap does not lather well in it; the soap reacts with the calcium and magnesium ions to form an insoluble greasy scum. However, synthetic soaps, known as detergents or syndets, are available; they have excellent cleaning qualities and do not form precipitates with hard water. Hard water is also undesirable because it causes "boiler scale" to form on the walls of water heaters and steam boilers, which greatly reduces their efficiency.

Four techniques used to "soften" hard water are distillation, chemical precipitation, ion exchange, and demineralization. In distillation the water is boiled, and the steam thus formed is condensed to a liquid again, leaving the minerals behind in the distilling vessel. Figure 13.9 illustrates a simple laboratory distillation apparatus. Commercial stills are available that are capable of producing hundreds of liters of distilled water per hour.

Calcium and magnesium ions are precipitated from hard water by adding sodium carbonate and lime. Insoluble calcium carbonate and magnesium hydroxide are precipitated and are removed by filtration or sedimentation.

In the ion-exchange method, used in many households, hard water is

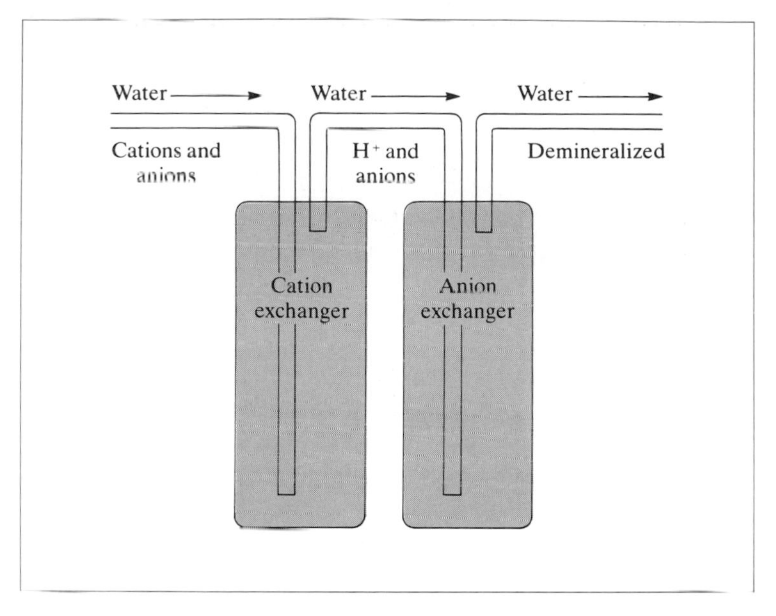

Figure 13.10

Demineralization of water. Water is passed through two beds of synthetic resin. In the cation exchanger metal ions are exchanged for hydrogen ions. In the anion exchanger anions are exchanged for hydroxide ions. The H^+ and OH^- ions react to form water, giving essentially pure, demineralized water.

effectively softened as it is passed through a bed or tank of zeolite. Zeolite is a complex sodium aluminum silicate. In this process sodium ions replace objectionable calcium and magnesium ions, and the water is thereby softened:

$$\mathrm{Na_2Zeolite}(s) + \mathrm{Ca^{2+}}(aq) \longrightarrow \mathrm{CaZeolite}(s) + 2\ \mathrm{Na^+}(aq)$$

The zeolite is regenerated by back-flushing with concentrated sodium chloride solution, reversing the foregoing reaction.

The sodium ions that are present in water softened either by chemical precipitation or by the zeolite process are not objectionable to most users of soft water.

In demineralization both cations and anions are removed by a two-stage ion-exchange system. Special synthetic organic resins are used in the ion-exchange beds. In the first stage metal cations are replaced by hydrogen ions. In the second stage anions are replaced by hydroxide ions. The hydrogen and hydroxide ions react, and essentially pure, mineral-free water leaves the second stage (see Figure 13.10).

The oceans are an inexhaustible source of water; however, seawater contains about 3.5 lb of salts per 100 lb of water. This 35,000 ppm of dissolved salts makes seawater unfit for agricultural and domestic uses. Water that contains less than

1000 ppm of salts is considered reasonably good for drinking, and potable (safe to drink) water is already being obtained from the sea in many parts of the world. Continuous research is being done in an effort to make usable water from the oceans more abundant and economical.

13.14 Water Pollution

Polluted water was formerly thought of as water that was unclear, had a bad odor or taste, and contained disease-causing bacteria. However, such factors as increased population, industrial requirements for water, atmospheric pollution, toxic waste dumps, and use of pesticides have greatly modified the problem of water pollution.

Many of the newer pollutants are not removed or destroyed by the usual water-treatment processes. For example, among the 66 organic compounds found in the drinking water of a major city on the Mississippi River, 3 are labeled slightly toxic, 17 moderately toxic, 15 very toxic, 1 extremely toxic, and 1 supertoxic. Two are known carcinogens (cancer-producing agents), 11 are suspect, and 3 are metabolized to carcinogens. The United States Public Health Service classifies water pollutants under eight broad categories. These categories are shown in Table 13.5.

Many outbreaks of disease or poisoning such as typhoid, dysentery, and cholera have been attributed directly to drinking water. Rivers and streams are an easy means for municipalities to dispose of their domestic and industrial waste products. Much of this water is used again by people downstream, and then discharged back into the water source. Then another community still farther downstream draws the same water and discharges its own wastes. Thus, along waterways such as the Mississippi and Delaware rivers, water is withdrawn and discharged many times. If this water is not properly treated, harmful pollutants will build up, causing epidemics of various diseases.

Hazardous waste products are unavoidable in the manufacture of many products that we use in everyday life. One common way to dispose of these wastes is to place them in toxic waste dumps. What has been found after many years of disposing of wastes in this manner is that toxic substances have seeped into the ground-water deposits. As a result many people have become ill, and water wells have been closed until satisfactory methods of detoxifying this water are found. This problem is serious, because one-half the United States population gets its drinking water from ground water. To clean up the thousands of industrial dumps and to find and implement new and safe methods of disposing of wastes will be very costly.

Mercury and its compounds have long been known to be highly toxic. Mercury gets into the body primarily in the foods we eat. Although it is not an essential mineral for the body, mercury accumulates in the blood, kidneys, liver, and brain tissues. Mercury in the brain causes serious damage to the central nervous system.

Table 13.5 Classification of water pollutants

Type of pollutant	Examples
Oxygen-demanding wastes	Decomposable organic wastes from domestic sewage and industrial wastes of plant and animal origin
Infectious agents	Bacteria, viruses, and other organisms from domestic sewage, animal wastes, and animal process wastes
Plant nutrients	Principally compounds of nitrogen and phosphorus
Organic chemicals	Large numbers of chemicals synthesized by industry, pesticides, chlorinated organic compounds
Other minerals and chemicals	Inorganic chemicals from industrial operations, mining, oil field operations, and agriculture
Radioactive substances	Waste products from mining and processing of radioactive materials, airborne radioactive fallout, increased use of radioactive materials in hospitals and research
Heat from industry	Large quantities of heated water returned to water bodies from power plants and manufacturing facilities after use for cooling
Sediment from land erosion	Solid matter washed into streams and oceans by erosion, rain, and water runoff

The sequence of events that have led to incidents of mercury poisoning is as follows: Mercury and its compounds are used in many industries and in agriculture, primarily as a fungicide in the treatment of seeds. One of the largest uses is in the electrochemical conversion of sodium chloride brines to chlorine and sodium hydroxide, as represented by this equation:

$$2\ NaCl + 2\ H_2O \xrightarrow{\text{Electrolysis}} Cl_2 + 2\ NaOH + H_2$$

Although no mercury is shown in the chemical equation, it is used in the process for electrical contact, and small amounts are discharged along with spent brine solutions. Thus considerable quantities of mercury, in low concentrations, have been discharged into lakes and other surface waters from the effluents of these manufacturing plants. The mercury compounds discharged into the water are converted by bacterial action and other organic compounds to methyl mercury, $(CH_3)_2Hg$, which then accumulates in the bodies of fish. Several major episodes of mercury poisoning that have occurred in the past years were the result of eating mercury-contaminated fish. The best way to control this contaminant is at the source, and much has been done since 1970 to eliminate the discharge of mercury in industrial wastes. In 1976 the Environmental Protection Agency banned the use of all mercury-containing insecticides and fungicides.

Many other major water pollutants have been recognized and steps have been taken to eliminate them. Three that pose serious problems are lead, detergents, and chlorine-containing organic compounds. Lead poisoning, for example, has been responsible for many deaths in past years. The toxic action of lead in the body is the inhibition of the enzymc necessary for the production of hemoglobin in the blood. The usual intake of lead into the body is through food. However, extraordinary amounts of lead can be ingested from water running through lead pipes and by using lead-containing ceramic containers for storage of food and beverages.

It has been clearly demonstrated that waterways rendered so polluted that the water is neither fit for human use nor able to sustain marine life can be successfully restored. However, keeping our lakes and rivers free from pollution is a very costly and complicated process.

QUESTIONS

A. *Review the meanings of the new terms introduced in this chapter.*

1. Heat of fusion
2. Heat of vaporization
3. Hydrogen bond
4. Evaporation
5. Vaporization
6. Sublimation
7. Vapor pressure
8. Volatile
9. Boiling point
10. Standard or normal boiling point
11. Vapor pressure curve
12. Freezing or melting point
13. Basic anhydride
14. Acid anhydride
15. Hydrate
16. Water of hydration
17. Water of crystallization
18. Hygroscopic substances
19. Deliquescence
20. Efflorescence
21. Allotropy

B. *Information useful in answering the following questions will be found in tables and figures.*

1. Compare the potential energy of the two states of water shown in Figure 13.1.
2. In what state (solid, liquid, or gas) would H_2S, H_2Se, and H_2Te be at 0°C? (See Table 13.1.)
3. The two thermometers in Figure 13.2 read 100°C. What is the pressure of the atmosphere?
4. Draw a diagram of a water molecule and point out the areas that are the negative and positive ends of the dipole.
5. If the water molecule were linear, with all three atoms in a straight line rather than in the shape of a V, as shown in Figure 13.3, what effect would this have on the physical properties of water?
6. Based on Table 13.4, how do we specify 1, 2, 3, 4, 5, 6, 7, and 8 molecules of water in the formulas of hydrates?
7. Would the distillation setup in Figure 13.9 be satisfactory for separating salt and water? Ethyl alcohol and water? Explain.
8. If the liquid in the flask in Figure 13.9 is ethyl alcohol and the atmospheric pressure is 543 torr, what temperature will show on the thermometer? (Use Figure 13.7.)

9. If water were placed in both containers in Figure 13.5, would both have the same vapor pressure at the same temperature? Explain.
10. In Figure 13.5, in which case, (a) or (b), will the atmosphere above the liquid reach a point of saturation?
11. Suppose that a solution of ethyl ether and ethyl alcohol were placed in the closed bottle in Figure 13.5. Use Figure 13.7 for information on the substances.
    (a) Would both substances be present in the vapor?
    (b) If the answer to part (a) is yes, which would have more molecules in the vapor?
12. In Figure 13.6, if 50% more water had been added in part (b), what equilibrium vapor pressure would have been observed in (c)?
13. At approximately what temperature would each of the substances listed in Table 13.3 boil when the pressure is 30 torr? (See Figure 13.7.)
14. Use the graph in Figure 13.7 to find the following:
    (a) The boiling point of water at 500 torr pressure
    (b) The normal boiling point of ethyl alcohol
    (c) The boiling point of ethyl ether at 0.50 atm
15. Consider Figure 13.8.
    (a) Why is line *BC* horizontal? What is happening in this interval?
    (b) What phases are present in the interval *BC*?
    (c) When heating is continued after point *C*, another horizontal line, *DE*, is reached at a higher temperature. What does this line represent?

C. *Review questions*

1. List six physical properties of water.
2. What condition is necessary for water to have its maximum density? What is its maximum density?
3. Account for the fact that an ice–water mixture remains at 0°C until all the ice is melted, even though heat is applied to it.
4. Which contains less heat, ice at 0°C or water at 0°C? Explain.
5. Why does ice float in water? Would ice float in ethyl alcohol ($d = 0.789$ g/mL)? Explain.
6. If water molecules were linear instead of bent, would the heat of vaporization be higher or lower? Explain.
7. The heat of vaporization for ethyl ether is 351 J/g (83.9 cal/g) and that for ethyl alcohol is 855 J/g (204.3 cal/g). Which of these compounds has hydrogen bonding? Explain.
8. Would there be more or less H-bonding if water molecules were linear instead of bent? Explain.
9. Which would show hydrogen bonding, ammonia, NH_3, or methane, CH_4? Explain.
10. In which condition are there fewer hydrogen bonds between molecules: water at 40°C or water at 80°C?
11. Which compound, $H_2NCH_2CH_2NH_2$ or $CH_3CH_2CH_2NH_2$, would you expect to have the higher boiling point? Explain your answer. (Both compounds have similar molecular weights.)
12. The vapor pressure at 20°C is given for the following compounds:

Methyl alcohol	96 torr
Acetic acid	11.7 torr
Benzene	74.7 torr
Bromine	173 torr
Water	17.5 torr
Carbon tetrachloride	91 torr
Mercury	0.0012 torr
Toluene	23 torr

    (a) Arrange these compounds in their order of increasing rate of evaporation.
    (b) Which substance listed would have the highest boiling point, and which would have the lowest?
13. Explain why rubbing alcohol, warmed to body temperature, still feels cold when applied to your skin.
14. Suggest a method whereby water could be made to boil at 50°C.
15. If a dish of water initially at 20°C is placed in a living room maintained at 20°C, the water temperature will fall below 20°C. Explain.
16. Explain why a higher temperature is obtained in a pressure cooker than in an ordinary cooking pot.
17. What is the relationship between vapor pressure and boiling point?
18. On the basis of the Kinetic-Molecular Theory, explain why vapor pressure increases with temperature.
19. Why does water have such a relatively high boiling point?

20. The boiling point of ammonia, NH_3, is $-33.4°C$ and that of sulfur dioxide, SO_2, is $-10.0°C$. Which has the higher vapor pressure at $-40°C$?
21. Explain what is occurring physically when a substance is boiling.
22. Explain why HF (bp = 19.4°C) has a higher boiling point than HCl (bp = −85°C), whereas F_2 (bp = −188°C) has a lower boiling point than Cl_2 (bp = −34°C).
23. Can ice be colder than 0°C? Explain.
24. Why does a boiling liquid maintain a constant temperature when heat is continuously being added?
25. At what specific temperature will copper have a vapor pressure of 760 torr?
26. Why does a lake freeze from the top down?
27. What water temperature would you theoretically expect to find at the bottom of a very deep lake? Explain.
28. Write equations to show how the following metals react with water: aluminum, calcium, iron, sodium, zinc. State the conditions for each reaction.
29. Is the formation of hydrogen and oxygen from water an exothermic or an endothermic reaction? How do you know?
30. (a) What is an anhydride?
    (b) What type of compound will be an acid anhydride?
    (c) What type of compound will be a basic anhydride?
31. (a) Write the formulas for the anhydrides of the following acids:
    H_2SO_3, H_2SO_4, HNO_3, $HClO_4$, H_2CO_3, H_3PO_4
    (b) Write the formulas for the anhydrides of the following bases:
    $NaOH$, KOH, $Ba(OH)_2$, $Ca(OH)_2$, $Mg(OH)_2$
32. Complete and balance the following equations:
    (a) $Ba(OH)_2 \xrightarrow{\Delta}$
    (b) $\underset{\text{Methyl alcohol}}{CH_3OH} + O_2 \longrightarrow$
    (c) $Rb + H_2O \longrightarrow$
    (d) $SnCl_2 \cdot 2\,H_2O \xrightarrow{\Delta}$
    (e) $HNO_3 + NaOH \longrightarrow$
    (f) $LiO + H_2O \longrightarrow$
    (g) $KOH \xrightarrow{\Delta}$
    (h) $Ba + H_2O \longrightarrow$
    (i) $Cl_2 + H_2O \longrightarrow$
    (j) $SO_3 + H_2O \longrightarrow$
    (k) $H_2SO_3 + KOH \longrightarrow$
    (l) $CO_2 + H_2O \longrightarrow$
33. Is the conversion of oxygen to ozone an exothermic or an endothermic reaction? How do you know?
34. Write formulas for an oxygen atom, an oxygen molecule, an ozone molecule, and a peroxide ion. How many electrons are in a peroxide ion? How many in an oxygen molecule?
35. How does ozone in the stratosphere protect the earth from excessive ultraviolet radiation?
36. Name each of the following hydrates:
    (a) $BaBr_2 \cdot 2\,H_2O$ (d) $MgNH_4PO_4 \cdot 6\,H_2O$
    (b) $AlCl_3 \cdot 6\,H_2O$ (e) $FeSO_4 \cdot 7\,H_2O$
    (c) $FePO_4 \cdot 4\,H_2O$ (f) $SnCl_4 \cdot 5\,H_2O$
37. Explain how anhydrous copper(II) sulfate ($CuSO_4$) can act as an indicator for moisture.
38. Compare the types of bonds in metal dioxides and metal peroxides.
39. Distinguish between deionized water and:
    (a) Hard water (c) Distilled water
    (b) Soft water
40. How can soap function to make soft water from hard water? What objections are there to using soap for this purpose?
41. What substance is commonly used to destroy bacteria in water?
42. What chemical, other than chlorine or chlorine compounds, can be used to disinfect water for domestic use?
43. Some organic pollutants in water can be oxidized by dissolved molecular oxygen. What harmful effect can result from this depletion of oxygen in the water?
44. Why should you not drink liquids that are stored in ceramic containers, especially unglazed ones?
45. Write the chemical equation showing how magnesium ions are removed by a zeolite water softener.
46. Write an equation to show how hard water containing calcium chloride ($CaCl_2$) is softened by using sodium carbonate (Na_2CO_3).
47. Which of the following statements are correct?
    (a) The process of a substance changing directly from a solid to a gas is called sublimation.
    (b) When water is decomposed, the volume ratio of H_2 to O_2 is 2:1, but the mass ratio of H_2 to O_2 is 1:8.

(c) Hydrogen sulfide is a larger molecule than water.
(d) The changing of ice into water is an exothermic process.
(e) Water and hydrogen fluoride are both nonpolar molecules.
(f) Hydrogen bonding is stronger in H_2O than in H_2S because oxygen is more electronegative than sulfur.
(g) $H_2O_2 \longrightarrow 2\ H_2O + O_2$ represents a balanced equation for the decomposition of hydrogen peroxide.
(h) Steam at 100°C can cause more severe burns than liquid water at 100°C.
(i) The density of water is independent of temperature.
(j) Liquid A boils at a lower temperature than liquid B. This fact indicates that liquid A has a lower vapor pressure than liquid B at any particular temperature.
(k) Water boils at a higher temperature in the mountains than at sea level.
(l) No matter how much heat you put under an open pot of pure water on a stove, you cannot heat the water above its boiling point.
(m) The vapor pressure of a liquid at its boiling point is equal to the prevailing atmospheric pressure.
(n) The normal boiling temperature of water is 273°C.
(o) The pressure exerted by a vapor in equilibrium with its liquid is known as the vapor pressure of the liquid.
(p) Sodium, potassium, and calcium each react with water to form hydrogen gas and a metal hydroxide.
(q) Calcium oxide reacts with water to form calcium hydroxide and hydrogen gas.
(r) Carbon dioxide is the hydride of carbonic acid.
(s) Water in a hydrate is known as water of hydration or water of crystallization.
(t) A substance that spontaneously loses its water of hydration when exposed to the air is said to be efflorescent.
(u) A substance that absorbs water from the air until it forms a solution is deliquescent.
(v) Distillation is effective for softening water because the minerals boil away, leaving soft water behind.
(w) The original source of mercury-contaminated fish is industrial pollution.
(x) Disposal of toxic industrial wastes in toxic waste dumps has been found to be a very satisfactory long-term solution to the problem of what to do with these wastes.
(y) The amount of heat needed to change 1 mole of ice at 0°C to a liquid at 0°C is 6.02 kJ (1.44 kcal).
(z) $BaCl_2 \cdot 2\ H_2O$ has a higher percentage of water than does $CaCl_2 \cdot 2\ H_2O$.

D. *Review problems*

1. How many moles of compound are in 100 g of each of these hydrates?
   (a) $CoCl_2 \cdot 6\ H_2O$ (b) $FeI_2 \cdot 4\ H_2O$
2. How many moles of water can be obtained from 100 g of each of these hydrates?
   (a) $CoCl_2 \cdot 6\ H_2O$ (b) $FeI_2 \cdot 4\ H_2O$
3. When a person purchases epsom salts, $MgSO_4 \cdot 7\ H_2O$, what percentage of the compound is water?
4. Calculate the percentage weight of water in the hydrate $Al_2(SO_4)_3 \cdot 18\ H_2O$.
5. Sugar of lead, a hydrate of lead acetate, $Pb(C_2H_3O_2)_2$, contains 14.2% H_2O. What is the formula for the hydrate?
6. A 25.0 g sample of a hydrate of $FePO_4$ was heated until no more water was driven off. The anhydrous sample weighed 16.9 g. What is the formula of the hydrate?
7. How many joules are needed to change 120 g of water at 20°C to steam at 100°C?
8. How many joules of energy must be removed from 126 g of water at 24°C to form ice at 0°C?
9. How many calories are required to change 225 g of ice at 0°C to steam at 100°C?
10. The molar heat of vaporization is the number of joules required to change 1 mole of a substance from liquid to vapor at its boiling point. What is the molar heat of vaporization of water?
11. Suppose 100 g of ice at 0°C are added to 300 g of water at 25°C. Is this sufficient ice to lower the temperature of the system to 0°C and still have ice remaining? Show evidence for your answer.

*12. Suppose 35.0 g of steam at 100°C are added to 300 g of water at 25°C. Is there sufficient steam to heat all the water to 100°C and still have steam remaining? Show evidence for your answer.

13. How many joules of energy would be liberated by condensing 50.0 mol of steam and allowing the liquid to cool to 30.0°C?

14. How many kilojoules of energy are needed to convert 100 g of ice at −10.0°C to water at 20.0°C. (The specific heat of ice at −10.0°C is 2.01 J/g°C.)

15. What weight of water must be decomposed to produce 25.0 L of oxygen at STP?

16. Compare the volume occupied by 1.00 mol of liquid water at 0°C and 1.00 mol of water vapor at STP.

17. (a) How many moles of oxygen can be obtained by the decomposition of 12.0 mol of hydrogen peroxide?
    (b) What volume will this much oxygen occupy at 35°C and 710 torr pressure?

18. What volume of oxygen at STP can be obtained by decomposing 625 g of 3.0% hydrogen peroxide?

19. How many grams of water will react with each of the following?
    (a) 1.00 mol K (d) 1.00 mole SO_3
    (b) 1.00 mol Ca (e) 1.00 g MgO
    (c) 1.00 g Na (f) 1.00 g N_2O_5

20. Suppose 1.00 mol of water evaporates in 1.00 day. How many water molecules, on the average, leave the liquid each second?

*21. A quantity of sulfuric acid is added to 100 mL of water. The final volume of the solution is 122 mL and it has a density of 1.26 g/mL. What weight of acid was added? Assume the density of the water is 1.00 g/mL.

22. A mixture of 80.0 mL of hydrogen and 60.0 mL of oxygen is ignited by a spark to form water.
    (a) Does any gas remain unreacted? Which one, H_2 or O_2?
    (b) What volume of which gas (if any) remains unreacted? (Assume the same conditions before and after the reaction.)

CHAPTER 14 Solutions

After studying Chapter 14, you should be able to

1. Understand the terms listed in Question A at the end of the chapter.
2. Describe the different types of solutions that are possible, based on the three states of matter.
3. List the general properties of solutions.
4. Outline the solubility rules for common salts and hydroxides.
5. Describe and illustrate the process by which an ionic substance like sodium chloride dissolves in water.
6. Tell how temperature changes affect the solubilities of solids and gases in liquids.
7. Tell how changes of pressure and temperature affect the solubility of a gas in a liquid.
8. Identify and discuss the variables that affect the rate at which a solid dissolves in a liquid.
9. Determine by using a solubility graph or table whether a given solution is unsaturated, saturated, or supersaturated at a given temperature.
10. Calculate the weight percent or volume percent composition of a solution from appropriate data.
11. Calculate the amount of solute in a given quantity of a solution when given the weight percent or volume percent composition.
12. Calculate the molarity of a solution when given the volume of solution and the grams or moles of solute.
13. Calculate the weight of a substance needed to prepare a solution of specified volume and molarity.

14 Determine the resulting molarity when a given volume of a solution of known molarity is mixed with a specified volume of water or is mixed with a solution of different molarity.
15 Apply stoichiometric principles to chemical reactions in which the amounts of reactants are given in units of volume and molarity.
16 Understand the concepts of equivalent weight and normality and do calculations involving these concepts.
17 Relate the effect of a solute on the vapor pressure of a solvent to the freezing point and the boiling point of a solution.
18 Calculate the boiling point or freezing point of a solution from appropriate concentration data.
19 Calculate molality and molecular weight of a solute from boiling point or freezing point and weight concentration data.

14.1 Components of a Solution

solution

solute

solvent

The term **solution** is used in chemistry to describe a system in which one or more substances are homogeneously mixed or dissolved in another substance. A simple solution has two components, a solute and a solvent. The **solute** is the component that is dissolved or the least abundant component in the solution. The **solvent** is the dissolving agent or the most abundant component in the solution. For example, when salt is dissolved in water to form a solution, salt is the solute and water is the solvent. Complex solutions containing more than one solute and/or more than one solvent are common.

14.2 Types of Solution

From the three states of matter—solid, liquid, and gas—it is possible to have nine different types of solutions: solid dissolved in solid, solid dissolved in liquid, solid dissolved in gas, liquid dissolved in liquid, and so on. Of these, the most common solutions are solid dissolved in liquid, liquid dissolved in liquid, gas dissolved in liquid, and gas dissolved in gas.

14.3 General Properties of Solutions

A true solution is one in which the particles of dissolved solute are molecular or ionic in size, generally in the range of 0.1 to 1 nm (10^{-8} to 10^{-7} cm). The properties of a true solution are as follows:

1 It is a homogeneous mixture of two or more components, solute and solvent.

2 It has a variable composition; that is, the ratio of solute to solvent may be varied.
3 The dissolved solute is molecular or ionic in size.
4 It may be either colored or colorless but is usually transparent.
5 The solute remains uniformly distributed throughout the solution and will not settle out with time.
6 The solute generally can be separated from the solvent by purely physical means (for example, by evaporation).

These properties are illustrated by water solutions of sugar and of potassium permanganate. Suppose that we prepare two sugar solutions, the first containing 10 g of sugar added to 100 mL of water and the second containing 20 g of sugar added to 100 mL of water. Each solution is stirred until all the solute dissolves, demonstrating that we can vary the composition of a solution. Every portion of the solution has the same sweet taste because the sugar molecules are uniformly distributed throughout. If confined so that no solvent is lost, the solution will taste and appear the same a week or a month later. The properties of the solution are unaltered after the solution is passed through filter paper. But by carefully evaporating the water, we can recover the sugar from the solution.

To observe the dissolving of potassium permanganate ($KMnO_4$) we affix a few crystals of $KMnO_4$ to paraffin wax or rubber cement at the end of a glass rod and submerge the entire rod, with the wax–permanganate end up, in a cylinder of water. Almost at once the beautiful purple color of dissolved permanganate ions (MnO_4^-) appears at the top of the rod and streams to the bottom of the cylinder as the crystals dissolve. The purple color at first is mostly at the bottom of the cylinder because potassium permanganate is denser than water. But after a while the purple color disperses until it is evenly distributed throughout the solution. This dispersal demonstrates that molecules and ions move about freely and spontaneously (diffuse) in a liquid or solution.

Once formed, a solution is permanent; the solute particles do not settle out. Solution permanency is explained in terms of the Kinetic-Molecular Theory (see Section 12.2). According to the KMT both the solute and solvent particles (molecules and/or ions) are in constant random thermal motion. This motion is energetic enough to prevent the solute particles from settling out under the influence of gravity. This same ceaseless, random thermal motion is also responsible for diffusion in liquids as well as in gases.

14.4 Solubility

solubility

The term **solubility** describes the amount of one substance (solute) that will dissolve in a specified amount of another substance (solvent) under stated conditions. For example, 36.0 g of sodium chloride (NaCl) will dissolve in 100 g of water at 20°C. We say, then, that the solubility of NaCl in water is 36.0 g per 100 g of water at 20°C.

Solubility is often used in a relative way. We say that a substance is very

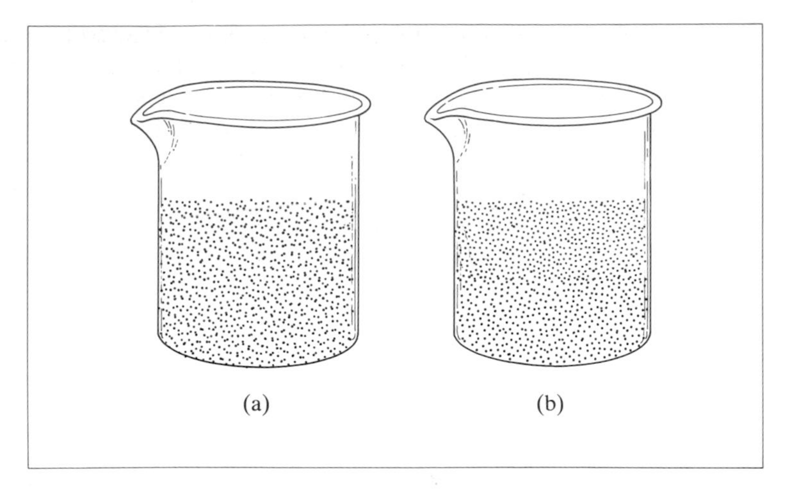

Figure 14.1

Miscible and immiscible systems: (a) miscible, H_2O and CH_3OH; (b) immiscible, H_2O and CCl_4. In a miscible system a solution is formed consisting of a single phase with the solute and solvent uniformly dispersed. An immiscible system is heterogeneous and, in the case of two liquids, forms two liquid layers.

Table 14.1 General solubility rules for common salts and hydroxides[a]

Class	Solubility in cold water[b]
Nitrates	Most nitrates are soluble.
Acetates	Most acetates are soluble.
Chlorides, Bromides, Iodides	Most chlorides, bromides, and iodides are soluble except those of Ag, Hg(I), and Pb(II); $PbCl_2$ and $PbBr_2$ are slightly soluble in hot water.
Sulfates	Most sulfates are soluble except those of Ba, Sr, and Pb; Ca and Ag sulfates are slightly soluble.
Carbonates, Phosphates	Most carbonates and phosphates are insoluble except those of Na, K, and NH_4^+. Many bicarbonates and acid phosphates are soluble.
Hydroxides	Most hydroxides are insoluble except those of the alkali metals and NH_4OH; $Ba(OH)_2$ and $Ca(OH)_2$ are slightly soluble.
Sodium salts, Potassium salts, Ammonium salts	Most common salts of these ions are soluble.
Sulfides	Most sulfides are insoluble except those of the alkali metals, ammonium, and the alkaline earth metals (Ca, Mg, Ba).

[a] When we say a substance is soluble, we mean that the substance is reasonably soluble. All substances have some solubility in water, although the amount of solubility may be very small; the solubility of silver iodide, for example, is about 1×10^{-8} mol AgI/liter H_2O.
[b] These rules have exceptions.

soluble, moderately soluble, slightly soluble, or insoluble. Although these terms do not accurately indicate how much solute will dissolve, they are frequently used to describe the solubility of a substance qualitatively.

Two other terms often used to describe solubility are miscible and immiscible. Liquids that are capable of mixing and forming a solution are **miscible**; those that do not form solutions or are generally insoluble in each other are **immiscible**. Methyl alcohol and water are miscible in each other in all proportions. Carbon tetrachloride and water are immiscible, forming two separate layers when they are mixed. Miscible and immiscible systems are illustrated in Figure 14.1.

miscible

immiscible

The general rules for the solubility of common salts and hydroxides are given in Table 14.1. The solubilities of over 200 compounds are given in the Solubility Table in Appendix IV. Solubility data for thousands of compounds can be found by consulting standard reference sources.*

The quantitative expression of the amount of dissolved solute in a particular quantity of solvent is known as the **concentration of a solution**. Several methods of expressing concentration will be described in Section 14.8.

concentration of a solution

14.5 Factors Related to Solubility

Predicting solubilities is complex and difficult. Many variables, such as size of ions, charge on ions, interaction between ions, interaction between solute and solvent, and temperature, bear upon the problem. Because of the factors involved, the general rules of solubility given in Table 14.1 have many exceptions. However, the rules are very useful, because they do apply to many of the more common compounds that we encounter in the study of chemistry. Keep in mind that these are rules, not laws, and are therefore subject to exceptions. Fortunately the solubility of a solute is relatively easy to determine experimentally. Factors related to solubility are discussed in the following paragraphs.

1. The Nature of the Solute and Solvent The old adage "like dissolves like" has merit, in a general way. Polar or ionic substances tend to be more miscible, or soluble, with other polar substances. Nonpolar substances tend to be miscible with other nonpolar substances and less miscible with polar substances. Thus, mineral acids, bases, and salts, which are polar, tend to be much more soluble in water, which is polar, than in solvents such as ether, carbon tetrachloride, or benzene, which are essentially nonpolar. Sodium chloride, an ionic substance, is soluble in water, slightly soluble in ethyl alcohol (less polar than water), and insoluble in ether and benzene. Pentane (C_5H_{12}), a nonpolar substance, is only slightly soluble in water but is very soluble in benzene and ether.

* Two commonly used handbooks are *Lange's Handbook of Chemistry*, 13th ed. (New York: McGraw-Hill, 1985), and *Handbook of Chemistry and Physics*, 66th ed. (Cleveland: Chemical Rubber Co., 1985).

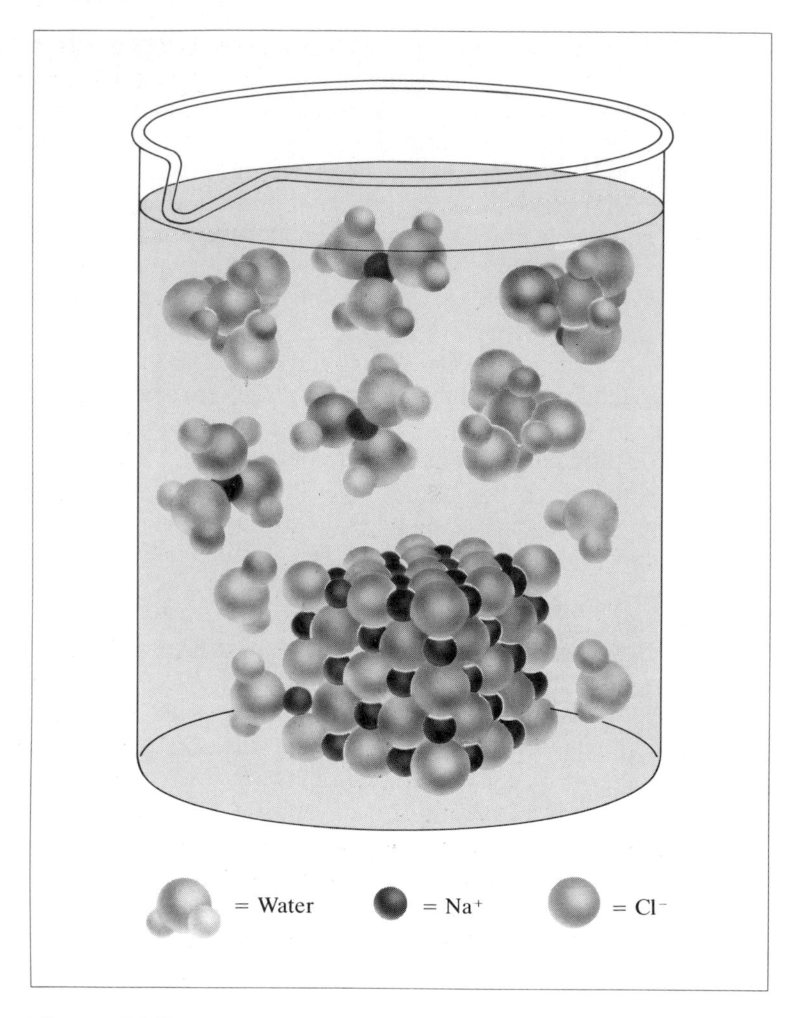

Figure 14.2

Dissolution of sodium chloride in water. Polar water molecules are attracted to Na^+ and Cl^- ions in the salt crystal, weakening the attraction between the ions. As the attraction between the ions weakens, the ions move apart and become surrounded by water dipoles. The hydrated ions slowly diffuse away from the crystal to become dissolved in solution.

At the molecular level the formation of a solution from two nonpolar substances, such as carbon tetrachloride and benzene, can be visualized as a process of simple mixing. The nonpolar molecules, having little tendency to either attract or repel one another, easily intermingle to form a homogeneous mixture.

Solution formation between polar substances is much more complex. For example, the process by which sodium chloride dissolves in water is illustrated in

Table 14.2 Solubility of alkali metal halides in water

Salt	Solubility (g salt/100 g H_2O) 0°C	100°C
LiF	0.12	0.14 (at 35°C)
LiCl	67	127.5
LiBr	143	266
LiI	151	481
NaF	4	5
NaCl	35.7	39.8
NaBr	79.5	121
NaI	158.7	302
KF	92.3 (at 18°C)	Very soluble
KCl	27.6	57.6
KBr	53.5	104
KI	127.5	208

Figure 14.2. Water molecules are very polar and are attracted to other polar molecules or ions. When salt crystals (NaCl) are put into water, polar water molecules become attracted to the sodium and chloride ions on the crystal surfaces and weaken the attraction between Na^+ and Cl^- ions. The positive end of the water dipole is attracted to the Cl^- ions, and the negative end of the water dipole to the Na^+ ions. The weaker attraction permits the ions to move apart, making room for more water dipoles. Thus, the surface ions are surrounded by water molecules, becoming hydrated ions, $Na^+(aq)$ and $Cl^-(aq)$, and slowly diffuse away from the crystals and dissolve in solution.

$$\text{NaCl(crystal)} \xrightarrow{H_2O} Na^+(aq) + Cl^-(aq)$$

Examination of the data in Table 14.2 reveals some of the complex questions relating to solubility. For example: Why are lithium halides, except for lithium fluoride (LiF), more soluble than sodium and potassium halides? Why are the solubilities of LiF and sodium fluoride (NaF) so low in comparison with those of the other salts? Why does not the solubility of LiF, NaF, and NaCl increase proportionately with temperature, as do the solubilities of the other salts? Sodium chloride is appreciably soluble in water but is insoluble in concentrated hydrochloric acid (HCl) solution. On the other hand, LiF and NaF are not very soluble in water but are quite soluble in hydrofluoric acid (HF) solution—why? These questions will not be answered directly here, but it is hoped that your curiosity will be aroused to the point that you will do some reading and research on the properties of solutions.

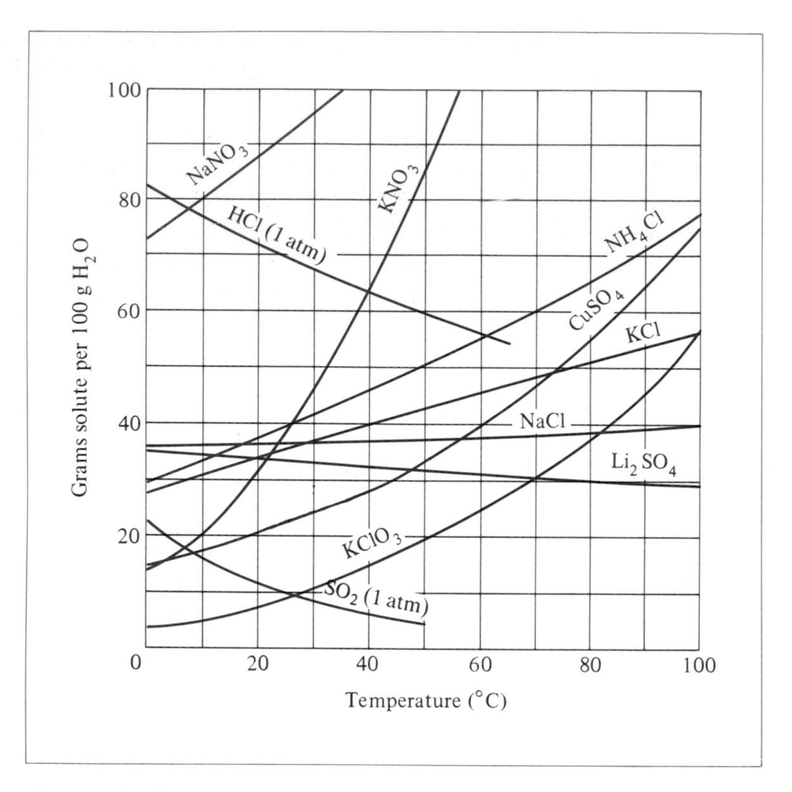

Figure 14.3
Solubility of various compounds in water

2. The Effect of Temperature on Solubility Temperature has major effects on the solubility of most substances, and most solutes have a limited solubility in a specific solvent at a fixed temperature. For most solids dissolved in a liquid, an increase in temperature results in increased solubility (see Figure 14.3). However, no completely valid general rule governs the solubility of solids in liquids with change in temperature. Some solids increase in solubility only slightly with increasing temperature (see NaCl in Figure 14.3); other solids decrease in solubility with increasing temperature (see Li_2SO_4 in Figure 14.3).

On the other hand, the solubility of a gas in a liquid always decreases with increasing temperature (see HCl and SO_2 in Figure 14.3). The tiny bubbles that are formed when water is heated are due to the decreased solubility of air at higher temperatures. The decreased solubility of gases at higher temperatures is explained in terms of the KMT by assuming that, in order to dissolve, the gas molecules must form "bonds" of some sort with the molecules of the liquid. An increase in temperature decreases the solubility of the gas because it increases the kinetic energy (speed) of the gas molecules and thereby decreases their ability to form "bonds" with the liquid molecules.

3. The Effect of Pressure on Solubility Small changes in pressure have little effect on the solubility of solids in liquids or liquids in liquids but have a marked effect on the solubility of gases in liquids. The solubility of a gas in a liquid is directly proportional to the pressure of that gas above the solution. Thus, the amount of a gas that is dissolved in solution will double if the pressure of that gas over the solution is doubled. For example, carbonated beverages contain dissolved carbon dioxide at pressures greater than atmospheric pressure. When a bottle of carbonated soda is opened, the pressure is immediately reduced to the atmospheric pressure, and the excess dissolved carbon dioxide bubbles out of the solution.

14.6 Rate of Dissolving Solids

The rate at which a solid dissolves is governed by (1) the size of the solute particles, (2) the temperature, (3) the concentration of the solution, and (4) agitation or stirring.

1. Particle Size A solid can dissolve only at the surface that is in contact with the solvent. Because the surface to volume ratio increases as size decreases, smaller crystals dissolve faster than large ones. For example, if a salt crystal 1 cm on a side (6 cm^2 surface area) is divided into 1000 cubes, each 0.1 cm on a side, the total surface of the smaller cubes is 60 cm^2—a tenfold increase in surface area (see Figure 14.4).

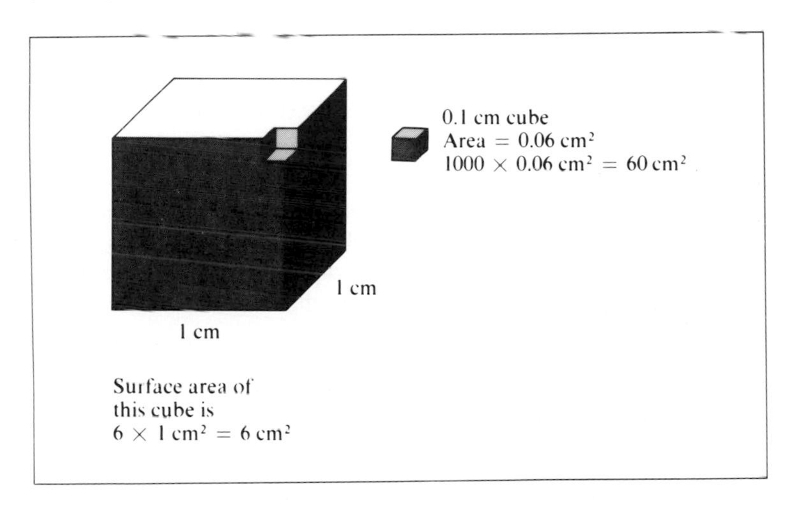

Figure 14.4

Surface area of crystals. A crystal 1 cm on a side has a surface area of 6 cm^2. Subdivided into 1000 smaller crystals, each 0.1 cm on a side, the total surface area is increased to 60 cm^2.

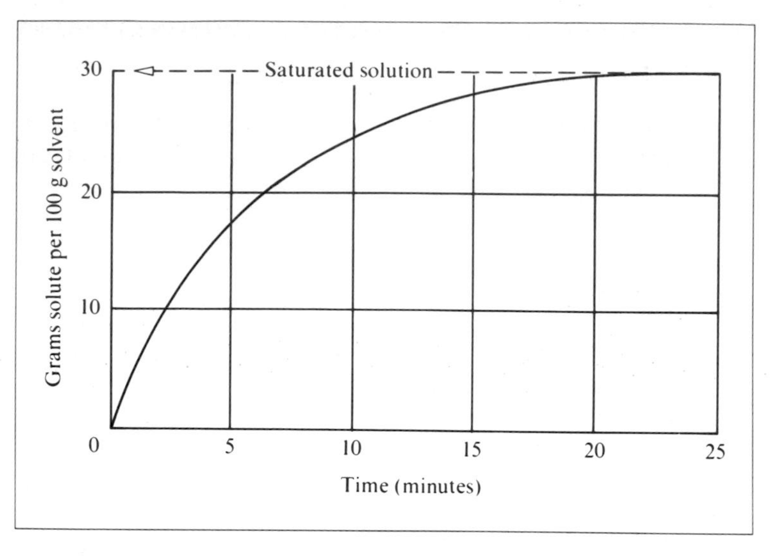

Figure 14.5

Rate of dissolution of a solid solute in a solvent. The rate is maximum at the beginning and decreases as the concentration approaches saturation.

2. Temperature In most cases the rate of dissolving of a solid increases with temperature. This increase is due to kinetic effects: The solvent molecules move more rapidly at higher temperatures and strike the solid surfaces more often and harder, causing the rate of dissolving to increase.

3. Concentration of the Solution When the solute and solvent are first mixed, the rate of dissolving is at its maximum. As the concentration of the solution increases and the solution becomes more nearly saturated with the solute, the rate of dissolving decreases greatly. The rate of dissolving is pictured graphically in Figure 14.5. Note that about 17 g dissolve in the first 5 minute interval, but only about 1 g dissolves in the fourth 5 minute interval. Although different solutes show different rates, the rate of dissolving always becomes very slow as the concentration approaches the saturation point.

4. Agitation or Stirring The effect of agitation or stirring is kinetic. When a solid is first put into water, the only solvent with which it comes in contact is in the immediate vicinity. As the solid dissolves, the amount of dissolved solute around the solid becomes more and more concentrated, and the rate of dissolving slows down. If the mixture is not stirred, the dissolved solute diffuses very slowly through the solution; weeks may pass before the solid is entirely dissolved. Stirring distributes the dissolved solute rapidly through the solution, and more solvent is brought into contact with the solid, causing it to dissolve more rapidly.

14.7 Solutions: A Reaction Medium

Many solids must be put into solution in order to undergo appreciable chemical reaction. We can write the equation for the double-displacement reaction between sodium chloride and silver nitrate:

$$NaCl + AgNO_3 \longrightarrow AgCl + NaNO_3$$

But suppose we mix solid NaCl and solid $AgNO_3$ and look for a chemical change. If any reaction occurs, it is slow and virtually undetectable. In fact, the crystalline structures of NaCl and $AgNO_3$ are so different that we could separate them by tediously picking out each kind of crystal from the mixture. But if we dissolve the sodium chloride and silver nitrate separately in water and mix the two solutions, we observe the immediate formation of a white, curdy precipitate of silver chloride.

Molecules or ions must come into intimate contact or collide with one another in order to react. In the foregoing example, the two solids did not react because the ions were securely locked within their crystal structures. But when the sodium chloride and silver nitrate are dissolved, their crystal lattices are broken down and the ions become mobile. When the two solutions are mixed, the mobile Ag^+ and Cl^- ions come into contact and react to form insoluble AgCl, which precipitates out of solution. The soluble Na^+ and NO_3^- ions remain mobile in solution but form the crystalline salt $NaNO_3$ when the water is evaporated:

$$NaCl(aq) + AgNO_3(aq) \longrightarrow AgCl\downarrow + NaNO_3(aq)$$

$$\underset{\text{Sodium chloride solution}}{(Na^+ + Cl^-)} + \underset{\text{Silver nitrate solution}}{(Ag^+ + NO_3^-)} \xrightarrow{H_2O} \underset{\text{Silver chloride}}{AgCl\downarrow} + \underset{\text{Sodium nitrate in solution}}{Na^+ + NO_3^-}$$

The mixture of the two solutions provides a medium or space in which the Ag^+ and Cl^- ions can react. (See Chapter 15 for further discussion of ionic reactions.)

Solutions also function as diluents (diluting agents) in reactions in which the undiluted reactants would combine with each other too violently. Moreover, a solution of known concentration provides a convenient method for delivering specific amounts of reactants.

14.8 Concentration of Solutions

The concentration of a solution expresses the amount of solute dissolved in a given quantity of solvent or solution. Because reactions are often conducted in solution, it is important to understand the methods of expressing concentration and to know how to prepare solutions of particular concentrations.

Dilute and Concentrated Solutions When we say that a solution is *dilute* or *concentrated*, we are expressing, in a relative way, the amount of solute present. One gram of salt and 2 g of salt in solution are both dilute solutions when compared with the same volume of a solution containing 20 g of salt. Ordinary concentrated hydrochloric acid (HCl) contains 12 moles of HCl per liter of solution. In some laboratories the dilute acid is made by mixing equal volumes of water and the concentrated acid. In other laboratories the concentrated acid is diluted with two or three volumes of water, depending on its use. The term **dilute solution,** then, describes a solution that contains a relatively small amount of dissolved solute. Conversely, a **concentrated solution** contains a relatively large amount of dissolved solute.

dilute solution

concentrated solution

Saturated, Unsaturated, and Supersaturated Solutions At a specific temperature there is a limit to the amount of solute that will dissolve in a given amount of solvent. When this limit is reached, the resulting solution is said to be *saturated*. For example, when we put 40.0 g of KCl into 100 g of H_2O at 20°C, we find that 34.0 g of KCl dissolve and 6.0 g of KCl remain undissolved. The solution formed is a saturated solution of KCl.

Two processes are occurring simultaneously in a saturated solution. The solid is dissolving into solution, and at the same time the dissolved solute is crystallizing out of solution. This may be expressed as

$$\text{solute (undissolved)} \rightleftharpoons \text{solute (dissolved)}$$

When these two opposing processes are occurring at the same rate, the amount of solute in solution is constant, and a condition of equilibrium is established between dissolved and undissolved solute. A **saturated solution** contains dissolved solute in equilibrium with undissolved solute.

saturated solution

It is important to state the temperature of a saturated solution, because a solution that is saturated at one temperature may not be saturated at another. If the temperature of a saturated solution is changed, the equilibrium is disturbed, and the amount of dissolved solute will change to reestablish equilibrium.

A saturated solution may be either dilute or concentrated, depending on the solubility of the solute. A saturated solution can be conveniently prepared by dissolving a little more than the saturated amount of solute at a temperature somewhat higher than room temperature. Then the amount of solute in solution will be in excess of its solubility at room temperature; and, when the solution cools, the excess solute will crystallize, leaving the solution saturated. (In this case, the solute must be more soluble at higher temperatures and must not form a supersaturated solution.) Examples expressing the solubility of saturated solutions at two different temperatures are given in Table 14.3.

unsaturated solution

An **unsaturated solution** contains less solute per unit of volume than does its corresponding saturated solution. In other words, additional solute can be dissolved in an unsaturated solution without altering any other conditions. Consider a solution made by adding 40 g of KCl to 100 g of H_2O at 20°C (see Table 14.3). The solution formed will be saturated and will contain about 6 g of

Table 14.3 Saturated solutions at 20°C and 50°C

Solute	Solubility (g solute/100 g H_2O) 20°C	50°C
NaCl	36.0	37.0
KCl	34.0	42.6
$NaNO_3$	88.0	114.0
$KClO_3$	7.4	19.3
$AgNO_3$	222.0	455.0
$C_{12}H_{22}O_{11}$	203.9	260.4

undissolved salt, because the maximum amount of KCl that can dissolve in 100 g of H_2O at 20°C is 34 g. If the solution is now heated and maintained at 50°C, all the salt will dissolve and, in fact, even more can be dissolved. Thus the solution at 50°C is unsaturated.

In some circumstances, solutions can be prepared that contain more solute than that needed for a saturated solution at a particular temperature. Such solutions are said to be **supersaturated**. However, we must qualify this definition by noting that a supersaturated solution is unstable. Disturbances such as jarring, stirring, scratching the walls of the container, or dropping in a "seed" crystal cause the supersaturation to break. When a supersaturated solution is disturbed, the excess solute crystallizes out rapidly, returning the solution to a saturated state.

supersaturated solution

Supersaturated solutions are not easy to prepare but may be made from certain substances by dissolving, in warm solvent, an amount of solute greater than that needed for a saturated solution at room temperature. The warm solution is then allowed to cool very slowly. With the proper solute and careful work, a supersaturated solution will result. Two substances commonly used to demonstrate this property are sodium thiosulfate pentahydrate, $Na_2S_2O_3 \cdot 5\,H_2O$, and sodium sulfate, Na_2SO_4 (from a saturated solution at 30°C).

PROBLEM 14.1 Will a solution made by adding 2.5 g of $CuSO_4$ to 10 g of H_2O be saturated or unsaturated at 20°C?

To answer this question we first need to know the solubility of $CuSO_4$ at 20°C. From Figure 14.3 we see that the solubility of $CuSO_4$ at 20°C is about 21 g per 100 g of H_2O. This amount is equivalent to 2.1 g of $CuSO_4$ per 10 g of H_2O.

Since 2.5 g per 10 g of H_2O is greater than 2.1 g per 10 g of H_2O, the solution will be saturated and 0.4 g of $CuSO_4$ will be undissolved.

Weight Percent Solution This method expresses concentration as the percentage of solute in a given mass of solution. It says that for a given mass of solution a certain percentage of that mass is solute. Suppose that we take a bottle from the reagent shelf that reads "Sodium hydroxide, NaOH, 10%." This statement means that for every 100 g of this solution, 10 g will be NaOH and 90 g will be water. (Note that this amount of solution is 100 g and not 100 mL.) We could also make this same concentration of solution by dissolving 2.0 g of NaOH in 18 g of water. Weight percent concentrations are most generally used for solids dissolved in liquids.

$$\text{weight percent} = \frac{\text{g solute}}{\text{g solute} + \text{g solvent}} \times 100\% = \frac{\text{g solute}}{\text{g solution}} \times 100\%$$

PROBLEM 14.2 What is the weight percent of sodium hydroxide in a solution that is made by dissolving 8.00 g of NaOH in 50.0 g of H_2O?

$$\text{grams of solute (NaOH)} = 8.00\ \text{g}$$

$$\text{grams of solvent (H}_2\text{O)} = 50.0\ \text{g}$$

$$\frac{8.00\ \text{g NaOH}}{8.00\ \text{g NaOH} + 50.0\ \text{g H}_2\text{O}} \times 100\% = 13.8\%\ \text{NaOH solution}$$

PROBLEM 14.3 What weights of potassium chloride (KCl) and water are needed to make 250 g of 5.00% solution?

The percentage expresses the weight of the solute.

$$250\ \text{g} = \text{total weight of solution}$$

$$5.00\%\ \text{of}\ 250\ \text{g} = 0.0500 \times 250\ \text{g} = 12.5\ \text{g KCl (solute)}$$

$$250\ \text{g} - 12.5\ \text{g} = 237.5\ \text{g H}_2\text{O}$$

Dissolving 12.5 g of KCl in 237.5 g of H_2O gives a 5.00% KCl solution.

PROBLEM 14.4 A 34.0% sulfuric acid solution has a density of 1.25 g/mL. How many grams of H_2SO_4 are contained in 1.00 L of this solution?

Since H_2SO_4 is the solute, we first solve the weight percent equation for grams of solute:

$$\text{weight percent} = \frac{\text{g solute}}{\text{g solution}} \times 100\%$$

$$\text{g solute} = \frac{\text{weight percent} \times \text{g solution}}{100\%}$$

The weight percent is given in the problem. We need to determine the grams of solution. The weight of the solution can be calculated from the density data.

Convert density (g/mL) to grams:

$$1.00 \text{ L} = 1000 \text{ mL}$$

$$\frac{1.25 \text{ g}}{\text{mL}} \times 1000 \text{ mL} = 1250 \text{ g} \quad \text{(g of solution)}$$

Now we have all the figures to calculate the grams of solute.

$$\text{g solute} = \frac{34.0\% \times 1250 \text{ g}}{100\%} = 425 \text{ g } H_2SO_4$$

1.00 L of 34.0% H_2SO_4 solution contains 425 g of H_2SO_4.

The student should note that the concentration expressed as weight percent is independent of the formula of the solute.

Weight/Volume Percent (w/v) This method expresses concentration as grams of solute per 100 mL of solution. With this system, a 10.0% (w/v) glucose solution is made by dissolving 10.0 g of glucose in water, diluting to 100 mL, and mixing. The 10.0% (w/v) solution could also be made by diluting 20.0 g to 200 mL, 50.0 g to 500 mL, and so on. Of course, any other appropriate dilution ratio may be used.

$$\text{weight/volume percent} = \frac{\text{g solute}}{\text{mL solution}} \times 100\%$$

Volume Percent Solutions that are formulated from two liquids are often expressed as *volume percent* with respect to the solute. The volume percent is the volume of a liquid in 100 mL of solution. The label on a bottle of ordinary rubbing alcohol reads "Isopropyl alcohol, 70% by volume." Such a solution could be made by mixing 70 mL of alcohol with water to make a total volume of 100 mL. We cannot use 30 mL of water because the two volumes are not necessarily additive.

$$\text{volume percent} = \frac{\text{volume of liquid in question}}{\text{total volume of solution}} \times 100\%$$

Molarity Weight percent solutions do not equate or express the number of formula or molecular weights of the solute in solution. For example, 1000 g of 10% NaOH solution contain 100 g of NaOH; 1000 g of 10% KOH solution contain 100 g of KOH. In terms of moles of NaOH and KOH, these solutions contain

$$\text{moles NaOH} = 100 \text{ g NaOH} \times \frac{1 \text{ mol NaOH}}{40.0 \text{ g NaOH}} = 2.50 \text{ mol NaOH}$$

$$\text{moles KOH} = 100 \text{ g KOH} \times \frac{1 \text{ mol KOH}}{56.1 \text{ g KOH}} = 1.78 \text{ mol KOH}$$

From these figures we see that the two 10% solutions do not contain the same number of moles of NaOH and KOH. Yet one mole of each of these bases will neutralize the same amount of acid. As a result we find that a 10% NaOH solution has more reactive alkali than a 10% KOH solution.

We need a method of expressing concentration that will easily indicate how many moles or formula weights of solute are present per unit volume of solution. For this purpose the molar method of expressing concentration is used.

1 molar solution

A **1 molar solution** contains 1 mole, or 1 molecular weight, or 1 formula weight of solute per liter of solution. For example, to make a 1 molar solution of sodium hydroxide (NaOH), we dissolve 40.0 g of NaOH (1 mole) in water and dilute the solution with more water to a volume of 1 liter. The solution contains 1 mole of the solute in 1 liter of solution and is said to be 1 molar (1 *M*) in concentration. Figure 14.6 illustrates the preparation of a 1 molar solution. Note that the volume of the solute and the solvent together is 1 liter.

molarity

The concentration of a solution can, of course, be varied by using more or less solute or solvent; but in any case the **molarity** of a solution is the number of moles of solute per liter of solution. A capital *M* is the abbreviation for molarity. The units of molarity are moles per liter. The expression "2.0 *M* NaOH" means a 2.0 molar solution of NaOH (2.0 mol, or 80 g, of NaOH dissolved in 1 liter of solution).

$$\text{molarity} = M = \frac{\text{number of moles of solute}}{\text{liter of solution}} = \frac{\text{moles}}{\text{liter}}$$

Flasks that are calibrated to contain specific volumes at a particular temperature are used to prepare solutions of a desired concentration. These *volumetric flasks* have a calibration mark on the neck to indicate accurately the measured volume. Molarity is based on a specific volume of solution and therefore will vary slightly with temperature because volume varies with temperature (1000 mL of H_2O at 20°C = 1001 mL at 25°C).

Suppose we want to make 500 mL of 1 *M* solution. This solution can be prepared by weighing 0.5 mol of the solute and diluting with water in a 500 mL (0.5 L) volumetric flask. The molarity will be

$$M = \frac{0.5 \text{ mol solute}}{0.5 \text{ L solution}} = 1 \text{ molar}$$

Thus you can see that it is not necessary to have a liter of solution to express molarity. All we need to know is the number of moles of dissolved solute and the volume of solution. Thus 0.001 mol of NaOH in 10 mL of solution is 0.1 *M*:

$$\frac{0.001 \text{ mol}}{10 \cancel{\text{mL}}} \times \frac{1000 \cancel{\text{mL}}}{1 \text{ L}} = 0.1\ M$$

When we stop to think that a balance is not calibrated in moles but in grams, we can incorporate grams into the molarity formula. We do so by using the relationship

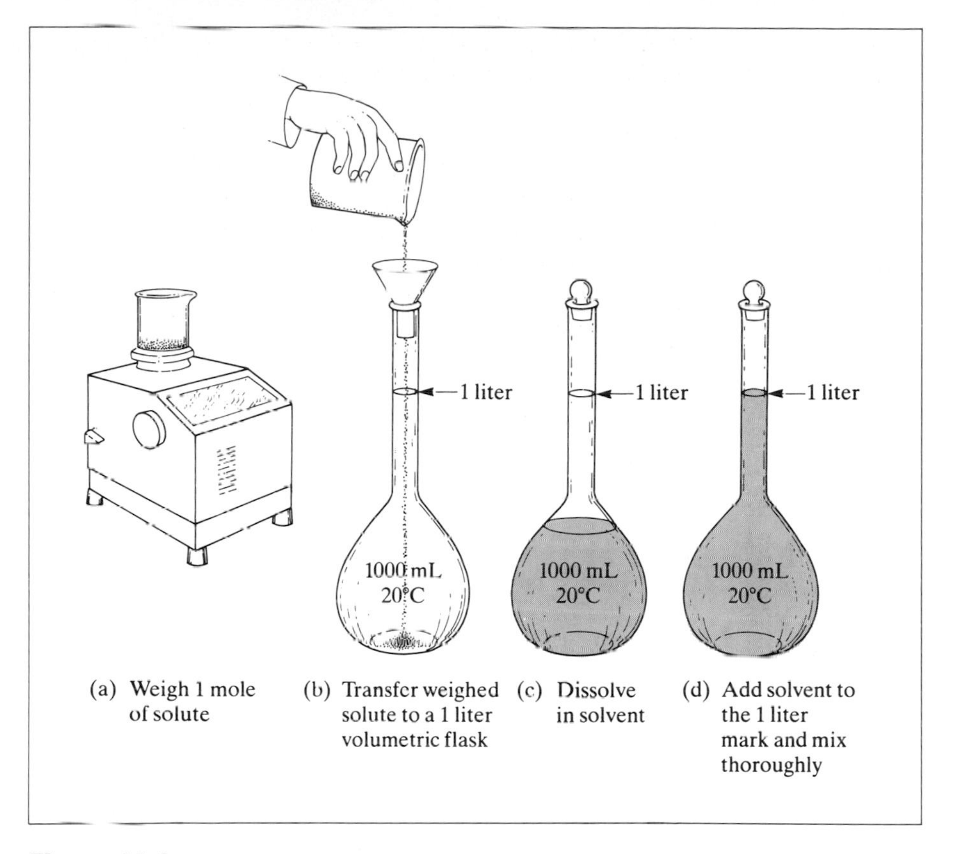

Figure 14.6
Preparation of a 1 molar solution

$$\text{moles} = \frac{\text{grams of solute}}{\text{formula wt}}$$

Substituting this relationship into our expression for molarity, we get

$$M = \frac{\text{mol}}{\text{L}} = \frac{\text{g solute}}{\text{form. wt solute} \times \text{L solution}} = \frac{\text{g}}{\text{form. wt} \times \text{L}}$$

We can now weigh any amount of a solute that has a known formula, dilute it to any volume, and calculate the molarity of the solution using this formula.

The molarities of the concentrated acids commonly used in the laboratory are

HCl 12 M $HC_2H_3O_2$ 17 M HNO_3 16 M H_2SO_4 18 M

PROBLEM 14.5 What is the molarity of a solution containing 1.4 mol of acetic acid ($HC_2H_3O_2$) in 250 mL of solution?

Substitute the data, 1.4 mol and 250 mL (0.250 L), directly into the equation for molarity.

$$M = \frac{\text{mol}}{\text{L}} = \frac{1.4\ \text{mol}}{0.250\ \text{L}} = \frac{5.6\ \text{mol}}{\text{L}} = 5.6\ M \quad \text{(Answer)}$$

By the unit conversion method we note that the concentration given in the problem statement is 1.4 mol per 250 mL (mol/mL). Since molarity = mol/L, the needed conversion is

$$\frac{\text{mol}}{\text{mL}} \longrightarrow \frac{\text{mol}}{\text{L}} = M$$

$$\frac{1.40\ \text{mol}}{250\ \cancel{\text{mL}}} \times \frac{1000\ \cancel{\text{mL}}}{1\ \text{L}} = \frac{5.6\ \text{mol}}{1\ \text{L}} = 5.6\ M \quad \text{(Answer)}$$

PROBLEM 14.6 What is the molarity of a solution made by dissolving 2.00 g of potassium chlorate ($KClO_3$) in enough water to make 150 mL of solution?

This problem can be solved using the unit conversion method. The steps in the conversions must lead to units of moles/liter.

$$\frac{\text{g } KClO_3}{\text{mL}} \longrightarrow \frac{\text{g } KClO_3}{\text{L}} \longrightarrow \frac{\text{mol } KClO_3}{\text{L}} = M$$

The data are

$$\text{g} = 200\ \text{g} \qquad \text{form. wt } KClO_3 = 122.6\ \text{g/mol} \qquad \text{volume} = 150\ \text{mL}$$

$$\frac{2.00\ \cancel{\text{g } KClO_3}}{150\ \cancel{\text{mL}}} \times \frac{1000\ \cancel{\text{mL}}}{1\ \text{L}} \times \frac{1\ \text{mol } KClO_3}{122.6\ \cancel{\text{g } KClO_3}} = \frac{0.109\ \text{mol}}{1\ \text{L}} = 0.109\ M$$

PROBLEM 14.7 How many grams of potassium hydroxide are required to prepare 600 mL of 0.450 *M* KOH solution?

The conversion is

milliliters ⟶ liters ⟶ moles ⟶ grams

The data are

$$\text{volume} = 600\ \text{mL} \qquad M = \frac{0.450\ \text{mol}}{\text{L}} \qquad \text{form. wt KOH} = \frac{56.1\ \text{g KOH}}{\text{mol}}$$

The calculation is

$$600\ \cancel{\text{mL}} \times \frac{1\ \cancel{\text{L}}}{1000\ \cancel{\text{mL}}} \times \frac{0.450\ \cancel{\text{mol}}}{\cancel{\text{L}}} \times \frac{56.1\ \text{g KOH}}{\cancel{\text{mol}}} = 15.1\ \text{g KOH}$$

PROBLEM 14.8 How many milliliters of 2.00 *M* HCl will react with 28.0 g of NaOH?

Step 1 Write and balance the equation for the reaction:

$$HCl(aq) + NaOH(aq) \longrightarrow NaCl(aq) + H_2O(aq)$$

The equation states that 1 mol of HCl reacts with 1 mol of NaOH.

Step 2 Find the number of moles of NaOH in 28.0 g of NaOH:

$$\text{g NaOH} \longrightarrow \text{mol NaOH}$$

$$28.0 \text{ g NaOH} \times \frac{1 \text{ mol}}{40.0 \text{ g}} = 0.700 \text{ mol NaOH}$$

$$28.0 \text{ g NaOH} = 0.700 \text{ mol NaOH}$$

Step 3 Solve for moles and volume of HCl needed. From Steps 1 and 2 we see that 0.700 mol of HCl will react with 0.700 mol of NaOH, because the ratio of moles reacting is 1:1. We know that 2.00 *M* HCl contains 2.00 mol of HCl per liter; therefore, the volume that contains 0.700 mol of HCl will be less than 1 L.

$$\text{mol NaOH} \longrightarrow \text{mol HCl} \longrightarrow \text{L HCl} \longrightarrow \text{mL HCl}$$

$$0.700 \cancel{\text{mol NaOH}} \times \frac{1 \cancel{\text{mol HCl}}}{1 \cancel{\text{mol NaOH}}} \times \frac{1 \text{ L HCl}}{2.00 \cancel{\text{mol HCl}}} = 0.350 \text{ L HCl}$$

$$0.350 \cancel{\text{L}} \text{ HCl} \times \frac{1000 \text{ mL}}{1 \cancel{\text{L}}} = 350 \text{ mL HCl}$$

Therefore, 350 mL of 2.00 *M* HCl contain 0.700 mol of HCl and will react with 0.700 mol, or 28.0 g, of NaOH.

PROBLEM 14.9 What volume of 0.250 *M* solution can be prepared from 16.0 g of potassium carbonate (K_2CO_3)?

We are starting with 16.0 g of K_2CO_3 and need to find the volume of 0.250 *M* solution that can be prepared from this K_2CO_3.

The conversion therefore is

$$\text{g } K_2CO_3 \longrightarrow \text{mol } K_2CO_3 \longrightarrow \text{L solution}$$

The data are

$$16.0 \text{ g } K_2CO_3 \qquad M = \frac{0.250 \text{ mol}}{1 \text{ L}} \qquad \text{form. wt } K_2CO_3 = \frac{138.2 \text{ g } K_2CO_3}{1 \text{ mol}}$$

$$16.0 \cancel{\text{g } K_2CO_3} \times \frac{1 \cancel{\text{mol } K_2CO_3}}{138.2 \cancel{\text{g } K_2CO_3}} \times \frac{1 \text{ L}}{0.250 \cancel{\text{mol } K_2CO_3}} = 0.463 \text{ L (463 mL)}$$

Thus, a 0.250 *M* solution can be made by dissolving 16.0 g of K_2CO_3 in water and diluting to 463 mL.

PROBLEM 14.10 Calculate the number of moles of nitric acid in 325 mL of 16 M HNO_3 solution.

Use the equation $\text{moles} = \text{liters} \times M$

Substituting the data given in the problem and solve:

$$\text{moles} = 0.325 \not{L} \times \frac{16 \text{ mol } HNO_3}{1 \not{L}} = 5.2 \text{ mol } HNO_3 \quad \text{(Answer)}$$

Dilution Problems Chemists often find it necessary to dilute solutions from one concentration to another by adding more solvent to the solution. If a solution is diluted by adding pure solvent, the volume of the solution increases, but the number of moles of solute in the solution remains the same. Thus, the moles/liter (molarity) of the solution decreases. It is important to read a problem carefully to distinguish between (1) how much solvent must be added to dilute a solution to a particular concentration and (2) to what volume a solution must be diluted to prepare a solution of a particular concentration.

PROBLEM 14.11 Calculate the molarity of a sodium hydroxide solution that is prepared by mixing 100 mL of 0.20 M NaOH with 150 mL of water.

This problem is a dilution problem. If we double the volume of a solution by adding water, we cut the concentration in half. Therefore, the concentration of the above solution should be less than 0.10 M. In the dilution, the moles of NaOH remain constant; the molarity and volume change. The final volume is (100 mL + 150 mL) or 250 mL.

To solve this problem, (1) calculate the moles of NaOH in the original solution, and (2) divide the moles of NaOH by the final volume of the solution to obtain the new molarity.

Step 1 Calculate the moles of NaOH in the original solution.

$$M = \frac{\text{mol}}{\text{L}} \qquad \text{mol} = \text{L} \times M$$

$$0.100 \not{L} \times \frac{0.20 \text{ mol NaOH}}{1 \not{L}} = 0.020 \text{ mol NaOH}$$

Step 2 Solve for the new molarity, taking into account that the total volume of the solution after dilution is 250 mL (0.250 L).

$$M = \frac{0.020 \text{ mol NaOH}}{0.250 \text{ L}} = 0.080\ M \text{ NaOH} \quad \text{(Answer)}$$

Alternate Solution. When the moles of solute in a solution before and after dilution are the same, then the moles before and after dilution may be set equal to each other:

$$\text{mol}_1 = \text{mol}_2$$

where mol_1 = moles before dilution, and mol_2 = moles after dilution. Then

$$\text{mol}_1 = \text{L}_1 \times M_1 \qquad \text{mol}_2 = \text{L}_2 \times M_2$$

$$\text{L}_1 \times M_1 = \text{L}_2 \times M_2$$

When both volumes are in the same units, a more general statement can be made:

$$V_1 \times M_1 = V_2 \times M_2$$

For this problem

$$V_1 = 100 \text{ mL} \qquad M_1 = 0.20\ M$$

$$V_2 = 250 \text{ mL} \qquad M_2 = M_2 \quad \text{(unknown)}$$

Then

$$100 \text{ mL} \times 0.20\ M = 250 \text{ mL} \times M_2$$

Solving for M_2, we get

$$M_2 = \frac{100 \cancel{\text{mL}} \times 0.20\ M}{250 \cancel{\text{mL}}} = 0.080\ M \text{ NaOH}$$

PROBLEM 14.12 How many grams of silver chloride, AgCl, will be precipitated by adding sufficient silver nitrate, $AgNO_3$, to react with 1500 mL of 0.400 M $BaCl_2$ (barium chloride) solution?

$$2\ AgNO_3(aq) + \underset{1 \text{ mol}}{BaCl_2(aq)} \longrightarrow \underset{2 \text{ mol}}{2\ AgCl\downarrow} + Ba(NO_3)_2(aq)$$

This problem is a stoichiometry problem. The fact that $BaCl_2$ is in solution means that we need to consider the volume and concentration of the solution in order to know the number of moles of $BaCl_2$ reacting.

Step 1 Determine the number of moles of $BaCl_2$ in 1500 mL of 0.400 M solution:

$$M = \frac{\text{mol}}{\text{L}} \qquad \text{mol} = \text{L} \times M \qquad 1500 \text{ mL} = 1.500 \text{ L}$$

$$1.500 \cancel{\text{L}} \times \frac{0.400 \text{ mol } BaCl_2}{\cancel{\text{L}}} = 0.600 \text{ mol } BaCl_2$$

Step 2 Calculate the moles and grams of AgCl formed by the mole-ratio method:

$$\text{mol } BaCl_2 \longrightarrow \text{mol AgCl} \longrightarrow \text{g AgCl}$$

$$0.600 \cancel{\text{mol } BaCl_2} \times \frac{2 \cancel{\text{mol AgCl}}}{1 \cancel{\text{mol } BaCl_2}} \times \frac{143.3 \text{ g AgCl}}{\cancel{\text{mol AgCl}}} = 172 \text{ g AgCl}$$

---

normality

Normality Normality is another way of expressing the concentration of a solution. It is based on an alternate chemical unit of mass called the *equivalent weight*. The **normality** of a solution is the concentration expressed as the number of equivalent weights (equivalents, abbreviated equiv) of solute per liter of

1 normal solution

solution. A **1 normal (1 *N*) solution** contains 1 equivalent weight of solute per liter of solution. Normality is widely used in analytical chemistry because it simplifies many of the calculations involving solution concentration.

$$\text{normality} = N = \frac{\text{number of equivalents of solute}}{\text{1 liter of solution}} = \frac{\text{equivalents}}{\text{liter}}$$

where

$$\text{number of equivalents of solute} = \frac{\text{grams of solute}}{\text{equivalent weight of solute}}$$

Every substance may be assigned an equivalent weight. The equivalent weight may be equal either to the formula weight of the substance or to an integral fraction of the formula weight (that is, the formula weight divided by 2, 3, 4, and so on). To gain an understanding of the meaning of equivalent weight, let us start by considering these two reactions:

$$\underset{\substack{1\ \text{mol} \\ (36.5\ \text{g})}}{HCl(aq)} + \underset{\substack{1\ \text{mol} \\ (40.0\ \text{g})}}{NaOH(aq)} \longrightarrow NaCl(aq) + H_2O$$

$$\underset{\substack{1\ \text{mol} \\ (98.1\ \text{g})}}{H_2SO_4(aq)} + \underset{\substack{2\ \text{mol} \\ (80.0\ \text{g})}}{2\ NaOH(aq)} \longrightarrow Na_2SO_4(aq) + 2\ H_2O$$

We note first that 1 mol of hydrochloric acid (HCl) reacts with 1 mol of sodium hydroxide (NaOH) and 1 mol of sulfuric acid (H_2SO_4) reacts with 2 mol of NaOH. If we make 1 molar solutions of these substances, 1 L of 1 *M* HCl will react with 1 L of 1 *M* NaOH, and 1 L of 1 *M* H_2SO_4 will react with 2 L of 1 *M* NaOH. From this reaction, we can see that H_2SO_4 has twice the chemical capacity of HCl when reacting with NaOH. We can, however, adjust these acid solutions to be equivalent in reactivity by dissolving only 0.5 mol of H_2SO_4 per liter of solution. By doing so, we find that we are required to use 49.0 g of H_2SO_4 per liter (instead of 98.1 g of H_2SO_4 per liter) to make a solution that is equivalent to one made from 36.5 g of HCl per liter. These weights, 49.0 g of H_2SO_4 and 36.5 g of HCl, are chemically equivalent and are known as the equivalent weights of these substances, because each will react with the same amount of NaOH (40.0 g). The equivalent weight of HCl is equal to its formula weight, but that of H_2SO_4 is one-half its formula weight. Table 14.4 summarizes these relationships.

Thus, 1 L of solution containing 36.5 g of HCl would be 1 *N*, and 1 L of solution containing 49.0 g of H_2SO_4 would also be 1 *N*. A solution containing 98.1 g of H_2SO_4 (1 mol) per liter would be 2 *N* when reacting with NaOH in the given equation.

equivalent weight

The **equivalent weight** is the weight of a substance that will react with, combine with, contain, replace, or in any other way be equivalent to 1 mole of hydrogen atoms or hydrogen ions.

Normality and molarity can be interconverted in the following manner.

Table 14.4 Comparison of molar and normal solutions of HCl and H_2SO_4 reacting with NaOH

	Formula weight	Concentration	Volumes that react	Equivalent weight	Concentration	Volumes that react
HCl	36.5	1 *M*	1 liter	36.5	1 *N*	1 liter
NaOH	40.0	1 *M*	1 liter	40.0	1 *N*	1 liter
H_2SO_4	98.1	1 *M*	1 liter	49.0	1 *N*	1 liter
NaOH	40.0	1 *M*	2 liters	40.0	1 *N*	1 liter

$$N = \frac{\text{equiv}}{\text{L}} \qquad M = \frac{\text{mol}}{\text{L}}$$

$$N = M \times \frac{\text{equiv}}{\text{mol}} = \frac{\cancel{\text{mol}}}{\text{L}} \times \frac{\text{equiv}}{\cancel{\text{mol}}} = \frac{\text{equiv}}{\text{L}}$$

$$M = N \times \frac{\text{mol}}{\text{equiv}} = \frac{\cancel{\text{equiv}}}{\text{L}} \times \frac{\text{mol}}{\cancel{\text{equiv}}} = \frac{\text{mol}}{\text{L}}$$

Thus a 2.0 *N* H_2SO_4 solution is 1.0 *M*.

One application of normality and equivalents is in acid–base neutralization reactions. The equivalent weight of an acid is that mass of the acid that will furnish 1 mole of H^+ ions. The equivalent weight of a base is that mass of base that will furnish 1 mole of OH^- ions. Using concentrations in normality, one equivalent of acid (A) will react with one equivalent of base (B).

$$N_A = \frac{\text{equiv}_A}{\text{L}_A} \quad \text{and} \quad N_B = \frac{\text{equiv}_B}{\text{L}_B}$$

$$\text{equiv}_A = \text{L}_A \times N_A \quad \text{and} \quad \text{equiv}_B = \text{L}_B \times N_B$$

Since $\text{equiv}_A = \text{equiv}_B$,

$$\text{L}_A \times N_A = \text{L}_B \times N_B$$

When both volumes are in the same units, we can write a more general equation:

$$\text{V}_A N_A = \text{V}_B N_B$$

which states that the volume of acid times the normality of the acid equals the volume of base times the normality of the base.

PROBLEM 14.13 (a) What is the normality of an H_2SO_4 solution if 25.00 mL of the solution requires 22.48 mL of 0.2018 *N* NaOH for complete neutralization? (b) What is the molarity of the H_2SO_4 solution?

(a) Solve for N_A by substituting the data into

$$V_A N_A = V_B N_B$$

$$25.00 \text{ mL} \times N_A = 22.48 \text{ mL} \times 0.2018\ N$$

$$N_A = \frac{22.48 \cancel{\text{mL}} \times 0.2018\ N}{25.00 \cancel{\text{mL}}} = 0.1815\ N\ H_2SO_4$$

(b) When H_2SO_4 is completely neutralized it furnishes 2 equivalents of H^+ ions per mole of H_2SO_4. The conversion from N to M is

$$\frac{\text{equiv}}{\text{L}} \longrightarrow \frac{\text{mol}}{\text{L}} \qquad H_2SO_4 = 0.1815\ N$$

$$\frac{0.1815 \cancel{\text{equiv}}}{1 \text{ L}} \times \frac{1 \text{ mol}}{2 \cancel{\text{equiv}}} = 0.09075 \text{ mol/L}$$

The H_2SO_4 solution is 0.09075 M.

The equivalent weight of a substance may be variable; its value is dependent on the reaction that the substance is undergoing. Consider the reactions represented by these equations:

$$NaOH + H_2SO_4 \longrightarrow NaHSO_4 + H_2O$$

$$2\ NaOH + H_2SO_4 \longrightarrow Na_2SO_4 + 2\ H_2O$$

In the first reaction 1 mol of sulfuric acid furnishes 1 mol of hydrogen. Therefore the equivalent weight of sulfuric acid is the formula weight, namely 98.1 g. But in the second reaction 1 mol of H_2SO_4 furnishes 2 mol of hydrogen. Therefore, the equivalent weight of the sulfuric acid is one-half the formula weight, or 49.0 g.

14.9 Colligative Properties of Solutions

Two solutions, one containing 1 mol (60.0 g) of urea (NH_2CONH_2) and the other containing 1 mol (342 g) of sucrose ($C_{12}H_{22}O_{11}$) in 1 kg of water, both have a freezing point of $-1.86°C$, not 0°C as for pure water. Urea and sucrose are distinctly different substances, yet they lower the freezing point of the water by the same amount. The only thing apparently common to these two solutions is that each contains 1 mol (6.022×10^{23} molecules) of solute and 1 kg of solvent. In fact, if we dissolve one mole of any nonionizable solute in 1 kg of water, the freezing point of the resulting solution will be $-1.86°C$.

These results lead us to conclude that the freezing point depression for a solution containing 6.022×10^{23} solute molecules (particles) and 1 kg of water is a constant, namely, 1.86°C. Freezing point depression is a general property of solutions. Furthermore the amount by which the freezing point is depressed is the

Table 14.5 Freezing point depression and boiling point elevation constants of selected solvents

Solvent	Freezing point of pure solvent (°C)	Freezing point depression constant, K_f $\left(\frac{°C, \text{kg solvent}}{\text{mol solute}}\right)$	Boiling point of pure solvent (°C)	Boiling point elevation constant, K_b $\left(\frac{°C, \text{kg solvent}}{\text{mol solute}}\right)$
Water	0.00	1.86	100.0	0.512
Acetic acid	16.6	3.90	118.5	3.07
Benzene	5.5	5.1	80.1	2.53
Camphor	178	40	208.2	5.95

same for all solutions made with a given solvent; that is, each solvent shows a characteristic *freezing point depression constant.* Freezing point depression constants for several solvents are given in Table 14.5.

The solution formed by the addition of a nonvolatile solute to a solvent has a lower freezing point, a higher boiling point, and a lower vapor pressure than that of the pure solvent. All these effects are related and are known as colligative properties. The **colligative properties** are properties that depend only upon the number of solute particles in a solution and not on the nature of those particles. Freezing point depression, boiling point elevation, and vapor pressure lowering are colligative properties of solutions.

colligative properties

The colligative properties of a solution can be considered in terms of vapor pressure. The vapor pressure of a pure liquid depends on the tendency of molecules to escape from its surface. Thus, if 10% of the molecules in a solution are nonvolatile solute molecules, the vapor pressure of the solution is 10% lower than that of the pure solvent. The vapor pressure is lower because the surface of the solution contains 10% nonvolatile molecules and 90% of the volatile solvent molecules. A liquid boils when its vapor pressure equals the pressure of the atmosphere. Thus, we can see that the solution just described as having a lower vapor pressure will have a higher boiling point than the pure solvent. The solution with a lowered vapor pressure does not boil until it has been heated above the boiling point of the solvent (see Figure 14.7). Each solvent has its own characteristic boiling point elevation constant (see Table 14.5). The boiling point elevation constant is based on a solution that contains 1 mole of solute particles per kilogram of solvent. For example, the boiling point elevation constant for a solution containing 1 mole of solute particles per kilogram of water is 0.512°C, which means that this water solution will boil at 100.512°C.

The freezing behavior of a solution can also be considered in terms of lowered vapor pressure. Figure 14.7 shows the vapor pressure relationships of ice, water, and a solution containing 1 mole of solute per kilogram of water. The freezing point of water is at the intersection of the water and ice vapor pressure curves—that is, at the point where water and ice have the same vapor pressure. Because the vapor pressure of water is lowered by the solute, the vapor pressure curve of the solution does not intersect the vapor pressure curve of ice until the

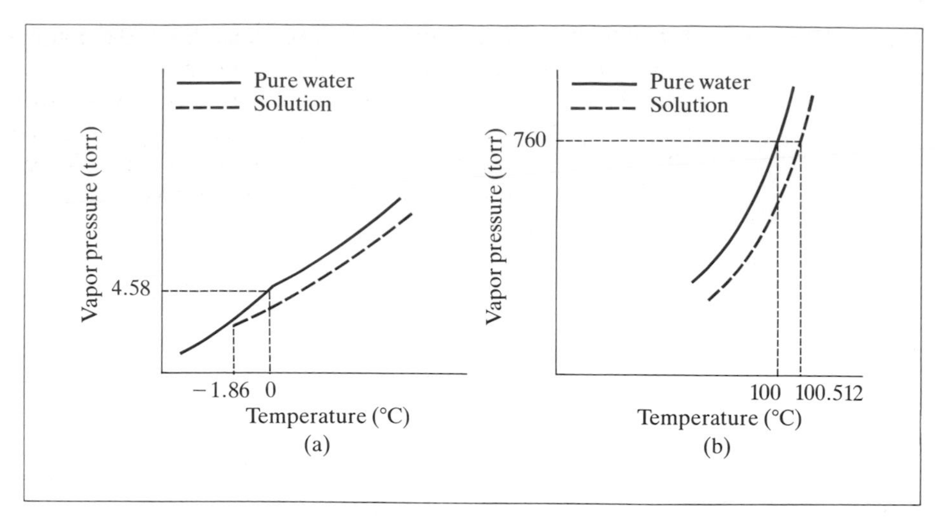

Figure 14.7

Vapor pressure curves of pure water and water solutions, showing (a) freezing point depression and (b) boiling point elevation effects. (Concentration: 1 mole of solute per kilogram of water.)

solution has been cooled below the freezing point of pure water. Thus it is necessary to cool the solution below 0°C in order to freeze out ice.

The foregoing discussion dealing with freezing point depressions is restricted to *un-ionized* substances. The discussion of boiling point elevations is restricted to *nonvolatile* and un-ionized substances. The colligative properties of ionized substances (Electrolytes, Chapter 15) are not under consideration at this point.

Some practical applications involving colligative properties are (1) use of salt–ice mixtures to provide low freezing temperatures for homemade ice cream, (2) use of sodium chloride or calcium chloride to melt ice from streets, and (3) use of ethylene glycol and water mixtures as antifreeze in automobile radiators (ethylene glycol also raises the boiling point of radiator fluid and thus allows the engine to operate at a higher temperature).

Both the freezing point depression and the boiling point elevation are directly proportional to the number of moles of solute per kilogram of solvent. When we deal with the colligative properties of solutions, another concentration expression, *molality*, is used. The **molality** (m) of a solute is the number of moles of solute per kilogram of solvent:

molality

$$m = \frac{\text{mol solute}}{\text{kg solvent}}$$

Note that a lowercase m is used for molality concentrations and a capital M for molarity. The difference between molality and molarity is that molality refers to moles of solute *per kilogram of solvent*, whereas molarity refers to moles of solute *per liter of solution.* For un-ionized substances, the colligative properties of a solution are directly proportional to its molality.

Molality is independent of volume. It is a mass-to-mass relationship of solute to solvent and allows for experiments, such as freezing point depression and boiling point elevation, to be conducted at variable temperatures.

The following equations are used in calculations involving colligative properties and molality.

$$\Delta t_f = mK_f \qquad \Delta t_b = mK_b \qquad m = \frac{\text{mol solute}}{\text{kg solvent}}$$

m = molality; mol solute/kg solvent

Δt_f = freezing point depression; °C

Δt_b = boiling point elevation; °C

K_f = freezing point depression constant; °C, kg solvent/mol solute

K_b = boiling point elevation constant; °C, kg solvent/mol solute

---

PROBLEM 14.14 What is the molality (m) of a solution prepared by dissolving 2.70 g of CH_3OH in 25.0 g of H_2O?

Since $m = \dfrac{\text{mol solute}}{\text{kg solvent}}$ the conversion is

$$\frac{2.70 \text{ g } CH_3OH}{25.0 \text{ g } H_2O} \longrightarrow \frac{\text{mol } CH_3OH}{25.0 \text{ g } H_2O} \longrightarrow \frac{\text{mol } CH_3OH}{1 \text{ kg } H_2O}$$

The molecular weight of CH_3OH is (12.0 + 4.0 + 16.0) or 32.0 g/mol.

$$\frac{2.70 \cancel{\text{g } CH_3OH}}{25.0 \cancel{\text{g } H_2O}} \times \frac{1 \text{ mol } CH_3OH}{32.0 \cancel{\text{g } CH_3OH}} \times \frac{1000 \cancel{\text{g } H_2O}}{1 \text{ kg } H_2O} = \frac{3.38 \text{ mol } CH_3OH}{1 \text{ kg } H_2O}$$

The molality is 3.38 m.

PROBLEM 14.15 A solution is made by dissolving 100 g of ethylene glycol ($C_2H_6O_2$) in 200 g of water. What is the freezing point of this solution?

To calculate the freezing point of the solution, we first need to calculate Δt_f, the change in freezing point. Use the equation

$$\Delta t_f = mK_f = \frac{\text{mol solute}}{\text{kg solvent}} \times K_f$$

K_f (for water): $\dfrac{1.86°\text{C kg solvent}}{\text{mol solute}}$ (from Table 14.5)

mol solute: $100 \cancel{\text{g } C_2H_6O_2} \times \dfrac{1 \text{ mol } C_2H_6O_2}{62.0 \cancel{\text{g } C_2H_6O_2}} = 1.61 \text{ mol } C_2H_6O_2$

$$\text{kg solvent:} \quad 200\ \cancel{g}\ H_2O \times \frac{1\ kg}{1000\ \cancel{g}} = 0.200\ kg\ H_2O$$

$$\Delta t_f = \frac{1.61\ \cancel{mol\ C_2H_6O_2}}{0.200\ \cancel{kg\ H_2O}} \times \frac{1.86°C\ \cancel{kg\ H_2O}}{1\ \cancel{mol\ C_2H_6O_2}} = 15.0°C$$

The freezing point depression, 15.0°C, must be subtracted from 0°C, the freezing point of the pure solvent.

$$\begin{aligned}\text{freezing point of solution} &= \text{freezing point of solvent} - \Delta t_f \\ &= 0.0°C - 15.0°C = -15°C\end{aligned}$$

Therefore, the freezing point of the solution is −15°C.

The calculation can also be done using the equation

$$\Delta t_f = K_f \times \frac{\text{g solute}}{\text{mol. wt solute}} \times \frac{1}{\text{kg solvent}}$$

PROBLEM 14.16 A solution made by dissolving 4.71 g of a compound of unknown molecular weight in 100.0 g of water has a freezing point of −1.46°C. What is the molecular weight of the compound?

First substitute the data in $\Delta t_f = mK_f$ and solve for m.

$\Delta t_f = +1.46$ since the solvent, water, freezes at 0°C.

$$K_f = \frac{1.86°C\ kg\ H_2O}{\text{mol solute}}$$

$$1.46°C = mK_f = m \times \frac{1.86°C\ kg\ H_2O}{\text{mol solute}}$$

$$m = \frac{1.46\cancel{°C} \times \text{mol solute}}{1.86\cancel{°C} \times kg\ H_2O} = \frac{0.785\ \text{mol solute}}{kg\ H_2O}$$

Now convert the data, 4.71 g solute/100.0 g H_2O, to g/mol.

$$\frac{4.71\ g\ \cancel{solute}}{100.0\ \cancel{g\ H_2O}} \times \frac{1000\ \cancel{g\ H_2O}}{1\ \cancel{kg\ H_2O}} \times \frac{1\ \cancel{kg\ H_2O}}{0.785\ mol\ \cancel{solute}} = 60.0\ g/mol$$

The molecular weight of the compound is 60.0 g/mol.

QUESTIONS

A. *Review the meanings of the new terms introduced in this chapter.*

1. Solution
2. Solute
3. Solvent
4. Solubility
5. Miscible
6. Immiscible
7. Concentration of a solution
8. Dilute solution

9. Concentrated solution
10. Saturated solution
11. Unsaturated solution
12. Supersaturated solution
13. 1 molar solution
14. Molarity
15. Normality
16. 1 normal solution
17. Equivalent weight
18. Colligative properties
19. Molality

B. *Information useful in answering the following questions will be found in the tables and figures.*

1. Make a sketch indicating the orientation of water molecules (a) about a single sodium ion and (b) about a single chloride ion in solution.
2. Which of the substances listed below are reasonably soluble and which are insoluble in water?
   (a) KOH (f) PbI_2
   (b) $NiCl_2$ (g) $MgCO_3$
   (c) ZnS (h) $CaCl_2$
   (d) $AgC_2H_3O_2$ (i) $Fe(NO_3)_3$
   (e) Na_2CrO_4 (j) $BaSO_4$
3. Estimate the number of grams of sodium fluoride that would dissolve in 100 g of water at 50°C.
4. What is the solubility at 25°C of each of the substances listed below? (See Figure 14.3.)
   (a) Potassium chloride (c) Potassium nitrate
   (b) Potassium chlorate
5. What is different in the solubility trend of the potassium halides compared with that of the lithium halides and the sodium halides? (See Table 14.2.)
6. What is the solubility, in grams of solute per 100 g of H_2O, of (a) $KClO_3$ at 60°C, (b) HCl at 20°C, (c) Li_2SO_4 at 80°C, and (d) KNO_3 at 0°C? (See Figure 14.3.)
7. Which substance, KNO_3 or NH_4Cl, shows the greater increase in solubility with increased temperature? (See Figure 14.3.)
8. Does a 2 molal solution in benzene or a 1 molal solution in camphor show the greater freezing point depression? (See Table 14.5.)
9. What would be the total surface area if the 1 cm cube in Figure 14.4 were cut into cubes 0.01 cm on a side?
10. At which temperatures—10°C, 20°C, 30°C, 40°C, or 50°C—would you expect a solution made from 63 g of ammonium chloride and 150 g of water to be unsaturated? (See Figure 14.3.)
11. Explain why the rate of dissolving decreases as shown in Figure 14.5.
12. Would the volumetric flasks in Figure 14.6 be satisfactory for preparing normal solutions? Explain.

C. *Review questions*

1. Name and distinguish between the two components of a solution.
2. Is it always apparent in a solution which component is the solute, for example, in a solution of a liquid in a liquid?
3. Explain why the solute does not settle out of a solution.
4. Is it possible to have one solid dissolved in another? Explain.
5. An aqueous solution of KCl is colorless, $KMnO_4$ is purple, and $K_2Cr_2O_7$ is orange. What color would you expect of an aqueous solution of $Na_2Cr_2O_7$? Explain.
6. Explain why carbon tetrachloride will dissolve benzene but will not dissolve sodium chloride.
7. Some drinks like tea are consumed either hot or cold, whereas others like Coca Cola are drunk only cold. Why?
8. Why is air considered to be a solution?
9. In which will a teaspoonful of sugar dissolve more rapidly, 200 mL of iced tea or 200 mL of hot coffee? Explain in terms of the KMT.
10. What is the effect of pressure on the solubility of gases in liquids? Solids in liquids?
11. Why do smaller particles dissolve faster than large ones?
12. In a saturated solution containing undissolved solute, solute is continuously dissolving, but the concentration of the solution remains unchanged. Explain.
13. Explain why there is no apparent reaction when crystals of $AgNO_3$ and NaCl are mixed, but a reaction is apparent immediately when solutions of $AgNO_3$ and NaCl are mixed.
14. What do we mean when we say that concentrated nitric acid (HNO_3) is 16 molar?
15. Will 1 liter of 1 molar NaCl contain more chloride ions than 0.5 liter of 1 molar $MgCl_2$? Explain.

16. What disadvantages are there in expressing the concentration of solutions as dilute or concentrated?
17. Explain how concentrated H_2SO_4 can be both 18 molar and 36 normal in concentration.
18. Describe how you would prepare 750 mL of 5 molar NaCl solution.
19. Arrange the following bases (in descending order) according to the volume of each that will react with 1 liter of 1 *M* HCl: (a) 1 *M* NaOH, (b) 1.5 *M* $Ca(OH)_2$, (c) 2 *M* KOH, and (d) 0.6 *M* $Ba(OH)_2$.
20. Explain in terms of vapor pressure why the boiling point of a solution containing a nonvolatile solute is higher than that of the pure solvent.
21. Explain why the freezing point of a solution is lower than the freezing point of the pure solvent.
22. Which would be colder, a glass of water and crushed ice or a glass of Seven-Up and crushed ice? Explain.
23. When water and ice are mixed, the temperature of the mixture is 0°C. But, if methyl alcohol and ice are mixed, a temperature of −10°C is readily attained. Explain why the two mixtures show such different temperature behavior.
24. Which would be more effective in lowering the freezing point of 500 g of water?
    (a) 100 g of sucrose ($C_{12}H_{22}O_{11}$) or 100 g of ethyl alcohol (C_2H_5OH)
    (b) 100 g of sucrose or 20.0 g of ethyl alcohol
    (c) 20.0 g of ethyl alcohol or 20.0 g of methyl alcohol (CH_3OH)
25. Is the molarity of a 5 molal aqueous solution of NaCl greater or less than 5 molar? Explain.
26. Express, in terms of its molarity, the normality of an H_2SO_4 solution used to titrate NaOH solutions. (Assume both hydrogens react.)
27. Which of the following statements are correct?
    (a) A solution is a homogeneous mixture.
    (b) It is possible for the same substance to be the solvent in one solution and the solute in another.
    (c) A solute can be removed from a solution by filtration.
    (d) Saturated solutions are always concentrated solutions.
    (e) If a solution of sugar in water is allowed to stand undisturbed for a long time, the sugar will gradually settle to the bottom of the container.
    (f) It is not possible to prepare an aqueous 1.0 *M* AgCl solution.
    (g) Gases are generally more soluble in hot water than in cold water.
    (h) It is impossible to prepare a two-phase liquid mixture from two liquids that are miscible with each other in all proportions.
    (i) A solution that is 10% NaCl by weight always contains 10 g of NaCl.
    (j) Small changes in pressure have little effect on the solubility of solids in liquids but a marked effect on the solubility of gases in liquids.
    (k) How fast a solute dissolves depends mainly on the size of the solute particles, the temperature of the solvent, and the degree of agitation or stirring taking place.
    (l) In order to have a 1 molar solution you must have 1 mol of solute dissolved in sufficient solvent to give 1 L of solution.
    (m) Dissolving 1 mole of NaCl in 1 liter of water will give a 1 molar solution.
    (n) One mole of solute in 1 L of solution has the same concentration as 0.1 mol of solute in 100 mL of solution.
    (o) When 100 mL of 0.200 *M* HCl is diluted to 200 mL volume by the addition of water, the resulting solution is 0.100 *M* and contains one-half the number of moles of HCl as were in the original solution.
    (p) Fifty milliliters of 0.1 *M* H_2SO_4 will neutralize the same volume of 0.1 *M* NaOH as 100 mL of 0.1 *M* HCl.
    (q) Fifty milliliters of 0.1 *N* H_2SO_4 will neutralize the same volume of 0.1 *M* NaOH as 100 mL of 0.1 *M* HCl.
    (r) The molarity of a solution will vary slightly with temperature.
    (s) The equivalent weight of $Ca(OH)_2$ is one-half its formula weight.
    (t) Gram for gram methyl alcohol, CH_3OH, is more effective than ethyl alcohol, C_2H_5OH, in lowering the freezing point of water.
    (u) An aqueous solution that freezes below 0°C will have a normal boiling point below 100°C.
    (v) The colligative properties of a solution depend on the number of solute particles dissolved in solution.

D. *Review problems*

Percent Solutions

1. Calculate the weight percent of the following solutions.
   (a) 25.0 g NaBr + 100 g H_2O
   (b) 1.20 g K_2SO_4 + 10.0 g H_2O
   (c) 40.0 g $Mg(NO_3)_2$ + 500 g H_2O
2. How many grams of a solution that is 12.5% by weight $AgNO_3$ would contain the following?
   (a) 30.0 g of $AgNO_3$ (b) 0.400 mol of $AgNO_3$
3. Calculate the weight percent of the following solutions:
   (a) 60.0 g NaCl + 200.0 g H_2O
   (b) 0.25 mol $HC_2H_3O_2$ + 3.0 mol H_2O
   (c) 1.0 molal solution of $C_6H_{12}O_6$ in water
4. How much solute is present in each of the following?
   (a) 65 g of 5.0% KCl solution
   (b) 250 g of 15.0% K_2CrO_4 solution
   *(c) A solution that contains 100.0 g of water and is 6.0% by weight sodium bicarbonate, $NaHCO_3$
5. What weight of 5.50% solution can be prepared from 25.0 g of KCl?
6. Physiological saline (NaCl) solutions used in intravenous injections have a concentration of 0.90% NaCl by weight.
   (a) How many grams of NaCl are needed to prepare 500 g of this solution?
   *(b) How much water must evaporate from this solution to give a solution that is 9.0% NaCl by weight?
7. A solution is made from 50.0 g of KNO_3 and 175 g of H_2O. How many grams of water must evaporate to give a saturated solution of KNO_3 in water at 20°C? (See Figure 14.3.)
8. Calculate the weight/volume percent of a solution made by dissolving
   (a) 22.0 g of CH_3OH dissolved in C_2H_5OH to make 100 mL of solution
   (b) 4.20 g of NaCl dissolved in H_2O to make 12.5 mL of solution
9. What is the volume percent of these solutions?
   (a) 10.0 mL of CH_3OH dissolved in water to a volume of 40.0 mL.
   (b) 2.0 mL of CCl_4 dissolved in benzene to a volume of 9.0 mL.
10. What volume of 70% rubbing alcohol can you prepare if you have only 150 mL of pure isopropyl alcohol on hand?
11. At 20°C an aqueous solution of HNO_3 that is 35.0% HNO_3 by weight has a density of 1.21 g/mL.
    (a) How many grams of HNO_3 are present in 1.00 L of this solution?
    (b) What volume of this solution will contain 500 g of HNO_3?

Molarity Problems

12. Calculate the molarity of the following solutions:
    (a) 0.10 mol of solute in 250 mL of solution
    (b) 2.5 mol of NaCl in 0.650 L of solution.
    (c) 0.025 mol of HCl in 10 mL of solution
    (d) 0.35 mol $BaCl_2 \cdot 2H_2O$ in 593 mL of solution
13. Calculate the molarity of the following solutions:
    (a) 53.0 g of Na_2CrO_4 in 1.00 L of solution
    (b) 260 g of $C_6H_{12}O_6$ in 800 mL of solution
    (c) 1.50 g of $Al_2(SO_4)_3$ in 2.00 L of solution
    (d) 0.0282 g of $Ca(NO_3)_2$ in 1.00 mL of solution
14. Calculate the number of moles of solute in each of the following solutions:
    (a) 40.0 L of 1.0 *M* LiCl solution
    (b) 25.0 mL of 3.00 *M* H_2SO_4
    (c) 349 mL of 0.0010 *M* NaOH
    (d) 5000 mL of 3.1 *M* $CoCl_2$
15. Calculate the grams of solute in each of the following solutions:
    (a) 150 L of 1.0 *M* NaCl
    (b) 0.035 L of 10.0 *M* HCl
    (c) 260 mL of 18 *M* H_2SO_4
    (d) 8.00 mL of 8.00 *M* $Na_2C_2O_4$
16. How many milliliters of 0.256 *M* KCl solution will contain the following?
    (a) 0.430 mol of KCl
    (b) 10.0 mol of KCl
    (c) 20.0 g of KCl
    *(d) 71.0 g of chloride ion, Cl^-

*17. What is the molarity of a nitric acid solution if the solution is 35.0% HNO_3 by weight and has a density of 1.21 g/mL?

Dilution Problems

18. What will be the molarity of the resulting solutions made by mixing the following? (Assume volumes are additive.)
    (a) 200 mL 12 *M* HCl + 200 mL H_2O
    (b) 60.0 mL 0.60 *M* $ZnSO_4$ + 500 mL H_2O
    (c) 100 mL 1.0 *M* HCl + 150 mL 2.0 *M* HCl

19. Calculate the volume of concentrated reagent required to prepare the diluted solutions indicated:
    (a) 12 *M* HCl to prepare 400 mL of 6.0 *M* HCl
    (b) 15 *M* NH_3 to prepare 50 mL of 6.0 *M* NH_3
    (c) 16 *M* HNO_3 to prepare 100 mL of 2.5 *M* HNO_3
    (d) 18 *M* H_2SO_4 to prepare 250 mL of 10.0 *N* H_2SO_4
20. To what volume must a solution of 80.0 g of H_2SO_4 in 500 mL of solution be diluted to give a 0.10 *M* solution?
21. What will be the molarity of each of the solutions made by mixing 250 mL of 0.75 *M* H_2SO_4 with (a) 150 mL of H_2O, (b) 250 mL of 0.70 *M* H_2SO_4, and (c) 400 mL of 2.50 *M* H_2SO_4?
22. How many milliliters of water must be added to 300 mL of 1.40 *M* HCl to make a solution that is 0.500 *M* HCl?
23. A 10.0 mL sample of 16 *M* HNO_3 solution is diluted to 500 mL. What is the molarity of the final solution?
24. Given a 5.00 *M* KOH solution, how would you prepare 250 mL of 0.625 *M* KOH?

Stoichiometry Problems

25. $BaCl_2(aq) + K_2CrO_4(aq) \longrightarrow BaCrO_4(s) + 2\,KCl(aq)$

    Using the above equation, calculate the following:
    (a) The grams of $BaCrO_4$ that can be obtained from 100 mL of 0.300 *M* $BaCl_2$
    (b) The volume of 1.0 *M* $BaCl_2$ solution needed to react with 50.0 mL of 0.300 *M* K_2CrO_4 solution
26. $3\,MgCl_2(aq) + 2\,Na_3PO_4(aq) \longrightarrow Mg_3(PO_4)_2(s) + 6\,NaCl(aq)$

    Using the above equation, calculate:
    (a) The milliliters of 0.250 *M* Na_3PO_4 that will react with 50.0 mL of 0.250 *M* $MgCl_2$.
    (b) The grams of $Mg_3(PO_4)_2$ that will be formed from 50.0 mL of 0.250 *M* $MgCl_2$.
27. (a) How many moles of hydrogen will be liberated from 200 mL of 3.00 *M* HCl reacting with an excess of magnesium? The equation is

    $$Mg(s) + 2\,HCl(aq) \longrightarrow MgCl_2(aq) + H_2(g)$$

    (b) How many liters of hydrogen gas, H_2, measured at 27°C and 720 torr, will be obtained? [*Hint*: Use the ideal gas equation.]
*28. What is the molarity of an HCl solution, 150 mL of which, when treated with excess magnesium, liberates 3.50 L of H_2 gas measured at STP?
29. Given the balanced equation

    $$6\,FeCl_2 + K_2Cr_2O_7 + 14\,HCl \longrightarrow 6\,FeCl_3 + 2\,CrCl_3 + 2\,KCl + 7\,H_2O$$

    (a) How many moles of KCl will be produced from 2.0 mol of $FeCl_2$?
    (b) How many moles of $CrCl_3$ will be produced from 1.0 mol of $FeCl_2$?
    (c) How many moles of $FeCl_2$ will react with 0.050 mol of $K_2Cr_2O_7$?
    (d) How many milliliters of 0.060 *M* $K_2Cr_2O_7$ will react with 0.025 mol of $FeCl_2$?
    (e) How many milliliters of 6.0 *M* HCl will react with 15.0 mL of 6.0 *M* $FeCl_2$?
30. $2\,KMnO_4 + 16\,HCl \longrightarrow 2\,MnCl_2 + 5\,Cl_2 + 8\,H_2O + 2\,KCl$

    Calculate the following using the above equation:
    (a) The moles of Cl_2 produced from 0.50 mol of $KMnO_4$
    (b) The moles of HCl required to react with 1.0 L of 2.0 *M* $KMnO_4$
    (c) The milliliters of 6.0 *M* HCl required to react with 200 mL of 0.50 *M* $KMnO_4$
    (d) The liters of Cl_2 gas at STP produced by the reaction of 75.0 mL of 6.0 *M* HCl

Equivalent Weight and Normality Problems

31. Calculate the equivalent weight of the acid and base in each of the following reactions:
    (a) $HCl + NaOH \longrightarrow NaCl + H_2O$
    (b) $2\,HCl + Ba(OH)_2 \longrightarrow BaCl_2 + 2\,H_2O$
    (c) $H_2SO_4 + Ca(OH)_2 \longrightarrow CaSO_4 + 2\,H_2O$
    (d) $H_2SO_4 + KOH \longrightarrow KHSO_4 + H_2O$
    (e) $H_3PO_4 + 2\,LiOH \longrightarrow Li_2HPO_4 + 2\,H_2O$
32. What is the normality of the following solutions? Assume complete neutralization.
    (a) 4.0 *M* HCl
    (b) 0.243 *M* HNO_3
    (c) 3.0 *M* H_2SO_4
    (d) 1.85 *M* H_3PO_4
    (e) 0.250 *M* $HC_2H_3O_2$
33. What is the normality of an H_2SO_4 solution if 36.26 mL are required to neutralize 2.50 g of $Ca(OH)_2$?

34. Which will be more effective in neutralizing stomach acid, HCl, a tablet containing 12.0 g of $Mg(OH)_2$ or a tablet containing 10.0 g of $Al(OH)_3$? Show evidence for your answer.
35. What volume of 0.2550 *N* NaOH is required to neutralize
    (a) 20.22 mL of 0.1254 *N* HCl
    (b) 14.86 mL of 0.1246 *N* H_2SO_4
    (c) 18.00 mL of 0.1430 *M* H_2SO_4

Molality and Colligative Properties Problems

36. Calculate the molality of these solutions.
    (a) 14.0 g of CH_3OH in 100 g of H_2O
    (b) 2.50 mol of benzene (C_6H_6) in 250 g of CCl_4
    (c) 1.0 g of $C_6H_{12}O_6$ in 1.0 g of H_2O
37. Which would be more effective as an antifreeze in an automobile radiator? (a) 10 kg of methyl alcohol, CH_3OH, or 10 kg of ethyl alcohol, C_2H_5OH (b) 10 *m* solution of methyl alcohol or 10 *m* solution of ethyl alcohol
38. Automobile battery acid is 38% H_2SO_4 and has a density of 1.29 g/mL. Calculate the molality and the molarity of this solution.
39. A sugar solution made to feed humming birds contains 1.00 lb of sugar to 4.00 lb of water. Can this solution be put outside, without freezing, where the temperature falls to 20.0°F at night? Show evidence for your answer.
*40. What would be (a) the boiling point and (b) the molality of an aqueous sugar ($C_{12}H_{22}O_{11}$) solution that freezes at −5.4°C?
41. (a) What is the molality of a solution containing 100.0 g of ethylene glycol, $C_2H_6O_2$, in 150.0 g of water?
    (b) What is the boiling point of this solution?
    (c) What is the freezing point of this solution?
42. What is (a) the molality, (b) the freezing point, and (c) the boiling point of a solution containing 2.68 g of naphthalene, $C_{10}H_8$, in 38.4 g of benzene (C_6H_6)?
*43. The freezing point of a solution of 8.00 g of an unknown compound dissolved in 60.0 g of acetic acid is 13.2°C. Calculate the molecular weight of the compound.
44. What is the molecular weight of a compound if 4.80 g of the compound dissolved in 22.0 g of H_2O gives a solution that freezes at −2.50°C?
45. A solution of 6.20 g of $C_2H_6O_2$ in water has a freezing point of −0.372°C. How many grams of H_2O are in the solution?
46. What (a) weight and (b) volume of ethylene glycol ($C_2H_6O_2$, density = 1.11 g/mL) should be added to 12.0 L of water in an automobile radiator to protect it from freezing at −20.0°C? (c) To what temperature Fahrenheit will the radiator be protected?

Additional Problems

47. How many grams of solution, 10% NaOH by weight, are required to neutralize 150 mL of a 1.0 *M* HCl solution?
*48. How many grams of solution, 10% NaOH by weight, are required to neutralize 250 g of a 1.0 molal solution of HCl?
*49. A sugar syrup solution contains 15.0% sugar, $C_{12}H_{22}O_{11}$, by weight and has a density of 1.06 g/mL.
    (a) How many grams of sugar are in 1.0 L of this syrup?
    (b) What is the molarity of this solution?
    (c) What is the molality of this solution?
*50. A solution of 3.84 g of C_4H_2N (empirical formula) in 250 g of benzene depresses the freezing point of benzene 0.614°C. What is the molecular formula of the compound?

CHAPTER **15**

Ionization: Acids, Bases, Salts

After studying Chapter 15, you should be able to

1. Understand the terms listed in Question A at the end of the chapter.
2. State the general characteristics of acids and bases.
3. Define an acid and a base in terms of Arrhenius, Brønsted–Lowry, and Lewis theories.
4. Identify conjugate acid–base pairs in a reaction.
5. When given the reactants, complete and balance equations for the reactions of acids with bases, metals, metal oxides, and carbonates.
6. When given the reactants, complete and balance equations for the reaction of an amphoteric hydroxide with either a strong acid or a strong base.
7. Write balanced equations for the reaction of sodium hydroxide or potassium hydroxide with zinc and with aluminum.
8. Classify common compounds as electrolytes or nonelectrolytes.
9. Distinguish between strong and weak electrolytes.
10. Understand the process of dissociation and ionization.
11. Write equations for the dissociation or ionization of acids, bases, and salts in water.
12. Describe and write equations for the ionization of water.
13. Understand pH as an expression of hydrogen ion or hydronium ion concentration.
14. Given pH as an integer, give the H^+ molarity and vice versa.
15. Use the simplified log scale given in the chapter to estimate pH values from corresponding H^+ molarities.

16 Understand the process of acid–base neutralization.
17 Calculate the molarity, normality, or volume of an acid or base solution from appropriate titration data.
18 Write balanced un-ionized, total ionic, and net ionic equations for neutralization reactions.

15.1 Acids and Bases

The word *acid* is derived from the Latin *acidus*, meaning "sour" or "tart," and is also related to the Latin word *acetum*, meaning "vinegar." Vinegar has been known since antiquity as the product of the fermentation of wine and apple cider. The sour constituent of vinegar is acetic acid ($HC_2H_3O_2$).

Some of the characteristic properties commonly associated with acids are the following: Water solutions of acids taste sour and change the color of litmus, a vegetable dye, from blue to red. Water solutions of nearly all acids react with (1) metals such as zinc and magnesium to produce hydrogen gas, (2) bases to produce water and a salt, and (3) carbonates to produce carbon dioxide. These properties are due to the hydrogen ions, H^+, released by acids in a water solution.

Classically, a *base* is a substance capable of liberating hydroxide ions, OH^-, in water solution. Hydroxides of the alkali metals (Group IA) and alkaline earth metals (Group IIA), such as LiOH, NaOH, KOH, $Ca(OH)_2$, and $Ba(OH)_2$, are the most common inorganic bases. Water solutions of bases are called *alkaline solutions* or *basic solutions*. They have a bitter or caustic taste, a slippery, soapy feeling, the ability to change litmus from red to blue, and the ability to interact with acids to form a salt and water.

Several theories have been proposed to answer the question "What is an acid and a base?" One of the earliest, most significant of these theories was advanced in a doctoral thesis in 1884 by Svante Arrhenius (1859–1927), a Swedish scientist, who stated that an acid is a hydrogen-containing substance that dissociates to produce hydrogen ions, and that a base is a hydroxide-containing substance that dissociates to produce hydroxide ions in aqueous solutions. Arrhenius postulated that the hydrogen ions were produced by the dissociation of acids in water, and that the hydroxide ions were produced by the dissociation of bases in water:

$$\underset{\text{Acid}}{HA} \longrightarrow H^+ + A^-$$

$$\underset{\text{Base}}{MOH} \longrightarrow M^+ + OH^-$$

Thus, an acid solution contains an excess of hydrogen ions and a base an excess of hydroxide ions.

In 1923 the Brønsted–Lowry proton transfer theory was introduced by J. N. Brønsted (1897–1947), a Danish chemist, and T. M. Lowry (1847–1936), an

English chemist. This theory states that an acid is a proton donor and a base is a proton acceptor.

Consider the reaction of hydrogen chloride gas with water to form hydrochloric acid:

$$HCl(g) + H_2O(l) \longrightarrow H_3O^+(aq) + Cl^-(aq) \quad (1)$$

In the course of the reaction, HCl donates, or gives up, a proton to form a Cl^- ion, and H_2O accepts a proton to form the H_3O^+ ion. Thus, HCl is an acid and H_2O is a base, according to the Brønsted–Lowry theory.

hydronium ion

A hydrogen ion, H^+, is nothing more than a bare proton and does not exist by itself in an aqueous solution. In water a proton combines with a polar water molecule to form a hydrated hydrogen ion, H_3O^+ [that is, $H(H_2O)^+$], commonly called a **hydronium ion**. The proton is attracted to a polar water molecule, forming a coordinate-covalent bond with one of the two pairs of unshared electrons:

$$H^+ + H:\ddot{\underset{\cdot\cdot}{O}}:\underset{}{} \longrightarrow \left[H:\ddot{\underset{\cdot\cdot}{O}}:H\right]^+$$
$$\qquad\quad H \qquad\qquad\quad H$$

Hydronium ion

Note the electron structure of the hydronium ion. For simplicity of expression in equations, we often use H^+ instead of H_3O^+, with the explicit understanding that H^+ is always hydrated in solution.

Whereas the Arrhenius theory is restricted to aqueous solutions, the Brønsted–Lowry approach has applications in all media and has become the more important theory when the chemistry of substances in solutions other than water is studied. Ammonium chloride (NH_4Cl) is a salt, yet its water solution has an acidic reaction. From this test we must conclude that NH_4Cl has acidic properties. The Brønsted–Lowry explanation shows that the ammonium ion, NH_4^+, is a proton donor, and water is the proton acceptor:

$$\underset{\text{Acid}}{NH_4^+} \rightleftharpoons \underset{\text{Base}}{NH_3} + \underset{\text{Acid}}{H^+} \quad (2)$$

$$\underset{\text{Acid}}{NH_4^+} + \underset{\text{Base}}{H_2O} \longrightarrow \underset{\text{Acid}}{H_3O^+} + \underset{\text{Base}}{NH_3} \quad (3)$$

A Brønsted–Lowry acid is a proton (H^+) donor.
A Brønsted–Lowry base is a proton (H^+) acceptor.

The Brønsted–Lowry theory also applies to certain cases where no solution is involved. For example, in the reaction of hydrogen chloride and ammonia gases, HCl is the proton donor and NH_3 is the base. [Remember that (g) after a formula in equations stands for a gas.]

$$\underset{\text{Acid}}{HCl(g)} + \underset{\text{Base}}{NH_3(g)} \longrightarrow \underset{\text{Acid}}{NH_4^+} + \underset{\text{Base}}{Cl^-} \qquad (4)$$

When a Brønsted–Lowry acid donates a proton, it forms the conjugate base of that acid. When a base accepts a proton, it forms the conjugate acid of that base.

In equations (1), (3), and (4) a conjugate acid and base are produced as products. The formulas of a conjugate acid–base pair differ by one proton (H^+). In equation (1) the conjugate acid–base pairs are HCl–Cl^- and H_3O^+–H_2O. Cl^- is the conjugate base of HCl, and HCl is the conjugate acid of Cl^-. H_2O is the conjugate base of H_3O^+, and H_3O^+ is the conjugate acid of H_2O.

conjugate acid–base pair

$$HCl(g) + H_2O(l) \longrightarrow Cl^-(aq) + H_3O^+(aq)$$

conjugate acid–base pair

In equation (3) the conjugate acid–base pairs are NH_4^+–NH_3 and H_3O^+–H_2O; in equation (4) they are HCl–Cl^- and NH_4^+–NH_3.

PROBLEM 15.1 Write the formula for (a) the conjugate base of H_2O and of HNO_3, and (b) the conjugate acid of SO_4^{2-} and of $C_2H_3O_2^-$. The difference between an acid or a base and its conjugate is one proton, H^+.

(a) To write the conjugate base of an acid, remove one proton from the acid formula. Thus,

$$H_2O \xrightarrow{-H^+} OH^- \quad \text{(Conjugate base)}$$

$$HNO_3 \xrightarrow{-H^+} NO_3^- \quad \text{(Conjugate base)}$$

Note that, by removing an H^+, the conjugate base becomes more negative than the acid by one minus charge.

(b) To write the conjugate acid of a base, add one proton to the formula of the base. Thus,

$$SO_4^{2-} \xrightarrow{+H^+} HSO_4^- \quad \text{(Conjugate acid)}$$

$$C_2H_3O_2^- \xrightarrow{+H^+} HC_2H_3O_2 \quad \text{(Conjugate acid)}$$

In each case the conjugate acid becomes more positive than the base by one positive charge due to the addition of H^+.

A more general concept of acids and bases was introduced by Gilbert N. Lewis. The Lewis theory deals with the way in which a substance with an unshared pair of electrons reacts in an acid–base type of reaction. According to this theory a base is any substance that has an unshared pair of electrons (electron-pair donor), and an acid is any substance that will attach itself to or accept a pair of electrons. In the reaction

$$\mathrm{H^+} + \begin{matrix} & \mathrm{H} & \\ & \cdot\cdot & \\ : & \mathrm{N} & :\mathrm{H} \\ & \cdot\cdot & \\ & \mathrm{H} & \end{matrix} \longrightarrow \begin{matrix} & \mathrm{H} & \\ & \cdot\cdot & \\ \mathrm{H}: & \mathrm{N} & :\mathrm{H^+} \\ & \cdot\cdot & \\ & \mathrm{H} & \end{matrix}$$

Acid Base

H^+ is a Lewis acid and $:NH_3$ is a Lewis base. According to the Lewis theory, substances other than proton donors (for example, BF_3) behave as acids:

$$\begin{matrix} & \mathrm{F} & \\ & \cdot\cdot & \\ \mathrm{F}: & \mathrm{B} & \\ & \cdot\cdot & \\ & \mathrm{F} & \end{matrix} + \begin{matrix} & \mathrm{H} & \\ & \cdot\cdot & \\ : & \mathrm{N} & :\mathrm{H} \\ & \cdot\cdot & \\ & \mathrm{H} & \end{matrix} \longrightarrow \begin{matrix} & \mathrm{F} & & \mathrm{H} & \\ & \cdot\cdot & & \cdot\cdot & \\ \mathrm{F}: & \mathrm{B} & : & \mathrm{N} & :\mathrm{H} \\ & \cdot\cdot & & \cdot\cdot & \\ & \mathrm{F} & & \mathrm{H} & \end{matrix}$$

Acid Base

The Lewis and Brønsted–Lowry bases are identical because, to accept a proton, a base must have an unshared pair of electrons.

The three theories are summarized in Table 15.1. These theories explain how acid–base reactions occur. We will generally use the theory that best explains the reaction that is under consideration. Most of our examples will refer to aqueous solutions. It is important to realize that in an aqueous acidic solution the H^+ ion concentration is always greater than the OH^- ion concentration. And, vice versa, in an aqueous basic solution the OH^- ion concentration is always greater than the H^+ ion concentration. When the H^+ and OH^- ion concentrations in a solution are equal, the solution is neutral; that is, it is neither acidic nor basic.

Table 15.1 Summary of acid–base definitions according to Arrhenius, Brønsted–Lowry, and G. N. Lewis theories

Theory	Acid	Base
Arrhenius	A hydrogen-containing substance that produces hydrogen ions in aqueous solution	A hydroxide-containing substance that produces hydroxide ions in aqueous solution
Brønsted–Lowry	A proton (H^+) donor	A proton (H^+) acceptor
Lewis	Any species that will bond to an unshared pair of electrons (electron-pair acceptor)	Any species that has an unshared pair of electrons (electron-pair donor)

15.2 Reactions of Acids

In aqueous solutions the H^+ or H_3O^+ ions are responsible for the characteristic reactions of acids. All the following reactions are in an aqueous medium.

(a) Reaction with Metals Acids react with metals that lie above hydrogen in the activity series of elements to produce hydrogen and a salt (see Section 17.5).

$$\text{acid} + \text{metal} \longrightarrow \text{hydrogen} + \text{salt}$$
$$2\,HCl(aq) + Ca(s) \longrightarrow H_2\uparrow + CaCl_2(aq)$$
$$H_2SO_4(aq) + Mg(s) \longrightarrow H_2\uparrow + MgSO_4(aq)$$
$$6\,HC_2H_3O_2(aq) + 2\,Al(s) \longrightarrow 3\,H_2\uparrow + 2\,Al(C_2H_3O_2)_3(aq)$$

Acids such as nitric acid (HNO_3) are oxidizing substances (see Chapter 17) and react with metals to produce water instead of hydrogen. For example,

$$3\,Zn(s) + 8\,HNO_3(\text{dilute}) \longrightarrow 3\,Zn(NO_3)_2(aq) + 2\,NO(g) + 4\,H_2O$$

(b) Reaction with Bases The interaction of an acid and a base is called a *neutralization reaction.* In aqueous solutions, the products of this reaction are a salt and water.

$$\text{acid} + \text{base} \longrightarrow \text{salt} + \text{water}$$
$$HBr(aq) + KOH(aq) \longrightarrow KBr(aq) + H_2O$$
$$2\,HNO_3(aq) + Ca(OH)_2(aq) \longrightarrow Ca(NO_3)_2(aq) + 2\,H_2O$$
$$2\,H_3PO_4(aq) + 3\,Ba(OH)_2(aq) \longrightarrow Ba_3(PO_4)_2\downarrow + 6\,H_2O$$

(c) Reaction with Metal Oxides This reaction is closely related to that of an acid with a base. With an aqueous acid, the products are a salt and water.

$$\text{acid} + \text{metal oxide} \longrightarrow \text{salt} + \text{water}$$
$$2\,HCl(aq) + Na_2O(s) \longrightarrow 2\,NaCl(aq) + H_2O$$
$$H_2SO_4(aq) + MgO(s) \longrightarrow MgSO_4(aq) + H_2O$$
$$6\,HCl(aq) + Fe_2O_3(s) \longrightarrow 2\,FeCl_3(aq) + 3\,H_2O$$

(d) Reaction with Carbonates Many acids react with carbonates to produce carbon dioxide, water, and a salt. Carbonic acid (H_2CO_3) is not the product, because it is unstable and decomposes into water and carbon dioxide.

$$\text{acid} + \text{carbonate} \longrightarrow \text{salt} + \text{water} + \text{carbon dioxide}$$
$$2\,HCl(aq) + Na_2CO_3(aq) \longrightarrow 2\,NaCl(aq) + H_2O + CO_2\uparrow$$
$$H_2SO_4(aq) + MgCO_3(s) \longrightarrow MgSO_4(aq) + H_2O + CO_2\uparrow$$

15.3 Reactions of Bases

The OH^- ions are responsible for the characteristic reactions of bases. All the following reactions are in an aqueous medium.

(a) Reaction with Acids Bases react with acids to produce a salt and water. See reaction of acids with bases in Section 15.2(b).

amphoteric

(b) Amphoteric Hydroxides Hydroxides of certain metals, such as zinc, aluminum, and chromium, are **amphoteric**; that is, they are capable of reacting as either an acid or a base. When treated with a strong acid, they behave like bases; when reacted with a strong base, they behave like acids.

$$Zn(OH)_2(s) + 2\ HCl(aq) \longrightarrow ZnCl_2(aq) + 2\ H_2O$$
$$Zn(OH)_2(s) + 2\ NaOH(aq) \longrightarrow Na_2Zn(OH)_4(aq)$$

(c) Reaction of NaOH and KOH with Certain Metals Some amphoteric metals react directly with the strong bases, sodium hydroxide and potassium hydroxide, to produce hydrogen.

$$\text{base} + \text{metal} + \text{water} \longrightarrow \text{salt} + \text{hydrogen}$$
$$2\ NaOH(aq) + Zn(s) + 2\ H_2O \longrightarrow Na_2Zn(OH)_4(aq) + H_2\uparrow$$
$$2\ KOH(aq) + 2\ Al(s) + 6\ H_2O \longrightarrow 2\ KAl(OH)_4(aq) + 3\ H_2\uparrow$$

(d) Reaction with Salts Bases will react with many salts in solution due to the formation of insoluble metal hydroxides.

$$\text{base} + \text{salt} \longrightarrow \text{metal hydroxide}\downarrow + \text{salt}$$
$$2\ NaOH(aq) + MnCl_2(aq) \longrightarrow Mn(OH)_2\downarrow + 2\ NaCl(aq)$$
$$3\ Ca(OH)_2(aq) + 2\ FeCl_3(aq) \longrightarrow 2\ Fe(OH)_3\downarrow + 3\ CaCl_2(aq)$$
$$2\ KOH(aq) + CuSO_4(aq) \longrightarrow Cu(OH)_2\downarrow + K_2SO_4(aq)$$

15.4 Salts

Salts are very abundant in nature. Most of the rocks and minerals of the earth's mantle are salts of one kind or another. Huge quantities of dissolved salts also exist in the oceans. Salts may be considered to be compounds that have been derived from acids and bases. They consist of positive metal or ammonium ions (H^+ excluded) combined with negative nonmetal ions (OH^- and O^{2-} excluded). The positive ion is the base counterpart and the nonmetal ion is the acid counterpart:

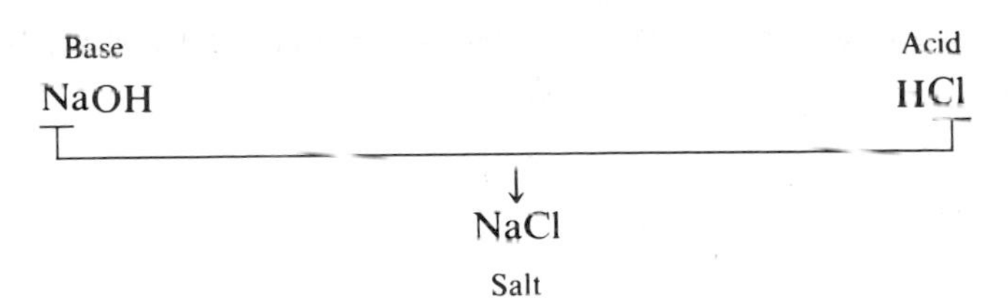

Salts are usually crystalline and have high melting and boiling points.

From a single acid such as hydrochloric acid (HCl), we can produce many chloride salts by replacing the hydrogen with metal ions (for example, NaCl, KCl, RbCl, $CaCl_2$, $NiCl_2$). Hence, the number of known salts greatly exceeds the number of known acids and bases. Salts are ionic compounds. If the hydrogen atoms of a binary acid are replaced by a nonmetal, the resulting compound has covalent bonding and is therefore not considered to be a salt (for example, PCl_3, S_2Cl_2, Cl_2O, NCl_3, ICl).

A review of Chapter 8 on the nomenclature of acids, bases, and salts may be beneficial at this point.

15.5 Electrolytes and Nonelectrolytes

Some of the most convincing evidence as to the nature of chemical bonding within a substance is the ability (or lack of ability) of a water solution of the substance to conduct electricity.

We can show that solutions of certain substances are conductors of electricity by using a simple conductivity apparatus consisting of a pair of electrodes connected to a voltage source through a light bulb and switch (see Figure 15.1). If the medium between the electrodes is a conductor of electricity, the light bulb will glow when the switch is closed. When chemically pure water is placed in the beaker and the switch is closed, the light does not glow, indicating that water is a virtual nonconductor. When we dissolve a small amount of sugar in the water and test the solution, the light still does not glow, showing that a sugar solution is also a nonconductor. But, when a small amount of salt, NaCl, is dissolved in water and this solution is tested, the light glows brightly. Thus, the salt solution conducts electricity. A fundamental difference exists between the chemical bonding of sugar and that of salt. Sugar is a covalently bonded (molecular) substance; salt is a substance with ionic bonds.

electrolyte

nonelectrolyte

Substances whose aqueous solutions are conductors of electricity are called **electrolytes**. Substances whose solutions are nonconductors are known as **nonelectrolytes**. The classes of compounds that are electrolytes are acids, bases, and salts. Solutions of certain oxides also are conductors because the oxides form an acid or a base when dissolved in water. One major difference between electrolytes and nonelectrolytes is that electrolytes are capable of producing ions in solution, whereas nonelectrolytes do not have this property. Solutions that contain a sufficient number of ions will conduct an electric current. Although

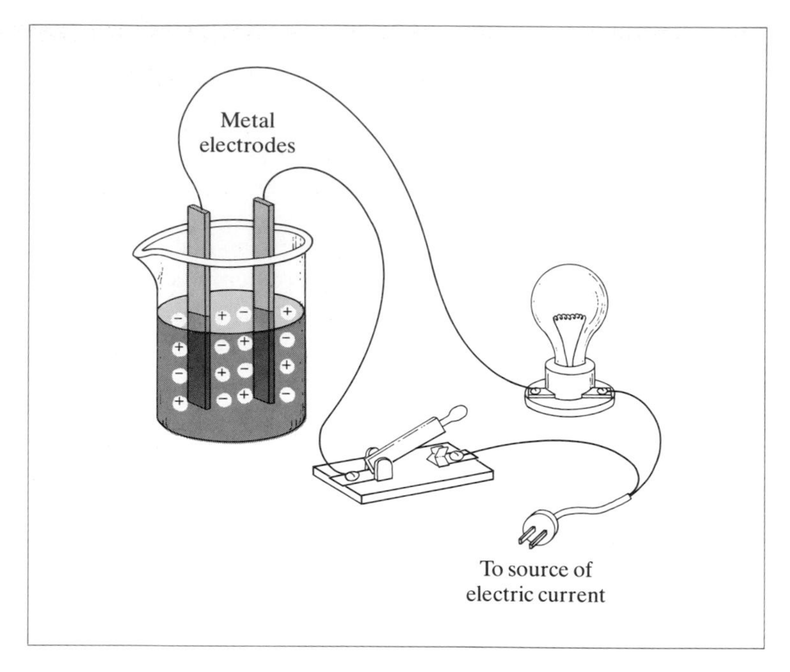

Figure 15.1
A simple conductivity apparatus for testing electrolytes and nonelectrolytes in solution. If the solution contains an electrolyte, the light will glow when the switch is closed.

Table 15.2 Representative electrolytes and nonelectrolytes

Electrolytes	Nonelectrolytes
H_2SO_4	$C_{12}H_{22}O_{11}$ (sugar)
HCl	C_2H_5OH (ethyl alcohol)
HNO_3	$C_2H_4(OH)_2$ (ethylene glycol)
$NaOH$	$C_3H_5(OH)_3$ (glycerol)
$HC_2H_3O_2$	CH_3OH (methyl alcohol)
NH_4OH	$CO(OH_2)_2$ (urea)
K_2SO_4	O_2
$NaNO_3$	H_2O

pure water is essentially a nonconductor, many city water supplies contain enough dissolved ionic matter to cause the light to glow dimly when the water is tested in a conductivity apparatus. Table 15.2 lists some common electrolytes and nonelectrolytes.

Acids, bases, and salts are electrolytes.

15.6 Dissociation and Ionization of Electrolytes

Arrhenius received the 1903 Nobel Prize in chemistry for his work on electrolytes. He stated that a solution conducts electricity because the solute dissociates immediately upon dissolving into electrically charged particles called *ions*. The movement of these ions toward oppositely charged electrodes causes the solution to be a conductor. According to his theory, solutions that are relatively poor conductors contain electrolytes that are only partly dissociated. Arrhenius also believed that ions exist in solution whether or not an electric current is present. In other words, the electric current does not cause the formation of ions. Positive ions, attracted to the cathode, are cations; negative ions, attracted to the anode, are anions.

We have seen that sodium chloride crystals consist of sodium and chloride ions held together by ionic bonds. When placed in water, the sodium and chloride ions are attracted by the polar water molecules, which surround each ion as it dissolves. In water, the salt dissociates, forming hydrated sodium and chloride ions (see Figure 15.2). The sodium and chloride ions in solution are bonded to a specific number of water dipoles and have less attraction for each other than they had in the crystalline state. The equation representing this dissociation is

$$NaCl(s) + (x + y)\,H_2O \longrightarrow Na^+(H_2O)_x + Cl^-(H_2O)_y$$

A simplified dissociation equation in which the water is omitted but understood to be present is

$$NaCl \longrightarrow Na^+ + Cl^-$$

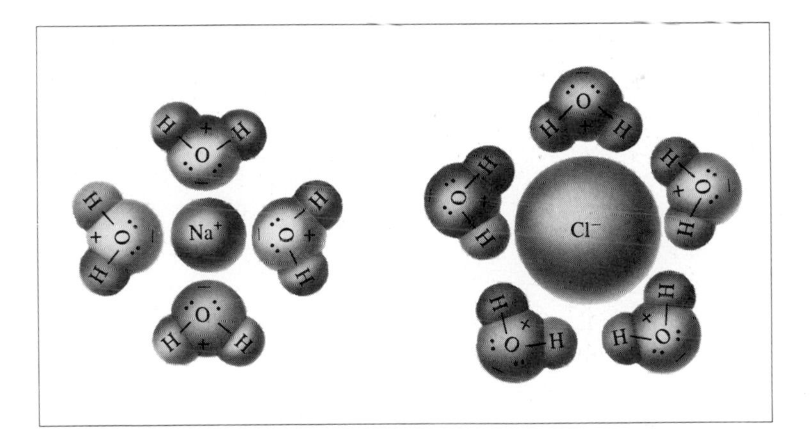

Figure 15.2

Hydrated sodium and chloride ions. When sodium chloride dissolves in water, each Na^+ and Cl^- ion becomes surrounded by water molecules. The negative end of the water dipole is attracted to the Na^+ ion, and the positive end is attracted to the Cl^- ion.

It is important to remember that sodium chloride exists in an aqueous solution as hydrated ions and not as NaCl units, even though the formula NaCl or $Na^+ + Cl^-$ is often used in equations.

The chemical reactions of salts in solution are the reactions of their ions. For example, when sodium chloride and silver nitrate react and form a precipitate of silver chloride, only the Ag^+ and Cl^- ions participate in the reaction. The Na^+ and NO_3^- remain as ions in solution.

$$Ag^+ + Cl^- \longrightarrow AgCl\downarrow$$

In many cases the number of molecules of water associated with a particular ion is known. For example, the blue color of the copper(II) ion is due to the hydrated ion $Cu(H_2O)_4^{2+}$. The hydration of ions can be demonstrated in a striking way with cobalt(II) chloride. When cobalt(II) chloride hexahydrate is dissolved in water, a pink solution forms due to the $Co(H_2O)_6^{2+}$ ions. If concentrated hydrochloric acid is added to this pink solution, the color gradually changes to blue. If water is then added to the blue solution, the color changes to pink again. These color changes are due to the exchange of water molecules and chloride ions on the cobalt ion. The $CoCl_4^{2-}$ ion is blue. Thus, the hydration of the cobalt ion is a reversible or equilibrium reaction (see Chapter 16). The equilibrium equation representing these changes is

$$\underset{\text{Pink}}{Co(H_2O)_6^{2+}} + 4\,Cl^- \rightleftharpoons \underset{\text{Blue}}{CoCl_4^{2-}} + 6\,H_2O$$

Ionization is the formation of ions; it may occur as a result of chemical reaction of certain substances with water. Glacial acetic acid (100% $HC_2H_3O_2$) is a liquid that behaves as a nonelectrolyte when tested by the method described in Section 15.5. But a water solution of acetic acid conducts an electric current (as indicated by the dull-glowing light of the conductivity apparatus). The equation for the reaction with water, which forms hydronium and acetate ions, is

$$\underset{\text{Acid}}{HC_2H_3O_2} + \underset{\text{Base}}{H_2O} \rightleftharpoons \underset{\text{Acid}}{H_3O^+} + \underset{\text{Base}}{C_2H_3O_2^-}$$

or, in the simplified equation,

$$HC_2H_3O_2 \rightleftharpoons H^+ + C_2H_3O_2^-$$

In this ionization reaction, water serves not only as a solvent but also as a base according to the Brønsted–Lowry theory.

Hydrogen chloride is predominantly covalently bonded, but when dissolved in water it reacts to form hydronium and chloride ions:

$$HCl(g) + H_2O \longrightarrow H_3O^+ + Cl^-$$

When a hydrogen chloride solution is tested for conductivity, the light glows brilliantly, indicating many ions in the solution.

Ionization occurs in each of the above two reactions with water, producing ions in solution. The necessity for water in the ionization process can be demonstrated by dissolving hydrogen chloride in a nonpolar solvent such as benzene, and testing the solution for conductivity. The solution fails to conduct electricity, indicating that no ions are produced.

dissociation

ionization

The terms *dissociation* and *ionization* are often used interchangeably to describe processes taking place in water. But, strictly speaking, the two are different. In the **dissociation** of a salt, the salt already exists as ions; when it dissolves in water, the ions separate, or dissociate, and increase in mobility. In the **ionization** process, ions are produced by the reaction of a compound with water.

15.7 Strong and Weak Electrolytes

strong electrolyte

weak electrolyte

Electrolytes are classified as strong or weak depending on the degree, or extent, of dissociation or ionization. **Strong electrolytes** are essentially 100% ionized in solution; **weak electrolytes** are much less ionized (based on comparing 0.1 *M* solutions). Most electrolytes are either strong or weak, with a few classified as moderately strong or weak. Most salts are strong electrolytes. Acids and bases that are strong electrolytes (highly ionized) are called *strong acids* and *strong bases*. Acids and bases that are weak electrolytes (slightly ionized) are called *weak acids* and *weak bases*.

For equivalent concentrations, solutions of strong electrolytes contain many more ions than do solutions of weak electrolytes. As a result, solutions of strong electrolytes are better conductors of electricity. Consider the two solutions, 1 *M* HCl and 1 *M* $HC_2H_3O_2$. Hydrochloric acid is almost 100% ionized; acetic acid is about 1% ionized. Thus HCl is a strong acid, and $HC_2H_3O_2$ is a weak acid. Hydrochloric acid has about 100 times as many hydronium ions in solution as acetic acid, making the HCl solution much more acidic.

One can distinguish between strong and weak electrolytes experimentally using the apparatus described in Section 15.5. A 1 *M* HCl solution causes the light to glow brilliantly, but a 1 *M* $HC_2H_3O_2$ solution causes only a dim glow. In a similar fashion the strong base sodium hydroxide (NaOH) may be distinguished from the weak base ammonium hydroxide (NH_4OH). The ionization of a weak electrolyte in water is represented by an equilibrium equation showing that both the un-ionized and ionized forms are present in solution. In the equilibrium equation of $HC_2H_3O_2$ and its ions, we say that the equilibrium lies "far to the left" because relatively few hydrogen and acetate ions are present in solution:

$$HC_2H_3O_2(aq) \rightleftarrows H^+ + C_2H_3O_2^-$$

We have previously used a double arrow in an equation to represent reversible processes in the equilibrium between dissolved and undissolved solute in a saturated solution. A double arrow ($\rightleftarrows$) is also used in the ionization equation of soluble weak electrolytes to indicate that the solution contains a considerable

Table 15.3 Selected list of strong and weak electrolytes

Strong electrolytes	Weak electrolytes
Most soluble salts	$HC_2H_3O_2$
H_2SO_4	H_2CO_3
HNO_3	HNO_2
HCl	H_2SO_3
HBr	H_2S
$HClO_4$	$H_2C_2O_4$
NaOH	H_3BO_3
KOH	HClO
$Ca(OH)_2$	NH_4OH
$Ba(OH)_2$	HF

amount of the un-ionized compound in equilibrium with its ions in solution. (See Section 16.1 for a discussion of reversible reactions.) A single arrow is used to indicate that the electrolyte is essentially all in the ionic form in the solution. For example, nitric acid is a strong acid; nitrous acid is a weak acid. Their ionization equations in water may be indicated as

$$HNO_3(aq) \xrightarrow{H_2O} H^+ + NO_3^-$$
$$HNO_2(aq) \xrightleftharpoons{H_2O} H^+ + NO_2^-$$

Practically all soluble salts; acids such as sulfuric, nitric, and hydrochloric acids; and bases such as sodium, potassium, calcium, and barium hydroxides are strong electrolytes. Weak electrolytes include numerous other acids and bases such as acetic acid, nitrous acid, carbonic acid, and ammonium hydroxide. The terms *strong acid*, *strong base*, *weak acid*, and *weak base* refer to whether an acid or base is a strong or weak electrolyte. A brief list of strong and weak electrolytes is given in Table 15.3.

Electrolytes yield two or more ions per formula unit upon dissociation, the actual number being dependent on the compound. Dissociation is complete or nearly complete for nearly all soluble salts and for certain other strong electrolytes such as those given in Table 15.3. The following are dissociation equations for several strong electrolytes. In all cases the ions are actually hydrated.

$$NaOH \xrightarrow{H_2O} Na^+ + OH^- \qquad \text{2 ions in solution per formula unit}$$
$$Na_2SO_4 \xrightarrow{H_2O} 2\ Na^+ + SO_4^{2-} \qquad \text{3 ions in solution per formula unit}$$
$$AlCl_3 \xrightarrow{H_2O} Al^{3+} + 3\ Cl^- \qquad \text{4 ions in solution per formula unit}$$
$$Fe_2(SO_4)_3 \xrightarrow{H_2O} 2\ Fe^{3+} + 3\ SO_4^{2-} \qquad \text{5 ions in solution per formula unit}$$

One mole of NaCl will give 1 mol of Na^+ ions and 1 mol of Cl^- ions in solution, assuming complete dissociation of the salt. One mole of $CaCl_2$ will give 1 mol of Ca^{2+} ions and 2 mol of Cl^- ions in solution.

$$\underset{1\text{ mol}}{NaCl} \xrightarrow{H_2O} \underset{1\text{ mol}}{Na^+} + \underset{1\text{ mol}}{Cl^-}$$

$$\underset{1\text{ mol}}{CaCl_2} \xrightarrow{H_2O} \underset{1\text{ mol}}{Ca^{2+}} + \underset{2\text{ mol}}{2\,Cl^-}$$

PROBLEM 15.2 What is the molarity of each ion in a solution of (a) 2.0 M NaCl, and (b) 0.40 M K_2SO_4? (Assume complete dissociation.)

(a) According to the dissociation equation,

$$\underset{1\text{ mol}}{NaCl} \xrightarrow{H_2O} \underset{1\text{ mol}}{Na^+} + \underset{1\text{ mol}}{Cl^-}$$

the concentration of Na^+ is equal to that of NaCl (1 mol NaCl $\longrightarrow$ 1 mol Na^+) and the concentration of Cl^- is also equal to that of NaCl. Therefore, the concentrations of the ions in 2.0 M NaCl are 2.0 M Na^+ and 2.0 M Cl^-.

(b) According to the dissociation equation,

$$\underset{1\text{ mol}}{K_2SO_4} \xrightarrow{H_2O} \underset{2\text{ mol}}{2\,K^+} + \underset{1\text{ mol}}{SO_4^{2-}}$$

The concentration of K^+ is twice that of K_2SO_4 and the concentration of SO_4^{2-} is equal to that of K_2SO_4. Therefore, the concentrations of the ions in 0.40 M K_2SO_4 are 0.80 M K^+ and 0.40 M SO_4^{2-}.

Colligative Properties of Electrolyte Solutions We have learned that, when 1 mol of sucrose, a nonelectrolyte, is dissolved in 1000 g of water, the solution freezes at $-1.86°C$. When 1 mol of NaCl is dissolved in 1000 g of water, the freezing point of the solution is not $-1.86°C$, as might be expected, but is closer to $-3.72°C$ (-1.86×2). The reason for the lower freezing point is that 1 mol of NaCl in solution produces 2 mol of particles ($2 \times 6.022 \times 10^{23}$ ions) in solution. Thus, the freezing point depression produced by 1 mol of NaCl is essentially equivalent to that produced by 2 mol of a nonelectrolyte. An electrolyte such as $CaCl_2$, which yields three ions in water, gives a freezing point depression of about three times that of a nonelectrolyte. These freezing point data provide additional evidence that electrolytes dissociate when dissolved in water. The other colligative properties are similarly affected by substances that yield ions in aqueous solutions.

15.8 Ionization of Water

The more we study chemistry, the more intriguing the little molecule of water becomes. Two equations commonly used to show how water ionizes are

$$\underset{\text{Acid}}{H_2O} + \underset{\text{Base}}{H_2O} \rightleftharpoons \underset{\text{Acid}}{H_3O^+} + \underset{\text{Base}}{OH^-}$$

and

$$H_2O \rightleftharpoons H^+ + OH^-$$

The first equation represents the Brønsted–Lowry concept, with water reacting as both an acid and a base, forming a hydronium ion and a hydroxide ion. The second equation is a simplified version, indicating that water ionizes to give a hydrogen and a hydroxide ion. Actually, the proton, H^+, is hydrated and exists as a hydronium ion. In either case equal molar amounts of acid and base are produced so that water is neutral, having neither H^+ nor OH^- ions in excess. The ionization of water at 25°C produces an H^+ ion concentration of 1.0×10^{-7} mole per liter and an OH^- ion concentration of 1.0×10^{-7} mole per liter. These concentrations are usually expressed as

$$[H^+] \text{ or } [H_3O^+] = 1.0 \times 10^{-7} \text{ mol/L}$$
$$[OH^-] = 1.0 \times 10^{-7} \text{ mol/L}$$

These figures mean that about two out of every billion water molecules are ionized. This amount of ionization, small as it is, is a significant factor in the behavior of water in many chemical reactions.

The square brackets, [ ], indicate that the concentration is in moles per liter. Thus $[H^+]$ means the concentration of H^+ in moles per liter.

15.9 Introduction to pH

The acidity of an aqueous solution depends on the concentration of hydrogen or hydronium ions. The acidity of solutions involved in a chemical reaction is often critically important, especially for biochemical reactions. The pH scale of acidity was devised to fill the need for a simple, convenient numerical way to state the acidity of a solution. Values on the pH scale are obtained by mathematical conversion of H^+ ion concentrations to pH by these expressions:

$$\text{pH} = \log \frac{1}{[H^+]} \qquad \text{or} \qquad \text{pH} = -\log[H^+]$$

pH

where $[H^+] = H^+$ or H_3O^+ ion concentration in moles per liter. The **pH** is defined as the logarithm (log) of the reciprocal of the H^+ or H_3O^+ ion

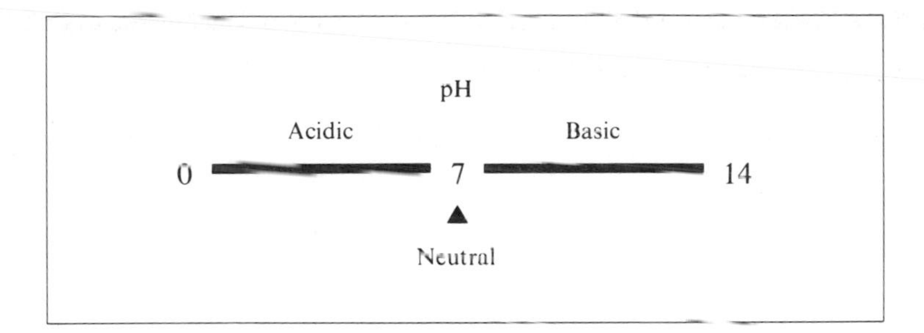

Figure 15.3

The pH scale of acidity and basicity

concentration in moles per liter. The scale itself is based on the H^+ concentration in water at 25°C. At this temperature, water has an H^+ concentration of 1×10^{-7} mole per liter and is calculated to have a pH of 7.

$$\mathrm{pH} = \log \frac{1}{[\mathrm{H}^+]} = \log \frac{1}{[1 \times 10^{-7}]} = \log 1 \times 10^7 = 7$$

By an alternate and mathematically equivalent definition, pH is the *negative* logarithm of the H^+ or H_3O^+ concentration in moles per liter:

$$\mathrm{pH} = -\log[\mathrm{H}^+] = -\log[1 \times 10^{-7}] = -(-7) = 7$$

The pH of pure water at 25°C is 7 and is said to be neutral; that is, it is neither acidic nor basic, because the concentrations of H^+ and OH^- are equal. Solutions that contain more H^+ ions than OH^- ions have pH values less than 7, and solutions that contain less H^+ ions than OH^- ions have values greater than 7.

pH < 7.00 Acidic solution
pH = 7.00 Neutral solution
pH > 7.00 Basic solution

When $[\mathrm{H}^+] = 1 \times 10^{-5}$ mol/L, pH = 5 (acidic)
When $[\mathrm{H}^+] = 1 \times 10^{-9}$ mol/L, pH = 9 (basic)

Instead of saying that the hydrogen ion concentration in the solution is 1×10^{-5} mole per liter, it is customary to say that the pH of the solution is 5. The smaller the pH value, the more acidic the solution (see Figure 15.3).

At a given molarity a strong acid is more acidic (has a higher H^+ concentration) than a weak acid. For example, in a 0.100 *M* concentration, the pH of HCl is 1.00 and that of $HC_2H_3O_2$ is 2.87. As a weak acid is made more dilute, greater percentages of its molecules ionize, and the pH tends to approach that of a strong acid at comparable dilutions. This behavior is illustrated by the following comparative data for hydrochloric acid (100% ionized) and acetic acid.

HCl solution			$HC_2H_3O_2$ solution		
M	**pH**	**% ionized**	**M**	**pH**	**% ionized**
0.100	1.00	100	0.100	2.87	1.35
0.0100	2.00	100	0.0100	3.37	4.27
0.00100	3.00	100	0.00100	3.90	12.6

The pH scale, along with its interpretation, is given in Table 15.4, and Table 15.5 lists the pH of some common solutions. Note that a change of only 1 pH unit means a tenfold increase or decrease in H^+ ion concentration. A simplified method of determining pH from $[H^+]$ follows:

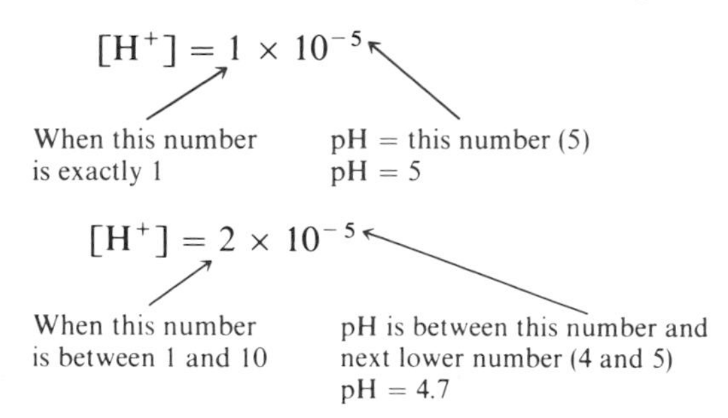

Calculation of the pH value corresponding to any H^+ ion concentration requires the use of logarithms. Logarithms are exponents. The logarithm (log) of a number is simply the power to which 10 must be raised to give that number. Thus the log of 100 is 2, and the log of 1000 is 3 ($100 = 10^2$, $1000 = 10^3$). The log of 500 is 2.70, but you cannot readily determine this value without a log table or a calculator with log capability. However, even though you may not be familiar with logarithms, the simplified log scale of Figure 15.4 can be used to estimate the logarithms of various numbers. For example, let us use this log scale to calculate the pH of a solution with $[H^+] = 2 \times 10^{-5}$:

$$[H^+] = ② \times 10^{-⑤}$$

pH = This number (5) minus the log of this number (2) (which must be between 1 and 10)

$\text{pH} = 5 - \log 2$

$\log 2 = 0.30$ (from the log scale)

$\text{pH} = 5 - 0.30 = 4.7$ (Answer)

Table 15.4 The pH scale for expressing acidity

$[H^+]$ (mol/L)	pH	
1×10^{-14}	14	↑
1×10^{-13}	13	
1×10^{-12}	12	Increasing basicity
1×10^{-11}	11	
1×10^{-10}	10	
1×10^{-9}	9	
1×10^{-8}	8	
1×10^{-7}	7	Neutral
1×10^{-6}	6	
1×10^{-5}	5	
1×10^{-4}	4	Increasing acidity
1×10^{-3}	3	
1×10^{-2}	2	
1×10^{-1}	1	
1×10^{0}	0	↓

Table 15.5 The pH of some common solutions

Solution	pH
Gastric juice	1.0
0.1 *M* HCl	1.0
Lemon juice	2.3
Vinegar	2.8
0.1 *M* $HC_2H_3O_2$	2.9
Orange juice	3.7
Tomato juice	4.1
Coffee, black	5.0
Urine	6.0
Milk	6.6
Pure water (25°C)	7.0
Blood	7.4
Household ammonia	11.0
1 *M* NaOH	14.0

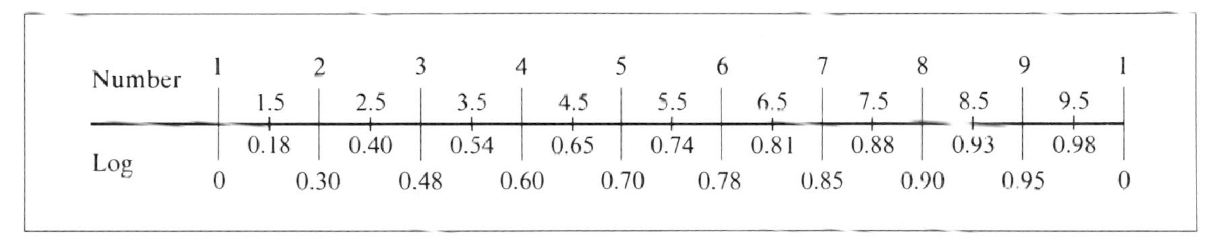

Figure 15.4
Simplified logarithm scale. For example, the logarithm (log) of 5 is 0.70; the logarithm of 7.5 is 0.88.

PROBLEM 15.3 What is the pH of a solution with an $[H^+]$ of (a) 1×10^{-11} and (b) 6×10^{-4}?

(a) $[H^+] = 1 \times 10^{-11}$

$pH = 11$

(b) $[H^+] = 6 \times 10^{-4}$

$pH = 4 - \log 6$

$\log 6 = 0.78$ (from Figure 15.4)

$pH = 4 - 0.78 = 3.22$ (Answer)

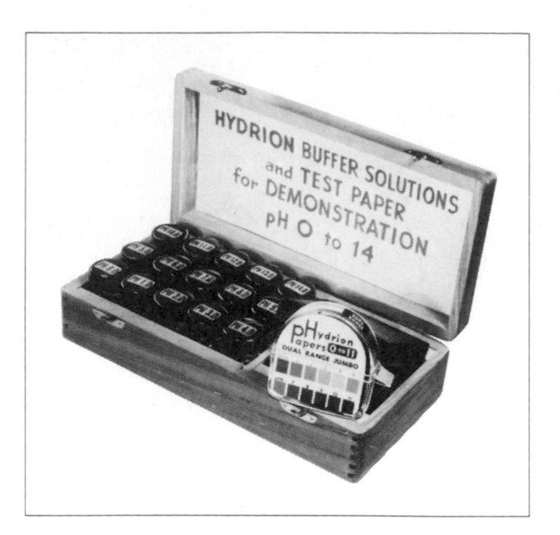

Figure 15.5

pH test paper for determining the approximate acidity of solutions. (*Courtesy Micro Essential Laboratory, Inc.*)

The measurement and control of pH is extremely important in many fields of science and technology. The proper soil pH is necessary to grow certain types of plants successfully. The pH of certain foods is too acid for some diets. Many biological processes are delicately controlled pH systems. The pH of human blood is regulated to very close tolerances through the uptake or release of H^+ by mineral ions such as HCO_3^-, HPO_4^{2-}, and $H_2PO_4^-$. Changes in the pH of the blood by as little as 0.4 pH unit result in death.

Compounds with colors that change at particular pH values are used as indicators in acid–base reactions. For example, phenolphthalein, an organic compound, is colorless in acid solution and changes to pink at a pH of 8.3. When a solution of sodium hydroxide is added to a hydrochloric acid solution containing phenolphthalein, the change in color (from colorless to pink) indicates that all the acid is neutralized. Commercially available pH test paper, such as shown in Figure 15.5, contains chemical indicators. The indicator in the paper takes on different colors when wetted with solutions of different pH. Thus the pH of a solution can be estimated by placing a drop on the test paper and comparing the color of the test paper with a color chart calibrated at different pH values. Electronic pH meters of the type shown in Figure 15.6 are used for making rapid and precise pH determinations.

15.10 Neutralization

neutralization

The reaction of an acid and a base to form a salt and water is known as **neutralization**. We have seen this reaction before; but now, in the light of what we have learned about ions and ionization, let us reexamine the process of neutralization.

Consider the reaction that occurs when solutions of sodium hydroxide and hydrochloric acid are mixed. The ions present initially are Na^+ and OH^- from the base and H^+ and Cl^- from the acid. The products, sodium chloride and water,

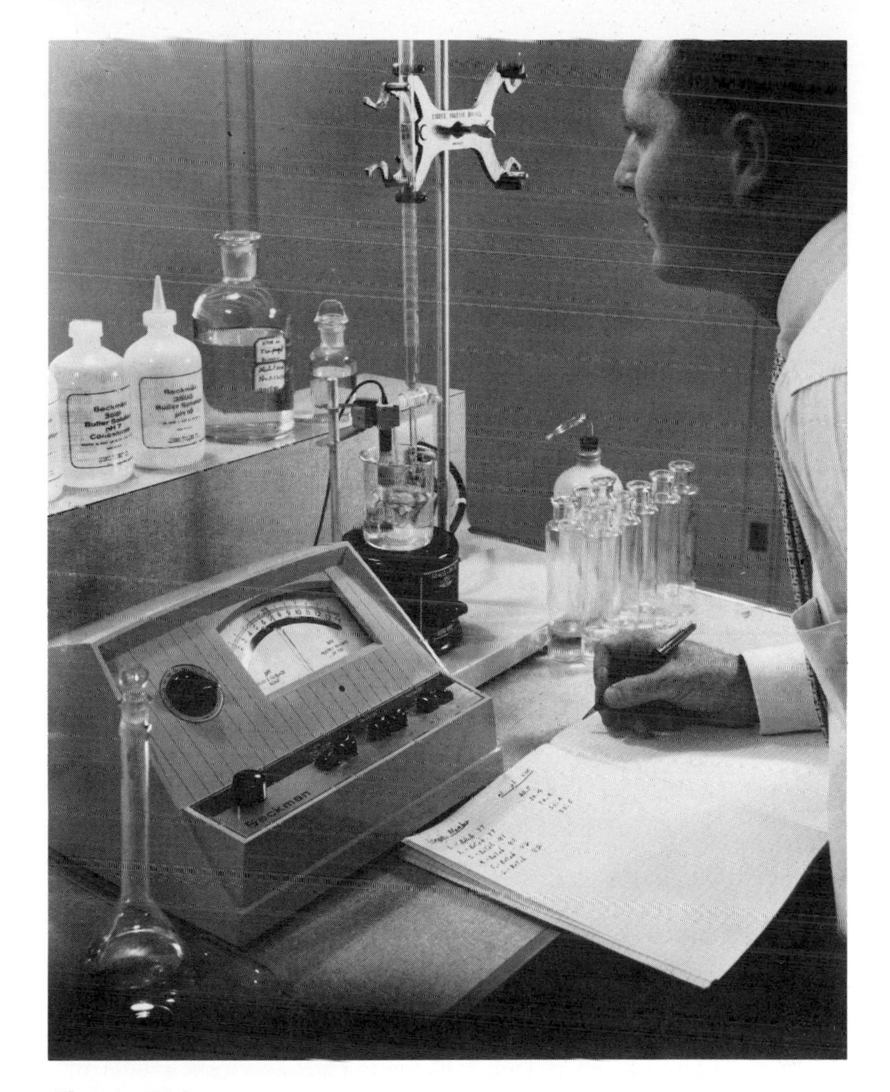

Figure 15.6

An electronic pH meter. Accurate measurements may be made by meters of this type. (*Courtesy Beckman Instruments, Inc. Zeromatic is a registered trademark.*)

exist as Na^+ and Cl^- ions and H_2O molecules. A chemical equation representing this reaction is

$$HCl(aq) + NaOH(aq) \longrightarrow NaCl(aq) + H_2O \quad (5)$$

This equation, however, does not show that HCl, NaOH, and NaCl exist as ions in solution. The following total ionic equation gives a better representation of the reaction:

$$(H^+ + Cl^-) + (Na^+ + OH^-) \longrightarrow Na^+ + Cl^- + H_2O \quad (6)$$

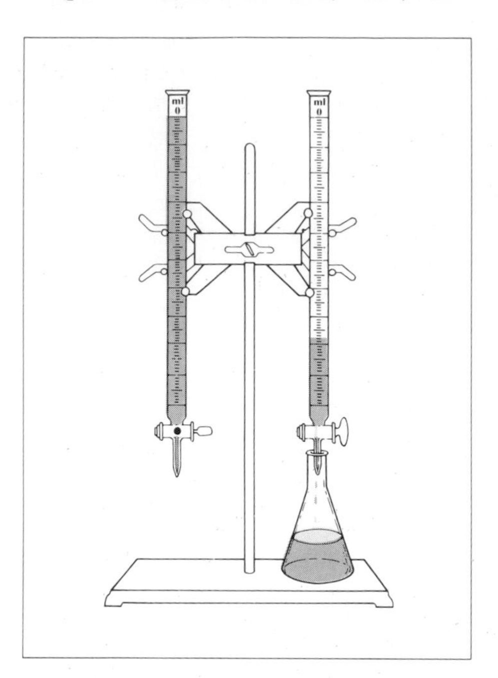

Figure 15.7

Graduated burets are used in titrations for neutralization of acids and bases as well as for many other volumetric determinations.

spectator ions

Equation (6) shows that the Na^+ and Cl^- ions did not react. These ions are called **spectator ions** because they were present but did not take part in the reaction. The only reaction that occurred was that between the H^+ and OH^- ions. Therefore, the equation for the neutralization can be written as this net ionic equation:

$$\underset{\text{Acid}}{H^+} + \underset{\text{Base}}{OH^-} \longrightarrow \underset{\text{Water}}{H_2O} \qquad (7)$$

This simple net ionic equation (7) represents not only the reaction of sodium hydroxide and hydrochloric acid but also the reaction of any acid with any base in an aqueous solution. The driving force of a neutralization reaction is the ability of an H^+ ion and an OH^- ion to react and form a molecule of un-ionized water.

titration

The amount of acid, base, or other species in a sample may be determined by titration. **Titration** is the process of measuring the volume of one reagent that is required to react with a measured weight or volume of another reagent. Let us consider the titration of an acid with a base. A measured volume of acid of unknown concentration is placed in a flask, and a few drops of an indicator solution are added. Base solution of known concentration is slowly added from a buret to the acid until the indicafor changes color (see Figure 15.7). The indicator selected is one that changes color when the stoichiometric quantity (according to the equation) of base has been added to the acid. At this point, known as the *end point of the titration*, the titration is complete, and the volume of base used to neutralize the acid is read from the buret. The concentration or amount of acid in solution can be calculated from the titration data and the chemical equation for the reaction. Illustrative problems follow.

PROBLEM 15.4 Suppose that 42.00 mL of 0.150 M NaOH solution is required to titrate 50.00 mL of hydrochloric acid solution. What is the molarity of the acid solution?

The equation for the reaction is

$$NaOH(aq) + HCl(aq) \longrightarrow NaCl(aq) + H_2O(l)$$

In this neutralization NaOH and HCl react in a 1:1 mole ratio. Therefore, the moles of HCl in solution are equal to the moles of NaOH required to react with it. First we calculate the moles of NaOH used, and from this value we determine the moles of HCl.

Data: 42.00 mL of 0.150 M NaOH 50.00 mL of HCl
Molarity of acid = M (unknown)

Moles of NaOH:

$$M = \text{mol/L} \qquad 42.00 \text{ mL} = 0.04200 \text{ L}$$

$$0.04200 \cancel{\text{L}} \times \frac{0.150 \text{ mol NaOH}}{1 \cancel{\text{L}}} = 0.00630 \text{ mol NaOH}$$

Since NaOH and HCl react in a 1:1 mole ratio, 0.00630 mol of HCl was present in the 50.00 mL of HCl solution. Therefore the molarity of the HCl is

$$M = \frac{\text{mol}}{\text{L}} = \frac{0.00630 \text{ mol HCl}}{0.05000 \text{ L}} = 0.126\ M \text{ HCl} \quad \text{(Answer)}$$

PROBLEM 15.5 Suppose that 42.00 mL of 0.150 M NaOH solution is required to titrate 50.00 mL of sulfuric acid (H_2SO_4) solution. What is the molarity of the acid solution?

The equation for the reaction is

$$2\ NaOH(aq) + H_2SO_4(aq) \longrightarrow Na_2SO_4(aq) + 2\ H_2O(l)$$

The same amount of base (0.00630 mol of NaOH) is used in this titration as in Problem 15.4, but the mole ratio of acid to base in the reaction is 1:2. The moles of H_2SO_4 reacted can be calculated by using the mole-ratio method.

Data: 42.00 mL of 0.150 M NaOH = 0.00630 mol NaOH

$$0.00630 \cancel{\text{mol NaOH}} \times \frac{1 \text{ mol } H_2SO_4}{2 \cancel{\text{mol NaOH}}} = 0.00315 \text{ mol } H_2SO_4$$

Therefore 0.00315 mol of H_2SO_4 was present in 50.00 mL of H_2SO_4 solution. The molarity of the H_2SO_4 is

$$M = \frac{\text{mol}}{\text{L}} = \frac{0.00315 \text{ mol } H_2SO_4}{0.05000} = 0.0630\ M\ H_2SO_4 \quad \text{(Answer)}$$

PROBLEM 15.6 A 25.00 mL sample of H_2SO_4 solution required 14.26 mL of 0.2240 N NaOH for complete neutralization. What is the normality and the molarity of the sulfuric acid?

The equation for the reaction is

$$2\ NaOH(aq) + H_2SO_4(aq) \longrightarrow Na_2SO_4(aq) + 2\ H_2O(l)$$

The normality of the acid can be calculated from

$$V_A N_A = V_B N_B$$

Substitute the data in the problem and solve for N_A.

$$25.00\ \text{mL} \times N_A = 14.26\ \text{mL} \times 0.2240\ N$$

$$N_A = \frac{14.26\ \text{mL} \times 0.2240\ N}{25.00\ \text{mL}} = 0.1278\ N\ H_2SO_4$$

The normality of the acid is 0.1278 N.

Because H_2SO_4 furnishes 2 equivalents of H^+ per mole, the conversion to molarity is

$$\frac{\text{equiv.}}{\text{L}} \times \frac{\text{mol}}{\text{equiv.}}$$

$$\frac{0.1278\ \cancel{\text{equiv. } H_2SO_4}}{1\ \text{L}} \times \frac{1\ \text{mol}\ H_2SO_4}{2\ \cancel{\text{equiv. } H_2SO_4}} = 0.06390\ \text{mol/L}$$

The H_2SO_4 solution is 0.06390 M.

15.11 Writing Ionic Equations

un-ionized equation

total ionic equation

net ionic equation

In Section 15.10 we wrote the reaction of hydrochloric acid and sodium hydroxide in three different equations. Equation (5) was the un-ionized equation; equation (6) was the total ionic equation; and equation (7) was the net ionic equation. In the **un-ionized equation**, compounds are written in their molecular, or normal, formula expressions. In the **total ionic equation**, compounds are written to show the form in which they are predominantly present: strong electrolytes as ions in solution; and nonelectrolytes, weak electrolytes, precipitates, and gases in their molecular (or un-ionized) forms. In the **net ionic equation**, only those molecules or ions that have changed are included in the equation; ions or molecules that do not change (spectators) are omitted.

Up to this point, when balancing an equation we have been concerned only with the atoms of the individual elements. Because ions are electrically charged, ionic equations often end up with a net electrical charge. A balanced equation must have the same net charge on each side, whether that charge is positive, negative, or zero. Therefore, when balancing an ionic equation, we must make sure that both the same number of each kind of atom and the same net electrical charge are present on each side.

Following is a list of rules for writing ionic equations:

1. Strong electrolytes in solution are written in their ionic form.
2. Weak electrolytes are written in their molecular (un-ionized) form.

3 Nonelectrolytes are written in their molecular form.
4 Insoluble substances, precipitates, and gases are written in their molecular forms.
5 The net ionic equation should include only those substances that have undergone a chemical change. Spectator ions are omitted from the net ionic equation.
6 Equations must be balanced, both in atoms and in electrical charge.

Study the examples below. Note that all reactions are in solution.

(a) $HNO_3(aq) + KOH(aq) \longrightarrow KNO_3(aq) + H_2O$ Un-ionized equation

$(H^+ + NO_3^-) + (K^+ + OH^-) \longrightarrow (K^+ + NO_3^-) + H_2O$ Total ionic equation

$H^+ + OH^- \longrightarrow H_2O$ Net ionic equation

HNO_3, KOH, and KNO_3 are soluble, strong electrolytes. K^+ and NO_3^- are spectator ions, have not changed, and are not included in the net ionic equation. Water is a nonelectrolyte and is written in the molecular form.

(b) $2\ AgNO_3(aq) + BaCl_2(aq) \longrightarrow 2\ AgCl\downarrow + Ba(NO_3)_2(aq)$

$(2\ Ag^+ + 2\ NO_3^-) + (Ba^{2+} + 2\ Cl^-) \longrightarrow 2\ AgCl\downarrow + (Ba^{2+} + 2\ NO_3^-)$

$Ag^+ + Cl^- \longrightarrow AgCl\downarrow$ Net ionic equation

Although silver chloride (AgCl) is an ionic salt, it is written in the un-ionized form on the right side of the ionic equations because most of the Ag^+ and Cl^- ions are no longer in solution but have formed a precipitate of AgCl. Ba^{2+} and NO_3^- are spectator ions.

(c) $Na_2CO_3(aq) + H_2SO_4(aq) \longrightarrow Na_2SO_4(aq) + H_2O + CO_2\uparrow$

$(2\ Na^+ + CO_3^{2-}) + (2\ H^+ + SO_4^{2-}) \longrightarrow (2\ Na^+ + SO_4^{2-}) + H_2O + CO_2\uparrow$

$CO_3^{2-} + 2\ H^+ \longrightarrow H_2O + CO_2\uparrow$ Net ionic equation

Carbon dioxide (CO_2) is a gas and evolves from the solution; Na^+ and SO_4^{2-} are spectator ions.

(d) $HC_2H_3O_2(aq) + NaOH(aq) \longrightarrow NaC_2H_3O_2(aq) + H_2O$

$HC_2H_3O_2 + (Na^+ + OH^-) \longrightarrow (Na^+ + C_2H_3O_2^-) + H_2O$

$HC_2H_3O_2 + OH^- \longrightarrow C_2H_3O_2^- + H_2O$ Net ionic equation

Acetic acid ($HC_2H_3O_2$), a weak acid, is written in the molecular form, but sodium acetate ($NaC_2H_3O_2$), a soluble salt, is written in the ionic form. The Na^+ ion is the only spectator ion in this reaction. Both sides of the net ionic equation have a -1 electrical charge.

(e) $Mg(s) + 2\ HCl(aq) \longrightarrow MgCl_2(aq) + H_2\uparrow$

$Mg + (2\ H^+ + 2\ Cl^-) \longrightarrow (Mg^{2+} + 2\ Cl^-) + H_2\uparrow$

$Mg + 2\ H^+ \longrightarrow Mg^{2+} + H_2\uparrow$ Net ionic equation

The net electrical charge on both sides of the equation is $+2$.

(f) $H_2SO_4(aq) + Ba(OH)_2(aq) \longrightarrow BaSO_4\downarrow + 2\,H_2O$

$(2\,H^+ + SO_4^{2-}) + (Ba^{2+} + 2\,OH^-) \longrightarrow BaSO_4\downarrow + 2\,H_2O$

$2\,H^+ + SO_4^{2-} + Ba^{2+} + 2\,OH^- \longrightarrow$

$BaSO_4\downarrow + 2\,H_2O$ Net ionic equation

Barium sulfate ($BaSO_4$) is a highly insoluble salt. If we conduct this reaction using the conductivity apparatus described in Section 15.5, the light glows brightly at first but goes out when the reaction is complete, because almost no ions are left in solution. The $BaSO_4$ precipitates out of solution, and water is a nonconductor of electricity.

QUESTIONS

A. *Review the meanings of the new terms introduced in this chapter.*

1. Hydronium ion
2. Amphoteric
3. Electrolyte
4. Nonelectrolyte
5. Dissociation
6. Ionization
7. Strong electrolyte
8. Weak electrolyte
9. pH
10. Neutralization
11. Spectator ions
12. Titration
13. Un-ionized equation
14. Total ionic equation
15. Net ionic equation

B. *Information useful in answering the following questions will be found in the tables and figures.*

1. Since a hydrogen ion and a proton are identical, what differences exist between the Arrhenius and Brønsted–Lowry definitions of an acid? (See Table 15.1.)
2. According to Figure 15.1, what type of substance must be in solution in order for the bulb to light?
3. Which of the following classes of compounds are electrolytes: acids, alcohols, bases, salts? (See Table 15.2.)
4. What two differences are apparent in the arrangement of water molecules about the hydrated ions as depicted in Figure 15.2?
5. The pH of a solution with a hydrogen ion concentration of 0.003 *M* is between what two whole numbers? (See Table 15.4.)
6. Which is the more acidic, tomato juice or blood? (See Table 15.5.)

C. *Review questions*

1. Using each of the three acid–base theories (Arrhenius, Brønsted–Lowry, and Lewis), define an acid and a base.
2. For each of the acid–base theories referred to in Question 1, write an equation illustrating the neutralization of an acid with a base.
3. Identify the conjugate acid–base pairs in the following equations:
   (a) $HCl + NH_3 \longrightarrow NH_4^+ + Cl^-$
   (b) $HCO_3^- + OH^- \rightleftharpoons CO_3^{2-} + H_2O$
   (c) $HCO_3^- + H_3O^+ \rightleftharpoons H_2CO_3 + H_2O$
   (d) $HC_2H_3O_2 + H_2O \rightleftharpoons H_3O^+ + C_2H_3O_2^-$
   (e) $HC_2H_3O_2 + H_2SO_4 \rightleftharpoons H_2C_2H_3O_2^+ + HSO_4^-$
   (f) The two-step ionization of sulfuric acid,
   $H_2SO_4 + H_2O \longrightarrow H_3O^+ + HSO_4^-$
   $HSO_4^- + H_2O \rightleftharpoons H_3O^+ + SO_4^{2-}$
   (g) $HClO_4 + H_2O \longrightarrow H_3O^+ + ClO_4^-$
   (h) $CH_3O^- + H_3O^+ \longrightarrow CH_3OH + H_2O$
4. Write the Lewis structure for (a) bromide ion, (b) hydroxide ion, and (c) cyanide ion. Why are these ions considered to be bases according to the Brønsted–Lowry and Lewis acid–base theories?
5. Complete and balance the following equations:
   (a) $Mg(s) + HCl(aq) \longrightarrow$
   (b) $BaO(s) + HBr(aq) \longrightarrow$
   (c) $Al(s) + H_2SO_4(aq) \longrightarrow$
   (d) $Na_2CO_3(aq) + HCl(aq) \longrightarrow$
   (e) $Fe_2O_3(s) + HBr(aq) \longrightarrow$

(f) $Ca(OH)_2(aq) + H_2CO_3(aq) \longrightarrow$
(g) $NaOH(aq) + HBr(aq) \longrightarrow$
(h) $KOH(aq) + HCl(aq) \longrightarrow$
(i) $Ca(OH)_2(aq) + HI(aq) \longrightarrow$
(j) $Al(OH)_3(s) + HBr(aq) \longrightarrow$
(k) $Na_2O(s) + HClO_4(aq) \longrightarrow$
(l) $LiOH(aq) + FeCl_3(aq) \longrightarrow$
(m) $NH_4OH(aq) + FeCl_2(aq) \longrightarrow$

6. Into what three classes of compounds do electrolytes generally fall?

7. Which of the following compounds are electrolytes? Consider each substance to be mixed with water.
(a) HCl
(b) CO_2
(c) $CaCl_2$
(d) $C_{12}H_{22}O_{11}$ (sugar)
(e) C_3H_7OH (rubbing alcohol)
(f) CCl_4 (insoluble)
(g) $NaHCO_3$ (baking soda)
(h) N_2 (insoluble gas)
(i) $AgNO_3$
(j) HCOOH (formic acid)
(k) RbOH
(l) K_2CrO_4

8. Name each compound listed in Table 15.3.

9. A solution of HCl in water conducts an electric current, but a solution of HCl in benzene does not. Explain this behavior in terms of ionization and chemical bonding.

10. How do salts exist in their crystalline structure? What occurs when they are dissolved in water?

11. An aqueous methyl alcohol, CH_3OH, solution does not conduct an electric current, but a solution of sodium hydroxide, NaOH, does. What does this information tell us about the OH group in the alcohol?

12. Why does molten sodium chloride conduct electricity?

13. Explain the difference between dissociation of ionic compounds and ionization of molecular compounds.

14. Distinguish between strong and weak electrolytes.

15. Explain why ions are hydrated in aqueous solutions.

16. Indicate, by simple equations, how the following substances dissociate or ionize in water:
(a) $Cu(NO_3)_2$
(b) $HC_2H_3O_2$
(c) HNO_2
(d) LiOH
(e) NH_4Br
(f) K_2SO_4
(g) $NaClO_3$
(h) K_3PO_4

17. What is the main distinction between water solutions of strong and weak electrolytes?

18. What are the relative concentrations of $H^+(aq)$ and $OH^-(aq)$ in (a) a neutral solution, (b) an acid solution, and (c) a basic solution?

19. Write the net ionic equation for the reaction of an acid with a base in an aqueous solution.

20. The solubility of hydrogen chloride gas in water, a polar solvent, is much greater than its solubility in benzene, a nonpolar solvent. How can you account for this difference?

21. Pure water, containing both acid and base ions, is neutral. Why?

22. Rewrite the following unbalanced equations, changing them into balanced net ionic equations. All reactions are in water solution.
(a) $K_2SO_4(aq) + Ba(NO_3)_2(aq) \longrightarrow KNO_3(aq) + BaSO_4(s)$
(b) $CaCO_3(s) + HCl(aq) \longrightarrow CaCl_2(aq) + CO_2(g) + H_2O$
(c) $Mg(s) + HC_2H_3O_2(aq) \longrightarrow Mg(C_2H_3O_2)_2(aq) + H_2(g)$
(d) $H_2S(g) + CdCl_2(aq) \longrightarrow CdS(s) + HCl(aq)$
(e) $Zn(s) + H_2SO_4(aq) \longrightarrow ZnSO_4(aq) + H_2(g)$
(f) $AlCl_3(aq) + Na_3PO_4(aq) \longrightarrow AlPO_4(s) + NaCl(aq)$

23. In each of the following pairs which solution is more acidic? (All are water solutions.)
(a) 1 molar HCl or 1 molar H_2SO_4?
(b) 1 molar HCl or 1 molar $HC_2H_3O_2$?
(c) 1 molar HCl or 2 molar HCl?
(d) 1 normal H_2SO_4 or 1 molar H_2SO_4?

24. How does a hydronium ion differ from a hydrogen ion?

25. Arrange, in decreasing order of freezing points, 1 molal aqueous solutions of HCl, $HC_2H_3O_2$, $C_{12}H_{22}O_{11}$ (sucrose), and $CaCl_2$. (List the one with the highest freezing point first.)

26. At 100°C the H^+ concentration in water is about 1×10^{-6} mol/L, about 10 times that of water at 25°C. At which of these temperatures is (a) the pH of water the greater, (b) the hydrogen ion (hydronium ion) concentration the higher, and (c) the water neutral?

27. What is the relative difference in H^+ concentration in solutions that differ by 1 pH unit?

28. A 1 molal solution of acetic acid in water freezes at a lower temperature than a 1 molal solution of ethyl alcohol, C_2H_5OH, in water. Explain.

29. At the same cost per pound, which alcohol, CH_3OH or C_2H_5OH, would be more economical to purchase as an antifreeze for your car?

30. Which of the following statements are correct?

(a) The Arrhenius theory of acids and bases is restricted to aqueous solutions.

(b) The Brønsted–Lowry theory of acids and bases is restricted to solutions other than aqueous solutions.

(c) All substances that are acids according to the Brønsted–Lowry theory will also be acids by the Lewis theory.

(d) All substances that are acids according to the Lewis theory will also be acids by the Bronsted–Lowry theory.

(e) An electron-pair donor is a Lewis acid.

(f) All Arrhenius acid–base neutralization reactions can be represented by a single net ionic equation.

(g) When an ionic compound dissolves in water, the ions separate; this process is called ionization.

(h) In the autoionization of water

$$2\,H_2O \rightleftharpoons H_3O^+ + OH^-$$

the H_3O^+ and the OH^- constitute a conjugate acid–base pair.

(i) In the reaction in part (h) H_2O is both the acid and the base.

(j) Most common Na, K, and NH_4^+ salts are soluble in water.

(k) A solution of pH 3 is 100 times more acidic than a solution of pH 5.

(l) In general, ionic substances when placed in water will give a solution capable of conducting an electric current.

(m) The terms *dissociation* and *ionization* are synonymous.

(n) A saturated solution may become unsaturated by raising the temperature of the solution.

(o) The terms *strong acid, strong base, weak acid,* and *weak base* refer to whether an acid or base solution is concentrated or dilute.

(p) pH is defined as the negative logarithm of the molar concentration of H^+ ions (or H_3O^+ ions).

(q) All reactions may be represented by net ionic equations.

(r) One mole of $CaCl_2$ contains more anions than cations.

(s) It is possible to boil seawater at a lower temperature than that required to boil pure water (both at the same pressure).

(t) It is possible to have a neutral aqueous solution whose pH is not 7.

D. *Review problems*

1. Calculate the molarity of the ions present in each of the following salt solutions. Assume each salt to be 100% dissociated.

(a) 0.015 *M* NaCl

(b) 4.25 *M* $NaKSO_4$

(c) 0.75 *M* $ZnBr_2$

(d) 1.65 *M* $Al_2(SO_4)_3$

(e) 0.20 *M* $CaCl_2$

(f) 22.0 g KI in 500 mL of solution

(g) 900 g $(NH_4)_2SO_4$ in 20.0 L of solution

(h) 0.0120 g $Mg(ClO_3)_2$ in 1.00 mL of solution

2. In Problem 1, how many grams of each ion would be present in 100 mL of each solution?

3. What is the concentration of Ca^{2+} ions in a solution of CaI_2 having an I^- ion concentration of 0.520 *M*?

4. What is the molar concentration of all ions present in a solution prepared by mixing the following?

(a) 30.0 mL of 1.0 *M* NaCl and 40.0 mL of 1.0 *M* NaCl

(b) 30.0 mL of 1.0 *M* HCl and 30.0 mL of 1.0 *M* NaOH

(c) 100.0 mL of 2.0 *M* KCl and 100.0 mL of 1.0 *M* $CaCl_2$

*(d) 100.0 mL of 0.40 *M* KOH and 100.0 mL of 0.80 *M* HCl

(e) 35.0 mL of 0.20 *M* $Ba(OH)_2$ and 35.0 mL of 0.20 *M* H_2SO_4

(f) 1.00 L of 1.0 *M* $AgNO_3$ and 500 mL of 2.0 *M* NaCl.

(Neglect the concentration of H^+ and OH^- from water. Also, assume volumes of solutions are additive.)

5. How many milliliters of 0.40 *M* HCl can be made by diluting 100 mL of 12 *M* HCl with water?

6. Given the data for the following six titrations, calculate the molarity of the HCl in titrations (a), (b), and (c), and the molarity of the NaOH in titrations (d), (e), and (f).

	mL HCl	Molarity HCl	mL NaOH	Molarity NaOH
(a)	40.13	*M* HCl	37.70	0.728
(b)	19.00	*M* HCl	33.66	0.306
(c)	27.25	*M* HCl	18.00	0.555
(d)	37.19	0.126	31.91	*M* NaOH
(e)	48.04	0.482	24.02	*M* NaOH
(f)	13.13	1.425	39.39	*M* NaOH

7. If 29.26 mL of 0.430 *M* HCl neutralizes 20.40 mL of $Ba(OH)_2$ solution, what is the molarity of the $Ba(OH)_2$ solution? The reaction is

$$Ba(OH)_2(aq) + 2\ HCl(aq) \longrightarrow BaCl_2(aq) + 2\ H_2O$$

8. What volume (in milliliters) of 0.245 *M* HCl will neutralize (a) 50.0 mL of 0.100 *M* $Ca(OH)_2$ and (b) 10.0 g of $Al(OH)_3$? The equations are
(a) $2\ HCl(aq) + Ca(OH)_2(aq) \longrightarrow CaCl_2(aq) + 2\ H_2O$
(b) $3\ HCl(aq) + Al(OH)_3(s) \longrightarrow AlCl_3(aq) + 3\ H_2O$

9. A sample of pure sodium carbonate weighing 0.452 g was dissolved in water and neutralized with 42.4 mL of hydrochloric acid. Calculate the molarity of the acid:

$$Na_2CO_3(aq) + 2\ HCl(aq) \longrightarrow 2\ NaCl(aq) + CO_2(g) + H_2O$$

10. What volume (mL) of 0.1234 *M* HCl is needed to neutralize 2.00 g $Ca(OH)_2$?

11. How many grams of KOH are required to neutralize 50.00 mL of 0.240 *M* HNO_3?

***12.** A 0.200 g sample of impure NaOH requires 18.25 mL of 0.2406 *M* HCl for neutralization. What is the weight percent of NaOH in the sample?

***13.** A batch of sodium hydroxide was found to contain sodium chloride as an impurity. To determine the amount of impurity, a 1.00 g sample was analyzed and found to require 49.90 mL of 0.466 *M* HCl for neutralization. What is the percentage of NaCl in the sample?

***14.** What volume of H_2 gas, measured at 27°C and 700 torr pressure, can be obtained by reacting 5.00 g of zinc metal with (a) 100 mL of 0.350 *M* HCl and (b) 200 mL of 0.350 *M* HCl? The equation is

$$Zn(s) + 2\ HCl(aq) \longrightarrow ZnCl_2(aq) + H_2(g)$$

15. Calculate the pH of solutions having the following H^+ ion concentrations:
(a) 0.01 *M* (d) 1×10^{-7} *M*
(b) 1.0 *M* (e) 0.50 *M*
(c) 6.5×10^{-9} *M* (f) 0.00010 *M*

16. Calculate the pH of the following:
(a) Orange juice, $3.7 \times 10^{-4}\ M\ H^+$
(b) Vinegar, $2.8 \times 10^{-3}\ M\ H^+$
(c) Black coffee, $5.0 \times 10^{-5}\ M\ H^+$
(d) Limewater, $3.4 \times 10^{-11}\ M\ H^+$

17. Two drops (0.1 mL) of 1.0 *M* HCl are added to water to make 1.0 L of solution. What is the pH of this solution if the HCl is 100% ionized?

18. What volume of concentrated (18.0 *M*) sulfuric acid must be used to prepare 50.0 L of 5.00 *M* solution?

19. Three (3.0) grams of NaOH are added to 500 mL of 0.10 *M* HCl. Will the resulting solution be acidic or basic? Show evidence for your answer.

20. A 10.00 mL sample of base solution requires 28.92 mL of 0.1240 *N* H_2SO_4 for neutralization. What is the normality of the base?

21. How many milliliters of 0.325 *N* HNO_3 are required to neutralize 32.8 mL of 0.225 *N* NaOH?

22. How many milliliters of 0.325 *N* H_2SO_4 are required to neutralize 32.8 mL of 0.225 *N* NaOH?

23. What is the normality and the molarity of a 25.00 mL sample of H_3PO_4 solution that requires 22.68 mL of 0.5000 *N* NaOH for complete neutralization?

Review Exercises for Chapters 13–15

CHAPTER 13 WATER AND THE PROPERTIES OF LIQUIDS

True–False. *Answer the following as either true or false.*

1. Sodium, potassium, and calcium each react with cold water to form hydrogen gas and a metal hydroxide.
2. Calcium oxide reacts with water to form calcium hydroxide and hydrogen gas.
3. Water in a hydrate is known as water of hydration or water of crystallization.
4. Substances that can spontaneously lose their water of hydration when exposed to air are said to be hygroscopic.
5. A substance that absorbs water from the air until it forms a solution is said to be deliquescent.
6. At pressures below 760 torr, water will boil above 100°C.
7. Evaporation is the escape of molecules from the liquid state to the vapor state.
8. Sublimation is the change from the vapor to the liquid state.
9. $CuSO_4 \cdot 5\,H_2O$ is named copper(II) sulfate pentahydrate.
10. The pressure exerted by a vapor in equilibrium with its liquid is known as the vapor pressure of the liquid.
11. The heat of vaporization of water is more than six times as large as its heat of fusion.
12. Of two liquids, the one having the higher vapor pressure at a given temperature will have the higher normal boiling point.
13. In treating city water supplies, fluorides are often added to prevent tooth decay.
14. The molar heat of fusion of water is $335\ J/g \times 18.0\ g/mol$.
15. In treating city water supplies, chlorine is injected into the water to kill harmful bacteria before it is distributed to the public.
16. Fluorocarbons, which were used in aerosol spray cans and are used in refrigeration and air conditioners, escape to the stratosphere and react to destroy part of the ozone layer.
17. Hard water contains relatively large amounts of Ca^{2+} and Mg^{2+} ions.
18. Substances that evaporate readily are said to be volatile.
19. The phenomenon in which an element can exist in two or more molecular or crystalline forms is known as allotropy.
20. Metal oxides that react with water to form bases are known as basic anhydrides.
21. Nonmetal oxides that react with water to form acids are known as acid anhydrides.
22. SO_2 is the anhydride of H_2SO_4.
23. When P_2O_5 reacts with water, it would be expected to make the solution basic.
24. Hydrogen bonds are stronger than covalent bonds.
25. Substances that readily absorb water from the atmosphere are said to be hygroscopic.
26. Ice at 0°C is at the same temperature as water at 0°C, but ice contains less heat energy than the water.
27. The vapor pressure of a liquid depends on the temperature and the atmospheric pressure.
28. The bond angle between the atoms in a water molecule is about 90°.
29. Ozone in the stratosphere aids ultraviolet radiation in reaching the earth's surface.
30. The formula for ozone is $3\,O_2$.

Multiple Choice. *Choose the correct answer to each of the following.*

1. The heat of fusion of water is:
   (a) 4.184 J/g (c) 2.26 kJ/g
   (b) 335 J/g (d) 2.26 kJ/mol
2. The heat of vaporization of water is:
   (a) 4.184 J/g (c) 2.26 kJ/g
   (b) 335 J/g (d) 2.26 kJ/mol
3. The specific heat of water is:
   (a) 4.184 J/g °C (c) 2.26 kJ/g °C
   (b) 335 J/g °C (d) 18 J/g °C
4. The density of water at 4°C is:
   (a) 1.0 g/mL (c) 18.0 g/mL
   (b) 80 g/mL (d) $14.7\ lb/in.^3$

5. SO_2 can be properly classified as a(n)
   (a) basic anhydride (c) anhydrous salt
   (b) hydrate (d) acid anhydride
6. When compared to H_2S, H_2Se, and H_2Te, water is found to have the highest boiling point because it:
   (a) has the lowest molecular weight
   (b) is the smallest molecule
   (c) has the highest bonding
   (d) will form hydrogen bonds better than the others
7. In which of the following molecules will hydrogen bonding be important?
   (a) H—F (c) H—Br

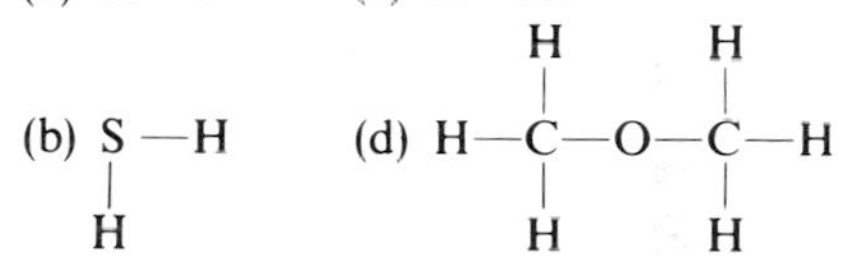

8. Which of the following is an incorrect equation?
   (a) $H_2SO_4 + 2\ NaOH \longrightarrow Na_2SO_4 + 2\ H_2O$
   (b) $C_2H_6 + O_2 \longrightarrow 2\ CO_2 + 3\ H_2$
   (c) $2\ H_2O \xrightarrow[H_2SO_4]{\text{Electrolysis}} 2\ H_2 + O_2$
   (d) $Ca + 2\ H_2O \longrightarrow H_2 + Ca(OH)_2$
9. Which of the following is an incorrect equation?
   (a) $C + H_2O(g) \xrightarrow{1000°C} CO(g) + H_2(g)$
   (b) $CaO + H_2O \longrightarrow Ca(OH)_2$
   (c) $2\ NO_2 + H_2O \longrightarrow 2\ HNO_3$
   (d) $Cl_2 + H_2O \longrightarrow HCl + HOCl$
10. A correct Lewis structure of hydrogen peroxide is:
   (a) :Ö:Ö:H (c) :Ö:O:H (with H below the second O)
   (b) :Ö:Ö:H (with H below the first O) (d) :Ö:H⁻
11. How many kilojoules are required to change 85 g of water at 25°C to steam at 100°C?
   (a) 219 kJ (b) 27 kJ (c) 590 kJ (d) 192 kJ
12. A chunk of 0°C ice weighing 145 g is dropped into 75 g of water at 62°C. The heat of fusion of water is 335 J/g. The result, after thermal equilibrium is attained, will be:
   (a) 87 g of ice and 133 g of liquid water, all at 0°C
   (b) 58 g of ice and 162 g of liquid water, all at 0°C
   (c) 220 g of water at 7°C
   (d) 220 g of water at 17°C
13. The formula for iron(II) sulfate heptahydrate is
   (a) $Fe_2SO_4 \cdot 7\ H_2O$ (c) $FeSO_4 \cdot 7\ H_2O$
   (b) $Fe(SO_4)_2 \cdot 6\ H_2O$ (d) $Fe_2(SO_4)_3 \cdot 7\ H_2O$
14. The process by which a solid changes directly to a vapor is called
   (a) vaporization (c) sublimation
   (b) evaporation (d) condensation
15. The anhydride of permanganic acid, $HMnO_4$, is
   (a) MnO_3 (c) MnO_4
   (b) Mn_2O_3 (d) Mn_2O_7
16. Hydrogen bonding
   (a) occurs only between water molecules
   (b) is stronger than covalent bonding
   (c) can occur between NH_3 and H_2O
   (d) results from strong attractive forces in ionic compounds
17. A liquid boils when
   (a) the vapor pressure of the liquid equals the external pressure above the liquid
   (b) the heat of vaporization exceeds the vapor pressure
   (c) the vapor pressure equals one atmosphere
   (d) the normal boiling temperature is reached
18. Consider two beakers, one containing 50 mL of liquid A and the other 50 mL of liquid B. The boiling point of A is 90°C and that of B is 72°C. Which of these statements is correct?
   (a) A will evaporate faster than B.
   (b) B will evaporate faster than A.
   (c) Both A and B evaporate at the same rate.
   (d) Insufficient data to answer the question.

CHAPTER 14 SOLUTIONS

True–False. *Answer the following as either true or false.*

1. Solubility describes the amount of one substance that will dissolve into another substance.
2. When solid NaCl dissolves in water, the sodium ions attract the positive end of the water dipole.
3. Bromine is more soluble in polar water than in nonpolar carbon tetrachloride because the polar water molecules help it form ions.

4. For most solids or gases dissolved in a liquid, an increase in temperature results in an increase in solubility.
5. An increase in temperature almost always increases the rate at which a solid will dissolve in a liquid.
6. A supersaturated solution contains dissolved solute in equilibrium with undissolved solute.
7. A 1 molar solution contains 1 mole of solute per liter of solution.
8. When a solute is dissolved in a solvent, the freezing point of the solution will be higher than that of the pure solvent.
9. Freezing point depression, boiling point elevation, and vapor pressure lowering are colligative properties of solutions.
10. Liquids that mix with water in all proportions are usually ionic in solution or are polar substances.
11. Liquids that do not mix are said to be miscible.
12. As a general rule, sodium and potassium salts of the common ions are soluble in water.
13. The vapor pressure of the solvent in a solution is lower than the vapor pressure of the pure solvent.
14. Salt water has a higher boiling point than distilled water.
15. One definition of equivalent weight is the weight of a substance that will combine with, contain, replace, or in any other way be equivalent to 1 mol of hydrogen atoms.
16. One mole of H_2SO_4 will react with twice as much NaOH as 1 mol of HCl.
17. A solution containing a nonvolatile solute has a lower boiling point than the pure solvent as a result of having a lower vapor pressure than the pure solvent.
18. A sulfuric acid solution with a molarity of 4.0 *M* would have a normality of 2.0 *N*.
19. Large crystals dissolve faster than do smaller ones because they expose a larger surface area to the solvent.
20. A major use of a solvent is as a medium for chemical reactions.
21. Molar solutions are temperature dependent, because the volume of the solution can vary with the temperature but the number of moles of solute remains the same.
22. A 15% by weight solution contains 15 g of solute per 100 mL of solution.
23. The molarity of a solution is the number of moles of solute per liter of solution.
24. A 1.0 *M* HCl solution will also be 1.0 *N* HCl.
25. A solution containing 0.001 mol of solute in 1 mL of solvent is 1 *M*.

Multiple Choice. *Choose the correct answer to each of the following.*

1. Which of the following is not a general property of solutions?
   (a) It is a homogeneous mixture of two or more substances.
   (b) It has a variable composition.
   (c) The dissolved solute breaks down to individual molecules.
   (d) The solution has the same chemical composition, the same chemical properties, and the same physical properties in every part.
2. If NaCl is soluble in water to the extent of 36.0 g NaCl/100 g H_2O at 20°C, then a solution at 20°C containing 45 g NaCl/150 g H_2O would be:
   (a) Dilute (c) Supersaturated
   (b) Saturated (d) Unsaturated
3. If 5.00 g of NaCl are dissolved in 25.0 g of water, the percentage of NaCl by weight is:
   (a) 16.7 (c) 0.20
   (b) 20.0 (d) No correct answer given
4. How many grams of 9.0% $AgNO_3$ solution will contain 5.3 g $AgNO_3$?
   (a) 47.7 (c) 59
   (b) 0.58 (d) No correct answer given
5. The molarity of a solution containing 2.5 mol of acetic acid, $HC_2H_3O_2$, in 400 mL of solution is:
   (a) 0.063 *M* (c) 0.103 *M*
   (b) 1.0 *M* (d) 6.3 *M*
6. What volume of 0.300 *M* KCl will contain 15.3 g of KCl?
   (a) 1.46 L (c) 61.5 mL
   (b) 684 mL (d) 4.60 L
7. What weight of $BaCl_2$ will be required to prepare 200 mL of 0.150 *M* solution?
   (a) 0.750 g (b) 156 g (c) 6.25 g (d) 31.2 g
8. If 15 g of H_2SO_4 are dissolved in water to make 80 mL of solution, the normality is:
   (a) 0.19 *N* (b) 1.9 *N* (c) 6.1 *N* (d) 3.8 *N*

Problems 9–11 relate to the reaction

$$CaCO_3 + 2\,HCl \longrightarrow CaCl_2 + H_2O + CO_2$$

9. What volume of 6.0 *M* HCl will be needed to react with 0.350 mol of $CaCO_3$?
(a) 42.0 mL (c) 117 mL
(b) 1.17 L (d) 583 mL
10. If 400 mL of 2.0 *M* HCl react with excess $CaCO_3$, the volume of CO_2 produced, measured at STP, is:
(a) 18 L (b) 5.6 L (c) 9.0 L (d) 56 L
11. If 5.3 g of $CaCl_2$ were produced in the reaction, what was the molarity of the HCl used if 25 mL of it reacted with excess $CaCO_3$?
(a) 3.8 *M* (b) 0.19 *M* (c) 0.38 *M* (d) 0.42 *M*
12. In the reaction $Ca + 2\,HCl \longrightarrow CaCl_2 + H_2$, the equivalent weight of Ca is:
(a) 36.5 g (b) 20.0 g (c) 40.1 g (d) 80.2 g
13. In the reaction

$$3\,HNO_3 + Cr(OH)_3 \longrightarrow Cr(NO_3)_3 + 3\,H_2O,$$

the equivalent weight of $Cr(OH)_3$ is:
(a) 34.3 g (b) 52 g (c) 103 g (d) 309 g
14. If 20.0 g of the nonelectrolyte urea, $CO(NH_2)_2$, are dissolved in 25.0 g of water, the freezing point of the solution will be:
(a) −2.48°C (c) −24.8°C
(b) −1.40°C (d) −3.72°C
15. When 256 g of a nonvolatile, nonelectrolyte unknown were dissolved in 500 g of H_2O, the freezing point was found to be −2.79°C. The molecular weight of the unknown solute is:
(a) 357 (b) 62.0 (c) 768 (d) 341
16. How many milliliters of 6.0 *M* H_2SO_4 must you use to prepare 500 mL of 0.20 *M* sulfuric acid solution?
(a) 30 (b) 17 (c) 12 (d) 100
17. How many milliliters of water must be added to 200 mL of 1.40 *M* HCl to make a solution that is 0.500 *M* HCl?
(a) 360 mL (c) 140 mL
(b) 560 mL (d) 280 mL
18. Which procedure is most likely to increase the solubility of most solids in liquids?
(a) Stirring
(b) Pulverizing the solid
(c) Heating the solution
(d) Increasing the pressure
19. The addition of a crystal of $NaClO_3$ to a solution of $NaClO_3$ causes additional crystals to precipitate. The original solution was
(a) Unsaturated (c) Saturated
(b) Dilute (d) Supersaturated
20. Which of the following anions will not form a precipitate with silver ions, Ag^+?
(a) Cl^- (b) NO_3^- (c) Br^- (d) CO_3^{2-}
21. Which of the following salts are considered to be soluble in water?
(a) $BaSO_4$ (b) NH_4Cl (c) AgI (d) PbS
22. A solution of ethyl alcohol and benzene is 40% alcohol by volume. Which statement is correct?
(a) The solution contains 40 mL of alcohol in 100 mL of solution.
(b) The solution contains 60 mL of benzene in 100 mL of solution.
(c) The solution contains 40 mL of alcohol in 100 g of solution.
(d) The solution is made by dissolving 40 mL of alcohol in 60 mL of benzene.

CHAPTER 15 IONIZATION: ACIDS, BASES, SALTS

True–False. *Answer the following as either true or false.*

1. Arrhenius defined a base as a hydroxide-containing substance that dissociates in water to produce hydroxide ions.
2. The Bronsted–Lowry theory defined an acid as a proton donor.
3. The Lewis theory defined an acid as an electron-pair donor.
4. In the reaction

$$HCl + NH_3 \longrightarrow NH_4^+ + Cl^-$$

according to the Brønsted–Lowry theory, the conjugate base of HCl is NH_4^+.
5. In the reaction

$$HNO_3 + H_2O \longrightarrow H_3O^+ + NO_3^-$$

the conjugate base of HNO_3 is NO_3^-.

6. When an acid reacts with a carbonate, the products are a salt, water, and carbon dioxide.
7. When electrolytes dissociate, they break into individual molecules.
8. The equation for the ionization of acetic acid is:

$$HC_2H_3O_2 + H_2O \rightleftharpoons H_3O^+ + C_2H_3O_2^-$$

9. In water, zinc nitrate dissociates as shown by the equation

$$Zn(NO_3)_2 \xrightarrow{H_2O} Zn^{2+} + 2\ NO_3^-$$

10. The equation for the dissociation of sodium carbonate in water is

$$Na_2CO_3 \xrightarrow{H_2O} Na_2^{2+} + CO_3^{2-}$$

11. When 1 mol of $CaCl_2$ dissolves in water, it will give 1 mol of Ca^{2+} and 2 mol of Cl^- ions in solution.
12. A solution with $[H^+] = 1.0 \times 10^{-9}\ M$ has a pH of -9.0.
13. A solution with $[H^+] = 7.8 \times 10^{-2}\ M$ has a pH between 1 and 2.
14. The reaction of an acid and a base to form a salt and water is known as neutralization.
15. In neutralizations between a strong acid and a strong base, the net ionic equation is

$$H^+ + OH^- \longrightarrow H_2O$$

16. For the aqueous reaction of NaOH with $FeCl_3$, the net ionic equation is

$$Fe^{3+} + 3\ OH^- \longrightarrow Fe(OH)_3(s)$$

17. A 1.0 M $HC_2H_3O_2$ solution will freeze at a lower temperature than a 1.0 M solution of KBr.
18. In writing net ionic equations, weak electrolytes are written in their un-ionized form.
19. Acids, bases, and salts are electrolytes.
20. Positive ions are called spectator ions.
21. Negative ions are called anions.
22. Titration is the process of measuring the volume of one reagent required to react with a measured weight or volume of another reagent.
23. A pH meter is an instrument used to measure the acidity of a solution.
24. As the acidity of a solution increases, the pH decreases.
25. When 50.0 mL of 0.20 M NaOH and 100 mL of 0.10 M HCl are mixed, the resulting solution will have a pH of 7.
26. The conjugate acid of HSO_4^- is SO_4^{2-}.
27. H_3O^+ is called a hydronium ion and NH_4^+ is called a nitronium ion.
28. A 0.1 M HCl and a 0.1 M $HC_2H_3O_2$ solution will have the same pH.
29. HNO_2, $HC_2H_3O_2$, HClO, and HBr are weak acids.
30. When water ionizes at 100°C it produces more H^+ ions than OH^- ions.

Multiple Choice. *Choose the correct answer to each of the following.*

1. When the reaction $Al + HCl \longrightarrow$ is completed and balanced, a term appearing in the balanced equation is:
(a) 3 HCl (b) $AlCl_2$ (c) 3 H_2 (d) 4 Al
2. When the reaction $CaO + HNO_3 \longrightarrow$ is completed and balanced, a term appearing in the balanced equation is:
(a) H_2 (b) 2 H_2 (c) 2 $CaNO_3$ (d) H_2O
3. When the reaction $H_3PO_4 + KOH \longrightarrow$ is completed and balanced, a term appearing in the balanced equation is:
(a) H_3PO_4 (c) KPO_4
(b) 6 H_2O (d) 2 H_3PO_4
4. When the reaction $HCl + Cr_2(CO_3)_3 \longrightarrow$ is completed and balanced, a term appearing in the balanced equation is:
(a) Cr_2Cl (b) 3 HCl (c) 3 CO_2 (d) H_2O
5. Which of the following is not a salt?
(a) $K_2Cr_2O_7$ (c) $Ca(OH)_2$
(b) $NaHCO_3$ (d) $Na_2C_2O_4$
6. Which of the following is not an acid?
(a) H_3PO_4 (b) H_2S (c) H_2SO_4 (d) NH_3
7. Which of the following is a weak electrolyte?
(a) NH_4OH (c) K_3PO_4
(b) $Ni(NO_3)_2$ (d) NaBr
8. Which of the following is a nonelectrolyte?
(a) $HC_2H_3O_2$ (c) $KMnO_4$
(b) $MgSO_4$ (d) CCl_4
9. Which of the following is a strong electrolyte?
(a) H_2CO_3 (c) NH_4OH
(b) HNO_3 (d) H_3BO_3
10. Which of the following is a weak electrolyte?
(a) NaOH (c) $HC_2H_3O_2$
(b) NaCl (d) H_2SO_4

11. A solution has a concentration of H^+ of 3.4×10^{-5} *M*. The pH is:
(a) 4.47 (b) 5.53 (c) 3.53 (d) 5.47

12. A solution with a pH of 5.85 has an H^+ concentration of:
(a) 7.1×10^{-5} *M* (c) 3.8×10^{-4} *M*
(b) 7.1×10^{-6} *M* (d) 1.4×10^{-6} *M*

13. 16.55 mL of 0.844 *M* NaOH is required to titrate 10.00 mL of a hydrochloric acid solution. The molarity of the acid solution is:
(a) 0.700 *M* (c) 1.40 *M*
(b) 0.510 *M* (d) 0.255 *M*

14. What volume of 0.462 *M* NaOH is required to titrate 20.00 mL of 0.391 *M* HNO_3?
(a) 23.6 mL (c) 9.03 mL
(b) 16.9 mL (d) 11.8 mL

15. 25.00 mL of H_2SO_4 solution required 18.92 mL of 0.1024 *N* NaOH for complete neutralization The normality of the acid is
(a) 0.1550 *N* (c) 0.07750 *N*
(b) 0.1353 *N* (d) 0.06765 *N*

16. Dilute hydrochloric acid is a typical acid, as shown by its:
(a) color (b) odor (c) solubility (d) taste

17. What is the pH of a 0.00015 *M* HCl solution?
(a) 4.0 (c) between 3 and 4
(b) 2.82 (d) No correct answer given

18. The chloride ion concentration in 300 mL of 0.10 *M* $AlCl_3$ is
(a) 0.30 *M* (c) 0.030 *M*
(b) 0.10 *M* (d) 0.90 *M*

19. The amount of $BaSO_4$ that will precipitate when 100 mL of 0.10 *M* $BaCl_2$ and 100 mL of 0.10 *M* Na_2SO_4 are mixed is:
(a) 0.010 mol (c) 23 g
(b) 0.10 mol (d) No correct answer given

20. The freezing point of a 0.50 molal NaCl aqueous solution will be about:
(a) -1.86°C (c) -2.79°C
(b) -0.93°C (d) No correct answer given

CHAPTER

16 Chemical Equilibrium

After studying Chapter 16, you should be able to

1 Understand the terms listed in Question A at the end of the chapter.
2 Describe a reversible reaction.
3 Explain why the rate of the forward reaction decreases and the rate of the reverse reaction increases as a chemical reaction approaches equilibrium.
4 State and understand the qualitative effect of Le Chatelier's principle.
5 Predict how the rate of a chemical reaction is affected by (a) changes in concentration of reactants, (b) changes in pressure on gaseous reactants, (c) changes in temperature, and (d) the presence of a catalyst.
6 Write the equilibrium constant expression for a chemical reaction from a balanced equation.
7 Explain the meaning of the numerical value of an equilibrium constant relative to the amounts of reactants and products in equilibrium.
8 Calculate the equilibrium constant, K_{eq}, when given the concentration of reactants and products in equilibrium.
9 Calculate the concentration of one substance in an equilibrium when given the equilibrium constant and the concentrations of all the other substances.
10 Calculate the ionization constant for a weak acid from appropriate data.
11 Calculate the concentrations of all the chemical species in a solution of a weak acid when given the percent ionization or the ionization constant.

12 Compare the relative strengths of acids by using their ionization constants.
13 Use the ion product constant for water, K_w, to calculate $[H^+]$, $[OH^-]$, pH, and pOH when given any one of these quantities.
14 Calculate the solubility product constant, K_{sp}, of a slightly soluble salt when given its solubility, or vice versa.
15 Compare the relative solubilities of salts when given their solubility products.
16 Discuss the common ion effect on a system in equilibrium.
17 Explain hydrolysis and why some salts form acidic or basic aqueous solutions.
18 Explain how a buffer solution is able to counteract the addition of small amounts of either H^+ or OH^- ions.
19 Draw the relative energy diagram of a reaction in terms of activation energy, exothermic or endothermic reaction, and the effect of a catalyst.

16.1 Reversible Reactions

In the preceding chapters we have treated chemical reactions mainly as reactants going to products. However, many reactions do not go to completion. Some reactions do not go to completion because they are reversible; that is, when the products are formed, they react to produce the starting reactants.

We have encountered reversible systems before. One is the vaporization of a liquid by heating and its subsequent condensation by cooling:

$$\text{liquid} + \text{heat} \longrightarrow \text{vapor}$$
$$\text{vapor} + \text{cooling} \longrightarrow \text{liquid}$$

The interconversion of nitrogen dioxide (NO_2) and dinitrogen tetroxide (N_2O_4) offers visible evidence of the reversibility of a reaction. NO_2 is a reddish-brown gas that changes, with cooling, to N_2O_4, a yellow liquid that boils at 21.2°C, and then to a colorless solid (N_2O_4) that melts at −11.2°C. The reaction is reversible by heating N_2O_4.

$$2\,NO_2(g) \xrightarrow{\text{Cooling}} N_2O_4(l)$$
$$N_2O_4(l) \xrightarrow{\text{Heating}} 2\,NO_2(g)$$

These two reactions may be represented by a single equation with a double arrow, $\rightleftarrows$, to indicate that the reactions are taking place in both directions at the same time.

$$2\,NO_2(g) \rightleftarrows N_2O_4(l)$$

This reversible reaction can be demonstrated by sealing samples of NO_2 in two tubes and placing one tube in warm water and the other in ice water (see Figure

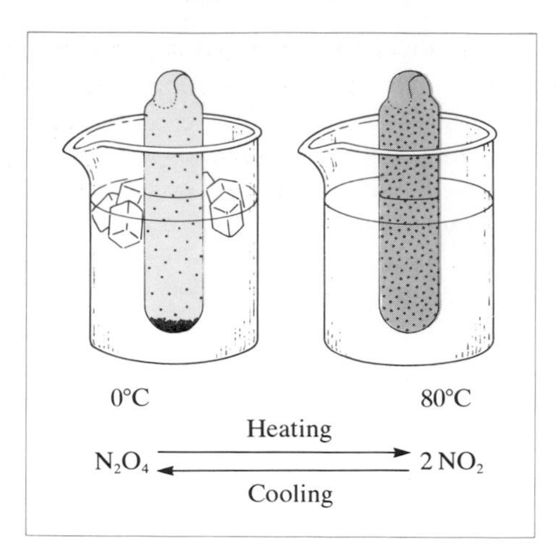

Figure 16.1
Reversible reaction of nitrogen dioxide (NO_2) and dinitrogen tetroxide (N_2O_4). More of the reddish-brown NO_2 molecules are visible in the tube that is heated than in the tube that is cooled.

16.1). Heating promotes disorder or randomness in a system, so we would expect more NO_2, a gas, to be present at higher temperatures.

reversible chemical reaction

A **reversible chemical reaction** is one in which the products formed react to produce the original reactants. Both the forward and reverse reactions occur simultaneously. The forward reaction is called *the reaction to the right*, and the reverse reaction is called *the reaction to the left*. A double arrow is used in the equation to indicate that the reaction is reversible.

16.2 Rates of Reaction

Every reaction has a rate, or speed, at which it proceeds. Some are fast and some are extremely slow. The study of reaction rates and reaction mechanisms is known as **chemical kinetics**.

chemical kinetics

The rate of a reaction is variable and depends on the concentration of the reacting species, the temperature, the presence or absence of catalytic agents, and the nature of the reactants. Consider the hypothetical reaction

$$A + B \longrightarrow C + D \quad \text{(forward reaction)}$$
$$C + D \longrightarrow A + B \quad \text{(reverse reaction)}$$

in which a collision between A and B is necessary for a reaction to occur. The rate at which A and B react depends on the concentration or the number of A and B molecules present; it will be fastest, for a fixed set of conditions, when they are first mixed. As the reaction proceeds, the number of A and B molecules available for reaction decreases, and the rate of reaction slows down. If the reaction is reversible, the speed of the reverse reaction is zero at first and gradually increases as the concentrations of C and D increase. As the number of A and B molecules decreases, the forward rate slows down because A and B cannot find one another as often in order to accomplish a reaction. To counteract this diminishing rate of reaction, an excess of one reagent is often used to keep the reaction from becoming impractically slow. Collisions between molecules may be likened to the scooters or "dodge'ems" at amusement parks. When many cars are on the floor, collisions occur frequently; but if only a few cars are present, collisions can usually be avoided.

16.3 Chemical Equilibrium

equilibrium

Any system at **equilibrium** represents a dynamic state in which two or more opposing processes are taking place at the same time and at the same rate. A chemical equilibrium is a dynamic system in which two or more opposing chemical reactions are going on at the same time and at the same rate. When the rate of the forward reaction is exactly equal to the rate of the reverse reaction, a condition of **chemical equilibrium** exists (see Figure 16.2). The concentrations of the products and the reactants are not changing, and the system appears to be at a

chemical equilibrium

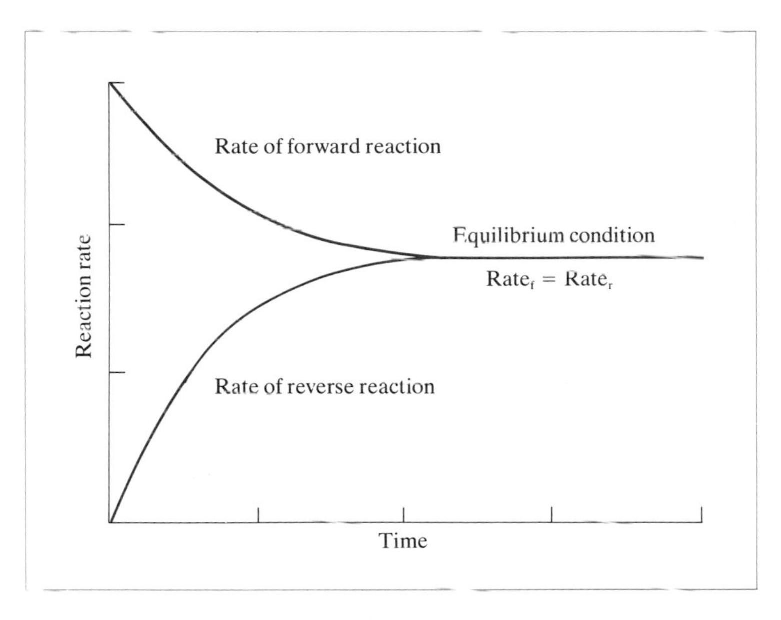

Figure 16.2

The graph illustrates that the rates of the forward and reverse reactions become equal at some point in time. The forward reaction rate decreases as a result of decreasing amounts of reactants. The reverse reaction rate starts at zero and increases as the amount of product increases. When the two rates become equal, a state of chemical equilibrium has been reached.

standstill because the products are reacting at the same rate at which they are being formed.

Chemical equilibrium:
rate of forward reaction = rate of reverse reaction

A saturated salt solution is in a condition of equilibrium:

$$NaCl(s) \rightleftharpoons Na^+(aq) + Cl^-(aq)$$

At equilibrium, salt crystals are continuously dissolving, and Na^+ and Cl^- ions are continuously crystallizing. Both processes are occurring at the same rate.

The ionization of weak electrolytes is another common chemical equilibrium system:

$$HC_2H_3O_2(aq) + H_2O(l) \rightleftharpoons H_3O^+(aq) + C_2H_3O_2^-(aq)$$

In this reaction, the equilibrium is established in a 1 *M* solution when the forward reaction has gone about 1%; that is, when only 1% of the acetic acid molecules in solution have ionized. Therefore, only a relatively few ions are present, and the acid behaves as a weak electrolyte. In any acid–base equilibrium system the position of equilibrium is toward the weaker conjugate acid and base. In the ionization of acetic acid, $HC_2H_3O_2$ is a weaker acid than H_3O^+, and H_2O is a weaker base than $C_2H_3O_2^-$.

The reactions represented by

$$H_2(g) + I_2(g) \rightleftharpoons 2\,HI(g)$$

provide another example of a chemical equilibrium. Theoretically, 1.00 mol of hydrogen should react with 1.00 mol of iodine to yield 2.00 mol of hydrogen iodide. Actually, when 1.00 mol of H_2 and 1.00 mol of I_2 are reacted at 700 K, only 1.58 mol of HI are present when equilibrium is attained. Since 1.58 is 79% of the theoretical yield of 2.00 mol of HI, the forward reaction is only 79% complete at equilibrium. The equilibrium mixture will also contain 0.21 mol each of unreacted H_2 and I_2 (1.00 mol − 0.79 mol = 0.21 mol).

$$\underset{\text{1.00 mol}}{H_2} + \underset{\text{1.00 mol}}{I_2} \xrightarrow{700\text{ K}} \underset{\text{2.00 mol}}{2\,HI}$$

(This equation would represent the condition if the reaction were 100% complete; 2.00 mol of HI would be formed and no H_2 and I_2 would be left unreacted.)

$$\underset{\text{0.21 mol}}{H_2} + \underset{\text{0.21 mol}}{I_2} \underset{}{\overset{700\text{ K}}{\rightleftharpoons}} \underset{\text{1.58 mol}}{2\,HI}$$

(This equation represents the actual equilibrium attained starting with 1.00 mol each of H_2 and I_2. It shows that the forward reaction is only 79% complete.)

16.4 Principle of Le Chatelier

principle of Le Chatelier

In 1888 the French chemist Henri Le Chatelier (1850–1936) set forth a simple, far-reaching generalization on the behavior of equilibrium systems. This generalization, known as the **principle of Le Chatelier**, states: If a stress is applied to a system in equilibrium, the system will respond in such a way as to relieve that stress and restore equilibrium under a new set of conditions. The application of Le Chatelier's principle helps us to predict the effect of changing conditions in chemical reactions. We will examine the effect of changes in concentration, temperature, and pressure.

16.5 Effect of Concentration on Reaction Rate and Equilibrium

The way in which the rate of a chemical reaction depends on the concentration of the reactants must be determined experimentally. Many simple, one-step reactions result from a collision between two molecules or ions. The rate of such one-step reactions can be altered by changing the concentration of the reactants or products. An increase in concentration of the reactants provides more individual reacting species for collisions and results in an increase in the rate of reaction.

An equilibrium is disturbed when the concentration of one or more of its components is changed. As a result the concentration of all the species will change, and a new equilibrium mixture will be established. Consider the hypothetical equilibrium represented by the equation

$$A + B \rightleftharpoons C + D$$

where A and B react in one step to form C and D. When the concentration of B is increased, the following occurs:

1. The rate of the reaction to the right (forward) increases. This rate is proportional to the concentration of A times the concentration of B.
2. The rate to the right becomes greater than the rate to the left.
3. Reactants A and B are used faster than they are produced; C and D are produced faster than they are used.
4. After a period of time, rates to the right and left become equal, and the system is again in equilibrium.
5. In the new equilibrium the concentration of A is less and the concentrations of B, C, and D are greater than in the original equilibrium.

Conclusion: The equilibrium has shifted to the right.

Applying this change in concentration to the equilibrium mixture of 1.00 mol of hydrogen and 1.00 mol of iodine from Section 16.3, we find that, when an additional 0.20 mol of I_2 is added, the yield of HI (based on H_2) is 85% (1.70 mol) instead of 79%. A comparison of the two systems, after the new equilibrium mixture is reached, follows.

Original equilibrium	New equilibrium
1.00 mol H_2 + 1.00 mol I_2	1.00 mol H_2 + 1.20 mol I_2
Yield: 79% HI	Yield: 85% HI (based on H_2)
Equilibrium mixture contains:	Equilibrium mixture contains:
1.58 mol HI	1.70 mol HI
0.21 mol H_2	0.15 mol H_2
0.21 mol I_2	0.35 mol I_2

Analyzing this new system, we see that, when the 0.20 mol I_2 was added, the equilibrium shifted to the right in order to counteract the increase in I_2 concentration. Some of the H_2 reacted with added I_2 and produced more HI, until an equilibrium mixture was established again. When I_2 was added, the concentration of I_2 increased, the concentration of H_2 decreased, and the concentration of HI increased. What do you think would be the effects of adding (a) more H_2 or (b) more HI?

The equation

$$\underset{\text{Pale yellow}}{Fe^{3+}(aq)} + \underset{\text{Colorless}}{SCN^-(aq)} \rightleftharpoons \underset{\text{Red}}{Fe(SCN)^{2+}(aq)}$$

represents an equilibrium that is used in certain analytical procedures as an indicator because of the readily visible, intense red color of the complex $Fe(SCN)^{2+}$ ion. A very dilute solution of iron(III), Fe^{3+}, and thiocyanate, SCN^-, is light red. When the concentration of either Fe^{3+} or SCN^- is increased, the equilibrium shift to the right is observed by an increase in the intensity of the color, resulting from the formation of additional $Fe(SCN)^{2+}$.

If either Fe^{3+} or SCN^- is removed from solution, the equilibrium will shift to the left, and the solution will become lighter in color. When Ag^+ is added to the solution, a white precipitate of silver thiocyanate (AgSCN) is formed, thus removing SCN^- ion from the equilibrium.

$$Ag^+(aq) + SCN^-(aq) \longrightarrow AgSCN\downarrow$$

The system accordingly responds to counteract the change in SCN^- concentration by shifting the equilibrium to the left. This shift is evident by a decrease in the intensity of the red color due to a decreased concentration of $Fe(SCN)^{2+}$.

Let us now consider the effect of changing the concentrations in the equilibrium mixture of chlorine water. The equilibrium equation is

$$Cl_2(aq) + 2\,H_2O \rightleftharpoons HOCl(aq) + H_3O^+(aq) + Cl^-(aq)$$

The variation in concentrations and the equilibrium shifts are tabulated in the following table. An X in the second or third column indicates the reagent that is increased or decreased. The fourth column indicates the direction of the equilibrium shift.

Reagent	Concentration		Equilibrium shift
	Increase	Decrease	
Cl_2	—	X	Left
H_2O	X	—	Right
$HOCl$	X	—	Left
H_3O^+	—	X	Right
Cl^-	X	—	Left

Consider the equilibrium in a 0.100 M acetic acid solution:

$$HC_2H_3O_2 + H_2O \rightleftharpoons H_3O^+ + C_2H_3O_2^-$$

In this solution the concentration of the hydronium ion (H_3O^+), which is a measure of the acidity, is 1.34×10^{-3} mol/L, corresponding to a pH of 2.87. What will happen to the acidity when 0.100 mol of sodium acetate ($NaC_2H_3O_2$) is added to 1 L of 0.100 M $HC_2H_3O_2$? When $NaC_2H_3O_2$ dissolves, it dissociates into sodium ions (Na^+) and acetate ions ($C_2H_3O_2^-$). The acetate ion from the salt is a common ion to the acetic acid equilibrium system and increases the total acetate ion concentration in the solution. As a result the equilibrium shifts to the left, decreasing the hydronium ion concentration and lowering the acidity of the solution. Evidence of this decrease in acidity is shown by the fact that the pH of a solution that is 0.100 M in $HC_2H_3O_2$ and 0.100 M in $NaC_2H_3O_2$ is 4.74. The pH of several different solutions of $HC_2H_3O_2$ and $NaC_2H_3O_2$ is shown in the table that follows. Each time the acetate ion is increased, the pH increases, indicating a further shift in the equilibrium toward un-ionized acetic acid.

Solution	pH
1 L 0.100 M $HC_2H_3O_2$	2.87
1 L 0.100 M $HC_2H_3O_2$ + 0.100 mol $NaC_2H_3O_2$	4.74
1 L 0.100 M $HC_2H_3O_2$ + 0.200 mol $NaC_2H_3O_2$	5.05
1 L 0.100 M $HC_2H_3O_2$ + 0.300 mol $NaC_2H_3O_2$	5.23

In summary, we can say that, when the concentration of a reagent on the left side of an equation is increased, the equilibrium shifts to the right. When the concentration of a reagent on the right side of an equation is increased, the equilibrium shifts to the left. In accordance with Le Chatelier's principle the equilibrium always shifts in the direction that tends to reduce the concentration of the added reactant.

16.6 Effect of Pressure on Reaction Rate and Equilibrium

Changes in pressure significantly affect the reaction rate only when one or more of the reactants or products is a gas. In these cases the effect of increasing the pressure of the reacting gases is equivalent to increasing their concentrations. In the reaction

$$CaCO_3(s) \overset{\Delta}{\rightleftharpoons} CaO(s) + CO_2(g)$$

calcium carbonate decomposes into calcium oxide and carbon dioxide when heated above 825°C. Increasing the pressure of the equilibrium system by adding CO_2 or by decreasing the volume, speeds up the reverse reaction and causes the equilibrium to shift to the left. The increased pressure gives the same effect as that caused by increasing the concentration of CO_2, the only gaseous substance in the reaction.

We have seen that, when the pressure on a gas is increased, its volume is decreased. In a system composed entirely of gases, an increase in pressure will cause the reaction and the equilibrium to shift to the side that contains the smaller volume or smaller number of moles. This shift occurs because the increase in pressure is partially relieved by the system's shifting its equilibrium toward the side in which the substances occupy the smaller volume.

Prior to World War I, Fritz Haber (1868–1934) in Germany invented the first major process for the fixation of nitrogen. In the Haber process nitrogen and hydrogen are reacted together in the presence of a catalyst at moderately high temperature and pressure to produce ammonia. The catalyst consists of iron and iron oxide with small amounts of potassium and aluminum oxides. For this process, Haber received the Nobel Prize in chemistry in 1918.

$$\underset{\substack{1\ \text{mol} \\ 1\ \text{volume}}}{N_2(g)} + \underset{\substack{3\ \text{mol} \\ 3\ \text{volumes}}}{3\,H_2(g)} \rightleftharpoons \underset{\substack{2\ \text{mol} \\ 2\ \text{volumes}}}{2\,NH_3(g)} + 92.5\ \text{kJ}(22.1\ \text{kcal}) \qquad (\text{at } 25°\text{C})$$

The left side of the equation in the Haber process represents four volumes of gas combining to give two volumes of gas on the right side of the equation. An increase in the total pressure on the system shifts the equilibrium to the right. This increase in pressure results in a higher concentration of both reactants and products. The equilibrium shifts to the right when the pressure is increased, because fewer moles of NH_3 than moles of N_2 and H_2 are in the equilibrium reaction.

Ideal conditions for the Haber process are 200°C and 1000 atm pressure. However, at 200°C the rate of reaction is very slow, and at 1000 atm extraordinarily heavy equipment is required. As a compromise the reaction is run at 400–600°C and 200–350 atm pressure, which gives a reasonable yield at a reasonable rate. The effect of pressure on the yield of ammonia at one particular temperature is shown in Table 16.1.

Table 16.1 The effect of pressure on the conversion of H_2 and N_2 to NH_3 at 450°C. The starting ratio of H_2 to N_2 is 3 moles to 1 mole.

Pressure (atm)	Yield of NH_3(%)	Pressure (atm)	Yield of NH_3(%)
10	2.04	300	35.5
30	5.8	600	53.4
50	9.17	1000	69.4
100	16.4		

When the total number of gaseous molecules on both sides of an equation is the same, a change in pressure does not cause an equilibrium shift. The following reaction is an example.

$$N_2(g) + O_2(g) \rightleftharpoons 2\,NO(g)$$

$N_2(g)$	$O_2(g)$	$2\,NO(g)$
1 mol	1 mol	2 mol
1 volume	1 volume	2 volumes
6.022×10^{23} molecules	6.022×10^{23} molecules	$2 \times 6.022 \times 10^{23}$ molecules

When the pressure on this system is increased, the rate of both the forward and the reverse reactions will increase because of the higher concentrations of N_2, O_2, and NO. But the equilibrium will not shift, because the increase in concentration of molecules is the same on both sides of the equation and the decrease in volume is the same on both sides of the equation.

PROBLEM 16.1 What effect would an increase in pressure have on the position of equilibrium in the following reactions?

(a) $2\,SO_2(g) + O_2(g) \rightleftharpoons 2\,SO_3(g)$
(b) $H_2(g) + Cl_2(g) \rightleftharpoons 2\,HCl(g)$
(c) $N_2O_4(l) \rightleftharpoons 2\,NO_2(g)$

(a) The equilibrium will shift to the right because the substance on the right has a smaller volume than those on the left.
(b) The equilibrium position will be unaffected because the volumes (or moles) of gases on both sides of the equation are the same.
(c) The equilibrium will shift to the left because $N_2O_4(l)$ occupies a much smaller volume than does $2\,NO_2(g)$.

16.7 Effect of Temperature on Reaction Rate and Equilibrium

An increase in temperature generally increases the rate of reaction. Molecules at elevated temperatures are more energetic and have more kinetic energy; thus, their collisions are more likely to result in a reaction. However, we cannot assume that the rate of a desired reaction will increase indefinitely as the temperature is raised. High temperatures may cause the destruction or decomposition of the reactants and products or may initiate reactions other than the one desired. For example, when calcium oxalate (CaC_2O_4) is heated to 500°C, it decomposes into calcium carbonate and carbon monoxide:

$$CaC_2O_4(s) \xrightarrow{500^\circ C} CaCO_3(s) + CO(g)$$

If calcium oxalate is heated to 850°C, the products are calcium oxide, carbon monoxide, and carbon dioxide:

$$CaC_2O_4(s) \xrightarrow{850^\circ C} CaO(s) + CO(g) + CO_2(g)$$

When heat is applied to a system in equilibrium, the reaction that absorbs heat is favored. When the process, as written, is endothermic, the forward reaction is increased. When the reaction is exothermic, the reverse reaction is favored. In this sense heat may be treated as a reactant in endothermic reactions or as a product in exothermic reactions. Therefore, temperature is analogous to concentration when applying Le Chatelier's principle to heat effects on a chemical reaction.

Hot coke (C) is a very reactive element. In the reaction

$$C(s) + CO_2(g) + \text{heat} \rightleftharpoons 2\ CO(g)$$

very little, if any, CO is formed at room temperature. At 1000°C the equilibrium mixture contains about an equal number of moles of CO and CO_2. At higher temperatures the equilibrium shifts to the right, increasing the yield of CO. The reaction is endothermic, and, as can be seen, the equilibrium is shifted to the right at higher temperatures.

Phosphorus trichloride reacts with dry chlorine gas to form phosphorus pentachloride. The reaction is exothermic:

$$PCl_3(l) + Cl_2(g) \rightleftharpoons PCl_5(s) + 88\ \text{kJ}(21\ \text{kcal})$$

Heat must continuously be removed during the reaction to obtain a good yield of the product. According to the principle of Le Chatelier, heat will cause the product, PCl_5, to decompose, re-forming PCl_3 and Cl_2. The equilibrium mixture at 200°C contains 52% PCl_5, and at 300°C it contains 3% PCl_5, verifying that heat causes the equilibrium to shift to the left.

When the temperature of a system is raised, the rate of reaction increases because of increased kinetic energy and more frequent collisions of the reacting species. In a reversible reaction the rate of both the forward and the reverse reactions is increased by an increase in temperature; however, the reaction that absorbs heat increases to a greater extent, and the equilibrium shifts to favor that reaction. The following examples illustrate these effects:

$$4\ HCl(g) + O_2(g) \rightleftharpoons 2\ H_2O(g) + 2\ Cl_2(g) + 95.4\ kJ\ (28.4\ kcal) \quad (1)$$

$$H_2(g) + Cl_2(g) \rightleftharpoons 2\ HCl(g) + 185\ kJ\ (44.2\ kcal) \quad (2)$$

$$CH_4(g) + 2\ O_2(g) \rightleftharpoons CO_2(g) + 2\ H_2O(g) + 890\ kJ\ (212.8\ kcal) \quad (3)$$

$$N_2O_4(l) + 58.6\ kJ\ (14\ kcal) \rightleftharpoons 2\ NO_2(g) \quad (4)$$

$$2\ CO_2(g) + 566\ kJ\ (135.2\ kcal) \rightleftharpoons 2\ CO(g) + O_2(g) \quad (5)$$

$$H_2(g) + I_2(g) + 51.9\ kJ\ (12.4\ kcal) \rightleftharpoons 2\ HI(g) \quad (6)$$

Reactions (1), (2), and (3) are exothermic; an increase in temperature will cause the equilibrium to shift to the left. Reactions (4), (5), and (6) are endothermic; an increase in temperature will cause the equilibrium to shift to the right.

16.8 Effect of Catalysts on Reaction Rate and Equilibrium

catalyst

A **catalyst** is a substance that influences the rate of a reaction and can be recovered essentially unchanged at the end of the reaction. A catalyst does not shift the equilibrium of a reaction; it affects only the speed at which the equilibrium is reached. If a catalyst does not affect the equilibrium, then it follows that it must affect the rate of both the forward and the reverse reactions equally.

The reaction between phosphorus trichloride (PCl_3) and sulfur is highly exothermic, but it is so slow that very little product, thiophosphoryl chloride ($PSCl_3$), is obtained, even after prolonged heating. When a catalyst, such as aluminum chloride ($AlCl_3$), is added, the reaction is complete in a few seconds:

$$PCl_3(l) + S(s) \xrightarrow{AlCl_3} PSCl_3(l)$$

In the laboratory preparation of oxygen, manganese dioxide is used as a catalyst to increase the rates of decomposition of both potassium chlorate and hydrogen peroxide.

$$2\ KClO_3(s) \xrightarrow[\Delta]{MnO_2} 2\ KCl(s) + 3\ O_2(g)$$

$$2\ H_2O_2(aq) \xrightarrow{MnO_2} 2\ H_2O(l) + O_2(g)$$

Catalysts are extremely important to industrial chemistry. Hundreds of chemical reactions that are otherwise too slow to be of practical value have been put to commercial use once a suitable catalyst was found. But in the area of biochemistry catalysts are of supreme importance because nearly all chemical reactions in all forms of life are completely dependent on biochemical catalysts known as *enzymes*.

16.9 Equilibrium Constants

In a reversible chemical reaction at equilibrium, the concentrations of the reactants and products are constant; that is, they are not changing. The rates of the forward and reverse reactions are constant, and an equilibrium constant expression can be written relating the products to the reactants. For the general reaction

$$\text{a A} + \text{b B} \rightleftharpoons \text{c C} + \text{d D}$$

at constant temperature, the following equilibrium constant expression can be written:

$$K_{eq} = \frac{[\text{C}]^{c}[\text{D}]^{d}}{[\text{A}]^{a}[\text{B}]^{b}}$$

equilibrium constant, K_{eq}

where K_{eq} is constant at a particular temperature and is known as the **equilibrium constant**. The quantities in brackets are the concentrations of each substance in moles per liter. The superscript letters a, b, c, and d are the coefficients of the substances in the balanced equation. The units for K_{eq} are not the same for every equilibrium reaction; however, the units are generally omitted. Observe that the concentration of each substance is raised to a power that is the same as the substance's numerical coefficient in the balanced equation. The convention is to place the concentrations of the products (the substances on the right side of the equation as written) in the numerator and the concentrations of the reactants in the denominator.

PROBLEM 16.2 Write equilibrium constant expressions for

(a) $3\,H_2(g) + N_2(g) \rightleftharpoons 2\,NH_3(g)$
(b) $CO(g) + 2\,H_2(g) \rightleftharpoons CH_3OH(g)$

(a) The only product, NH_3, has a coefficient of 2. Therefore the numerator of the equilibrium constant will be $[NH_3]^2$. Two reactants are present, H_2 with a coefficient of 3 and N_2 with a coefficient of 1. Therefore the denominator will be $[H_2]^3[N_2]$. The equilibrium constant expression is

$$K_{eq} = \frac{[NH_3]^2}{[H_2]^3[N_2]}$$

(b) For this equation the numerator is $[CH_3OH]$ and the denominator is $[CO][H_2]^2$. The equilibrium constant expression is

$$K_{eq} = \frac{[CH_3OH]}{[CO][H_2]^2}$$

The magnitude of an equilibrium constant indicates the extent to which the forward and reverse reactions take place. When K_{eq} is greater than 1, the amount of the products at equilibrium is greater than the amount of the reactants. When K_{eq} is less than 1, the amount of reactants at equilibrium is greater than the amount of the products. A very large value for K_{eq} indicates that the forward reaction goes essentially to completion. A very small K_{eq} means that the reverse reaction goes nearly to completion and that the equilibrium is far to the left (toward the reactants). Two examples follow:

$$H_2(g) + I_2(g) \rightleftharpoons 2\ HI(g) \qquad K_{eq} = 54.8 \text{ at } 425^\circ C$$

This K_{eq} indicates that considerably more product than reactants is present at equilibrium.

$$COCl_2(g) \rightleftharpoons CO(g) + Cl_2(g) \qquad K_{eq} = 7.6 \times 10^{-4} \text{ at } 400^\circ C$$

This K_{eq} indicates that $COCl_2$ is stable and that very little decomposition to CO and Cl_2 occurs at 400°C. The equilibrium is far to the left.

When the molar concentrations of all the species in an equilibrium reaction are known, the K_{eq} can be calculated by substituting the concentrations into the equilibrium constant expression.

PROBLEM 16.3 Calculate the K_{eq} for the following reaction based on concentrations of: PCl_5 = 0.030 mol/L; PCl_3 = 0.97 mol/L; Cl_2 = 0.97 mol/L at 300°C.

$$PCl_5(g) \rightleftharpoons PCl_3(g) + Cl_2(g)$$

First write the K_{eq} expression; then substitute the respective concentrations into this equation and solve:

$$K_{eq} = \frac{[PCl_3][Cl_2]}{[PCl_5]} = \frac{[0.97][0.97]}{[0.030]} = 31$$

This K_{eq} is considered to be a fairly large value, indicating that at 300°C the decomposition of PCl_5 proceeds far to the right.

16.10 Ionization Constants

As a first application of an equilibrium constant, let us consider the constant for acetic acid in solution. Because it is a weak acid, an equilibrium is established between molecular $HC_2H_3O_2$ and its ions in solution. The constant is called the

acid ionization constant, K_a

acid ionization constant, K_a, a special type of equilibrium constant. The concentration of water in the solution is large compared to the other concentrations and does not change appreciably, so we may use the following simplified equation to set up the constant:

$$HC_2H_3O_2 \rightleftharpoons H^+ + C_2H_3O_2^-$$

The ionization constant expression is the concentration of the products divided by the concentration of the reactants:

$$K_a = \frac{[H^+][C_2H_3O_2^-]}{[HC_2H_3O_2]}$$

It states that the ionization constant, K_a, is equal to the product of the hydrogen ion $[H^+]$ concentration and the acetate ion $[C_2H_3O_2^-]$ concentration divided by the concentration of the un-ionized acetic acid $[HC_2H_3O_2]$.

At 25°C a 0.1 M $HC_2H_3O_2$ solution is 1.34% ionized and has a hydrogen ion concentration of 1.34×10^{-3} mol/L. From this information we can calculate the ionization constant for acetic acid.

A 0.10 M solution initially contains 0.10 mol of acetic acid per liter. Of this 0.10 mol, only 1.34%, or 1.34×10^{-3} mol, is ionized, which gives an H^+ ion concentration of 1.34×10^{-3} mol/L. Because each molecule of acid that ionizes yields one H^+ and one $C_2H_3O_2^-$, the concentration of $C_2H_3O_2^-$ ions is also 1.34×10^{-3} mol/L. This ionization leaves $0.10 - 0.00134 = 0.09866$ mol/L of un-ionized acetic acid.

	Initial concentration (mol/L)	Equilibrium concentration (mol/L)
$[HC_2H_3O_2]$	0.10	0.09866
$[H^+]$	0	0.00134
$[C_2H_3O_2^-]$	0	0.00134

Substituting these concentrations in the equilibrium expression, we obtain the value for K_a:

$$K_a = \frac{[H^+][C_2H_3O_2^-]}{[HC_2H_3O_2]} = \frac{[1.34 \times 10^{-3}][1.34 \times 10^{-3}]}{[0.09866]} = 1.8 \times 10^{-5}$$

The K_a for acetic acid, 1.8×10^{-5}, is small and indicates that the position of the equilibrium is far toward the un-ionized acetic acid. In fact, a 0.10 M acetic acid solution is 98.66% un-ionized.

Once the K_a for acetic acid is established, it can be used to describe other systems containing H^+, $C_2H_3O_2^-$, and $HC_2H_3O_2$ in equilibrium at 25°C. The ionization constants for several other weak acids are listed in Table 16.2.

Table 16.2 Ionization constants (K_a) of weak acids at 25°C

Acid	Formula	K_a
Acetic	$HC_2H_3O_2$	1.8×10^{-5}
Benzoic	$HC_7H_5O_2$	6.3×10^{-5}
Carbolic (phenol)	HC_6H_5O	1.3×10^{-10}
Cyanic	$HCNO$	2.0×10^{-4}
Formic	$HCHO_2$	1.8×10^{-4}
Hydrocyanic	HCN	4.0×10^{-10}
Hypochlorous	$HClO$	3.5×10^{-8}
Nitrous	HNO_2	4.5×10^{-4}
Hydrofluoric	HF	6.5×10^{-4}

PROBLEM 16.4 What is the H^+ ion concentration in a 0.50 M $HC_2H_3O_2$ solution? The ionization constant, K_a, for $HC_2H_3O_2$ is 1.8×10^{-5}.

To solve this problem, first write the equilibrium equation and the K_a expression:

$$HC_2H_3O_2 \rightleftharpoons H^+ + C_2H_3O_2^- \qquad K_a = \frac{[H^+][C_2H_3O_2^-]}{[HC_2H_3O_2]} = 1.8 \times 10^{-5}$$

We know that the initial concentration of $HC_2H_3O_2$ is 0.50 M. We also know from the ionization equation that one $C_2H_3O_2^-$ is produced for every H^+ produced; that is, the $[H^+]$ and the $[C_2H_3O_2^-]$ are equal. To solve, let $Y = [H^+]$, which also equals the $[C_2H_3O_2^-]$. The un-ionized $[HC_2H_3O_2]$ remaining will then be $0.50 - Y$, the starting concentration minus the amount that ionized.

$$[H^+] = [C_2H_3O_2^-] = Y \qquad [HC_2H_3O_2] = 0.50 - Y$$

Substituting these values into the K_a expression, we obtain

$$K_a = \frac{(Y)(Y)}{0.50 - Y} = \frac{Y^2}{0.50 - Y} = 1.8 \times 10^{-5}$$

An exact solution of this equation for Y requires the use of a mathematical equation known as the quadratic equation. However, an approximate solution is obtained if we assume that Y is small and can be neglected compared with 0.50. Then $0.50 - Y$ will be equal to approximately 0.50. The equation now becomes

$$\frac{Y^2}{0.50} = 1.8 \times 10^{-5}$$

$$Y^2 = 0.50 \times 1.8 \times 10^{-5} = 0.90 \times 10^{-5} = 9.0 \times 10^{-6}$$

Taking the square root of both sides of the equation, we obtain

$$Y = \sqrt{9.0 \times 10^{-6}} = 3.0 \times 10^{-3} \text{ mol/L}$$

Thus, $[H^+]$ is approximately 3.0×10^{-3} mol/L in a 0.50 *M* $HC_2H_3O_2$ solution. The exact solution to this problem, using the quadratic equation, gives a value of 2.99×10^{-3} mol/L for $[H^+]$, showing that we were justified in neglecting *Y* compared with 0.50.

PROBLEM 16.5 Calculate the percent ionization in a 0.50 *M* $HC_2H_3O_2$ solution.

The percent ionization of a weak acid, $HA(aq) \rightleftharpoons H^+ + A^-$, is found by dividing the concentration of the H^+ or A^- ions at equilibrium by the initial concentration of HA. For acetic acid

$$\frac{\text{concentration of } [H^+] \text{ or } [C_2H_3O_2^-]}{\text{initial concentration of } [HC_2H_3O_2]} \times 100\% = \text{percent ionized}$$

To solve this problem we first need to calculate $[H^+]$. This calculation has already been done in Problem 16.4 for a 0.50 *M* solution.

$$[H^+] = 3.0 \times 10^{-3} \text{ mol/L in a 0.50 } M \text{ solution} \quad \text{(from Problem 16.4)}$$

This $[H^+]$ represents a fractional amount of the initial 0.50 *M* $HC_2H_3O_2$. Therefore

$$\frac{3.0 \times 10^{-3} \text{ mol/L}}{0.50 \text{ mol/L}} \times 100\% = 0.60\% \text{ ionized}$$

A 0.50 *M* $HC_2H_3O_2$ solution is 0.60% ionized.

16.11 Ion Product Constant for Water

We have seen that water ionizes to a slight degree. This ionization is represented by these equilibrium equations:

$$H_2O + H_2O \rightleftharpoons H_3O^+ + OH^- \quad (7)$$

$$H_2O \rightleftharpoons H^+ + OH^- \quad (8)$$

Equation (7) is the more accurate representation of the equilibrium, since free protons (H^+) do not exist in water. Equation (8) is a simplified and often-used representation of the water equilibrium. The actual concentration of H^+ produced in pure water is very minute and amounts to only 1.00×10^{-7} mol/L at 25°C. In pure water

$$[H^+] = [OH^-] = 1.00 \times 10^{-7} \text{ mol/L}$$

since both ions are produced in equal molar amounts, as shown in equation (8).

ion product constant for water, K_w

The $H_2O \rightleftharpoons H^+ + OH^-$ equilibrium exists in water and in all water solutions. A special equilibrium constant called the **ion product constant for water**, K_w, applies to this equilibrium. The constant K_w is defined as the product of the H^+ ion concentration and the OH^- ion concentration, each in moles per liter:

Table 16.3 Relationship of H^+ and OH^- concentrations in water solutions

$[H^+]$	$[OH^-]$	K_w	pH	pOH
1.00×10^{-2}	1.00×10^{-12}	1.00×10^{-14}	2.00	12.0
1.00×10^{-4}	1.00×10^{-10}	1.00×10^{-14}	4.00	10.0
2.00×10^{-6}	5.00×10^{-9}	1.00×10^{-14}	5.70	8.30
1.00×10^{-7}	1.00×10^{-7}	1.00×10^{-14}	7.00	7.00
1.00×10^{-9}	1.00×10^{-5}	1.00×10^{-14}	9.00	5.00

$$K_w = [H^+][OH^-]$$

The numerical value of K_w is 1.00×10^{-14}, since for pure water at 25°C

$$K_w = [H^+][OH^-] = [1.00 \times 10^{-7}][1.00 \times 10^{-7}] = 1.00 \times 10^{-14}$$

The value of K_w for all water solutions at 25°C is the constant 1.00×10^{-14}. It is important to realize that, as the concentration of one of these ions, H^+ or OH^-, increases, the other decreases. However, the product of $[H^+]$ and $[OH^-]$ always equals the constant 1.00×10^{-14}. This relationship can be seen in the examples shown in Table 16.3. If the concentration of one ion is known, the concentration of the other can be calculated from the K_w expression.

$$K_w = [H^+][OH^-] \qquad [H^+] = \frac{K_w}{[OH^-]} \qquad [OH^-] = \frac{K_w}{[H^+]}$$

PROBLEM 16.6 What is the concentration of (a) H^+ and (b) OH^- in a 0.001 *M* HCl solution? Assume that HCl is 100% ionized.

(a) Since all the HCl is ionized, $H^+ = 0.001$ mol/L.

$$\underset{0.001\,M}{HCl} \longrightarrow \underset{0.001\,M}{H^+} + \underset{0.001\,M}{Cl^-}$$

$[H^+] = 1 \times 10^{-3}$ mol/L (Answer)

(b) To calculate the $[OH^-]$ in this solution, use the following equation and substitute the values for K_w and $[H^+]$:

$$[OH^-] = \frac{K_w}{[H^+]}$$

$$[OH^-] = \frac{1.00 \times 10^{-14}}{1 \times 10^{-3}} = 1 \times 10^{-11} \text{ mol/L} \quad \text{(Answer)}$$

PROBLEM 16.7 What is the pH of a 0.010 M NaOH solution? Assume that NaOH is 100% ionized.

Since all the NaOH is ionized, $OH^- = 0.010$ mol/L or 1.0×10^{-2} mol/L.

$$\underset{0.010\,M}{NaOH} \longrightarrow \underset{0.010\,M}{Na^+} + \underset{0.010\,M}{OH^-}$$

To find the pH of the solution we first calculate the H^+ concentration. Use the following equation and substitute the values for K_w and $[OH^-]$.

$$[H^+] = \frac{K_w}{[OH^-]} = \frac{1.00 \times 10^{-14}}{1.0 \times 10^{-2}} = 1.0 \times 10^{-12} \text{ mol/L}$$

$$pH = -\log[H^+] = -\log 1 \times 10^{-12} = 12 \quad \text{(Answer)}$$

Just as pH is used to express the acidity of a solution, pOH is used to express the basicity of an aqueous solution. The pOH is related to the OH^- ion concentration in the same way that the pH is related to the H^+ ion concentration:

$$pOH = \log\frac{1}{[OH^-]} \qquad \text{or} \qquad pOH = -\log[OH^-]$$

Thus, a solution in which $[OH^-] = 1.0 \times 10^{-2}$, as in Problem 16.7, will have $pOH = 2.0$.

In pure water, where $[H^+] = 1 \times 10^{-7}$ and $[OH^-] = 1 \times 10^{-7}$, the pH is 7, and the pOH is 7. The sum of the pH and pOH is always 14.

$$pH + pOH = 14$$

In Problem 16.7 the pH can also be found by first calculating the pOH (2) from the OH^- ion concentration and then subtracting from 14.

$$pH = 14 - pOH = 14 - 2 = 12$$

16.12 Solubility Product Constant

solubility product constant, K_{sp}

The **solubility product constant**, abbreviated K_{sp}, is another application of the equilibrium constant. It is the equilibrium constant of a slightly soluble salt. The following example illustrates how K_{sp} is evaluated.

The solubility of silver chloride (AgCl) in water is 1.3×10^{-5} mol/L at 25°C. The equation for the equilibrium between AgCl and its ions in solution is

$$AgCl(s) \rightleftharpoons Ag^+ + Cl^-$$

The equilibrium constant expression is

$$K_{eq} = \frac{[Ag^+][Cl^-]}{[AgCl(s)]}$$

The amount of solid AgCl does not affect the equilibrium system provided that some is present. In other words, the concentration of solid silver chloride is constant whether 1 mg or 10 g of the salt are present. Therefore, the product obtained by multiplying the two constants K_{eq} and $[AgCl(s)]$ is also a constant. This constant is the solubility product constant, K_{sp}.

$$K_{eq} \times [AgCl(s)] = [Ag^+][Cl^-] = K_{sp}$$
$$K_{sp} = [Ag^+][Cl^-]$$

The K_{sp} is equal to the product of the Ag^+ ion and the Cl^- ion concentrations, each in moles per liter. When 1.3×10^{-5} mol/L of AgCl dissolves, it produces 1.3×10^{-5} mol/L each of Ag^+ and Cl^-. From these concentrations the K_{sp} can be evaluated.

$$[Ag^+] = 1.3 \times 10^{-5}\ \text{mol/L} \qquad [Cl^-] = 1.3 \times 10^{-5}\ \text{mol/L}$$
$$K_{sp} = [Ag^+][Cl^-] = [1.3 \times 10^{-5}][1.3 \times 10^{-5}] = 1.7 \times 10^{-10}$$

Once the K_{sp} value for AgCl is established, it can be used to describe other systems containing Ag^+ and Cl^-.

The K_{sp} expression does not have a denominator. It consists only of the concentrations (mol/L) of the ions in solution. As in other equilibria expressions, each of these concentrations is raised to a power that is the same number as its coefficient in the balanced equation. The equilibrium equations and the K_{sp} expressions for several other substances follow.

$$AgBr(s) \rightleftharpoons Ag^+ + Br^- \qquad K_{sp} = [Ag^+][Br^-]$$
$$BaSO_4(s) \rightleftharpoons Ba^{2+} + SO_4^{2-} \qquad K_{sp} = [Ba^{2+}][SO_4^{2-}]$$
$$Ag_2CrO_4(s) \rightleftharpoons 2\ Ag^+ + CrO_4^{2-} \qquad K_{sp} = [Ag^+]^2[CrO_4^{2-}]$$
$$CuS(s) \rightleftharpoons Cu^{2+} + S^{2-} \qquad K_{sp} = [Cu^{2+}][S^{2-}]$$
$$Mn(OH)_2(s) \rightleftharpoons Mn^{2+} + 2\ OH^- \qquad K_{sp} = [Mn^{2+}][OH^-]^2$$
$$Fe(OH)_3(s) \rightleftharpoons Fe^{3+} + 3\ OH^- \qquad K_{sp} = [Fe^{3+}][OH^-]^3$$

Table 16.4 lists K_{sp} values for these and several other substances.

When the product of the molar concentration of the ions in solution, each raised to its proper power, is greater than the K_{sp} for that substance, precipitation should occur. If the ion product is less than the K_{sp} value, no precipitation will occur.

PROBLEM 16.8 Write K_{sp} expressions for AgI and PbI_2, both of which are slightly soluble salts.

First write the equilibrium equations:

$$AgI(s) \rightleftharpoons Ag^+ + I^-$$
$$PbI_2(s) \rightleftharpoons Pb^{2+} + 2\ I^-$$

Table 16.4 Solubility product constants (K_{sp}) at 25°C

Compound	K_{sp}
$AgCl$	1.7×10^{-10}
$AgBr$	5×10^{-13}
AgI	8.5×10^{-17}
$AgC_2H_3O_2$	2×10^{-3}
Ag_2CrO_4	1.9×10^{-12}
$BaCrO_4$	8.5×10^{-11}
$BaSO_4$	1.5×10^{-9}
CaF_2	3.9×10^{-11}
CuS	9×10^{-45}
$Fe(OH)_3$	6×10^{-38}
PbS	7×10^{-29}
$PbSO_4$	1.3×10^{-8}
$Mn(OH)_2$	2.0×10^{-13}

Since the concentration of the solid crystals is constant, the K_{sp} equals the product of the molar concentrations of the ions in solution. In the case of PbI_2, the $[I^-]$ must be squared.

$$K_{sp} = [Ag^+][I^-]$$
$$K_{sp} = [Pb^{2+}][I^-]^2$$

PROBLEM 16.9 The K_{sp} value for lead sulfate is 1.3×10^{-8}. Calculate the solubility of $PbSO_4$ in grams per liter.

First write the equilibrium equation and the K_{sp} expression:

$$PbSO_4 \rightleftharpoons Pb^{2+} + SO_4^{2-}$$
$$K_{sp} = [Pb^{2+}][SO_4^{2-}] = 1.3 \times 10^{-8}$$

Since the lead sulfate that is in solution is completely dissociated, the concentration of $[Pb^{2+}]$ or $[SO_4^{2-}]$ is equal to the solubility of $PbSO_4$ in moles per liter.

Let

$$Y = [Pb^{2+}] = [SO_4^{2-}]$$

Substitute Y into the K_{sp} equation and solve.

$$[Pb^{2+}][SO_4^{2-}] = [Y][Y] = 1.3 \times 10^{-8}$$
$$Y^2 = 1.3 \times 10^{-8}$$
$$Y = 1.1 \times 10^{-4} \text{ mol/L}$$

The solubility of $PbSO_4$, therefore, is 1.1×10^{-4} mol/L. Now convert mol/L to g/L:

1 mol of $PbSO_4$ weighs (207.2 + 32.1 + 64.0) or 303.3 g

$$\frac{1.1 \times 10^{-4}\ \text{mol}}{\text{L}} \times \frac{303.3\ \text{g}}{\text{mol}} = 3.3 \times 10^{-2}\ \text{g/L}$$

The solubility of $PbSO_4$ is 3.3×10^{-2} g/L.

---

common ion effect

An ion added to a solution that already contains that ion is called a common ion. When a common ion is added to an equilibrium solution of a weak electrolyte or a slightly soluble salt, the equilibrium shifts according to Le Chatelier's principle. For example, when silver nitrate, $AgNO_3$, is added to a saturated solution of silver chloride, AgCl ($AgCl(s) \rightleftharpoons Ag^+ + Cl^-$), the equilibrium shifts to the left due to the increase in the Ag^+ concentration. As a result, the Cl^- concentration and the solubility of AgCl decreases. AgCl and $AgNO_3$ have the common ion Ag^+. The effect of the common ion causing the equilibrium to shift is known as the **common ion effect**.

PROBLEM 16.10 Silver nitrate, $AgNO_3$, is added to a saturated AgCl solution until the Ag^+ concentration is 0.10 M. What will be the Cl^- concentration remaining in solution?

This problem is an example of the common ion effect. The addition of $AgNO_3$ puts more Ag^+ in solution; the Ag^+ combines with Cl^- and causes the equilibrium to shift to the left, reducing the Cl^- concentration in solution.

We use the K_{sp} to calculate the Cl^- ion concentration remaining in solution. The K_{sp} is constant at a particular temperature and remains the same no matter how we change the concentration of the species involved.

$$K_{sp} = [Ag^+][Cl^-] = 1.7 \times 10^{-10} \qquad [Ag^+] = 0.10\ \text{mol/L}$$

We then substitute the concentration of Ag^+ ion into the K_{sp} expression and calculate the Cl^- concentration.

$$[0.10][Cl^-] = 1.7 \times 10^{-10}$$

$$[Cl^-] = \frac{1.7 \times 10^{-10}}{0.10} = 1.7 \times 10^{-9}\ \text{mol/L}$$

This calculation shows a 10,000-fold reduction of Cl^- ions in solution. It illustrates that Cl^- ions may be quantitatively removed from solution with an excess of Ag^+ ions.

16.13 Hydrolysis

hydrolysis

Hydrolysis is the term used for the general reaction in which a water molecule is split. For example, the net ionic hydrolysis reaction for a sodium acetate solution is

$$C_2H_3O_2^- + H_2O(l) \rightleftharpoons HC_2H_3O_2(aq) + OH^-$$

In the reaction above the water molecule is split, with the H^+ combining with $C_2H_3O_2^-$ to give the weak acid $HC_2H_3O_2$ and the OH^- going into solution, making the solution more basic.

Salts that contain an ion of a weak acid or a weak base undergo hydrolysis. For example, a 0.10 *M* NH_4Cl solution has a pH of 5.1, and a 0.10 *M* NaCN solution has a pH of 11.1. The hydrolysis reactions that cause these solutions to become acidic or basic are

$$NH_4^+ + H_2O(l) \rightleftharpoons NH_4OH(aq) + H^+$$
$$CN^- + H_2O(l) \rightleftharpoons HCN(aq) + OH^-$$

The ions of a salt derived from a strong acid and a strong base, such as NaCl, do not undergo hydrolysis and thus form neutral solutions. Table 16.5 lists the ionic composition of various salts and the nature of the aqueous solutions that they form.

Table 16.5 Ionic composition of salts and the nature of the aqueous solutions they form

Type of salt	Nature of aqueous solution	Examples
Weak base–strong acid	Acidic	NH_4Cl, NH_4NO_3
Strong base–weak acid	Basic	$NaC_2H_3O_2$, K_2CO_3
Weak base–weak acid	Depends on the salt	$NH_4C_2H_3O_2$, NH_4NO_2
Strong base–strong acid	Neutral	NaCl, KBr

16.14 Buffer Solutions: The Control of pH

The control of pH within narrow limits is critically important in many chemical applications and vitally important in many biological systems. For example, human blood must be maintained between pH 7.35 and 7.45 for the efficient transport of oxygen from the lungs to the cells. This narrow pH range is maintained by buffer systems in the blood.

buffer solution

A **buffer solution** resists changes in pH when diluted or when small amounts of acid or base are added. Two common types of buffer solutions are (1) a weak acid mixed with a salt of that weak acid and (2) a weak base mixed with a salt of that weak base.

The action of a buffer system can be understood by considering a solution of acetic acid and sodium acetate. The weak acid, $HC_2H_3O_2$, is mostly un-ionized and is in equilibrium with its ions in solution. The salt, $NaC_2H_3O_2$, is completely ionized.

$$HC_2H_3O_2(aq) \rightleftharpoons H^+(aq) + C_2H_3O_2^-(aq)$$
$$NaC_2H_3O_2(aq) \longrightarrow Na^+(aq) + C_2H_3O_2^-(aq)$$

Because the salt is completely ionized, the solution contains a much higher concentration of acetate ions than would be present if only acetic acid were in solution. The acetate ion represses the ionization of acetic acid and also reacts with water, causing the solution to have a higher pH (be more basic) than an acetic acid solution (see Section 16.5). Thus, a 0.1 M acetic acid solution has a pH of 2.87, but a solution that is 0.1 M in acetic acid and 0.1 M in sodium acetate has a pH of 4.74. This difference in pH is the result of the common ion effect.

A buffer solution has a built-in mechanism that counteracts the effect of adding acid or base. Consider the effect of adding HCl or NaOH to an acetic acid–sodium acetate buffer. When a small amount of HCl is added, the acetate ions of the buffer combine with the H^+ ions from HCl to form un-ionized acetic acid, thus neutralizing the added acid and maintaining the approximate pH of the solution. When NaOH is added, the OH^- ions react with acetic acid to neutralize the added base and thus maintain the approximate pH. The equations for these reactions are

$$H^+ + C_2H_3O_2^- \rightleftharpoons HC_2H_3O_2(aq)$$
$$OH^- + HC_2H_3O_2(aq) \rightleftharpoons H_2O(l) + C_2H_3O_2^-$$

Data comparing the changes in pH caused by adding HCl and NaOH to pure water and to an acetic acid–sodium acetate buffer solution are shown in Table 16.6.

The human body has a number of buffer systems. One of these, the bicarbonate–carbonic acid buffer, HCO_3^-–H_2CO_3, maintains the blood plasma at a pH of 7.4. The phosphate system, HPO_4^{2-}–$H_2PO_4^-$, is an important buffer in the red blood cells as well as in other places in the body.

Table 16.6 Changes in pH caused by the addition of HCl and NaOH to pure water and to an acetic acid–sodium acetate buffer solution.

Solution	pH	Change in pH
H_2O (1000 mL)	7	—
H_2O + 0.010 mol HCl	2	5
H_2O + 0.010 mol NaOH	12	5
Buffer solution (1000 mL)		
0.10 M $HC_2H_3O_2$ + 0.10 M $NaC_2H_3O_2$	4.74	—
Buffer + 0.010 mol HCl	4.66	0.08
Buffer + 0.010 mol NaOH	4.83	0.09

16.15 Mechanism of Reactions

mechanism of a reaction

How a reaction occurs—that is, the manner in which it proceeds—is known as the **mechanism of the reaction**. The mechanism is the path, or route, the atoms and molecules take to arrive at the products. Our aim here is not to study the mechanisms themselves but to show that chemical reactions occur by specific routes.

activated complex

activation energy

When hydrogen and iodine are mixed at room temperature, we observe no appreciable reaction. In this case, the reaction takes place as a result of collisions between H_2 and I_2 molecules, but at room temperature the collisions do not result in reaction because the molecules lack sufficient energy to react. We say that an energy barrier to reaction exists. If heat is added, the kinetic energy of the molecules increases. When molecules of H_2 and I_2 with sufficient energy collide, an intermediate product, known as the **activated complex**, is formed. The amount of energy needed to form the activated complex is known as the **activation energy**. The activated complex, H_2I_2, is in a metastable form and has an energy level higher than that of the reactants or the product. It can decompose to form either the reactants or the product. Three steps constitute the mechanism of the reaction: (1) collision of an H_2 and an I_2 molecule; (2) formation of the activated complex, H_2I_2; and (3) decomposition to the product, HI. The various steps in the formation of HI are shown in Figure 16.3. Figure 16.4 illustrates the energy relationships in this reaction.

The reaction of hydrogen and chlorine proceeds by a different mechanism. When H_2 and Cl_2 are mixed and kept in the dark, essentially no product is formed. But, if the mixture is exposed to sunlight or ultraviolet radiation, it reacts very rapidly. The overall reaction is

$$H_2(g) + Cl_2(g) \longrightarrow 2\ HCl(g)$$

free radical

This reaction proceeds by what is known as a *free radical mechanism.* A **free radical** is a neutral atom or group of atoms containing one or more unpaired

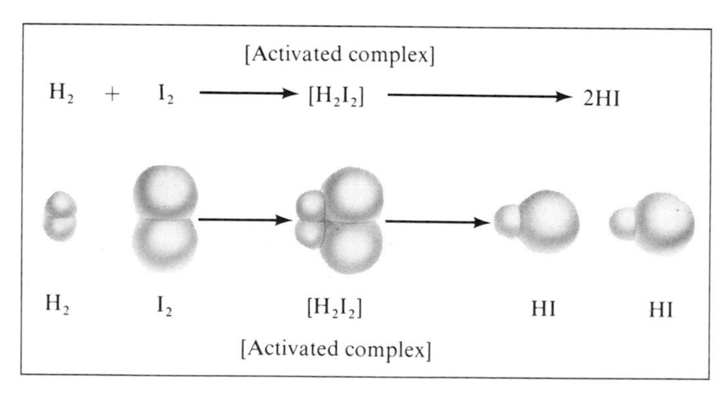

Figure 16.3

Mechanism of the reaction between hydrogen and iodine. H_2 and I_2 molecules of sufficient energy unite, forming the intermediate activated complex that decomposes to the product, hydrogen iodide.

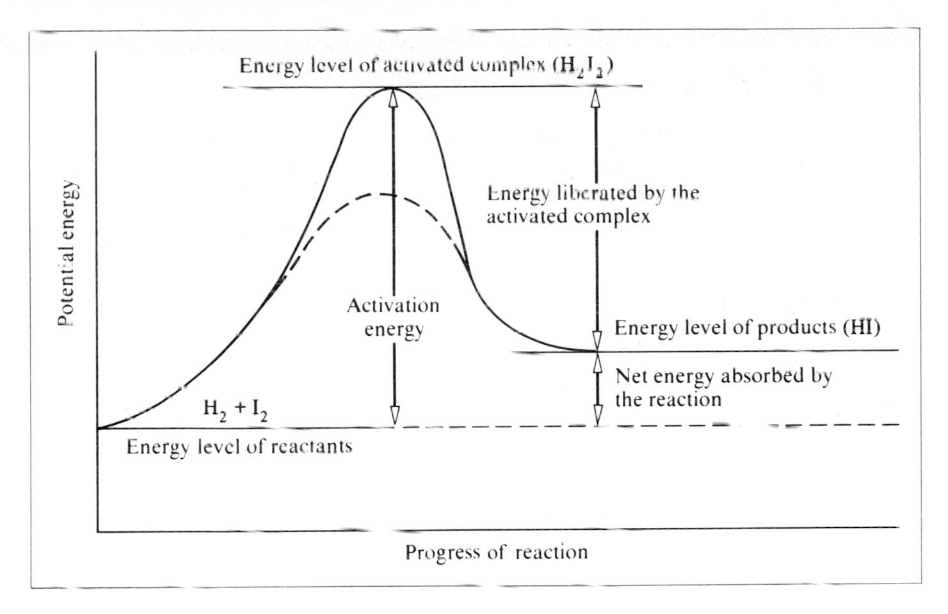

Figure 16.4
Relative energy diagram for the reaction between hydrogen and iodine.

$$H_2 + I_2 \longrightarrow [H_2I_2] \longrightarrow 2\ HI$$

Energy equal to the activation energy is put into the system to form the activated complex, H_2I_2. When this complex decomposes, it liberates energy, forming the product. In this case, the product is at a higher energy level than the reactants, indicating that the reaction is endothermic and that energy is absorbed during the reaction. The dotted line represents the effect that a catalyst would have on the reaction. The catalyst lowers the activation energy, thereby increasing the rate of the reaction.

electrons. Both atomic chlorine ($:\ddot{\underset{..}{Cl}}\cdot$) and atomic hydrogen ($H\cdot$) have an unpaired electron and are free radicals. The reaction occurs in three steps:

Step 1 *Initiation*

$$:\ddot{\underset{..}{Cl}}:\ddot{\underset{..}{Cl}}: + h\nu \longrightarrow \underset{\text{Chlorine free radicals}}{:\ddot{\underset{..}{Cl}}\cdot + :\ddot{\underset{..}{Cl}}\cdot}$$

In this step a chlorine molecule absorbs energy in the form of a photon, $h\nu$, of light or ultraviolet radiation. The energized chlorine molecule then splits into two chlorine free radicals.

Step 2 *Propagation*

$$:\ddot{\underset{..}{Cl}}\cdot + H:H \longrightarrow HCl + \underset{\text{Hydrogen free radical}}{H\cdot}$$

$$H\cdot + :\ddot{\underset{..}{Cl}}:\ddot{\underset{..}{Cl}}: \longrightarrow HCl + :\ddot{\underset{..}{Cl}}\cdot$$

This step begins when a chlorine free radical reacts with a hydrogen molecule to produce a molecule of hydrogen chloride and a hydrogen free radical. The hydrogen radical then reacts with another chlorine molecule to form hydrogen chloride and another chlorine free radical. This chlorine radical can repeat the process by reacting with another hydrogen molecule, and the reaction continues to propagate itself in this manner until one or both of the reactants are used up. Almost all of the product is formed in this step.

Step 3 *Termination*

$$:\ddot{\underset{..}{Cl}}\cdot + :\ddot{\underset{..}{Cl}}\cdot \longrightarrow Cl_2$$
$$H\cdot + H\cdot \longrightarrow H_2$$
$$H\cdot + :\ddot{\underset{..}{Cl}}\cdot \longrightarrow HCl$$

Hydrogen and chlorine free radicals can react in any of the three ways shown. Unless further activation occurs, the formation of hydrogen chloride will terminate when the radicals form molecules. In an exothermic reaction such as that between hydrogen and chlorine, usually enough heat and light energy is available to maintain the supply of free radicals, and the reaction will continue until at least one reactant is exhausted.

QUESTIONS

A. *Review the meanings of the new terms introduced in this chapter.*

1. Reversible chemical reaction
2. Chemical kinetics
3. Equilibrium
4. Chemical equilibrium
5. Principle of Le Chatelier
6. Catalyst
7. Equilibrium constant, K_{eq}
8. Acid ionization constant, K_a
9. Ion product constant for water, K_w
10. Solubility product constant, K_{sp}
11. Common ion effect
12. Hydrolysis
13. Buffer solution
14. Mechanism of a reaction
15. Activated complex
16. Activation energy
17. Free radical

B. *Information useful in answering the following questions will be found in tables and figures.*

1. How would you expect the two tubes in Figure 16.1 to appear if both are at 25°C?
2. Is the reaction $N_2O_4 \rightleftharpoons 2\ NO_2$ (Figure 16.1) exothermic or endothermic?
3. At equilibrium how do the forward and reverse reaction rates compare? (See Figure 16.2.)
4. Would the reaction of 30 mol of H_2 and 10 mol of N_2 produce a greater yield of NH_3 carried out in a 1 L or a 2 L vessel? (See Table 16.1.)
5. Of the acids listed in Table 16.2, which ones are stronger than acetic acid and which are weaker?
6. For each of the solutions in Table 16.3, what is the sum of the pH plus the pOH? What would be the pOH of a solution whose pH was -1?
7. Tabulate the relative order of molar solubilities of AgCl, AgBr, AgI, $AgC_2H_3O_2$, $PbSO_4$, $BaSO_4$, $BaCrO_4$, and PbS. (Use Table 16.4.) List the most soluble first.

8. Which compound in each of the following pairs has the greater molar solubility? (See Table 16.4.)
   (a) $Mn(OH)_2$ or Ag_2CrO_4
   (b) $BaCrO_4$ or Ag_2CrO_4
9. Using Table 16.6, explain how the acetic acid–sodium acetate buffer system maintains its pH when 0.010 mol of HCl is added to 1 L of the buffer solution.
10. How would Figure 16.4 be altered if the reaction were exothermic?

C. *Review questions*

1. Express the following reversible systems in equation form:
   (a) A mixture of ice and liquid water at 10°C
   (b) Liquid water and vapor at 100°C in a pressure cooker
   (c) Crystals of Na_2SO_4 in a saturated aqueous solution of Na_2SO_4
   (d) A closed system containing boiling sulfur dioxide, SO_2
2. Explain why a precipitate of NaCl forms when hydrogen chloride gas is passed into a saturated aqueous solution of NaCl.
3. Why does the rate of a reaction usually increase when the concentration of one of the reactants is increased?
4. Consider the following system at equilibrium:

$$4\ NH_3(g) + 3\ O_2(g) \rightleftharpoons 2\ N_2(g) + 6\ H_2O(g) + 1531\ kJ$$

   (a) Is the reaction exothermic or endothermic?
   (b) If the system's state of equilibrium is disturbed by the addition of O_2, in which direction, left or right, must reaction occur to reestablish equilibrium? After the new equilibrium has been established, how will the final molar concentrations of NH_3, O_2, N_2, and H_2O compare (increase or decrease) with their concentrations before the addition of the O_2?
   (c) If the system's state of equilibrium is disturbed by the addition of heat, in which direction will reaction occur, left or right, to reestablish equilibrium?
5. Consider the following system at equilibrium:

$$N_2(g) + 3\ H_2(g) \rightleftharpoons 2\ NH_3(g) + 92.5\ kJ$$

Complete the following table. Indicate changes in moles by entering I, D, N, or ? in the table. (I = increase, D = decrease, N = no change, ? = insufficient information to determine.)

Change or stress imposed on the system at equilibrium	Direction of reaction, left or right, to reestablish equilibrium	Change in number of moles		
		N_2	H_2	NH_3
(a) Add N_2				
(b) Remove H_2				
(c) Decrease volume of reaction vessel				
(d) Increase volume of reaction vessel				
(e) Increase temperature				
(f) Add catalyst				
(g) Add both H_2 and NH_3				

6. If pure hydrogen iodide, HI, is placed in a vessel at 700 K, will it decompose? Explain.
7. For each of the equations that follow, tell in which direction, left or right, the equilibrium will shift when the following changes are made: The temperature is increased; the pressure is increased by decreasing the volume of the reaction vessel; a catalyst is added.
   (a) $3\ O_2(g) + 271\ kJ \rightleftharpoons 2\ O_3(g)$
   (b) $CH_4(g) + Cl_2(g) \rightleftharpoons CH_3Cl(g) + HCl(g) + 110\ kJ$
   (c) $2\ NO(g) + 2\ H_2(g) \rightleftharpoons N_2(g) + 2\ H_2O(g) + 665\ kJ$
   (d) $2\ SO_3(g) + 197\ kJ \rightleftharpoons 2\ SO_2(g) + O_2(g)$
   (e) $4\ NH_3(g) + 3\ O_2(g) \rightleftharpoons 2\ N_2(g) + 6\ H_2O(g) + 1531\ kJ$
8. Explain why an increase in temperature causes the rate of reaction to increase.
9. Give a word description of how equilibrium is reached when the substances A and B are first mixed and react as

$$A + B \rightleftharpoons C + D$$

10. With dilution, aqueous solutions of acetic acid, $HC_2H_3O_2$, show increased ionization. For example, a 1.0 *M* solution of acetic acid is 0.42% ionized, whereas a 0.10 *M* solution is 1.34% ionized. Explain this behavior using the ionization equation and equilibrium principles.

11. A 1.0 *M* solution of acetic acid ionizes less and has a higher concentration of H^+ ions than a 0.10 *M* acetic acid solution. Explain this behavior. (See Question 10 for data.)

12. Write the equilibrium constant expression for each of the following reactions:
(a) $4\ HCl(g) + O_2(g) \rightleftharpoons 2\ Cl_2(g) + 2\ H_2O(g)$
(b) $N_2(g) + 3\ H_2(g) \rightleftharpoons 2\ NH_3(g)$
(c) $PCl_5(g) \rightleftharpoons PCl_3(g) + Cl_2(g)$
(d) $HClO_2(aq) \rightleftharpoons H^+(aq) + ClO_2^-(aq)$
(e) $NH_4OH(aq) \rightleftharpoons NH_4^+(aq) + OH^-(aq)$
(f) $4\ NH_3(g) + 5\ O_2(g) \rightleftharpoons 4\ NO(g) + 6\ H_2O(g)$

13. What would cause two separate samples of pure water to have slightly different pH values?

14. Why are the pH and pOH equal in pure water?

15. What effect will increasing the H^+ ion concentration of a solution have on (a) pH, (b) pOH, (c) $[OH^-]$, and (d) K_w?

16. Write the solubility product expression (K_{sp}) for each of the following substances:
(a) CuS (e) $Fe(OH)_3$
(b) $BaSO_4$ (f) Sb_2S_5
(c) $PbBr_2$ (g) CaF_2
(d) Ag_3AsO_4 (h) $Ba_3(PO_4)_2$

17. Explain why silver acetate, $AgC_2H_3O_2$, is more soluble in nitric acid than in water. [*Hint:* Write the equilibrium equation first and then consider the effect of the acid on the acetate ion.] What would happen if hydrochloric acid, HCl, were used in place of nitric acid?

18. Decide whether each of the following salts forms an acidic, a basic, or a neutral aqueous solution.
(a) KCl (d) $(NH_4)_2SO_4$ (g) $NaNO_2$
(b) Na_2CO_3 (e) $Ca(CN)_2$ (h) NaF
(c) K_2SO_4 (f) $BaBr_2$

19. Dissolution of sodium acetate, $NaC_2H_3O_2$, in pure water gives a basic solution. Why? [*Hint:* A small amount of $HC_2H_3O_2$ is formed.]

20. Write hydrolysis equations for aqueous solutions of these salts.
(a) KNO_2 (c) NH_4NO_3
(b) $Mg(C_2H_3O_2)_2$ (d) Na_2SO_3

21. Write hydrolysis equations for the following ions.
(a) HCO_3^- (b) NH_4^+ (c) OCl^- (d) ClO_2^-

22. Describe why the pH of a buffer solution remains almost constant when a small amount of acid or base is added to it.

23. One of the important pH-regulating systems in the blood consists of a carbonic acid–sodium bicarbonate buffer.

$$H_2CO_3(aq) \rightleftharpoons H^+(aq) + HCO_3^-(aq)$$
$$NaHCO_3(aq) \longrightarrow Na^+(aq) + HCO_3^-(aq)$$

Explain how this buffer resists changes in pH when (a) excess acid and (b) excess base get into the bloodstream.

24. Which of the following statements are correct?
(a) In a reaction at equilibrium the concentrations of reactants and products are equal.
(b) A catalyst increases the concentrations of products present at equilibrium.
(c) Enzymes are the catalysts in living systems.
(d) A catalyst lowers the activation energy of a reaction by equal amounts for both the forward and the reverse reactions.
(e) If an increase in temperature causes an increase in the concentration of products present at equilibrium, then the reaction is exothermic.
(f) The magnitude of an equilibrium constant is independent of the reaction temperature.
(g) A large equilibrium constant for a reaction indicates that the reaction, at equilibrium, favors products over reactants.
(h) The amount of product obtained at equilibrium is proportional to how fast equilibrium is attained.
(i) The study of reaction rates and reaction mechanisms is known as chemical kinetics.
(j) For the reaction

$$CaCO_3(s) \overset{\Delta}{\rightleftharpoons} CaO(s) + CO_2(g)$$

increasing the pressure of CO_2 present at equilibrium will cause the reaction to shift left.
(k) The larger the value of the equilibrium constant, the greater the proportion of products present at equilibrium.
(l) At chemical equilibrium the rate of the reverse reaction is equal to the rate of the forward reaction.

Statements (m)–(s) pertain to the equilibrium system

$$2\ NO(g) + O_2(g) \rightleftharpoons 2\ NO_2(g) + \text{heat}$$

(m) The reaction as shown is endothermic.
(n) Increasing the temperature will cause the equilibrium to shift left.
(o) Increasing the temperature will increase the magnitude of the equilibrium constant, K_{eq}.
(p) Decreasing the volume of the reaction vessel will shift the equilibrium to the right and decrease the concentrations of the reactants.
(q) Removal of some of the O_2 will cause an increase in the concentration of NO.
(r) High temperatures and pressures favor increased yields of NO_2.
(s) The equilibrium constant expression for the reaction is

$$K_{eq} = \frac{[NO_2]^2}{[NO]^2[O_2]}$$

(t) A solution with an H^+ ion concentration of 1×10^{-5} mol/L has a pOH of 9.
(u) An aqueous solution that has an OH^- ion concentration of 1×10^{-4} mol/L has an H^+ ion concentration of 1×10^{-10} mol/L.
(v) $K_w = [H^+][OH^-] = 1.00 \times 10^{-14}$, and pH + pOH = 14.
(w) As solid $BaSO_4$ is added to a saturated solution of $BaSO_4$, the magnitude of its K_{sp} increases.
(x) A solution of pOH 10 is basic.
(y) The pH increases as the $[H^+]$ increases.
(z) The pH of 0.050 *M* $Ca(OH)_2$ is 13.

D. *Review problems*

Equilibrium Constants

1. What is the maximum amount (moles) of HI that can be obtained from a reaction mixture containing 2.30 mol of I_2 and 2.10 mol of H_2?
2. (a) How many moles of hydrogen iodide, HI, will be produced when 2.00 mol of H_2 and 2.00 mol of I_2 are reacted at 700 K? (Reaction is 79% complete.)
   (b) Addition of 0.27 mol of I_2 to the system increases the yield of HI to 85%. How many moles of H_2, I_2, and HI are now present?
   (c) From the data in part (a), calculate K_{eq} for the reaction at 700 K.
*3. 6.00 g of hydrogen, H_2, and 200 g of iodine, I_2, are reacted at 500 K. After equilibrium is reached, analysis shows that the flask contains 64.0 g of HI. How many moles of H_2, I_2, and HI are present in this equilibrium mixture?
4. What is the equilibrium constant of the reaction shown if a 20 L flask contains 0.10 mol of PCl_3, 1.50 mol of Cl_2, and 0.22 mol of PCl_5?

$$PCl_3(g) + Cl_2(g) \rightleftharpoons PCl_5$$

5. If the rate of a reaction doubles for every 10° rise in temperature, how much faster will the reaction go at 100°C than at 30°C?

Ionization Constants

6. Calculate the ionization constant for each of the following monoprotic acids. Each acid ionizes as follows: $HA \rightleftharpoons H^+ + A^-$.

Acid	Acid concentration	$[H^+]$
Hypochlorous, HOCl	0.10 *M*	5.9×10^{-5} mol/L
Propanoic, $HC_3H_5O_2$	0.15 *M*	1.4×10^{-3} mol/L
Hydrocyanic, HCN	0.20 *M*	8.9×10^{-6} mol/L

7. Calculate (a) the H^+ ion concentration, (b) the pH, and (c) the percent ionization of a 0.25 *M* solution of $HC_2H_3O_2$. ($K_a = 1.8 \times 10^{-5}$.)
8. A 1.0 *M* solution of a weak acid, HA, is 0.52% ionized. Calculate the ionization constant, K_a, for the acid.
9. A 0.15 *M* solution of a weak acid, HA, has a pH of 5. Calculate the ionization constant, K_a, for the acid.
10. Calculate the percent ionization and pH of solutions of $HC_2H_3O_2$ having the following molarities: (a) 1.0 *M*, (b) 0.10 *M*, and (c) 0.010 *M*. ($K_a = 1.8 \times 10^{-5}$.)
11. A 0.37 *M* solution of a weak acid, HA, has a pH of 3.7. What is the K_a for this acid?
12. A 0.23 *M* solution of a weak acid, HA, has a pH of 2.89. What is the K_a for this acid?
13. A common laboratory reagent is 6.0 *M* HCl. Calculate the $[H^+]$, $[OH^-]$, pH, and pOH of this solution.

14. Calculate the pH and the pOH of the following solutions:
(a) 0.00010 M HCl
(b) 0.010 M NaOH
(c) 0.0025 M NaOH
(d) 0.10 M HClO ($K_a = 3.5 \times 10^{-8}$)
* (e) Saturated $Fe(OH)_2$ solution ($K_{sp} = 8.0 \times 10^{-16}$)

15. Calculate the $[OH^-]$ in each of these solutions:
(a) $[H^+] = 1.0 \times 10^{-4}$
(b) $[H^+] = 2.8 \times 10^{-6}$
(c) $[H^+] = 4.0 \times 10^{-9}$

16. Calculate the $[H^+]$ in each of these solutions:
(a) $[OH^-] = 6.0 \times 10^{-7}$
(b) $[OH^-] = 1 \times 10^{-8}$
(c) $[OH^-] = 4.5 \times 10^{-6}$

Solubility Product Constant

17. Given the following solubility data, calculate the solubility product constant for each substance.
(a) $BaSO_4$, 3.9×10^{-5} mol/L
(b) Ag_2CrO_4, 7.8×10^{-5} mol/L
(c) ZnS, 3.5×10^{-12} mol/L
(d) $Pb(IO_3)_2$, 4.0×10^{-5} mol/L
(e) Bi_2S_3, 4.9×10^{-15} mol/L
(f) AgCl, 0.0019 g/L
(g) $CaSO_4$, 0.67 g/L
(h) $Zn(OH)_2$, 2.33×10^{-4} g/L
(i) Ag_3PO_4, 6.73×10^{-3} g/L

18. Calculate the molar solubility for each of the following substances:
(a) $BaCO_3$, $K_{sp} = 2.0 \times 10^{-9}$
(b) $AlPO_4$, $K_{sp} = 5.8 \times 10^{-19}$
(c) Ag_2SO_4, $K_{sp} = 1.5 \times 10^{-5}$
(d) $Mg(OH)_2$, $K_{sp} = 7.1 \times 10^{-12}$

19. Calculate, for each of the substances in Question 18, the solubility in grams per 100 mL of water.

20. The K_{sp} of CaF_2 is 3.9×10^{-11}. Calculate (a) the molar concentrations of Ca^{2+} and F^- in a saturated solution, and (b) the grams of CaF_2 that will dissolve in 500 mL of water.

* **21.** The following pairs of solutions are mixed. Show by calculation whether or not a precipitate will form.
(a) 100 mL of 0.010 M Na_2SO_4 and 100 mL of 0.001 M $Pb(NO_3)_2$
(b) 50.0 mL of 1.0×10^{-4} M $AgNO_3$ and 100 mL of 1.0×10^{-4} M NaCl
(c) 1.0 g $Ca(NO_3)_2$ in 150 mL H_2O and 250 mL of 0.01 M NaOH
K_{sp} $PbSO_4 = 1.3 \times 10^{-8}$
K_{sp} $AgCl = 1.7 \times 10^{-10}$
K_{sp} $Ca(OH)_2 = 1.3 \times 10^{-6}$

22. $BaCl_2$ is added to a saturated $BaSO_4$ solution until the Ba^{2+} concentration is 0.050 M. (a) What concentration of SO_4^{2-} remains in solution? (b) How many grams of $BaSO_4$ remain dissolved in 100 mL of the solution? ($K_{sp} = 1.5 \times 10^{-9}$ for $BaSO_4$.)

23. The concentration of a solution is 0.10 M Ba^{2+} and 0.10 M Sr^{2+}. Which sulfate, $BaSO_4$ or $SrSO_4$, will precipitate first when a dilute solution of H_2SO_4 is added dropwise to the solution? Show evidence for your answer. ($K_{sp} = 1.5 \times 10^{-9}$ for $BaSO_4$ and $K_{sp} = 3.5 \times 10^{-7}$ for $SrSO_4$.)

24. How many moles of AgBr will dissolve in 1.0 L of (a) 0.10 M NaBr and (b) 0.10 M $MgBr_2$? ($K_{sp} = 5.0 \times 10^{-13}$ for AgBr.)

25. The K_{sp} for $PbCl_2$ is 2.0×10^{-5}. Will a precipitate form when 0.050 mol of $Pb(NO_3)_2$ and 0.010 mol of NaCl are dissolved in 1.0 L H_2O? Show evidence for your answer.

Buffer Solutions

26. Calculate the H^+ ion concentration and the pH of buffer solutions that are 0.20 M in $HC_2H_3O_2$ and contain sufficient sodium acetate to make the $C_2H_3O_2^-$ ion concentration equal to (a) 0.10 M and (b) 0.20 M. (K_a $HC_2H_3O_2 = 1.8 \times 10^{-5}$.)

27. (a) When 1.0 mL of 1.0 M HCl is added to 50 mL of 1.0 M NaCl, the H^+ ion concentration changes from 1×10^{-7} M to 2.0×10^{-2} M.
(b) When 1.0 mL of 1.0 M HCl is added to 50 mL of a buffer solution that is 1.0 M in $HC_2H_3O_2$ and 1.0 M in $NaC_2H_3O_2$, the H^+ ion concentration changes from 1.8×10^{-5} M to 1.9×10^{-5} M.
Calculate the initial pH and the pH change in each solution (log 1.8 = 0.26; log 1.9 = 0.28; log 2.0 = 0.30).

CHAPTER 17

Oxidation–Reduction

After studying Chapter 17, you should be able to

1. Understand the terms listed in Question A at the end of the chapter.
2. Assign oxidation numbers to all the elements in a given compound or ion.
3. Determine which element is being oxidized and which element is being reduced in an oxidation–reduction reaction.
4. Identify the oxidizing agent and the reducing agent in an oxidation–reduction reaction.
5. Balance oxidation–reduction equations in molecular and ionic form.
6. Outline the general principles concerning the activity series of the metals.
7. Use the activity series to determine whether a proposed single-displacement reaction will occur.
8. Distinguish between an electrolytic and a voltaic cell.
9. Draw a voltaic cell that will produce electric current from an oxidation–reduction reaction involving two metals and their salts.
10. Identify the anode reaction and the cathode reaction in a given electrolytic or voltaic cell.
11. Write equations for the overall chemical reaction and for the oxidation and reduction reactions involved in the discharging or charging of a lead storage battery.
12. Explain how the charge condition of a lead storage battery can be estimated with the aid of a hydrometer.

17.1 Oxidation Number

The oxidation number of an atom (sometimes called its oxidation state) can be considered to represent the number of electrons lost, gained, or unequally shared by the atom. Oxidation numbers can be zero, positive, or negative. When the oxidation number of an atom is zero, the atom has the same number of electrons assigned to it as there are in the free neutral atom. When the oxidation number is positive, the atom has fewer electrons assigned to it than there are in the neutral atom. When the oxidation number is negative, the atom has more electrons assigned to it than there are in the neutral atom.

The oxidation number of an atom that has lost or gained electrons to form an ion is the same as the plus or minus charge of the ion. In the ionic compound NaCl the oxidation numbers are clearly established to be +1 for the Na^+ ion and −1 for the Cl^- ion. The Na^+ ion has one less electron than the neutral Na atom; and the Cl^- ion has one more electron than the neutral Cl atom. In $MgCl_2$ two electrons have transferred from the Mg atom to the Cl atoms; thus, the oxidation number of Mg is +2.

In covalently bonded substances, where electrons are shared between two atoms, oxidation numbers are assigned by a somewhat arbitrary system based on relative electronegativities. For symmetrical covalent molecules, such as H_2 and Cl_2, each atom is assigned an oxidation number of zero because the bonding pair of electrons is shared equally between two like atoms, neither of which is more electronegative than the other.

H:H   :C̤̈l:C̤̈l:

When the covalent bond is between two unlike atoms, the bonding electrons are shared unequally because the more electronegative element has a greater attraction for them. In this case the oxidation numbers are determined by assigning both electrons to the more electronegative element.

Thus in compounds with covalent bonds, such as NH_3 and H_2O,

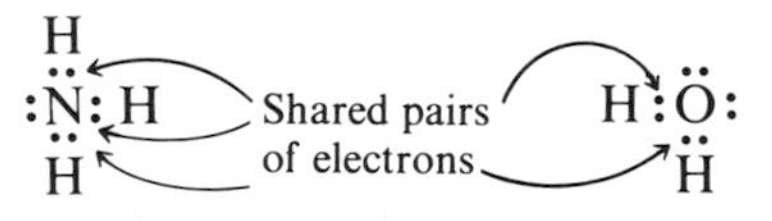

the pairs of electrons are unequally shared between the atoms and are attracted toward the more electronegative elements, N and O. This unequal sharing causes the N and O atoms to be relatively negative with respect to the H atoms. At the same time it causes the H atoms to be relatively positive with respect to the N and O atoms. In H_2O both pairs of shared electrons are assigned to the O atom, giving it two electrons more than the neutral O atom. At the same time, each H atom is assigned one electron less than the neutral H atom. Therefore, the O atom is assigned an oxidation number of −2, and each H atom is assigned an oxidation number of +1. In NH_3 the three pairs of shared electrons are assigned to the N atom, giving it three electrons more than the neutral N atom. At the same time, each H atom has one electron less than the neutral atom. Therefore, the N

Table 17.1 Arbitrary rules for assigning oxidation numbers

1. All elements in their free state (uncombined with other elements) have an oxidation number of zero (for example, Na, Cu, Mg, H_2, O_2, Cl_2, N_2).
2. H is +1, except in metal hydrides, where it is −1 (for example, NaH, CaH_2).
3. O is −2, except in peroxides, where it is −1, and in OF_2, where it is +2.
4. The metallic element in an ionic compound has a positive oxidation number.
5. In covalent compounds the negative oxidation number is assigned to the most electronegative atom.
6. The algebraic sum of the oxidation numbers of the elements in a compound is zero.
7. The algebraic sum of the oxidation numbers of the elements in a polyatomic ion is equal to the charge of the ion.

Table 17.2 Oxidation numbers of atoms in selected compounds

Ion or compound	Oxidation number	
H_2O	H, +1;	O, −2
SO_2	S, +4;	O, −2
CH_4	C, −4;	H, +1
CO_2	C, +4;	O, −2
$KMnO_4$	K, +1;	Mn, +7; O, −2
Na_3PO_4	Na, +1;	P, +5; O, −2
$Al_2(SO_4)_3$	Al, +3;	S, +6; O, −2
NO	N, +2;	O, −2
BCl_3	B, +3;	Cl, −1
SO_4^{2-}	S, +6;	O, −2
NO_3^-	N, +5;	O, −2
CO_3^{2-}	C, +4;	O, −2

atom is assigned an oxidation number of −3, and each H atom is assigned an oxidation number of +1.

The assignment of correct oxidation numbers to elements is essential for balancing oxidation–reduction equations. Review Sections 8.1, 8.2, and 8.4, regarding oxidation numbers, oxidation number tables, and the determination of oxidation numbers from formulas. Table 7.3 lists relative electronegativities of the elements. Rules for assigning oxidation numbers are given in Section 8.1 and are summarized in Table 17.1. Examples showing oxidation numbers in compounds and ions are given in Table 17.2.

PROBLEM 17.1 Determine the oxidation number of each element in (a) KNO_3 and (b) SO_4^{2-}.

(a) K is a Group IA metal; therefore it has an oxidation number of +1. The oxidation number of each O atom is −2 (Table 17.1, Rule 3). Using these values and the fact that the sum of the oxidation numbers of all the atoms in a compound is zero, we can determine the oxidation number of N.

$$KNO_3$$

$$+1 + N + 3(-2) = 0$$

$$N = +6 - 1 = +5$$

The oxidation numbers are K, +1; N, +5; O, −2.

(b) SO_4^{2-} is an ion; therefore, the sum of oxidation numbers of the S and the O atoms must be −2, the charge of the ion. The oxidation number of each O atom is −2 (Table 17.1, Rule 3). Then

$$SO_4^{2-}$$

$$S + 4(-2) = -2$$

$$S = -2 + 8 = +6$$

The oxidation numbers are S, +6; O, −2.

17.2 Oxidation–Reduction

oxidation–reduction

redox

oxidation

reduction

Oxidation–reduction, also known as **redox**, is a chemical process in which the oxidation number of an element is changed. The process may involve the complete transfer of electrons to form ionic bonds or only a partial transfer or shift of electrons to form covalent bonds.

Oxidation occurs whenever the oxidation number of an element increases as a result of losing electrons. Conversely, **reduction** occurs whenever the oxidation number of an element decreases as a result of gaining electrons. For example, a change in oxidation number from +2 to +3 or from −1 to 0 is oxidation; a change from +5 to +2 or from −2 to −4 is reduction (see Figure 17.1). Oxidation and reduction occur simultaneously in a chemical reaction; one cannot take place without the other.

Many combination, decomposition, and single-displacement reactions involve oxidation–reduction. Let us examine the combustion of hydrogen and oxygen from this point of view:

$$2\,H_2 + O_2 \longrightarrow 2\,H_2O$$

oxidizing agent

Both reactants, hydrogen and oxygen, are elements in the free state and have an oxidation number of zero. In the product, water, hydrogen has been oxidized to +1 and oxygen reduced to −2. The substance that causes an increase in the oxidation state of another substance is called an **oxidizing agent**. The substance that causes a decrease in the oxidation state of another substance is called a

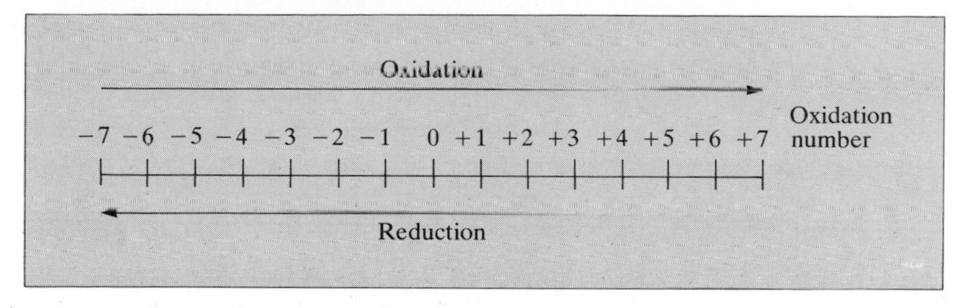

Figure 17.1

Oxidation and reduction. Oxidation results in an increase in the oxidation number, and reduction results in a decrease in the oxidation number.

reducing agent

reducing agent. In this reaction the oxidizing agent is free oxygen, and the reducing agent is free hydrogen. In the reaction

$$Zn(s) + H_2SO_4(aq) \longrightarrow ZnSO_4(aq) + H_2\uparrow$$

metallic zinc is oxidized, and hydrogen ions are reduced. Zinc is the reducing agent, and hydrogen ions, the oxidizing agent. Electrons are transferred from the zinc metal to the hydrogen ions. The reaction is better expressed as

$$Zn^0 + 2\ H^+ + SO_4^{2-} \longrightarrow Zn^{2+} + SO_4^{2-} + H_2^0\uparrow$$

Oxidation:	**Increase in oxidation number** **Loss of electrons**
Reduction:	**Decrease in oxidation number** **Gain of electrons**

The oxidizing agent is reduced and gains electrons. The reducing agent is oxidized and loses electrons. The loss and gain of electrons is characteristic of all redox reactions.

17.3 Balancing Oxidation–Reduction Equations

Many simple redox equations can be balanced readily by inspection, or trial and error.

$$Na + Cl_2 \longrightarrow NaCl \quad \text{(unbalanced)}$$
$$2\ Na + Cl_2 \longrightarrow 2\ NaCl \quad \text{(balanced)}$$

Balancing this equation is certainly not complicated. But as we study more complex reactions and equations, such as

$$P + HNO_3 + H_2O \longrightarrow NO + H_3PO_4 \quad \text{(unbalanced)}$$
$$3\,P + 5\,HNO_3 + 2\,H_2O \longrightarrow 5\,NO + 3\,H_3PO_4 \quad \text{(balanced)}$$

the trial-and-error method of finding the proper numbers to balance the equation would take an unnecessarily long time.

One systematic method for balancing oxidation–reduction equations is based on the transfer of electrons between the oxidizing and reducing agents. Consider the first equation again.

$$Na^0 + Cl_2^0 \longrightarrow Na^+Cl^- \quad \text{(unbalanced)}$$

In this reaction sodium metal loses one electron per atom when it changes to a sodium ion. At the same time chlorine gains one electron per atom. Because chlorine is diatomic, two electrons per molecule are needed to form a chloride ion from each atom. These electrons are furnished by two sodium atoms. Stepwise, the reaction may be written as two half-reactions, the oxidation half-reaction and the reduction half-reaction:

Oxidation half-reaction $\quad 2\,Na^0 \longrightarrow 2\,Na^+ + 2\,e^-$

Reduction half-reaction $\quad Cl_2^0 + 2\,e^- \longrightarrow 2\,Cl^-$

$$2\,Na^0 + Cl_2^0 \longrightarrow 2\,Na^+Cl^-$$

When the two half-reactions, each containing the same number of electrons, are added together algebraically, the electrons cancel out. In this reaction there are no excess electrons; the two electrons lost by the two sodium atoms are utilized by chlorine. In all redox reactions the loss of electrons by the reducing agent must equal the gain of electrons by the oxidizing agent. Sodium is oxidized; chlorine is reduced. Chlorine is the oxidizing agent; sodium is the reducing agent.

The following examples illustrate a systematic method of balancing more complicated redox equations by the change-in-oxidation-number method.

PROBLEM 17.2 Balance the equation

$$Sn + HNO_3 \longrightarrow SnO_2 + NO_2 + H_2O \quad \text{(unbalanced)}$$

Step 1 Assign oxidation numbers to each element to identify the elements that are being oxidized and those that are being reduced. Write the oxidation numbers below each element in order to avoid confusing them with the charge on an ion or radical.

$$\underset{0}{Sn} + \underset{+1}{H}\underset{+5}{N}\underset{-2}{O_3} \longrightarrow \underset{+4}{Sn}\underset{-2}{O_2} + \underset{+4}{N}\underset{-2}{O_2} + \underset{+1}{H_2}\underset{-2}{O}$$

Note that the oxidation numbers of Sn and N have changed.

Step 2 Now write two new equations, using only the elements that change in oxidation number. Then add electrons to bring the equations into electrical balance. One equation represents the oxidation step; the other represents the reduction step. The oxidation step produces electrons; the reduction step uses electrons.

oxidation $Sn^0 \longrightarrow Sn^{4+} + 4\,e^-$ (Sn^0 loses 4 electrons)

reduction $N^{5+} + 1\,e^- \longrightarrow N^{4+}$ (N^{5+} gains 1 electron)

Step 3 Now multiply the two equations by the smallest integral numbers that will make the loss of electrons by the oxidation step equal to the number of electrons gained in the reduction step. In this reaction the oxidation step is multiplied by 1 and the reduction step by 4. The equations become

oxidation $Sn^0 \longrightarrow Sn^{4+} + 4\,e^-$ (Sn^0 loses 4 electrons)

reduction $4\,N^{5+} + 4\,e^- \longrightarrow 4\,N^{4+}$ ($4\,N^{5+}$ gain 4 electrons)

We have now established the ratio of the oxidizing to the reducing agent as being four atoms of N to one atom of Sn.

Step 4 Now transfer the coefficient that appears in front of each substance in the balanced oxidation–reduction equations to the corresponding substance in the original equation. We need to use 1 Sn, 1 SnO_2, 4 HNO_3, and 4 NO_2:

$$Sn + 4\,HNO_3 \longrightarrow SnO_2 + 4\,NO_2 + H_2O \quad \text{(unbalanced)}$$

Step 5 In the usual manner, balance the remaining elements that are not oxidized or reduced to give the final balanced equation:

$$Sn + 4\,HNO_3 \longrightarrow SnO_2 + 4\,NO_2 + 2\,H_2O \quad \text{(balanced)}$$

In balancing the final elements, we must not change the ratio of the elements that were oxidized and reduced. We should make a final check to ensure that both sides of the equation have the same number of atoms of each element. The final balanced equation contains 1 atom of Sn, 4 atoms of N, 4 atoms of H, and 12 atoms of O on each side.

Because each new equation may present a slightly different problem and because proficiency in balancing equations requires practice, we will work through a few more problems.

PROBLEM 17.3 Balance the equation

$$I_2 + Cl_2 + H_2O \longrightarrow HIO_3 + HCl \quad \text{(unbalanced)}$$

Step 1 Assign oxidation numbers:

$$\underset{0}{I_2} + \underset{0}{Cl_2} + \underset{+1-2}{H_2O} \longrightarrow \underset{+1+5-2}{H\,I\,O_3} + \underset{+1-1}{HCl}$$

The oxidation numbers of I_2 and Cl_2 have changed, I_2 from 0 to +5 and Cl_2 from 0 to −1.

Step 2 Write oxidation and reduction steps. Balance the number of atoms and then balance the electrical charge using electrons.

oxidation $I_2 \longrightarrow 2\,I^{5+} + 10\,e^-$ (I_2 loses 10 electrons)

reduction $Cl_2 + 2\,e^- \longrightarrow 2\,Cl^-$ (Cl_2 gains 2 electrons)

Step 3 Adjust loss and gain of electrons so that they are equal. Multiply the oxidation step by 1 and the reduction step by 5.

oxidation $I_2 \longrightarrow 2\,I^{5+} + 10\,e^-$ (I_2 loses 10 electrons)

reduction $5\,Cl_2 + 10\,e^- \longrightarrow 10\,Cl^-$ ($5\,Cl_2$ gain 10 electrons)

Step 4 Transfer the coefficients from the balanced redox equations into the original equation. We need to use 1 I_2, 2 HIO_3, 5 Cl_2, and 10 HCl.

$$I_2 + 5\,Cl_2 + H_2O \longrightarrow 2\,HIO_3 + 10\,HCl \quad \text{(unbalanced)}$$

Step 5 Balance the remaining elements, H and O:

$$I_2 + 5\,Cl_2 + 6\,H_2O \longrightarrow 2\,HIO_3 + 10\,HCl \quad \text{(balanced)}$$

Check: The final balanced equation contains 2 atoms of I, 10 atoms of Cl, 12 atoms of H, and 6 atoms of O on each side.

PROBLEM 17.4 Balance the equation

$$K_2Cr_2O_7 + FeCl_2 + HCl \longrightarrow CrCl_3 + KCl + FeCl_3 + H_2O \quad \text{(unbalanced)}$$

Step 1 Assign oxidation numbers (Cr and Fe have changed):

$$\underset{+1\,+6\,-2}{K_2Cr_2O_7} + \underset{+2\,-1}{FeCl_2} + \underset{+1\,-1}{HCl} \longrightarrow \underset{+3\,-1}{CrCl_3} + \underset{+1\,-1}{KCl} + \underset{+3\,-1}{FeCl_3} + \underset{+1\,-2}{H_2O}$$

Step 2 Write the oxidation and reduction steps. Balance the number of atoms and then balance the electrical charge using electrons.

oxidation $Fe^{2+} \longrightarrow Fe^{3+} + 1\,e^-$ (Fe^{2+} loses 1 electron)

reduction $2\,Cr^{6+} + 6\,e^- \longrightarrow 2\,Cr^{3+}$ ($2\,Cr^{6+}$ gain 6 electrons)

Step 3 Balance the loss and gain of electrons. Multiply the oxidation step by 6 and the reduction step by 1 to equalize the transfer of electrons.

oxidation $6\,Fe^{2+} \longrightarrow 6\,Fe^{3+} + 6\,e^-$ ($6\,Fe^{2+}$ lose 6 electrons)

reduction $2\,Cr^{6+} + 6\,e^- \longrightarrow 2\,Cr^{3+}$ ($2\,Cr^{6+}$ gain 6 electrons)

Step 4 Transfer the coefficients from the balanced redox equations into the original equation. (Note that one formula unit of $K_2Cr_2O_7$ contains two Cr atoms.) We need to use 1 $K_2Cr_2O_7$, 2 $CrCl_3$, 6 $FeCl_2$, and 6 $FeCl_3$.

$$K_2Cr_2O_7 + 6\ FeCl_2 + HCl \longrightarrow 2\ CrCl_3 + KCl + 6\ FeCl_3 + H_2O \quad \text{(unbalanced)}$$

Step 5 Balance the remaining elements in this order: K, Cl, H, O.

$$K_2Cr_2O_7 + 6\ FeCl_2 + 14\ HCl \longrightarrow 2\ CrCl_3 + 2\ KCl + 6\ FeCl_3 + 7\ H_2O \quad \text{(balanced)}$$

Check: The final balanced equation contains 2 K atoms, 2 Cr atoms, 7 O atoms, 6 Fe atoms, 26 Cl atoms, and 14 H atoms on each side.

17.4 Balancing Ionic Redox Equations

The main difference between balancing ionic and balancing molecular redox equations is in the handling of ions. In addition to having the same number of each kind of element on both sides of the final equation, the net charges must also be equal. In assigning oxidation numbers we must be careful to consider the charge on the ions. In many respects, balancing ionic equations is much simpler than balancing molecular equations.

Several methods can be used to balance ionic redox equations. These methods include, with slight modification, the oxidation-number method just shown for molecular equations. But the most popular is probably the ion–electron method, which is explained in the following paragraphs.

The ion–electron method uses ionic charges and electrons to balance ionic redox equations. Oxidation numbers are not formally used, but it is necessary to determine what is being oxidized and what is being reduced. The method is as follows:

1. Write the two half-reactions that contain the elements being oxidized and reduced.
2. Balance the elements other than oxygen and hydrogen.
3. Balance oxygen and hydrogen: (a) acidic solutions; (b) basic solutions.
   (a) For reactions that occur in acidic solution, use H^+ and H_2O to balance oxygen and hydrogen. For each oxygen needed, use one H_2O. Then add H^+ as needed to balance the hydrogen atoms.
   (b) Balancing equations that occur in alkaline solutions is a bit more complicated. For reactions that occur in alkaline solutions, first balance as though the reaction were in an acidic solution, using Steps 1, 2, and 3(a). Then add as many OH^- ions to each side of the equation as there are H^+ ions in the equation. Now combine the H^+ and OH^- ions into water (for example, 4 H^+ and 4 OH^- give 4 H_2O). Rewrite the equation, cancelling equal numbers of water molecules that appear on opposite sides of the equation. Now proceed to Step 4 for electrical balance.

4 Add electrons (e^-) to each half-reaction to bring them into electrical balance.
5 Since the loss and gain of electrons must be equal, multiply each half-reaction by the appropriate number to make the number of electrons the same in each half-reaction.
6 Add the two half-reactions together, canceling electrons and any other identical substances that appear on opposite sides of the equation.

PROBLEM 17.5 Balance this equation using the ion–electron method:

$$MnO_4^- + S^{2-} \longrightarrow Mn^{2+} + S^0 \quad \text{(acidic solution)}$$

Step 1 Write two half-reactions, one containing the element being oxidized and the other, the element being reduced.

oxidation $S^{2-} \longrightarrow S^0$

reduction $MnO_4^- \longrightarrow Mn^{2+}$

Step 2 Balance elements other than oxygen and hydrogen (accomplished in Step 1 in this example—1 S and 1 Mn on each side).

Step 3 Balance O and H. The oxidation requires neither O nor H, but the reduction equation needs 4 H_2O on the right and 8 H^+ on the left [Step 3(a)].

$$S^{2-} \longrightarrow S^0$$

$$8\,H^+ + MnO_4^- \longrightarrow Mn^{2+} + 4\,H_2O$$

Step 4 Balance electrically with electrons.

$$S^{2-} \longrightarrow S^0 + 2\,e^- \quad \text{(net charge} = -2 \text{ on each side)}$$

$$5\,e^- + 8\,H^+ + MnO_4^- \longrightarrow Mn^{2+} + 4\,H_2O \quad \text{(net charge} = +2 \text{ on each side)}$$

Step 5 Equalize loss and gain of electrons. In this case multiply the oxidation equation by 5 and the reduction equation by 2.

$$5\,S^{2-} \longrightarrow 5\,S^0 + 10\,e^-$$

$$10\,e^- + 16\,H^+ + 2\,MnO_4^- \longrightarrow 2\,Mn^{2+} + 8\,H_2O$$

Step 6 Add the two half-reactions together, cancelling the 10 e^- from each side, to obtain the balanced equation.

$$5\,S^{2-} \longrightarrow 5\,S^0 + \cancel{10\,e^-}$$

$$\cancel{10\,e^-} + 16\,H^+ + 2\,MnO_4^- \longrightarrow 2\,Mn^{2+} + 8\,H_2O$$

$$16\,H^+ + 2\,MnO_4^- + 5\,S^{2-} \longrightarrow 2\,Mn^{2+} + 5\,S^0 + 8\,H_2O \quad \text{(balanced)}$$

Check: Both sides of the equation have a charge of +4 and contain the same number of atoms of each element.

PROBLEM 17.6 Balance the following equation.

$$CrO_4^{2-} + Fe(OH)_2 \longrightarrow Cr(OH)_3 + Fe(OH)_3 \quad \text{(basic solution)}$$

Step 1 Write the two half-reactions.

oxidation $Fe(OH)_2 \longrightarrow Fe(OH)_3$

reduction $CrO_4^{2-} \longrightarrow Cr(OH)_3$

Step 2 Balance elements other than H and O (accomplished in Step 1).

Step 3 Balance O and H as though the solution were acidic (Step 3a). Use H_2O and H^+. To balance O and H in the oxidation equation, add 1 H_2O on the left and 1 H^+ on the right side of the equation.

$$Fe(OH)_2 + H_2O \longrightarrow Fe(OH)_3 + H^+$$

Add 1 OH^- to each side.

$$Fe(OH)_2 + H_2O + OH^- \longrightarrow Fe(OH)_3 + H^+ + OH^-$$

Combine H^+ and OH^- as H_2O and rewrite, cancelling H_2O on each side [Step 3(b)].

$$Fe(OH)_2 + \cancel{H_2O} + OH^- \longrightarrow Fe(OH)_3 + \cancel{H_2O}$$

$$\boxed{Fe(OH)_2 + OH^- \longrightarrow Fe(OH)_3}$$

To balance O and H in the reduction equation, add 1 H_2O on the right and 5 H^+ on the left.

$$CrO_4^{2-} + 5\,H^+ \longrightarrow Cr(OH)_3 + H_2O$$

Add 5 OH^- to each side.

$$CrO_4^{2-} + 5\,H^+ + 5\,OH^- \longrightarrow Cr(OH)_3 + H_2O + 5\,OH^-$$

Combine $5\,H^+ + 5\,OH^- \longrightarrow 5\,H_2O$.

$$CrO_4^{2-} + 5\,H_2O \longrightarrow Cr(OH)_3 + H_2O + 5\,OH^-$$

Rewrite, cancelling 1 H_2O from each side.

$$\boxed{CrO_4^{2-} + 4\,H_2O \longrightarrow Cr(OH)_3 + 5\,OH^-}$$

Step 4 Balance electrically with electrons.

$$Fe(OH)_2 + OH^- \longrightarrow Fe(OH)_3 + e^- \quad \text{(balanced oxidation equation)}$$

$$CrO_4^{2-} + 4\,H_2O + 3\,e^- \longrightarrow Cr(OH)_3 + 5\,OH^- \quad \text{(balanced reduction equation)}$$

Step 5 Equalize the loss and gain of electrons. Multiply the oxidation reaction by 3.

$$3\ Fe(OH)_2 + 3\ OH^- \longrightarrow 3\ Fe(OH)_3 + 3\ e^-$$
$$CrO_4^{2-} + 4\ H_2O + 3\ e^- \longrightarrow Cr(OH)_3 + 5\ OH^-$$

Step 6 Add the two half-reactions together, cancelling the 3 e⁻ and 3 OH^- from each side of the equation.

$$3\ Fe(OH)_2 + 3\ OH^- \longrightarrow 3\ Fe(OH)_3 + 3\ e^-$$
$$CrO_4^{2-} + 4\ H_2O + 3\ e^- \longrightarrow Cr(OH)_3 + 5\ OH^-$$
$$CrO_4^{2-} + 3\ Fe(OH)_2 + 4\ H_2O \longrightarrow Cr(OH)_3 + 3\ Fe(OH)_3 + 2\ OH^- \quad \text{(balanced)}$$

Check: Each side of the equation has a charge of −2 and contains the same number of atoms of each element.

Ionic equations can also be balanced by using the electron-transfer method shown in Section 17.3. Steps 1, 2, 3, and 4 are the same. In Step 5 the electrical charges are balanced with H^+ in acidic solutions and with OH^- in basic solutions. H_2O is then used to complete the balancing of the equation. Let us use the same equation as in Problem 17.6 as an example of this method.

PROBLEM 17.7 Balance the following equation using the electron-transfer method.

$$CrO_4^{2-} + Fe(OH)_2 \longrightarrow Cr(OH)_3 + Fe(OH)_3 \quad \text{(basic solution)}$$

Steps 1 and 2 Assign oxidation numbers and balance the charges with electrons.

reduction $Cr^{6+} + 3\ e^- \longrightarrow Cr^{3+}$ (Cr^{6+} gains 3 e^-)

oxidation $Fe^{2+} \longrightarrow Fe^{3+} + e^-$ (Fe^{2+} loses 1 e^-)

Step 3 Equalize the loss and gain of electrons. Multiply the oxidation step by 3.

$3\ Fe^{2+} \longrightarrow 3\ Fe^{3+} + 3\ e^-$ (3 Fe^{2+} lose 3 e^-)

$Cr^{6+} + 3\ e^- \longrightarrow Cr^{3+}$ (Cr^{6+} gains 3 e^-)

Step 4 Transfer coefficients back to the original equation.

$$CrO_4^{2-} + 3\ Fe(OH)_2 \longrightarrow Cr(OH)_3 + 3\ Fe(OH)_3$$

Step 5 Balance electrically. Because the solution is basic, use OH^- to balance charges. The charge on the left side is −2, and on the right side is 0. Add 2 OH^- ions to the right side of the equation.

$$CrO_4^{2-} + 3\ Fe(OH)_2 \longrightarrow Cr(OH)_3 + 3\ Fe(OH)_3 + 2\ OH^-$$

Adding 4 H_2O to the left side balances the equation.

$$CrO_4^{2-} + 3\ Fe(OH)_2 + 4\ H_2O \longrightarrow Cr(OH)_3 + 3\ Fe(OH)_3 + 2\ OH^- \quad \text{(balanced)}$$

Check: Each side of the equation has a charge of -2 and contains the same number of atoms of each element.

17.5 Activity Series of Metals

Knowledge of the relative chemical reactivities of the elements helps us predict the course of many chemical reactions.

Calcium reacts with cold water to produce hydrogen, and magnesium reacts with steam to produce hydrogen. Therefore, calcium is considered to be a more reactive metal than magnesium.

$$Ca(s) + 2\ H_2O(l) \longrightarrow Ca(OH)_2(aq) + H_2(g)$$

$$Mg(s) + \underset{\text{Steam}}{H_2O(g)} \longrightarrow MgO(s) + H_2(g)$$

The difference in their activity is attributed to the fact that calcium loses its two valence electrons more easily than does magnesium and is therefore more reactive and/or more readily oxidized than magnesium.

When a strip of copper is placed in a solution of silver nitrate ($AgNO_3$), free silver begins to plate out on the copper. After the reaction has continued for some time, we can observe a blue color in the solution, indicating the presence of copper(II) ions. If a strip of silver is placed in a solution of copper(II) nitrate, $Cu(NO_3)_2$, no reaction is visible. The equations are

$$Cu^0 + 2\ AgNO_3(aq) \longrightarrow 2\ Ag^0 + Cu(NO_3)_2(aq)$$

$$Cu^0 + 2\ Ag^+ \longrightarrow 2\ Ag^0 + Cu^{2+} \qquad \text{net ionic equation}$$

$$Cu^0 \longrightarrow Cu^{2+} + 2\ e^- \qquad \text{oxidation of } Cu^0$$

$$Ag^+ + e^- \longrightarrow Ag^0 \qquad \text{reduction of } Ag^+$$

$$Ag^0 + Cu(NO_3)_2(aq) \longrightarrow \text{no reaction}$$

In the reaction between Cu and $AgNO_3$, electrons are transferred from Cu^0 atoms to Ag^+ ions in solution. Copper has a greater tendency than silver to lose electrons, so an electrochemical force is exerted upon silver ions to accept electrons from copper atoms. When a Ag^+ ion accepts an electron, it is reduced to a Ag^0 atom and is no longer soluble in solution. At the same time, Cu^0 is oxidized and goes into solution as Cu^{2+} ions. From this reaction we can conclude that copper is more reactive than silver.

Metals such as sodium, magnesium, zinc, and iron, which react with solutions of acids to liberate hydrogen, are more reactive than hydrogen. Metals such as copper, silver, and mercury, which do not react with solutions of acids to

Table 17.3 Activity Series of Metals

Ease of oxidation (increasing upward ↑)
$K \longrightarrow K^+ + e^-$
$Ba \longrightarrow Ba^{2+} + 2e^-$
$Ca \longrightarrow Ca^{2+} + 2e^-$
$Na \longrightarrow Na^+ + e^-$
$Mg \longrightarrow Mg^{2+} + 2e^-$
$Al \longrightarrow Al^{3+} + 3e^-$
$Zn \longrightarrow Zn^{2+} + 2e^-$
$Cr \longrightarrow Cr^{3+} + 3e^-$
$Fe \longrightarrow Fe^{2+} + 2e^-$
$Ni \longrightarrow Ni^{2+} + 2e^-$
$Sn \longrightarrow Sn^{2+} + 2e^-$
$Pb \longrightarrow Pb^{2+} + 2e^-$
$H_2 \longrightarrow 2H^+ + 2e^-$
$Cu \longrightarrow Cu^{2+} + 2e^-$
$As \longrightarrow As^{3+} + 3e^-$
$Ag \longrightarrow Ag^+ + e^-$
$Hg \longrightarrow Hg^{2+} + 2e^-$
$Au \longrightarrow Au^{3+} + 3e^-$

liberate hydrogen, are less reactive than hydrogen. By studying a series of reactions such as those given above, we can list metals according to their chemical activity, placing the most active at the top and the least active at the bottom. This list is called the **Activity Series of Metals**. Table 17.3 shows some of the common metals in the series. The arrangement corresponds to the ease with which the elements listed are oxidized or lose electrons, with the most easily oxidizable element listed first. More extensive tables are available in chemistry reference books.

Activity Series of Metals

The general principles governing the arrangement and use of the Activity Series are as follows:

1. The reactivity of the metals listed decreases from top to bottom.
2. A free metal can displace the ion of a second metal from solution provided that the free metal is above the second metal in the Activity Series.
3. Free metals above hydrogen react with nonoxidizing acids in solution to liberate hydrogen gas.
4. Free metals below hydrogen do not liberate hydrogen from acids.
5. Conditions such as temperature and concentration may affect the relative position of some of these elements.

Two examples of the application of the Activity Series are given in the following problems.

PROBLEM 17.8 Will zinc metal react with dilute sulfuric acid?

From Table 17.3 we see that zinc is above hydrogen; therefore zinc atoms will lose electrons more readily than hydrogen atoms. Hence, zinc atoms will reduce hydrogen ions from the acid to form hydrogen gas and zinc ions. In fact, these reagents are commonly used for the laboratory preparation of hydrogen. The equation is

$$Zn(s) + H_2SO_4(aq) \longrightarrow ZnSO_4(aq) + H_2(g)$$

$$Zn + 2\,H^+ \longrightarrow Zn^{2+} + H_2(g) \qquad \text{net ionic equation}$$

PROBLEM 17.9 Will a reaction occur when copper metal is placed in an iron(II) sulfate solution?

No, copper lies below iron in the series, loses electrons less easily than iron, and therefore will not displace iron(II) ions from solution. In fact, the reverse is true. When an iron nail is dipped into a copper(II) sulfate solution, it becomes coated with free copper. The equations are

$$Cu(s) + FeSO_4(aq) \longrightarrow \text{no reaction}$$

$$Fe(s) + CuSO_4(aq) \longrightarrow FeSO_4(aq) + Cu\downarrow$$

From Table 17.3 we may abstract the following pair in their relative position to each other.

$$Fe \longrightarrow Fe^{2+} + 2\,e^-$$

$$Cu \longrightarrow Cu^{2+} + 2\,e^-$$

According to the second principle listed above on the use of the Activity Series, we can predict that free iron will react with copper(II) ions in solution to form free copper metal and iron(II) ions in solution.

$$Fe(s) + Cu^{2+}(aq) \longrightarrow Fe^{2+}(aq) + Cu(s) \qquad \text{net ionic equation}$$

17.6 Electrolytic and Voltaic Cells

electrolysis

electrolytic cell

cathode

anode

The process in which electrical energy is used to bring about chemical change is known as **electrolysis.** An **electrolytic cell** uses electrical energy to produce a nonspontaneous chemical reaction. The use of electrical energy has many applications in the chemical industry—for example, in the production of sodium, sodium hydroxide, chlorine, fluorine, magnesium, aluminum, and pure hydrogen and oxygen, and in the purification and electroplating of metals.

What happens when an electric current is passed through a solution? Let us consider a hydrochloric acid solution in a simple electrolytic cell, as shown in Figure 17.2. The cell consists of a source of direct current (a battery) connected to two electrodes that are immersed in a solution of hydrochloric acid. The negative electrode is called the **cathode** because cations are attracted to it. The positive electrode is called the **anode** because anions are attracted to it. The cathode is

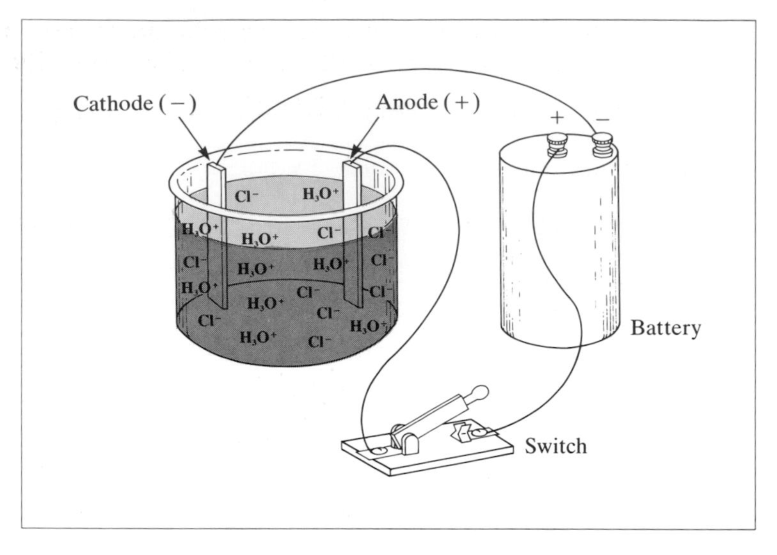

Figure 17.2

Electrolysis. During the electrolysis of a hydrochloric acid solution, positive hydronium ions are attracted to the cathode, where they gain electrons and form hydrogen gas. Chloride ions migrate to the anode, where they lose electrons and form chlorine gas. The equation for this process is $2\ HCl(aq) \longrightarrow H_2(g) + Cl_2(g)$.

attached to the negative pole and the anode to the positive pole of the battery. The battery supplies electrons to the cathode.

When the switch is closed, the electric circuit is completed; positive hydronium ions (H_3O^+) migrate to the cathode, where they pick up electrons and evolve hydrogen gas. At the same time the negative chloride ions (Cl^-) migrate to the anode, where they lose electrons and evolve chlorine gas.

Reaction at the cathode	$H_3O^+ + 1\ e^- \longrightarrow H^0 + H_2O$
(reduction)	$H^0 + H^0 \longrightarrow H_2(g)$
Reaction at the anode	$Cl^- \longrightarrow Cl^0 + 1\ e^-$
(oxidation)	$Cl^0 + Cl^0 \longrightarrow Cl_2(g)$
Net reaction	$2\ HCl(aq) \xrightarrow{\text{Electrolysis}} H_2(g) + Cl_2(g)$

Note that oxidation–reduction has taken place. Chloride ions lose electrons (are oxidized) at the anode, and hydronium ions gain electrons (are reduced) at the cathode.

Oxidation always occurs at the anode and reduction at the cathode.

When concentrated sodium chloride solutions (brines) are electrolyzed, the products are sodium hydroxide, hydrogen, and chlorine. The overall reaction is

$$2\ Na^+ + 2\ Cl^- + 2\ H_2O \xrightarrow{\text{Electrolysis}} 2\ Na^+ + 2\ OH^- + H_2(g) + Cl_2(g)$$

The net ionic equation is

$$2\ Cl^-(aq) + 2\ H_2O(l) \longrightarrow 2\ OH^-(aq) + H_2(g) + Cl_2(g)$$

During the electrolysis, Na^+ ions move toward the cathode and Cl^- ions move toward the anode. The anode reaction is similar to that of hydrochloric acid; chlorine is liberated.

$$2\ Cl^-(aq) \longrightarrow Cl_2(g) + 2\ e^-$$

Even though Na^+ ions are attracted by the cathode, the facts show that hydrogen is liberated there. No evidence of metallic sodium is found, but the area around the cathode tests alkaline from accumulated OH^- ions. The reaction at the cathode is

$$2\ H_2O(l) + 2\ e^- \longrightarrow H_2(g) + 2\ OH^-(aq)$$

If the electrolysis is allowed to continue until all the chloride is reacted, the solution remaining will contain only sodium hydroxide, which on evaporation yields solid NaOH. Large tonnages of sodium hydroxide and chlorine are made by this process.

When molten sodium chloride (without water) is subjected to electrolysis, metallic sodium and chlorine gas are formed:

$$2\ Na^+(l) + 2\ Cl^-(l) \xrightarrow{\text{Electrolysis}} 2\ Na(l) + Cl_2(g)$$

An important electrochemical application is the electroplating of metals. Electroplating is the art of covering a surface or an object with a thin adherent electrodeposited metal coating. Electroplating is done for protection of the surface of the base metal or for a purely decorative effect. The layer deposited is surprisingly thin, varying from as little as 5×10^{-5} cm to 2×10^{-3} cm, depending on the metal and the intended use. The object to be plated is set up as the cathode and is immersed in a solution containing ions of the metal to be plated. When an electric current passes through the solution, metal ions that migrate to the cathode are reduced, depositing on the object as the free metal. In most cases the metal deposited on the object is replaced in the solution by using an anode of the same metal. The following equations show the chemical changes in the electroplating of nickel:

Reaction at the cathode	$Ni^{2+}(aq) + 2\ e^- \longrightarrow Ni(s)$	Ni plated out on an object
Reaction at the anode	$Ni(s) \longrightarrow Ni^{2+}(aq) + 2\ e^-$	Ni replenished in solution

Metals commonly used in commercial electroplating are copper, nickel, zinc, lead, cadmium, chromium, tin, gold, and silver.

In the electrolytic cell shown in Figure 17.2, electrical energy from the battery is used to bring about nonspontaneous redox reactions. The hydrogen and chlorine produced have more potential energy than was present in the hydrochloric acid before electrolysis.

Conversely, some spontaneous redox reactions can be made to supply useful amounts of electrical energy. When a piece of zinc is put in a copper(II) sulfate solution, the zinc quickly becomes coated with metallic copper. We expect this coating to happen because zinc is above copper in the Activity Series; copper(II) ions are therefore reduced by zinc atoms:

$$Zn^0(s) + Cu^{2+}(aq) \longrightarrow Zn^{2+}(aq) + Cu^0(s)$$

This reaction is clearly a spontaneous redox reaction, but simply dipping a zinc rod into a copper(II) sulfate solution will not produce useful electric current. However, when we carry out this reaction in the cell shown in Figure 17.3, an electric current is produced. The cell consists of a piece of zinc immersed in a zinc sulfate solution and connected by a wire through a voltmeter to a piece of copper immersed in copper(II) sulfate solution. The two solutions are connected by a salt

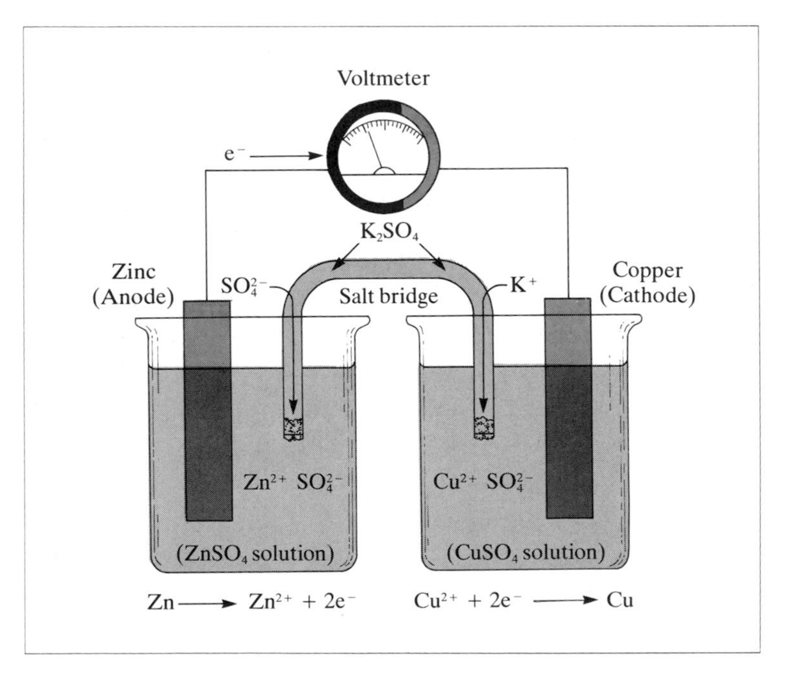

Figure 17.3

Zinc–copper voltaic cell. The cell has a potential of 1.1 volts when $ZnSO_4$ and $CuSO_4$ solutions are 1.0 *M*. The salt bridge provides electrical contact between the two half-cells.

bridge. Such a cell produces an electric current and a potential of about 1.1 volts when both solutions are 1.0 *M* in concentration. A cell that produces electric current from a spontaneous chemical reaction is called a **voltaic cell**. A voltaic cell is also known as a *galvanic cell.*

voltaic cell

The driving force responsible for the electric current in the zinc–copper cell originates in the great tendency of zinc atoms to lose electrons relative to the tendency of copper(II) ions to gain electrons. In the cell shown in Figure 17.3 zinc atoms lose electrons and are converted to zinc ions at the zinc electrode surface; the electrons flow through the wire (external circuit) to the copper electrode. Here copper(II) ions pick up electrons and are reduced to copper atoms, which plate out on the copper electrode. Sulfate ions flow from the $CuSO_4$ solution via the salt bridge into the $ZnSO_4$ solution (internal circuit) to complete the circuit. The equations for the reactions of this cell are

anode	$Zn^0(s) \longrightarrow Zn^{2+}(aq) + 2\,e^-$	(oxidation)
cathode	$Cu^{2+}(aq) + 2\,e^- \longrightarrow Cu^0(s)$	(reduction)
net ionic	$Zn^0(s) + Cu^{2+}(aq) \longrightarrow Zn^{2+}(aq) + Cu^0(s)$	
overall	$Zn(s) + CuSO_4(aq) \longrightarrow Cu(s) + ZnSO_4(aq)$	

The redox reaction, the movement of electrons in the metallic or external part of the circuit, and the movement of ions in the solution or internal part of the circuit of the copper–zinc cell are very similar to the actions that occur in the electrolytic cell of Figure 17.2. The only important difference is that the reactions of the zinc–copper cell are spontaneous. This spontaneity is the crucial difference between all voltaic and electrolytic cells.

Voltaic cells use chemical reactions to produce electrical energy, and electrolytic cells use electrical energy to produce chemical reactions.

Although the zinc–copper voltaic cell is no longer used commercially, it was used to energize the first transcontinental telegraph lines. Such cells were the direct ancestors of the many different kinds of "dry" cells that operate portable radio and television sets, automatic cameras, tape recorders, and so on.

One such "dry" cell, the alkaline zinc–mercury cell, is shown diagrammatically in Figure 17.4. The reactions occurring in this cell are

anode	$Zn^0 + 2\,OH^- \longrightarrow ZnO + H_2O + 2\,e^-$	(oxidation)
cathode	$HgO + H_2O + 2\,e^- \longrightarrow Hg^0 + 2\,OH^-$	(reduction)
net ionic	$Zn^0 + Hg^{2+} \longrightarrow Zn^{2+} + Hg^0$	
overall	$Zn^0 + HgO \longrightarrow ZnO + Hg^0$	

To offset the relatively high initial cost, this cell (a) provides current at a very steady potential of about 1.5 volts, (b) has an exceptionally long service life—that is, high energy output to weight ratio, (c) is completely self-contained, and (d) can be stored for relatively long periods of time when not in use.

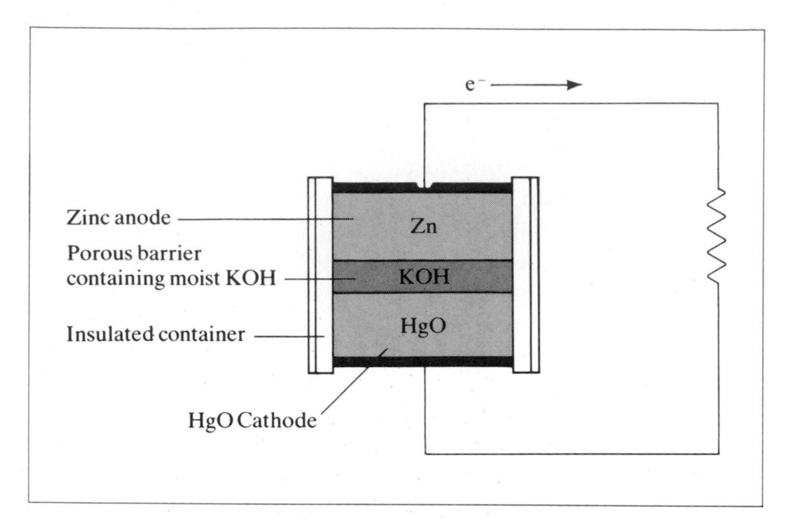

Figure 17.4
Diagram of an alkaline zinc–mercury cell

An automobile storage battery is an energy reservoir. The charged battery acts as a voltaic cell and through chemical reactions furnishes electrical energy to operate the starter, lights, radio, and so on. When the engine is running, a generator, or alternator, produces and forces an electric current through the battery and, by electrolytic chemical action, restores the battery to the charged condition.

The cell unit consists of a lead plate filled with spongy lead and a lead dioxide plate, both immersed in dilute sulfuric acid solution, which serves as the electrolyte (see Figure 17.5). When the cell is discharging, or acting as a voltaic cell, these reactions occur:

Pb plate (anode)	$Pb^0 \longrightarrow Pb^{2+} + 2\,e^-$	(oxidation)
PbO_2 plate (cathode)	$PbO_2 + 4\,H^+ + 2\,e^- \longrightarrow Pb^{2+} + 2\,H_2O$	(reduction)
Net ionic redox reaction	$Pb^0 + PbO_2 + 4\,H^+ \longrightarrow 2\,Pb^{2+} + 2\,H_2O$	
Precipitation reaction on plates	$Pb^{2+} + SO_4^{2-} \longrightarrow PbSO_4(s)$	

Because lead(II) sulfate is insoluble, the Pb^{2+} ions combine with SO_4^{2-} ions to form a coating of $PbSO_4$ on each plate. The overall chemical reaction of the cell is

$$Pb(s) + PbO_2(s) + 2\,H_2SO_4(aq) \xrightarrow[\text{cycle}]{\text{Discharge}} 2\,PbSO_4(s) + 2\,H_2O(l)$$

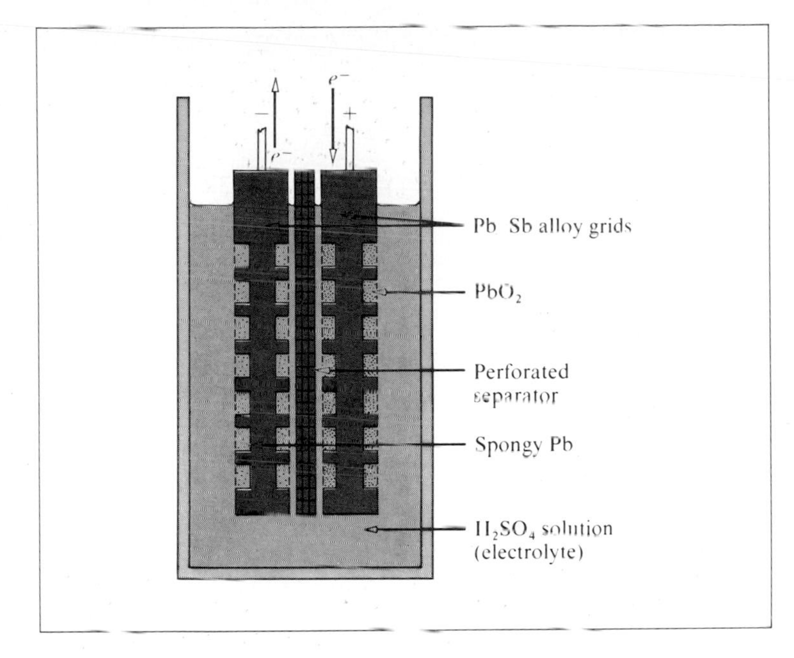

Figure 17.5
Cross-sectional diagram of a lead storage battery cell

The cell can be recharged by reversing the chemical reaction. This reversal is accomplished by forcing an electric current through the cell in the opposite direction. Lead sulfate and water are reconverted to lead, lead dioxide, and sulfuric acid:

$$2\ PbSO_4(s) + 2\ H_2O(l) \xrightarrow[\text{cycle}]{\text{Charge}} Pb(s) + PbO_2(s) + 2\ H_2SO_4(aq)$$

The electrolyte in a lead storage battery is a 38% by weight sulfuric acid solution having a density of 1.29 g/mL. As the battery is discharged, sulfuric acid is removed, thereby decreasing the density of the electrolyte solution. The state of charge or discharge of the battery can be estimated by measuring the density (or specific gravity) of the electrolyte solution with a hydrometer. When the density has dropped to about 1.05 g/mL, the battery needs recharging.

In a commercial battery, each cell consists of a series of cell units of alternating lead–lead dioxide plates separated and supported by wood, glass wool, or fiberglass. The energy storage capacity of a single cell is limited, and its electrical potential is only about 2 volts. Therefore, a bank of six cells is connected in series to provide the 12 volt output of the usual automobile battery.

QUESTIONS

A. *Review the meanings of the new terms introduced in this chapter.*

1. Oxidation–reduction
2. Redox
3. Oxidation
4. Reduction
5. Oxidizing agent
6. Reducing agent
7. Activity Series of Metals
8. Electrolysis
9. Electrolytic cell
10. Cathode
11. Anode
12. Voltaic cell

B. *Information useful in answering the following questions will be found in the tables and figures.*

1. In the equation

$$I_2 + 5\,Cl_2 + 6\,H_2O \longrightarrow 2\,HIO_3 + 10\,HCl$$

   (a) Has iodine been oxidized or has it been reduced?
   (b) Has chlorine been oxidized or has it been reduced? (See Figure 17.1.)
2. Based on Table 17.3, which element of each pair is more active?
   (a) Ag or Al (b) Na or Ba (c) Ni or Cu
3. Based on Table 17.3, will the following combinations react in aqueous solution?
   (a) $Zn + Cu^{2+}$ (e) $Ba + FeCl_2$
   (b) $Ag + H^+$ (f) $Pb + NaCl$
   (c) $Sn + Ag^+$ (g) $Ni + Hg(NO_3)_2$
   (d) $As + Mg^{2+}$ (h) $Al + CuSO_4$
4. The reaction between powdered aluminum and iron(III) oxide (in the Thermite process), producing molten iron, is very exothermic.
   (a) Write the equation for the chemical reaction that occurs.
   (b) Explain in terms of Table 17.3 why a reaction occurs.
   (c) Would you expect a reaction between powdered iron and aluminum oxide?
   (d) Would you expect a reaction between powdered aluminum and chromium(III) oxide?
5. Write equations for the chemical reaction of each of the following metals with dilute solutions of (a) hydrochloric acid and (b) sulfuric acid: aluminum, chromium, gold, iron, copper, magnesium, mercury, and zinc. If a reaction will not occur, write "no reaction" as the product. (See Table 17.3.)
6. A $NiCl_2$ solution is placed in the apparatus shown in Figure 17.2, instead of the HCl solution shown. Write equations for:
   (a) The anode reaction
   (b) The cathode reaction
   (c) The net electrochemical reaction
7. What is the major distinction between the reactions occurring in Figure 17.2 and those in Figure 17.3?
8. In the cell shown in Figure 17.3:
   (a) What would be the effect of removing the voltmeter and connecting the wires shown coming to the voltmeter?
   (b) What would be the effect of removing the salt bridge?

C. *Review questions*

1. What is the oxidation number of the underlined element in each compound:
   (a) $\underline{Na}Cl$ (e) $H_2\underline{S}O_3$ (i) $\underline{N}H_3$
   (b) $Fe\underline{Cl}_3$ (f) $\underline{N}H_4Cl$ (j) $K\underline{Cl}O_3$
   (c) $\underline{Pb}O_2$ (g) $K\underline{Mn}O_4$ (k) $K_2\underline{Cr}O_4$
   (d) $Na\underline{N}O_3$ (h) $\underline{I}_2$ (l) $K_2\underline{Cr}_2O_7$
2. What is the oxidation number of the underlined elements?
   (a) $\underline{S}^{2-}$ (d) $\underline{Mn}O_4^-$ (g) $\underline{As}O_4^{3-}$
   (b) $\underline{N}O_2^-$ (e) $\underline{Bi}^{3+}$ (h) $Fe(\underline{O}H)_3$
   (c) $Na_2\underline{O}_2$ (f) $\underline{O}_2$ (i) $\underline{I}O_3^-$
3. In the following half-reactions, which element is changing oxidation state? Is the half-reaction an oxidation or a reduction? Supply the proper number of electrons to the proper side to balance each equation.
   (a) $Zn^{2+} \longrightarrow Zn$
   (b) $2\,Br^- \longrightarrow Br_2$
   (c) $MnO_4^- + 8\,H^+ \longrightarrow Mn^{2+} + 4\,H_2O$
   (d) $Ni \longrightarrow Ni^{2+}$
   (e) $SO_3^{2-} + H_2O \longrightarrow SO_4^{2-} + 2\,H^+$
   (f) $NO_3^- + 4\,H^+ \longrightarrow NO + 2\,H_2O$

(g) $S_2O_4^{2-} + 2\,H_2O \longrightarrow 2\,SO_3^{2-} + 4\,H^+$
(h) $Fe^{2+} \longrightarrow Fe^{3+}$

4. In the following unbalanced equations,
(a) Identify the element that is oxidized and the element that is reduced.
(b) Identify the oxidizing agent and the reducing agent.
(1) $Cr + HCl \longrightarrow CrCl_3 + H_2$
(2) $SO_4^{2-} + I^- + H^+ \longrightarrow H_2S + I_2 + H_2O$
(3) $AsH_3 + Ag^+ + H_2O \longrightarrow H_3AsO_4 + Ag + H^+$
(4) $Cl_2 + NaBr \longrightarrow NaCl + Br_2$

5. Balance these equations by the change-in-oxidation-number method.
(a) $Zn + S \longrightarrow ZnS$
(b) $AgNO_3 + Pb \longrightarrow Pb(NO_3)_2 + Ag$
(c) $Fe_2O_3 + CO \longrightarrow Fe + CO_2$
(d) $H_2S + HNO_3 \longrightarrow S + NO + H_2O$
(e) $MnO_2 + HBr \longrightarrow MnBr_2 + Br_2 + H_2O$
(f) $Cl_2 + KOH \longrightarrow KCl + KClO_3$
(g) $Ag + HNO_3 \longrightarrow AgNO_3 + NO + H_2O$
(h) $CuO + NH_3 \longrightarrow N_2 + Cu + H_2O$
(i) $PbO_2 + Sb + NaOH \longrightarrow PbO + NaSbO_2 + H_2O$
(j) $H_2O_2 + KMnO_4 + H_2SO_4 \longrightarrow O_2 + MnSO_4 + K_2SO_4 + H_2O$

6. Balance the following ionic redox equations using the ion–electron method. All reactions occur in acidic solution.
(a) $Zn + NO_3^- \longrightarrow Zn^{2+} + NH_4^+$
(b) $NO_3^- + S \longrightarrow NO_2 + SO_4^{2-}$
(c) $PH_3 + I_2 \longrightarrow H_3PO_2 + I^-$
(d) $Cu + NO_3^- \longrightarrow Cu^{2+} + NO$
(e) $ClO_3^- + I^- \longrightarrow I_2 + Cl^-$
(f) $Cr_2O_7^{2-} + Fe^{2+} \longrightarrow Cr^{3+} + Fe^{3+}$
(g) $MnO_4^- + SO_2 \longrightarrow Mn^{2+} + SO_4^{2-}$
(h) $H_3AsO_3 + MnO_4^- \longrightarrow H_3AsO_4 + Mn^{2+}$
*(i) $ClO_3^- + Cl^- \longrightarrow Cl_2$
*(j) $Cr_2O_7^{2-} + H_3AsO_3 \longrightarrow Cr^{3+} + H_3AsO_4$

7. Balance the following ionic redox equations using the ion–electron method. All reactions occur in basic solutions.
(a) $Cl_2 + IO_3^- \longrightarrow Cl^- + IO_4^-$
(b) $MnO_4^- + ClO_2^- \longrightarrow MnO_2 + ClO_4^-$
(c) $Se \longrightarrow Se^{2-} + SeO_3^{2-}$
(d) $MnO_4^- + SO_3^{2-} \longrightarrow MnO_2 + SO_4^{2-}$
(e) $ClO_2(g) + SbO_2^- \longrightarrow ClO_2^- + Sb(OH)_6^-$
*(f) $Fe_3O_4 + MnO_4^- \longrightarrow Fe_2O_3 + MnO_2$
*(g) $BrO^- + Cr(OH)_4^- \longrightarrow Br^- + CrO_4^{2-}$
*(h) $P_4 \longrightarrow HPO_3^{2-} + PH_3$
*(i) $Al + OH^- \longrightarrow Al(OH)_4^- + H_2$
*(j) $Al + NO_3^- \longrightarrow NH_3 + Al(OH)_4$

8. Why are oxidation and reduction said to be complementary processes?

9. When molten $CaBr_2$ is electrolyzed, calcium metal and bromine are produced. Write equations for the two half-reactions that occur at the electrodes. Label the anode half-reaction and the cathode half-reaction.

10. Why is direct current used instead of alternating current in the electroplating of metals?

11. The chemical reactions taking place during discharge in a lead storage battery are

$$Pb + SO_4^{2-} \longrightarrow PbSO_4$$

$$PbO_2 + SO_4^{2-} + 4\,H^+ \longrightarrow PbSO_4 + 2\,H_2O$$

(a) Complete each half-reaction by supplying electrons.
(b) Which reaction is oxidation and which is reduction?
(c) Which reaction occurs at the anode of the battery?

12. What property of lead dioxide and lead(II) sulfate makes it unnecessary to have salt bridges in the cells of a lead storage battery?

13. Explain why the density of the electrolyte in a lead storage battery decreases during the discharge cycle.

14. In one type of alkaline cell used to power devices such as portable radios, Hg^{2+} ions are reduced to metallic mercury when the cell is being discharged. Does this reduction occur at the anode or the cathode? Explain.

15. Differentiate between an electrolytic cell and a voltaic cell.

16. Why is a porous barrier or a salt bridge necessary in some voltaic cells?

17. Which of the following statements are correct?
(a) An atom of an element in the uncombined state has an oxidation number of zero.
(b) The oxidation number of molybdenum in Na_2MoO_4 is +4.
(c) The oxidation number of an ion is the same as the electrical charge on the ion.
(d) The process in which an atom or an ion loses electrons is called reduction.
(e) The reaction $Fe^{3+} + e^- \longrightarrow Fe^{2+}$ is a reduction reaction.

(f) In the reaction

$$2\,Al + 3\,CuCl_2 \longrightarrow 2\,AlCl_3 + 3\,Cu$$

aluminum is the oxidizing agent.

(g) In a redox reaction the oxidizing agent is reduced, and the reducing agent is oxidized.

(h) $Cu^0 \longrightarrow Cu^{2+}$ is a balanced oxidation half-reaction.

(i) In the electrolysis of sodium chloride brine (solution), Cl_2 gas is formed at the cathode, and hydroxide ions are formed at the anode.

(j) In any cell, electrolytic or voltaic, reduction takes place at the cathode, and oxidation occurs at the anode.

(k) In the Zn–Cu voltaic cell, the reaction at the anode is $Zn \longrightarrow Zn^{2+} + 2\,e^-$.

The statements in (l) through (o) pertain to this Activity Series:

Ba Mg Zn Fe H Cu Ag

(l) The reaction $Zn + MgCl_2 \rightarrow Mg + ZnCl_2$ is a spontaneous reaction.

(m) Barium is more active than copper.

(n) Silver metal will react with acids to liberate hydrogen gas.

(o) Iron is a better reducing agent than zinc.

(p) Oxidation and reduction occur simultaneously in a chemical reaction; one cannot take place without the other.

(q) A free metal can displace from solution the ions of a metal that lies below the free metal in the Activity Series.

(r) In electroplating, the piece to be electroplated with a metal is attached to the cathode.

(s) In an automobile lead storage battery, the density of the sulfuric acid solution decreases as the battery discharges.

(t) In an electrolytic cell, chemical energy is used to produce electrical energy.

D. *Review problems*

1. How many moles of NO gas will be formed by the reaction of 25.0 g of silver with nitric acid? [See the equation given in Question C.5(g).]
2. What volume of chlorine gas, measured at STP, is required to react with excess KOH to form 0.300 mol of $KClO_3$? [See the equation given in Question C.5(f).]
3. What weight of $KMnO_4$ would be needed to react with 100 mL of H_2O_2 solution ($d = 1.031$ g/mL, 9.0% H_2O_2 by weight)? [See the equation given in Question C.5(j).]

*4. What volume of 0.200 M $K_2Cr_2O_7$ will be required to oxidize 5.00 g of H_3AsO_3? [See the equation given in Question C.6(j).]

*5. What volume of 0.200 M $K_2Cr_2O_7$ will be required to oxidize the Fe^{2+} ion in 60.0 mL of 0.200 M $FeSO_4$ solution? [See the equation given in Question C.6(f).]

6. A sample of crude potassium iodide was analyzed using the following reaction (not balanced):

$$I^- + SO_4^{2-} \longrightarrow I_2 + H_2S \quad \text{(acid solution)}$$

If a 4.00 g sample of crude KI produced 2.79 g of iodine, what is the percent purity of the KI?

7. What weight of copper is formed when 35.0 L of ammonia gas, measured at STP, reacts with copper(II) oxide? [See the equation given in Question C.5(h).]

*8. What volume of NO gas, measured at 28°C and 744 torr, will be formed by the reaction of 0.500 mol of Ag reacting with excess nitric acid? [See the equation given in Question C.5(g).]

9. How many moles of H_2 can be produced from 100 g of Al according to the following reaction?

$$Al + OH^- \longrightarrow Al(OH)_4^- + H_2 \quad \text{(basic solution)}$$

Review Exercises for Chapters 16–17

CHAPTER 16 CHEMICAL EQUILIBRIUM

True–False. *Answer the following as either true or false.*

1. A reversible reaction is one in which the products formed in a chemical reaction are reacting to produce the original reactants.
2. The study of reaction rates is known as chemical kinetics.
3. When the rate of the forward reaction is exactly equal to the rate of the reverse reaction, a condition of chemical equilibrium exists.
4. A statement of Le Chatelier's principle is that, if a stress is applied to a system in equilibrium, the system will behave in such a way as to relieve that stress and restore equilibrium but under a new set of conditions.
5. A catalyst will shift the point of equilibrium of a reaction but will not alter the reaction rates.
6. The reaction $CaCO_3(s) \rightleftharpoons CaO(s) + CO_2(g)$ will proceed to the right better in a closed container where the pressure of the CO_2 can build up than in an open container where the CO_2 can escape.
7. When heat is applied to a system in equilibrium, the reaction that absorbs the heat is favored.
8. The ionization constant expression for the ionization of the weak acid HCN is $K_a = [H^+][CN^-]$.
9. If K_a for acetic acid is 1.8×10^{-5}, and K_a for nitrous acid is 4.5×10^{-4}, then, at equal concentrations, acetic acid is a stronger acid.
10. If the K_{sp} for AgI is 1.6×10^{-16}, and the K_{sp} for CuS is 8×10^{-45}, then AgI is more soluble than CuS.
11. The amount of energy needed to form an activated complex is known as the activation energy of a reaction.
12. A catalyst can lower the activation energy, thus increasing the speed of a reaction.
13. A chemical reaction at equilibrium will have different equilibrium constants at different temperatures.
14. Generally, as the concentrations of the reactants increase in a chemical reaction, the speed of the reaction decreases.
15. When the temperature of an exothermic reaction is increased, the forward reaction is favored.
16. The ion product constant for water at 25°C is 1×10^{-14}.
17. A solution of pOH 12 has an H^+ concentration of 0.010 mol per liter.
18. A solution of pOH 3 will turn blue litmus red.
19. KNO_2 dissolved in water will give a solution with a pH less than 7.
20. A solution made from NaCl and HCl will act as a buffer solution.
21. A solution made from 100 mL of 0.1 *M* NaOH and 100 mL of 0.2 *M* $HC_2H_3O_2$ will act as a buffer solution.
22. KCN will hydrolyze to give an alkaline solution.
23. The equilibrium constant expression for

$$CH_4(g) + 2\,O_2(g) \rightleftharpoons CO_2(g) + 2\,H_2O(g)$$

is $\dfrac{[CO_2][H_2O]}{[CH_4][O_2]}$

Multiple Choice. *Choose the correct answer to each of the following.*

1. The equation $HC_2H_3O_2 + H_2O \rightleftharpoons H_3O^+ + C_2H_3O_2^-$ implies that
   (a) If you start with 1.0 mole of $HC_2H_3O_2$, 1.0 mole of H_3O^+ and 1.0 mole of $C_2H_3O_2^-$ will be produced.
   (b) An equilibrium exists between the forward reaction and the reverse reaction.
   (c) At equilibrium, equal molar amounts of all four substances will exist.

(d) The reaction proceeds all the way to the products, then reverses, going all the way back to the reactants.

2. If the reaction $A + B \rightleftharpoons C + D$ is initially at equilibrium, and then more A is added, which of the following is not true?
   (a) More collisions of A and B will occur, thus the rate of the forward reaction will be increased.
   (b) The equilibrium will shift toward the right.
   (c) The moles of B will be increased.
   (d) The moles of D will be increased.
3. What will be the H^+ concentration in a 1.0 M HCN solution? ($K_a = 4.0 \times 10^{-10}$)
   (a) 2.0×10^{-5} M (c) 4.0×10^{-10} M
   (b) 1.0 M (d) 2.0×10^{-10} M
4. What is the percent ionization of HCN in Exercise 3?
   (a) 100% (c) 2.0×10^{-3} %
   (b) 2.0×10^{-8} % (d) 4.0×10^{-8} %
5. If $[H^+] = 1 \times 10^{-5}$ M, which of the following is not true?
   (a) pH = 5 (c) $[OH^-] = 1 \times 10^{-5}$ M
   (b) pOH = 9 (d) The solution is acidic.
6. If $[H^+] = 2.0 \times 10^{-4}$ M, then $[OH^-]$ will be:
   (a) 5.0×10^{-9} M (c) 2.0×10^{-4} M
   (b) 3.70 (d) 5.0×10^{-11} M
7. The solubility product of $PbCrO_4$ is 2.8×10^{-13}. The solubility of $PbCrO_4$ is:
   (a) 5.3×10^{-7} M (c) 7.8×10^{-14} M
   (b) 2.8×10^{-13} M (d) 1.0 M
8. The solubility of AgBr is 6.3×10^{-7} M. The value of the solubility product is:
   (a) 6.3×10^{-7} (c) 4.0×10^{-48}
   (b) 4.0×10^{-13} (d) 4.0×10^{-15}
9. Which of the following solutions would be the best buffer solution?
   (a) 0.10 M $HC_2H_3O_2$ + 0.10 M $NaC_2H_3O_2$
   (b) 0.10 M HCl
   (c) 0.10 M HCl + 0.10 M NaCl
   (d) Pure water
10. For the reaction $H_2(g) + I_2(g) \rightleftharpoons 2\ HI(g)$, at 700 K, $K_{eq} = 56.6$. If an equilibrium mixture at 700 K was found to contain 0.55 M HI and 0.21 M H_2, the I_2 concentration must be:
    (a) 0.046 M (c) 22 M
    (b) 0.025 M (d) 0.21 M
11. The equilibrium constant for the reaction $2\ A + B \rightleftharpoons 3\ C + D$ is
    (a) $\frac{[C]^3[D]}{[A]^2[B]}$ (b) $\frac{[2A][B]}{[3C][D]}$
    (c) $\frac{[3C][D]}{[2A][B]}$ (d) $\frac{[A]^2[B]}{[C]^3[D]}$
12. In the equilibrium represented by

    $$N_2(g) + O_2(g) \rightleftharpoons 2\ NO_2(g)$$

    as the pressure is increased, the amount of NO_2 formed:
    (a) Increases
    (b) Decreases
    (c) Remains the same
    (d) Increases and decreases irregularly
13. Which factor will not increase the concentration of ammonia as represented by the following equation?

    $$3\ H_2(g) + N_2(g) \rightleftharpoons 2\ NH_3(g) + 92.5\ kJ$$

    (a) Increasing the temperature
    (b) Increasing the concentration of N_2
    (c) Increasing the concentration of H_2
    (d) Increasing the pressure
14. If $HCl(g)$ is added to a saturated solution of AgCl, the concentration of Ag^+ in solution:
    (a) Increases
    (b) Decreases
    (c) Remains the same
    (d) Increases and decreases irregularly
15. The solubility of $CaCO_3$ at 20°C is 0.013 g/L. What is the K_{sp} for $CaCO_3$?
    (a) 1.3×10^{-8} (c) 1.7×10^{-8}
    (b) 1.3×10^{-4} (d) 1.7×10^{-4}
16. The K_{sp} for $BaCrO_4$ is 8.5×10^{-11}. What is the solubility of $BaCrO_4$ in grams per liter?
    (a) 9.2×10^{-6} (c) 2.3×10^{-3}
    (b) 0.073 (d) 8.5×10^{-11}
17. What will be the $[Ba^{2+}]$ when 0.010 mol of Na_2CrO_4 is added to 1.0 L of saturated $BaCrO_4$ solution? See Exercise 16 for K_{sp}.
    (a) 8.5×10^{-11} M (c) 9.2×10^{-6} M
    (b) 8.5×10^{-9} M (d) 9.2×10^{-4} M
18. Which would occur if a small amount of sodium acetate crystals, $NaC_2H_3O_2$, were added to 100 mL of 0.1 M $HC_2H_3O_2$ at constant temperature?
    (a) The number of acetate ions in the solution would decrease.
    (b) The number of acetic acid molecules would decrease.
    (c) The number of sodium ions in solution would decrease.
    (d) The H^+ concentration in the solution would decrease.

CHAPTER 17 OXIDATION–REDUCTION

True–False. *Answer the following as either true or false.*

1. In oxidation, the oxidation number of an element increases in a positive direction as a result of gaining electrons.
2. The oxidation number of chlorine in Cl_2 is -1.
3. Oxidation and reduction occur simultaneously in a chemical reaction; one cannot take place without the other.
4. The negative electrode is called the cathode.
5. The cathode is the electrode at which oxidation takes place.
6. The change in the oxidation number of an element from -2 to 0 is reduction.
7. A free metal can displace from solution the ions of a metal that lies below the free metal in the Activity Series.
8. Metallic zinc will react with hydrochloric acid.
9. In electroplating, the piece to be electroplated with a metal is attached to the cathode.
10. In a lead storage battery, $PbSO_4$ is produced at both electrodes in the discharging cycle.
11. As a lead storage battery discharges, the electrolyte becomes less dense.
12. The algebraic sum of the oxidation numbers of all the atoms in $K_2Cr_2O_7$ is zero.
13. A reducing agent will always decrease in oxidation number.
14. Potassium is a better reducing agent than sodium.
15. In the reaction $2\,Cl^- \longrightarrow Cl_2 + 2\,e^-$, each chloride ion loses one electron.
16. $2\,Ag(s) + 2\,HCl(aq) \longrightarrow 2\,AgCl(s) + H_2(g)$
17. In a voltaic cell, reduction occurs at the cathode.
18. The oxidation number of P in $Mg_2P_2O_7$ is $+7$.
19. $CrO_4^{2-} + 4\,H_2O + 2\,e^- \longrightarrow Cr(OH)_3 + 5\,OH^-$ is a balanced reduction half-reaction.
20. In an electrolytic cell, electrical energy is used to bring about a chemical reaction.

Multiple Choice. *Choose the correct answer to each of the following.*

1. In K_2SO_4, the oxidation number of sulfur is:
(a) $+2$ (b) $+4$ (c) $+6$ (d) -2
2. In $Ba(NO_3)_2$, the oxidation number of N is:
(a) $+5$ (b) -3 (c) $+4$ (d) -1
3. In the reaction $H_2S + 4\,Br_2 + 4\,H_2O \longrightarrow H_2SO_4 + 8\,HBr$, the oxidizing agent is:
(a) H_2S (b) Br_2 (c) H_2O (d) H_2SO_4
4. In the reaction

$$VO_3 + Fe^{2+} + 4\,H^+ \rightarrow VO^{2+} + Fe^{3+} + 2\,H_2O,$$

the element reduced is:
(a) V (b) Fe (c) O (d) H

Questions 5, 6, and 7 pertain to the Activity Series

K Ca Mg Al Zn Fe H Cu Ag

5. Which of the following pairs will not react in water solution?
(a) Zn, $CuSO_4$ (c) Fe, $AgNO_3$
(b) Cu, $Al_2(SO_4)_3$ (d) Ca, $Al_2(SO_4)_3$
6. Which element is the most easily oxidized?
(a) K (b) Mg (c) Zn (d) Cu
7. Which element will reduce Cu^{2+} to Cu but will not reduce Zn^{2+} to Zn?
(a) Fe (b) Ca (c) Ag (d) Mg
8. In the electrolysis of fused (molten) $CaCl_2$, the product at the negative electrode is:
(a) Ca^{2+} (b) Cl^- (c) Cl_2 (d) Ca
9. In its reactions, a free element from Group IIA in the periodic table is most likely to:
(a) Be oxidized (c) Be unreactive
(b) Be reduced (d) Gain electrons
10. In the partially balanced redox equation

$$3\,Cu + HNO_3 \longrightarrow 3\,Cu(NO_3)_2 + 2\,NO + H_2O$$

the coefficient needed to balance H_2O is:
(a) 8 (b) 6 (c) 4 (d) 2
11. Which reaction does not involve oxidation–reduction?
(a) Burning sodium in chlorine
(b) Chemical union of Fe and S
(c) Decomposition of $KClO_3$
(d) Neutralization of NaOH with H_2SO_4
12. How many moles of Fe^{2+} can be oxidized to Fe^{3+} by 2.50 mol of Cl_2 according to the following equation?

$$Fe^{2+} + Cl_2 \longrightarrow Fe^{3+} + Cl^-$$

(a) 2.50 mol (c) 1.00 mol
(b) 5.00 mol (d) 22.4 mol

13. How many grams of sulfur can be produced from 100 mL of 6.00 *M* HNO_3?

$$HNO_3 + H_2S \longrightarrow S + NO + H_2O$$

(a) 28.9 g (b) 19.3 g (c) 32.1 g (d) 289 g

Balancing Oxidation–Reduction Equations. *Balance each of the following equations.*

1. $P + HNO_3 \longrightarrow HPO_3 + NO + H_2O$
2. $MnSO_4 + PbO_2 + H_2SO_4 \longrightarrow HMnO_4 + PbSO_4 + H_2O$
3. $Cr_2O_7^{2-} + Cl^- \longrightarrow Cr^{3+} + Cl_2$ (acidic solution)
4. $MnO_4^- + AsO_3^{3-} \longrightarrow Mn^{2+} + AsO_4^{3-}$ (acidic solution)
5. $S^{2-} + Cl_2 \longrightarrow SO_4^{2-} + Cl^-$ (basic solution)
6. $Zn + NO_3^- \longrightarrow Zn(OH)_4^{2-} + NH_3$ (basic solution)

APPENDIX I

Mathematical Review

1. Multiplication Multiplication is a process of adding any given number or quantity to itself a certain number of times. Thus, 4 times 2 means 4 added two times, or 2 added together four times, to give the product 8. Various ways of expressing multiplication are

$$ab \qquad a \times b \qquad a \cdot b \qquad a(b) \qquad (a)(b)$$

All mean a times b, or a multiplied by b, or b times a.

When $a = 16$ and $b = 24$, we have $16 \times 24 = 384$.

The expression $°F = (1.8 \times °C) + 32$ means that we are to multiply 1.8 times °C and add 32 to the product. When °C equal 50,

$$°F = (1.8 \times 50) + 32 = 90 + 32 = 122°F$$

The result of multiplying two or more numbers together is known as the *product*.

2. Division The word *division* has several meanings. As a mathematical expression, it is the process of finding how many times one number or quantity is contained in another. Various ways of expressing division are

$$a \div b \qquad \frac{a}{b} \qquad a/b$$

All mean a divided by b.

When $a = 15$ and $b = 3$, $\frac{15}{3} = 5$.

The number above the line is called the *numerator*; the number below the line is the *denominator*. Both the horizontal and the slanted (/) division signs also mean "per." For example, in the expression for density, the mass per unit volume:

$$\text{density} = \text{mass/volume} = \frac{\text{mass}}{\text{volume}} = \text{g/mL}$$

The diagonal line still refers to a division of grams by the number of milliliters occupied by that weight.

The result of dividing one number into another is called the *quotient.*

3. Fractions and Decimals A fraction is an expression of division, showing that the numerator is divided by the denominator. A *proper fraction* is one in which the numerator is smaller than the denominator. In an *improper fraction,* the numerator is the larger number. A decimal or a decimal fraction is a proper fraction in which the denominator is some power of 10. The decimal fraction is determined by carrying out the division of the proper fraction. Examples of proper fractions and their decimal fraction equivalents are shown in the following table.

Proper fraction		Decimal fraction		Proper fraction
$\frac{1}{8}$	=	0.125	=	$\frac{125}{1000}$
$\frac{1}{10}$	=	0.1	=	$\frac{1}{10}$
$\frac{3}{4}$	=	0.75	=	$\frac{75}{100}$
$\frac{1}{100}$	=	0.01	=	$\frac{1}{100}$
$\frac{1}{4}$	=	0.25	=	$\frac{25}{100}$

4. Addition of Numbers with Decimals To add numbers with decimals we use the same procedure as that used when adding whole numbers, but we always line up the decimal points in the same column. For example, add 8.21 + 143.1 + 0.325

$$\begin{array}{r} 8.21 \\ +143.1 \\ +\quad 0.325 \\ \hline 151.635 \end{array}$$

When adding numbers that express units of measurement, we must be certain that the numbers added together represent the same units. For example, what is the total length of three pieces of glass tubing: 10.0 cm, 125 mm, and 8.4 cm? If we simply add the numbers, we obtain a value of 143.4, but we are not certain what the unit of measurement is. To add these lengths correctly, first change 125 mm to 12.5 cm. Now all the lengths are expressed in the same units and can be added.

$$\begin{array}{r} 10.0 \text{ cm} \\ 12.5 \text{ cm} \\ 8.4 \text{ cm} \\ \hline 30.9 \text{ cm} \end{array}$$

5. Subtraction of Numbers with Decimals To subtract numbers containing decimals, we use the same procedure as for subtracting whole numbers, but we always line up the decimal points in the same column. For example, subtract 20.60 from 182.49.

$$\begin{array}{r} 182.49 \\ -\ 20.60 \\ \hline 161.89 \end{array}$$

6. Multiplication of Numbers with Decimals To multiply two or more numbers together that contain decimals, we first multiply as if they were whole numbers. Then, to locate the decimal point in the product, we add together the number of digits to the right of the decimal in all the numbers multiplied together. The product should have this same number of digits to the right of the decimal point.

Multiply 2.05×2.05 (total of four digits to the right of the decimal):

$$\begin{array}{r} 2.05 \\ \times\, 2.05 \\ \hline 1025 \\ 4100 \\ \hline 4.2025 \end{array} \quad \text{(four digits to the right of the decimal)}$$

Here are more examples:

$14.25 \times 6.01 \times 0.75 = 64.231875$ (six digits to the right of the decimal)

$39.26 \times 60 = 2355.60$ (two digits to the right of the decimal)

[*Note*: When at least one of the numbers that is multiplied is a measurement, the answer must be adjusted to contain the correct number of significant figures. (See Section 2.2 on significant figures.)]

7. Division of Numbers with Decimals To divide numbers containing decimals, we first relocate the decimal points of the numerator and denominator by moving them to the right as many places as needed to make the denominator a whole number. (Move the decimal of both the numerator and the denominator the same amount and in the same direction.) For example,

$$\frac{136.94}{4.1} = \frac{1369.4}{41}$$

The decimal point adjustment in this example is equivalent to multiplying both numerator and denominator by 10. Now we carry out the division normally, locating the decimal point immediately above its position in the dividend.

$$\begin{array}{r} 33.4 \\ 41\overline{)1369.4} \\ 123 \\ \hline 139 \\ 123 \\ \hline 164 \\ 164 \\ \hline \end{array} \qquad \frac{0.441}{26.25} = \frac{44.1}{2625} = \begin{array}{r} 0.0168 \\ 2625\overline{)44.1000} \\ 2625 \\ \hline 17850 \\ 15750 \\ \hline 21000 \\ 21000 \\ \hline \end{array}$$

[*Note*: When at least one of the numbers in the division is a measurement, the answer must be adjusted to contain the correct number of significant figures. (See Section 2.2 on significant figures.)]

The foregoing examples are merely guides to the principles used in performing the various mathematical operations illustrated. There are, no doubt, shortcuts and other methods, and the student will discover these with experience. Every student of chemistry should learn to use an electronic calculator for solving mathematical problems. The use of a calculator will save many hours of doing tedious longhand calculations. After solving a problem, the student should check for errors and evaluate the answer to see if it is logical and consistent with the data given.

8. Algebraic Equations Many mathematical problems that are encountered in chemistry fall into the following algebraic forms. Solutions to these problems are simplified by first isolating the desired term on one side of the equation. This rearrangement is accomplished by treating both sides of the equation in an identical manner (so as not to destroy the equality) until the desired term is isolated.

(a) $a = \dfrac{b}{c}$

To solve for a, divide b by c.
To solve for b, multiply both sides of the equation by c.

$$a \times c = \frac{b}{\not{c}} \times \not{c}$$

$$b = a \times c$$

To solve for c, multiply both sides of the equation by $\dfrac{c}{a}$.

$$\not{a} \times \frac{c}{\not{a}} = \frac{b}{\not{c}} \times \frac{\not{c}}{a}$$

$$c = \frac{b}{a}$$

(b) $\frac{a}{b} = \frac{c}{d}$

To solve for a, multiply both sides of the equation by b.

$$\frac{a}{\cancel{b}} \times \cancel{b} = \frac{c}{d} \times b$$

$$a = \frac{c \times b}{d}$$

To solve for b, multiply both sides of the equation by $\frac{b \times d}{c}$.

$$\frac{a}{\cancel{b}} \times \frac{\cancel{b} \times d}{c} = \frac{\cancel{c}}{\cancel{d}} \times \frac{b \times \cancel{d}}{\cancel{c}}$$

$$b = \frac{a \times d}{c}$$

(c) $a \times b = c \times d$

To solve for a, divide both sides of the equation by b.

$$\frac{a \times \cancel{b}}{\cancel{b}} = \frac{c \times d}{b}$$

$$a = \frac{c \times d}{b}$$

(d) $\frac{(b - c)}{a} = d$

To solve for b, first multiply both sides of the equation by a.

$$\frac{\cancel{a}(b - c)}{\cancel{a}} = d \times a$$

$$b - c = d \times a$$

Then add c to both sides of the equation.

$$b - \cancel{c} + \cancel{c} = d \times a + c$$

$$b = (d \times a) + c$$

When $a = 1.8$, $c = 32$, and $d = 35$,

$$b = (35 \times 1.8) + 32 = 63 + 32 = 95$$

9. Exponents, Powers of 10, Expression of Large and Small Numbers In scientific measurements and calculations, we often encounter very large and very small numbers—for example, 0.00000384 and 602,000,000,000,000,000,000,000. These numbers are troublesome to write and awkward to work with, especially in calculations. A convenient method of expressing these large and small numbers in a simplified form is by means of exponents or powers of 10. This method of expressing numbers is known as **scientific** or **exponential notation.**

An *exponent* is a number written as a superscript following another number; it is also called a *power* of that number, and it indicates how many times the number is used as a factor. In the number 10^2, 2 is the exponent, and the number means 10 squared, or 10 to the second power, or $10 \times 10 = 100$. Three other examples are

$$3^2 = 3 \times 3 = 9$$

$$3^4 = 3 \times 3 \times 3 \times 3 = 81$$

$$10^3 = 10 \times 10 \times 10 = 1000$$

For ease of handling, large and small numbers are expressed in powers of 10. Powers of 10 are used because multiplying or dividing by 10 coincides with moving the decimal point in a number by one place. Thus, a number multiplied by 10^1 would move the decimal point one place to the right; 10^2, two places to the right; 10^{-2}, two places to the left. To express a number in powers of 10, we move the decimal point in the original number to a new position, placing it so that the number is a value between 1 and 10. This new decimal number is multiplied by 10 raised to the proper power. For example, to write the number 42,389 in exponential form (powers of 10), the decimal point is placed between the 4 and the 2 (4.2389), and the number is multiplied by 10^4; thus, the number is 4.2380×10^4.

$$42{,}389 = 4.2389 \times 10^4$$

4 3 2 1

The exponent (power) of 10 (4) tells us the number of places that the decimal point has been moved from its original position. If the decimal point is moved to the left, the exponent is a positive number; if it is moved to the right, the exponent is a negative number. To express the number 0.00248 in exponential notation (as a power of 10), the decimal point is moved three places to the right; the exponent of 10 is -3, and the number is 2.48×10^{-3}.

$$0.00248 = 2.48 \times 10^{-3}$$

1 2 3

Study the following examples.

$$1237 = 1.237 \times 10^3$$

$$988 = 9.88 \times 10^2$$

$$147.2 = 1.472 \times 10^2$$

$$2{,}200{,}000 = 2.2 \times 10^6$$

$$0.0123 = 1.23 \times 10^{-2}$$

$$0.00005 = 5 \times 10^{-5}$$

$$0.000368 = 3.68 \times 10^{-4}$$

Exponents in multiplication and division. The use of powers of 10 in multiplication and division greatly simplifies locating the decimal point in the answer. In multiplication, first change all numbers to powers of 10, then multiply the numerical portion in the usual manner, and finally add the exponents of 10 algebraically, expressing them as a power of 10 in the product. In multiplication, the exponents (powers of 10) are added algebraically.

$$10^2 \times 10^3 = 10^{(2+3)} = 10^5$$

$$10^2 \times 10^2 \times 10^{-1} = 10^{(2+2-1)} = 10^3$$

Multiply: $40{,}000 \times 4200$

Change to powers of 10: $4 \times 10^4 \times 4.2 \times 10^3$

Rearrange: $4 \times 4.2 \times 10^4 \times 10^3$

$16.8 \times 10^{(4+3)}$

16.8×10^7 or 1.68×10^8 (Answer)

Multiply: 380×0.00020

$3.80 \times 10^2 \times 2.0 \times 10^{-4}$

$3.80 \times 2.0 \times 10^2 \times 10^{-4}$

$7.6 \times 10^{(2-4)}$

7.6×10^{-2} or 0.076 (Answer)

Multiply: $125 \times 284 \times 0.150$

$1.25 \times 10^2 \times 2.84 \times 10^2 \times 1.50 \times 10^{-1}$

$1.25 \times 2.84 \times 1.50 \times 10^2 \times 10^2 \times 10^{-1}$

$5.325 \times 10^{(2+2-1)}$

5.32×10^3 (Answer)

In division, after changing the numbers to powers of 10, move the 10 and its exponent from the denominator to the numerator, changing the sign of the exponent. Carry out the division in the usual manner and evaluate the power of 10. The following is a proof of the equality of moving the power of 10 from the denominator to the numerator.

$$1 \times 10^{-2} = 0.01 = \frac{1}{100} = \frac{1}{10^2} = 1 \times 10^{-2}$$

In division, change the sign(s) of the exponent(s) of 10 in the denominator and move the 10 and its exponent(s) to the numerator. Then add all the exponents of 10 together. For example,

$$\frac{10^5}{10^3} = 10^5 \times 10^{-3} = 10^{(5-3)} = 10^2$$

$$\frac{10^3 \times 10^4}{10^{-2}} = 10^3 \times 10^4 \times 10^2 = 10^{(3+4+2)} = 10^9$$

Divide: $\dfrac{2871}{0.0165}$

Change to powers of 10: $\dfrac{2.871 \times 10^{3}}{1.65 \times 10^{-2}}$

Move 10^{-2} to the numerator, changing the sign of the exponent. This is mathematically equivalent to multiplying both numerator and denominator by 10^2.

$$\frac{2.871 \times 10^{3} \times 10^{2}}{1.65}$$

$$\frac{2.871 \times 10^{(3+2)}}{1.65} = 1.74 \times 10^{5} \quad \text{(Answer)}$$

Divide: $\dfrac{0.000585}{0.00300}$

$$\frac{5.85 \times 10^{-4}}{3.00 \times 10^{-3}}$$

$$\frac{5.85 \times 10^{-4} \times 10^{3}}{3.00} = \frac{5.85 \times 10^{(-4+3)}}{3.00}$$

$$= 1.95 \times 10^{-1} \quad \text{or} \quad 0.195 \quad \text{(Answer)}$$

Calculate: $\dfrac{760 \times 300 \times 40.0}{700 \times 273}$

$$\frac{7.60 \times 10^{2} \times 3.00 \times 10^{2} \times 4.00 \times 10^{1}}{7.00 \times 10^{2} \times 2.73 \times 10^{2}}$$

$$\frac{7.60 \times 3.00 \times 4.00 \times 10^{2} \times 10^{2} \times 10^{1}}{7.00 \times 2.73 \times 10^{2} \times 10^{2}}$$

$$= 4.77 \times 10^{1} \quad \text{or} \quad 47.7 \quad \text{(Answer)}$$

10. Significant Figures in Calculations The result of a calculation based on experimental measurements cannot be more precise than the measurement that has the greatest uncertainty. (See Section 2.5 for additional discussion.)

Addition and subtraction. The result of an addition or subtraction should contain no more digits to the right of the decimal point than are contained in the quantity that has the least number of digits to the right of the decimal point.

Perform the operation indicated and then round off the number to the proper number of significant figures.

142.8 g	
18.843 g	93.45 mL
36.42 g	− 18.0 mL
198.063 g	75.45 mL
198.1 g (Answer)	75.5 mL (Answer)

Multiplication and division. In calculations involving multiplication or division, the answer should contain the same number of significant figures as the measurement that has the least number of significant figures. In multiplication or division the position of the decimal point has nothing to do with the number of significant figures in the answer. Study the following examples:

	Round off to
$2.05 \times 2.05 = 4.2025$	4.20
$18.48 \times 5.2 = 96.096$	96
$0.0126 + 0.020 - 0.000252$ or	
$1.26 \times 10^{-2} \times 2.0 \times 10^{-2} = 2.520 \times 10^{-4}$	2.5×10^{-4}
$\frac{1369.4}{41} = 33.4$	33
$\frac{2268}{4.20} - 540$	540

11. Dimensional Analysis Many problems of chemistry can be solved readily by dimensional analysis using the factor-label or conversion-factor method. Dimensional analysis involves the use of proper units of dimensions for all factors that are multiplied, divided, added, or subtracted in setting up and solving a problem. Dimensions are physical quantities such as length, mass, and time, which are expressed in such units as centimeters, grams, and seconds, respectively. In solving a problem, we treat these units mathematically just as though they were numbers, which gives us an answer that contains the correct dimensional units.

A measurement or quantity given in one kind of unit can be converted to any other kind of unit having the same dimension. To convert from one kind of unit to another, the original quantity or measurement is multiplied or divided by a conversion factor. The key to success lies in choosing the correct conversion factor. This general method of calculation is illustrated in the following examples.

Suppose we want to change 24 ft to inches. We need to multiply 24 ft by a conversion factor containing feet and inches. Two such conversion factors can be written relating inches to feet.

$$\frac{12\text{ in.}}{1\text{ ft}} \quad \text{or} \quad \frac{1\text{ ft}}{12\text{ in.}}$$

We choose the factor that will mathematically cancel feet and leave the answer in inches. Note that the units are treated in the same way we treat numbers, multiplying or dividing as required. Two possibilities then arise to change 24 ft to inches:

$$24\ \cancel{\text{ft}} \times \frac{12\text{ in.}}{1\ \cancel{\text{ft}}} \quad \text{or} \quad 24\text{ ft} \times \frac{1\text{ ft}}{12\text{ in.}}$$

In the first case (the correct method), feet in the numerator and the denominator cancel, giving us an answer of 288 in. In the second case, the units of the answer are ft^2/in., the answer being 2.0 ft^2/in. In the first case, the answer is reasonable since it is expressed in units having the proper dimensions. That is, the dimension of length expressed in feet has been converted to length in inches according to the mathematical expression

$$\cancel{\text{ft}} \times \frac{\text{in.}}{\cancel{\text{ft}}} = \text{in.}$$

In the second case, the answer is not reasonable since the units (ft^2/in.) do not correspond to units of length. The answer is therefore incorrect. The units are the guiding factor for the proper conversion.

The reason we can multiply 24 ft times 12 in./ft and not change the value of the measurement is that the conversion factor is derived from two equivalent quantities. Therefore, the conversion factor 12 in./ft is equal to unity. And when you multiply any factor by 1, it does not change the value.

$$12 \text{ in.} = 1 \text{ ft} \qquad \text{and} \qquad \frac{12 \text{ in.}}{1 \text{ ft}} = 1$$

Convert 16 kg to milligrams. In this problem it is best to proceed in this fashion:

$$\text{kg} \longrightarrow \text{g} \longrightarrow \text{mg}$$

The possible conversion factors are

$$\frac{1000 \text{ g}}{1 \text{ kg}} \quad \text{or} \quad \frac{1 \text{ kg}}{1000 \text{ g}} \qquad\qquad \frac{1000 \text{ mg}}{1 \text{ g}} \quad \text{or} \quad \frac{1 \text{ g}}{1000 \text{ mg}}$$

We use the conversion factor that leaves the proper unit at each step for the next conversion. The calculation is

$$16 \cancel{\text{kg}} \times \frac{1000 \cancel{\text{g}}}{1 \cancel{\text{kg}}} \times \frac{1000 \text{ mg}}{1 \cancel{\text{g}}} = 1.6 \times 10^7 \text{ mg}$$

Many problems can be solved by a sequence of steps involving unit conversion factors. This sound, basic approach to problem solving, together with neat and orderly setting up of data, will lead to correct answers having the right units, fewer errors, and considerable saving of time.

12. Graphical Representation of Data A graph is often the most convenient way to present or display a set of data. Various kinds of graphs have been devised, but the most common type uses a set of horizontal and vertical coordinates to show the relationship of two variables. It is called an x–y graph because the data of one variable are represented on the horizontal or x axis

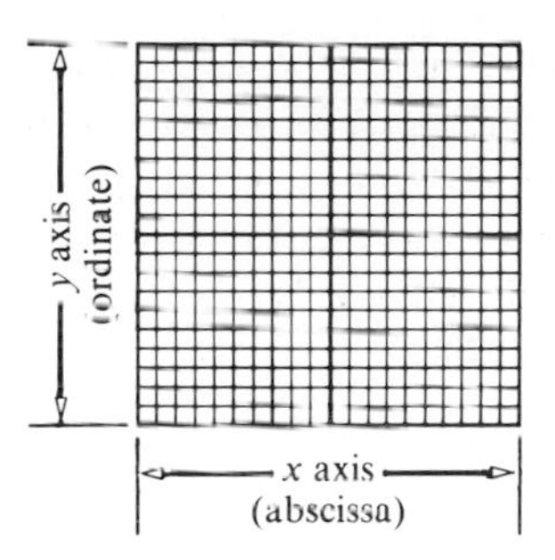

Figure I.1

(abscissa) and the data of the other variable are represented on the vertical or y axis (ordinate). (See Figure I.1.)

As a specific example of a simple graph, let us graph the relationship between Celsius and Fahrenheit temperature scales. Assume that initially we have only the information in the following table.

°C	°F
0	32
50	122
100	212

On a set of horizontal and vertical coordinates (graph paper), scale off at least 100 Celsius degrees on the x axis and at least 212 Fahrenheit degrees on the y axis. Locate and mark the three points corresponding to the three temperatures given and draw a line connecting these points (see Figure I.2).

Here is how a point is located on the graph: Using the 50°C–122°F data, trace a vertical line up from 50°C on the x axis and a horizontal line across from 122°F on the y axis and mark the point where the two lines intersect. This process is called *plotting*. The other two points are plotted on the graph in the same way. [*Note*: The number of degrees per scale division was chosen to give a graph of convenient size. In this case there are 5 Fahrenheit degrees per scale division and 2 Celsius degrees per scale division.]

The graph in Figure I.2 shows that the relationship between Celsius and Fahrenheit temperature is that of a straight line. The Fahrenheit temperature corresponding to any given Celsius temperature between 0 and 100° can be determined from the graph. For example, to find the Fahrenheit temperature corresponding to 40°C, trace a perpendicular line from 40°C on the x axis to the line plotted on the graph. Now trace a horizontal line from this point on the plotted line to the y axis and read the corresponding Fahrenheit temperature (104°F). See the dashed lines in Figure I.2. In turn, the Celsius temperature corresponding to any Fahrenheit temperature between 32 and 212° can be determined from the graph by tracing a horizontal line from the Fahrenheit

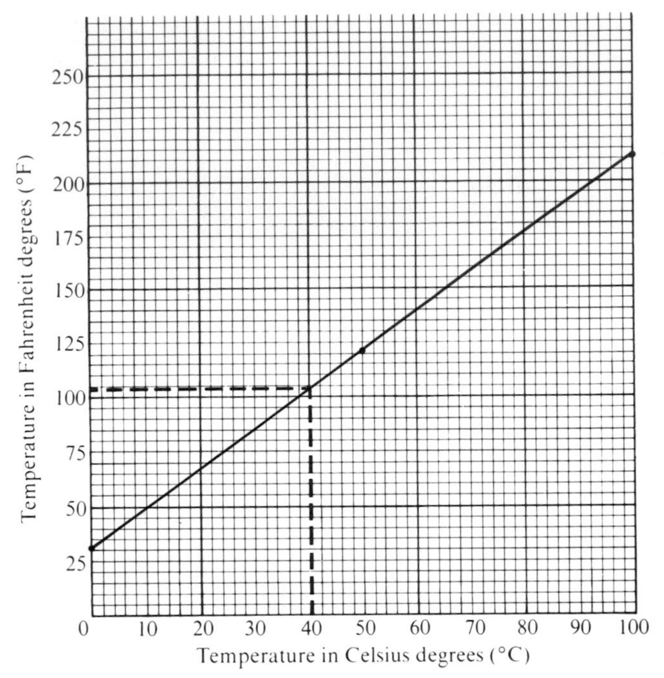

Figure I.2

temperature to the plotted line and reading the corresponding temperature on the Celsius scale directly below the point of intersection.

The mathematical relationship of Fahrenheit and Celsius temperatures is expressed by the equation $°F = 1.8 \times °C + 32$. Figure I.2 is a graph of this equation. Because the graph is a straight line, it can be extended indefinitely at either end. Any desired Celsius temperature can be plotted against the corresponding Fahrenheit temperature by extending the scales along both axes as necessary. Negative, as well as positive, values can be plotted on the graph (see Figure I.3).

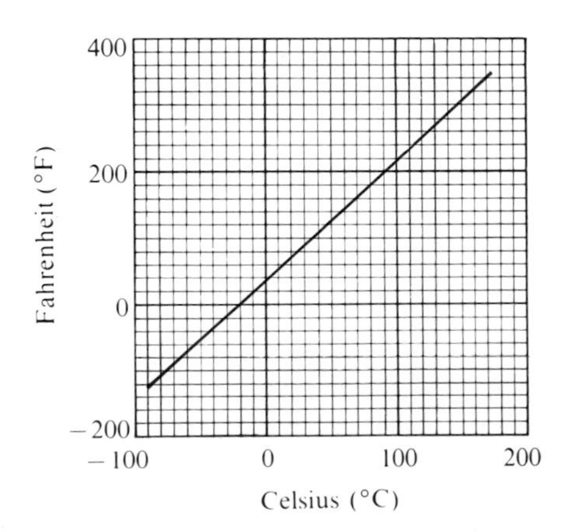

Figure I.3

Figure I.4 is a graph showing the solubility of potassium chlorate in water at various temperatures. The solubility curve on this graph was plotted from the data in the following table.

Temperature (°C)	Solubility (g $KClO_3$/100 g water)
10	5.0
20	7.4
30	10.5
50	19.3
60	24.5
80	38.5

In contrast to the Celsius–Fahrenheit temperature relationship, there is no known mathematical equation that describes the exact relationship between temperature and the solubility of potassium chlorate. The graph in Figure I.4 was

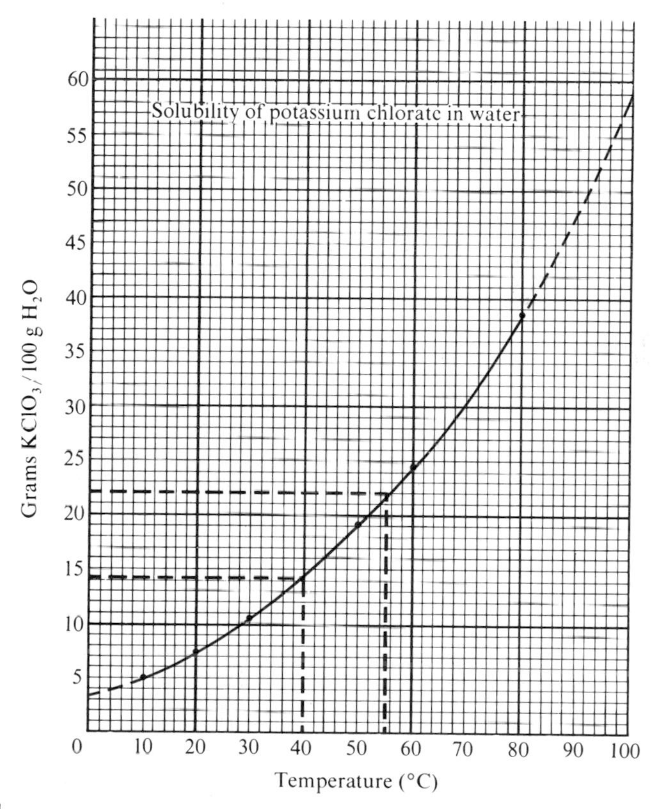

Figure I.4

constructed from experimentally determined solubilities at the six temperatures shown. These experimentally determined solubilities are all located on the smooth curve traced by the unbroken line portion of the graph. We are therefore confident that the unbroken line represents a very good approximation of the solubility data for potassium chlorate over the temperature range from 10 to 80°C. All points on the plotted curve represent the composition of saturated solutions. Any point below the curve represents an unsaturated solution.

The dashed-line portions of the curve are *extrapolations*; that is, they extend the curve above and below the temperature range actually covered by the plotted solubility data. Curves such as this one are often extrapolated a short distance beyond the range of the known data, although the extrapolated portions may not be highly accurate. Extrapolation is justified only in the absence of more reliable information.

The graph in Figure I.4 can be used with confidence to obtain the solubility of $KClO_3$ at any temperature between 10 and 80°C, but the solubilities between 0 and 10°C and between 80 and 100°C are less reliable. For example, what is the solubility of $KClO_3$ at 55°C, at 40°C, and at 100°C?

First draw a perpendicular line from each temperature to the plotted solubility curve. Now trace a horizontal line to the solubility axis from each point on the curve and read the corresponding solubilities. The values that we read from the graph are

55°	22.0 g $KClO_3$/100 g water
40°	14.2 g $KClO_3$/100 g water
100°	59 g $KClO_3$/100 g water

Of these solubilities, the one at 55°C is probably the most reliable because experimental points are plotted at 50°C and at 60°C. The 40°C solubility value is a bit less reliable because the nearest plotted points are at 30°C and 50°C. The 100°C solubility is the least reliable of the three values because it was taken from the extrapolated part of the curve, and the nearest plotted point is 80°C. Actual handbook solubility values are 14.0 and 57.0 g of $KClO_3$/100 g of water at 40°C and 100°C, respectively.

The graph in Figure I.4 can also be used to determine whether a solution is saturated or unsaturated. For example, a solution contains 15 g of $KClO_3$/100 g of water and is at a temperature of 55°C. Is the solution saturated or unsaturated? *Answer*: The solution is unsaturated because the point corresponding to 15 g and 55°C on the graph is below the solubility curve; all points below the curve represent unsaturated solutions.

APPENDIX II

Vapor Pressure of Water at Various Temperatures

Temperature (°C)	Vapor pressure (torr)	Temperature (°C)	Vapor pressure (torr)
0	4.6	26	25.2
5	6.5	27	26.7
10	9.2	28	28.3
15	12.8	29	30.0
16	13.6	30	31.8
17	14.5	40	55.3
18	15.5	50	92.5
19	16.5	60	149.4
20	17.5	70	233.7
21	18.6	80	355.1
22	19.8	90	525.8
23	21.2	100	760.0
24	22.4	110	1074.6
25	23.8	—	—

APPENDIX III

Units of Measurements

Prefixes and numerical values for SI units

Prefix	Symbol	Numerical value	Power of 10 equivalent
exa	E	1,000,000,000,000,000,000	10^{18}
peta	P	1,000,000,000,000,000	10^{15}
tera	T	1,000,000,000,000	10^{12}
giga	G	1,000,000,000	10^{9}
mega	M	1,000,000	10^{6}
kilo	k	1,000	10^{3}
hecto	h	100	10^{2}
deka	da	10	10^{1}
—	—	1	10^{0}
deci	d	0.1	10^{-1}
centi	c	0.01	10^{-2}
milli	m	0.001	10^{-3}
micro	μ	0.000001	10^{-6}
nano	n	0.000000001	10^{-9}
pico	p	0.000000000001	10^{-12}
femto	f	0.000000000000001	10^{-15}
atto	a	0.000000000000000001	10^{-18}

Length
1 in. = 2.54 cm
10 mm = 1 cm
100 cm = 1 m
1000 mm = 1 m
1000 m = 1 km
1 mile = 1.61 km
1 Å = 10^{-8} cm

Mass
1 lb = 453.6 g
1000 mg = 1 g
1000 g = 1 kg
1 ounce = 28.3 g
2.20 lb = 1 kg

Volume
1 mL = 1 cm^3
1000 mL = 1 L
1 fluid ounce = 29.6 mL
1 qt = 0.946 L
1 gal = 3.785 L

Temperature
$°F = 1.8 \times °C + 32$
$°C = \frac{(°F - 32)}{1.8}$
$K = °C + 273$
$°F = 1.8(°C + 40)$ 40
Absolute zero = −273.18°C or −459.72°F

APPENDIX IV

Solubility Table

	F^-	Cl^-	Br^-	I^-	O^{2-}	S^{2-}	OH^-	NO_3^-	CO_3^{2-}	SO_4^{2-}	$C_2H_3O_2^-$
H^+	S	S	S	S	S	s	S	S	s	S	S
Na^+	S	S	S	S	S	S	S	S	S	S	S
K^+	S	S	S	S	S	S	S	S	S	S	S
NH_4^+	S	S	S	S	—	S	S	S	S	S	S
Ag^+	S	I	I	I	I	I	—	S	I	I	I
Mg^{2+}	I	S	S	S	I	d	I	S	I	S	S
Ca^{2+}	I	S	S	S	I	d	I	S	I	I	S
Ba^{2+}	I	S	S	S	s	d	s	S	I	I	S
Fe^{2+}	s	S	S	S	I	I	I	S	s	S	S
Fe^{3+}	I	S	S	—	I	I	I	S	I	S	I
Co^{2+}	S	S	S	S	I	I	I	S	I	S	S
Ni^{2+}	s	S	S	S	I	I	I	S	I	S	S
Cu^{2+}	s	S	S	—	I	I	I	S	I	S	S
Zn^{2+}	s	S	S	S	I	I	I	S	I	S	S
Hg^{2+}	d	S	I	I	I	I	I	S	I	d	S
Cd^{2+}	s	S	S	S	I	I	I	S	I	S	S
Sn^{2+}	S	S	S	s	I	I	I	S	I	S	S
Pb^{2+}	I	I	I	I	I	I	I	S	I	I	S
Mn^{2+}	s	S	S	S	I	I	I	S	I	S	S
Al^{3+}	I	S	S	S	I	d	I	S	—	S	S

Key: S = soluble in water
s = slightly soluble in water
I = insoluble in water (less than 1 g/100 g H_2O)
d = decomposes in water

APPENDIX V

Glossary

absolute zero *See* Kelvin scale.

acid (1) A substance that produces H^+ (H_3O^+) when dissolved in water. (2) A proton donor. (3) An electron-pair acceptor. A substance that bonds to an electron pair.

acid anhydride A nonmetal oxide that reacts with water to form an acid.

activated complex The intermediate high-energy species formed when reactants collide. The complex can decompose to form either reactants or products.

activation energy The amount of energy needed to form the activated complex.

Activity Series of Metals A listing of metallic elements in descending order or reactivity.

alchemists Practitioners of chemistry during the Middle Ages whose aims were to change the baser metals into gold and to discover the "philosopher's stone," which was thought to bring eternal youth.

alkali metal An element (except H) from Group IA of the periodic table.

alkaline earth metal An element from Group IIA of the periodic table.

allotropy A phenomenon in which an element exists in two or more molecular or crystalline forms. Graphite and diamond are two allotropic forms of carbon.

amorphous A solid without definite crystalline form.

amphoteric substance A substance that can react as either an acid or a base.

anion A negatively charged ion.

anode The electrode where oxidation occurs in an electrochemical reaction.

aqueous solution A water solution.

atmospheric pressure The pressure experienced by objects on the earth as a result of the layer of air surrounding our planet. A pressure of 1 atmosphere (1 atm) is the pressure that will support a column of mercury 760 mm high at 0°C.

atom The smallest particle of an element that can enter into a chemical reaction.

atomic mass unit (amu) A unit of mass equal to one-twelfth the mass of a carbon-12 atom.

atomic number The number of protons in the nucleus of an atom.

atomic theory The theory that substances are composed of atoms, and that chemical reactions are explained by the properties and the interactions of these atoms.

atomic weight The average relative mass of the isotopes of an element referred to the atomic mass of carbon-12 as exactly 12 amu.

Avogadro's law Equal volumes of different gases at the same temperature and pressure contain equal numbers of molecules.

Avogadro's number 6.022×10^{23}; the number of formula units in 1 mole of whatever is indicated by the formula.

balanced equation A chemical equation having the same number and kind of atoms and the same electrical charge on each side of the equation.

barometer A device used to measure pressure.

base (1) A substance that produces OH^- when dissolved in water. (2) A proton acceptor. (3) An electron-pair donor.

basic anhydride A metal oxide that reacts with water to form a base.

binary compound A compound composed of two different elements.

biochemistry The branch of chemistry concerned with the chemical reactions occurring in living organisms.

boiling point The temperature at which the vapor pressure of a liquid is equal to the pressure above the liquid.

bond dissociation energy The energy required to break a covalent bond.

bond length The distance between two nuclei that are joined by a chemical bond.

Boyle's law At constant temperature, the volume of a given mass of gas is inversely proportional to the pressure ($PV =$ constant).

buffer solution A solution that resists changes in pH when diluted or when small amounts of a strong acid or strong base are added.

calorie (cal) One calorie is a quantity of heat energy that will raise the temperature of 1 gram of water 1°C (from 14.5 to 15.5°C).

catalyst A substance that influences the rate of a reaction and can be recovered in its original form at the end of the reaction.

cathode The electrode where reduction occurs in an electrochemical reaction.

cation A positively charged ion.

Celsius scale (°C) The temperature scale on which water freezes at 0°C and boils at 100°C at 1 atm pressure.

Charles' law At constant pressure, the volume of a gas is directly proportional to the absolute (K) temperature ($V/T =$ constant).

chemical bond The attractive force that holds atoms together in a compound.

chemical change A change producing products that differ in composition from the original substances.

chemical equation An expression showing the reactants and the products of a chemical change (for example, $2\,H_2 + O_2 \rightarrow 2\,H_2O$).

chemical equilibrium The state in which the rate of the forward reaction equals the rate of the reverse reaction for a chemical change.

chemical family *See* groups or families of elements.

chemical formula A shorthand method for showing the composition of a compound using symbols of the elements.

chemical kinetics The study of reaction rates or the speed of a particular reaction.

chemical properties Properties of a substance related to its chemical changes.

chemistry The science dealing with the composition of matter and the changes in composition that matter undergoes.

colligative properties Properties of a solution that depend on the number of solute particles in solution and not on the nature of the solute (for example, vapor pressure lowering, freezing point lowering, boiling point elevation).

combination reaction A direct union or combination of two substances to produce one new substance.

combustion In general, the process of burning or uniting a substance with oxygen, which is accompanied by the evolution of light and heat.

common ion effect The shift of an equilibrium caused by the addition of an ion common to the ions in the equilibrium.

compound A substance composed of two or more elements combined in a definite proportion by weight.

concentrated solution A solution containing a relatively large amount of solute.

concentration of a solution A quantitative expression of the amount of dissolved solute in a certain quantity of solvent or solution.

conjugate acid-base Two molecules or ions whose formulas differ by one H^+. (The acid is the species with the H^+, and the base is the species without the H^+.)

coordinate-covalent bond A covalent bond in

which the shared pair of electrons is furnished by only one of the bonded atoms.

covalent bond A chemical bond formed between two atoms by sharing a pair of electrons.

Dalton's atomic theory The first modern atomic theory to state that elements are composed of tiny, individual particles called atoms.

Dalton's law of partial pressures The total pressure in a mixture of gases is equal to the sum of the partial pressures of each gas in the mixture.

decomposition reaction A breaking down or decomposition of one substance into two or more different substances.

deliquescence The absorption of water by a compound beyond the hydrate stage to form a solution.

density The mass of an object divided by its volume.

diffusion The process by which gases and liquids mix spontaneously because of the random motion of their particles.

dilute solution A solution containing a relatively small amount of solute.

dipole A molecule with separation of charge causing it to be positive at one end and negative at the other end.

dissociation The process by which a salt separates into individual ions when dissolved in water.

double bond A covalent bond in which two pairs of electrons are shared.

double-displacement reaction A reaction of two compounds to produce two different compounds by exchanging the components of the reacting compounds.

efflorescence The spontaneous loss of water of hydration by a compound when exposed to air.

effusion The passage of gas through a tiny orifice from a region of high pressure to a region of lower pressure.

Einstein's mass-energy equation $E = mc^2$; the relationship between mass and energy.

electrolysis The process whereby electrical energy is used to bring about a chemical change.

electrolyte A substance whose aqueous solution conducts electricity.

electrolytic cell An electrolysis apparatus in which electrical energy from an outside source is used to produce a chemical change.

electron A subatomic particle that exists outside the nucleus and has an assigned electrical charge of -1.

electron affinity The energy released or absorbed when an electron is added to an atom or an ion.

electron-dot structure See Lewis structure

electronegativity The relative attraction that an atom has for the electrons in a covalent bond.

electron shell *See* energy levels of electrons.

element A basic building block of matter that cannot be broken down into simpler substances by ordinary chemical changes.

empirical formula A chemical formula that gives the smallest whole-number ratio of atoms in a compound.

endothermic reaction A chemical reaction that absorbs energy from the surroundings as it proceeds from reactants to products.

energy The capacity or ability of matter to do work.

energy levels of electrons Areas in which electrons are located at various distances from the nucleus.

energy sublevels The *s*, *p*, *d*, and *f* orbitals within a principal energy level occupied by electrons in an atom.

equilibrium A dynamic state in which two or more opposing processes are taking place at the same time and at the same rate.

equilibrium constant, K_{eq} A value representing the equilibrium state of a chemical reaction involving the concentrations of the reactants and the products.

equivalent weight That weight of a substance that will react with, combine with, contain, replace, or in any other way be equivalent to 1 mole of hydrogen atoms or hydrogen ions.

evaporation The escape of molecules from the liquid state to the gas or vapor state.

exothermic reaction A chemical reaction in which heat is released as a product.

Fahrenheit scale (°F) The temperature scale on

which water freezes at 32°F and boils at 212°F at 1 atm pressure.

formula weight The sum of the atomic weights of all the atoms in a chemical formula.

free radical A chemical species having one or more unpaired electrons.

freezing or melting point The temperature at which the solid and liquid states of a substance are in equilibrium.

galvanic cell *See* voltaic cell.

gas The state of matter that is the least compact of the three physical states; a gas has no shape or definite volume and completely fills its container.

Gay-Lussac's law of combining volumes of gases At constant temperature and pressure, the ratios of the volumes of reacting gases are small whole numbers.

Graham's law of effusion The rates of effusion of different gases are inversely proportional to the square root of their molecular weights or densities.

groups or families of elements Vertical groups of elements in the periodic table (IA, IIA, and so on). Families of elements have similar outer-orbital electron structures.

halogen family Group VIIA of the periodic table; consists of the elements fluorine, chlorine, bromine, iodine, and astatine.

heat A form of energy associated with the motion of small particles of matter.

heat of fusion The amount of heat required to change 1 gram of a solid into a liquid at its melting point.

heat of reaction The quantity of heat produced by a chemical reaction.

heat of vaporization The amount of heat required to change 1 gram of a liquid to a vapor at its normal boiling point.

heterogeneous Matter without uniform composition; two or more phases present.

homogeneous Matter having uniform properties throughout.

hydrate A substance that contains water molecules as a part of its crystalline structure; $CuSO_4 \cdot 5H_2O$ is an example.

hydrogen bond A chemical bond between polar molecules that contain hydrogen covalently bonded to the highly electronegative atoms F, O, or N:

```
          O
         / \
        H   H
       .
      O
     / \
    H   H
```

hydrolysis A chemical reaction with water in which the water molecule is split into H^+ and OH^-

hydrometer An instrument used to measure the specific gravity of a liquid. It consists of a weighted bulb at the end of a sealed calibrated tube.

hydronium ion H_3O^+; formed when an H^+ ion reacts with a water molecule.

hygroscopic substance A substance that readily absorbs and retains water vapor.

hypothesis A tentative explanation of the results of data to provide a basis for further experimentation.

ideal gas A gas that obeys the gas laws and the Kinetic-Molecular Theory exactly.

ideal gas equation $PV = nRT$; a single equation relating the four variables—P, V, T, and n—used in the gas laws. R is a proportionality constant known as the ideal or universal gas constant.

immiscible Incapable of mixing. Immiscible liquids do not form solutions with one another.

inorganic chemistry The chemistry of the elements and their compounds other than the carbon compounds.

ion An electrically charged atom or group of atoms. A positively charged (+) ion is called a *cation*, and a negatively charged (−) ion is called an *anion*.

ionic bond A chemical bond between a positively charged ion and a negatively charged ion.

ionization The formation of ions.

ionization constant (K_i) The equilibrium constant for the ionization of a weak electrolyte in water.

ionization energy The energy required to remove an electron from an atom, an ion, or a molecule.

ion product constant for water (K_w) $K_w = [H^+][OH^-] = 1 \times 10^{-14}$ at 25°C

isotopes Atoms of an element having the same atomic number but different atomic masses. Since the atomic numbers are identical, isotopes vary only in the number of neutrons in the nucleus.

IUPAC International Union of Pure and Applied Chemistry.

joule (J) The SI unit of energy; 4.184 J = 1 cal.

Kelvin (absolute) scale (K) Absolute temperature scale starting at absolute zero, the lowest temperature possible. Freezing and boiling points of water on this scale are 273 K and 373 K, respectively, at 1 atm pressure.

kilocalorie (kcal) 1000 cal; the kilocalorie is also known as the nutritional or large Calorie, used for measuring the energy produced by food.

kilogram (kg) The standard unit of mass in the metric system.

kinetic energy (KE) Energy of motion; $KE = \frac{1}{2}mv^2$.

Kinetic-Molecular Theory A group of assumptions used to explain the behavior and properties of ideal gas molecules.

law A statement of the occurrence of natural phenomena that occur with unvarying uniformity under the same conditions.

Law of Conservation of Energy Energy cannot be created or destroyed, but it may be transformed from one form to another.

Law of Conservation of Mass There is no detectable change in the total mass of the substances in a chemical reaction; the mass of the products equals the mass of the reactants.

Law of Definite Composition A compound always contains the same elements in a definite proportion by weight.

Le Chatelier's principle If the conditions of an equilibrium system are altered, the system will shift to establish a new equilibrium system under the new set of conditions.

Lewis structure A method of indicating the covalent bonds between atoms in a molecule or an ion such that a pair of electrons (:) represents the valence electrons forming the covalent bond.

limiting reactant A reactant in a chemical reaction that limits the amount of product formed. The limitation is imposed because an insufficient quantity of the reactant, compared to amounts of the other reactants, was used in the reaction.

liquid One of the three physical states of matter. The particles in a liquid move about freely while the liquid still retains a definite volume. Thus, liquids flow and take the shape of their containers.

liter (L) A unit of volume commonly used in chemistry; 1 L = 1000 mL; the volume of a kilogram of water at 4°C.

mass The quantity or amount of matter that an object possesses.

mass number (*A*) The sum of the number of protons and neutrons in the nucleus of a given isotope of an atom.

matter Anything that has mass and occupies space.

mechanism of a reaction The route or steps by which a reaction takes place. The mechanism describes the manner in which atoms or molecules are transformed from reactants into products.

metal An element that is lustrous, ductile, malleable, and a good conductor of heat and electricity. Metals tend to lose their valence electrons and become positive ions.

metalloid An element having properties that are intermediate between those of metals and nonmetals.

meter (m) The standard unit of length in the SI and metric systems.

metric system A decimal system of measurements.

miscible Capable of mixing and forming a solution.

mixture Matter containing two or more substances that can be present in variable amounts.

molality (m) The number of moles of solute dissolved in 1000 grams of solvent.

molarity (*M*) The number of moles of solute per liter of solution.

molar solution A solution containing 1 mole of solute per liter of solution.

molar volume of a gas The volume of 1 mole of a gas at STP, 22.4 L/mol.

mole The amount of a substance containing the same number of formula units (6.022×10^{23}) as there are in exactly 12 grams of carbon-12. One mole is equal to the formula weight in grams of any substance.

molecular formula The true formula representing the total number of atoms of each element present in one molecule of a compound.

molecular weight The sum of the atomic weights of all the atoms in a molecule.

molecule A small, uncharged individual unit of a compound formed by the union of two or more atoms.

mole ratio A ratio of the number of moles of any two species in a balanced chemical equation. The mole ratio can be used as a conversion factor in stoichiometric calculations.

net ionic equation A chemical equation that includes only those molecules and ions that have changed in the chemical reaction.

neutralization The reaction of an acid and a base to form water plus a salt.

neutron A subatomic particle that is electrically neutral and has an assigned mass of 1 amu.

noble gases A family of elements in the periodic table—helium, neon, argon, krypton, and xenon—that contain a particularly stable electron structure.

nonelectrolyte A substance whose aqueous solutions do not conduct electricity.

nonmetal Any of a number of elements that do not have the characteristics of metals. They are located mainly in the upper right-hand corner of the periodic table.

nonpolar covalent bond A covalent bond between two atoms with the same electronegativity value. Thus, the electrons are shared equally between the two atoms.

normal boiling point The temperature at which the vapor pressure of a liquid equals 1 atm or 760 torr pressure.

normality The number of equivalent weights (equivalents) of solute per liter of solution.

nucleons A general term for the neutrons and protons in the nucleus of an atom.

nucleus The central part of an atom where all the protons and neutrons of the atom are located. The nucleus is very dense and has a positive electrical charge.

octet rule An atom tends to lose or gain electrons until it has eight electrons in its outer shell.

orbital A cloudlike region around the nucleus where electrons are located. Orbitals are considered to be energy sublevels within the principal levels and are labeled *s*, *p*, *d*, and *f*.

organic chemistry The branch of chemistry that deals with carbon compounds.

oxidation An increase in the oxidation number of an atom as a result of losing electrons.

oxidation number (oxidation state) A small number representing the state of oxidation of an atom. For an ion, it is the positive or negative charge on the ion; for covalently bonded atoms, it is a positive or negative number assigned to the more electronegative atom; in free elements, it is zero.

oxidation-reduction A chemical reaction wherein electrons are transferred from one element to another.

oxidizing agent A substance that causes an increase in the oxidation state of another substance. The oxidizing agent is reduced during the course of the reaction.

percentage composition of a compound The weight percent of each element in a compound.

percent yield $\frac{\text{Actual yield}}{\text{Theoretical yield}} \times 100\%$

periodic law The properties of the chemical elements are a periodic function of their atomic numbers.

periodic table An arrangement of the elements according to their atomic numbers, illustrating the periodic law. The table consists of horizontal rows or periods and vertical columns or families of elements. Each period ends with a noble gas.

periods of elements The horizontal groupings of elements in the periodic table.

pH A method of expressing the H^+ concentration (acidity) of a solution. $pH = -\log[H^+]$; $pH = 7$ is a neutral solution, $pH < 7$ is acidic, and $pH > 7$ is basic.

phase A homogeneous part of a system separated from other parts by a physical boundary.

physical change A change in form (such as size, shape, physical state) without a change in composition.

physical properties Characteristics associated with the existence of a particular substance. Inherent characteristics such as color, taste, density, and melting point are physical properties of various substances.

physical states of matter Solids, liquids, and gases.

pOH A method of expressing the basicity of a solution. pOH = $-\log[OH^-]$. pOH = 7 is a neutral solution, pOH < 7 is basic, and pOH > 7 is acidic.

polar covalent bond A covalent bond between two atoms with differing electronegativity values resulting in unequal sharing of bonding electrons.

polyatomic ion An ion composed of more than one atom.

positron A particle with a +1 charge having the mass of an electron (a positive electron).

potential energy Stored energy or the energy an object has because of its relative position.

pressure Force per unit area; expressed in many units, such as mm Hg, atm, in./cm^2, torr.

product A chemical substance produced from reactants by a chemical change.

properties The characteristics, or traits, of substances. Properties are classified as physical or chemical.

proton A subatomic particle found in the nucleus of all atoms; has a charge of +1 and a mass of about 1 amu. An H^+ ion is a proton.

quantum mechanics or wave mechanics The modern theory of atomic structure based on the wave properties of matter.

rate of reaction The rate at which the reactants of a chemical reaction disappear and the products form.

reactant A chemical substance entering into a reaction.

redox An abbreviation for *oxidation–reduction.*

reducing agent A substance that causes a decrease in the oxidation state of another substance. The reducing agent is oxidized during the course of a reaction.

reduction A decrease in the oxidation number of an element as a result of gaining electrons.

representative element An element in one of the A groups in the periodic table.

reversible reaction A chemical reaction in which products can react to form the original reactants. A double arrow ($\rightleftarrows$) is used to indicate that a reaction is reversible.

salts Ionic compounds of cations and anions.

saturated solution A solution containing dissolved solute in equilibrium with undissolved solute.

scientific method A method of solving problems by observation; recording and evaluating data of an experiment; formulating hypotheses and theories to explain the behavior of nature; and devising additional experiments to test the hypotheses and theories to see if they are correct.

scientific notation A number between 1 and 10 (the decimal point after the first nonzero digit) multiplied by 10 raised to a power; for example, 6.022×10^{23}.

significant figures The number of digits that are known plus one that is uncertain are considered significant in a measured quantity.

simplest formula *See* empirical formula.

single bond A covalent bond in which one pair of electrons is shared between two atoms.

single-displacement reaction A reaction of an element and a compound to produce a different element and a different compound.

solid One of the three physical states of matter; matter in the solid state has a definite shape and a definite volume.

solubility An amount of solute that will dissolve in a specific amount of solvent.

solubility product constant (K_{sp}) The equilibrium constant for the solubility of a slightly soluble salt.

solute The substance that is dissolved in a solvent to form a solution.

solution A homogeneous mixture of two or more substances.

solvent The substance present to the largest extent in a solution. The solvent dissolves the solute.

specific gravity The ratio of the density of one substance to the density of another substance taken as a standard. Water is usually the standard for liquids and solids; air, for gases.

specific heat The quantity of heat required to change the temperature of 1 gram of any substance by 1°C.

spectator ion An ion in solution that does not undergo chemical change during a chemical reaction.

standard boiling point *See* normal boiling point.

standard conditions *See* STP.

stoichiometry The area of chemistry that deals with the quantitative relationships among reactants and products in a chemical reaction.

STP (standard temperature and pressure) 0°C (273 K) and 1 atm (760 torr).

strong electrolyte An electrolyte that is essentially 100% ionized in aqueous solution.

subatomic particles Mainly protons, neutrons, and electrons.

sublimation The process of going directly from the solid state to the vapor state without becoming a liquid.

substance Matter that is homogeneous and has a definite, fixed composition. Substances occur in two forms—as elements and as compounds.

supersaturated solution A solution containing more solute than a saturated solution at a particular temperature. Supersaturated solutions tend to be unstable; jarring the container or dropping in a "seed" crystal will cause crystallization of the excess solute.

symbol In chemistry, an abbreviation for the name of an element.

temperature A measure of the intensity of heat or how hot or cold a system is.

theoretical yield The maximum amount of product that can be produced according to a balanced equation.

theory An explanation of the general principles of certain phenomena with considerable evidence of facts to support it.

titration The process of measuring the volume of one reagent required to react with a measured weight or volume of another reagent.

torr A unit of pressure (1 torr = 1 mm Hg).

total ionic equation An equation that shows compounds in the form in which they actually exist. Strong electrolytes are written as ions in solution, whereas nonelectrolytes, weak electrolytes, precipitates, and gases are written in the un-ionized form.

transition elements The metallic elements characterized by increasing numbers of *d* and *f* electrons in an inner shell. These elements are located in Groups IB through VIIB and in Group VIII of the periodic table.

triple bond A covalent bond in which three pairs of electrons are shared between two atoms.

un-ionized equation (molecular equation) A chemical equation in which all the reactants and products are written in their molecular or normal formula expression.

unsaturated solution A solution containing less solute per unit volume than its corresponding saturated solution.

valence electrons Electrons in the outer energy level of an atom. These electrons are primarily involved in chemical reactions.

vaporization *See* evaporation.

vapor pressure The pressure exerted by a vapor in equilibrium with its liquid.

vapor pressure curve A graph generated by plotting the temperature of a liquid on the *x* axis and its vapor pressure on the *y* axis. Any point on the curve represents an equilibrium between the vapor and liquid.

volatile substance A substance that evaporates readily; a liquid with a high vapor pressure and a low boiling point.

voltaic cell An electrochemical cell that produces an electric current from a spontaneous chemical reaction.

volume The amount of space occupied by matter.

volume percent solution The volume of solute in 100 mL of solution.

water of crystallization or hydration Water molecules that are part of a crystalline structure, as in a hydrate.

weak electrolyte A substance that is ionized to a small extent in aqueous solution.

weight (mass) An extraneous property that an object possesses. The weight of an object depends on the gravitational attraction of the earth for that object. Therefore, an object's weight depends on its location in relation to the earth.

weight percent solution The grams of solute in 100 g of a solution.

yield The amount of product obtained from a chemical reaction.

APPENDIX VI
Answers to Selected Problems

Chapter 1

6. The following statements are correct: a, b, d, f, g, h, i.

Chapter 2

C.8 The following statements are correct: a, c, d, e, g, h, i, j, l, n, o, r, s.

D.1 (a) 2 (b) 3 (c) 3 (d) 2 (e) 3 (f) 6 (g) 3 (h) 4

D.2 (a) 93.2 (b) 8.87 (c) 0.0285 (d) 21.3 (e) 4.64 (f) 130 (g) 34.3 (h) 2.00×10^{6}

D.3 (a) 2.9×10^{6} (b) 4.56×10^{-2} (c) 5.8×10^{-1} (d) 4.0822×10^{3} (e) 8.40×10^{-3} (f) 4.030×10^{1} (g) 1.2×10^{7} (h) 5.5×10^{-6}

D.4 (a) 14.4 (b) 58.5 (c) 1.08×10^{8} (d) 40. (e) 2.0×10^{2} (f) 2009 (g) 0.54 (h) 1.79×10^{3} (i) 7.18×10^{-3} (j) 2.49×10^{-4}

D.5 (a) 0.833 (b) 0.429 (c) 0.750 (d) 0.500

D.6 (a) 1.9 (b) 86.7 (c) 2.1 (d) 51.3 (e) 8.93 (f) 0.030

D.7 (a) 100°C (b) 72°F (c) 318 K (d) 4.6 mL

D.8 (a) 0.280 m (b) 1.000 km (c) 92.8 mm (d) 1.5×10^{-4} km (e) 6.06×10^{-6} km (f) 4.5×10^{8} Å (g) 6.5×10^{3} Å (h) 1.21×10^{3} cm (i) 8.0×10^{3} m (j) 31.5 cm (k) 2.5×10^{7} mm (l) 1.2×10^{-6} cm (m) 5.20×10^{4} cm (n) 0.3884 nm (o) 107 cm (p) 4.1×10^{4} in. (q) 811 km (r) 12.9 cm^2 (s) 117 ft (t) 10.2 miles (u) 7.4×10^{4} mm^3 (v) 1.25×10^{19} mm^3

D.9 (a) 1.068×10^{4} mg (b) 6.8×10^{-2} kg (c) 8.54×10^{-3} kg (d) 94.3 lb (e) 0.164 g (f) 6.5×10^{5} mg (g) 5.5×10^{3} g (h) 4.3×10^{4} g

D.10 (a) 2.50×10^{-2} L (b) 2.24×10^{4} mL (c) 3.3×10^{3} mL (d) 1.3×10^{3} m^3 (e) 468 mL (f) 9.41 gal (g) 9.0×10^{-3} mL (h) 75.7 L

D.11 (a) 89 km/hr (b) 81 ft/s

D.12 (a) 33.4 ft/s (b) 22.8 miles/hr

D.13 297 g salt remaining

D.14 (a) 4.7×10^{4} miles/hr (b) 7.5×10^{4} km/hr

D.15 77.2 kg

D.16 0.32 g

D.17 5.0×10^{2} sec

D.18 3×10^{4} mg

D.19 $0.392/kg

D.20 2.52×10^{3}

D.21 3.0×10^{3} times heavier

D.22 $21

D.23 56 L

D.24 2.3×10^{2} L

D.25 7.6×10^{4} drops

D.26 1.6×10^{2} L

D.27 5.0×10^{-3} mL

D.28 2.83×10^{4} mL

D.29 4×10^{5} m^2

D.30 (a) 72.2°C (b) −18°C (c) 255 K (d) −0.40°F (e) 90°F (f) −23°C (g) 546 K (h) −61°C

D.31 98.6°F

D.32 −100°C is 10°F colder than −138°F

D.33 (a) −40°C = −40°F (b) 11.4°F = −11.4°C

D.34 1.7×10^{4} J

D.35 1.9×10^{3} J; 4.5×10^{2} cal

D.36 0.30 J/g °C

D.37 0.49 J/g °C
D.38 17°C
D.39 41.6°C
D.40 1.565 g/mL
D.41 3.12 g/mL
D.42 7.1 g/mL
D.43 595 g
D.44 1.28 g/mL
D.45 340 g
D.46 76.9 g
D.47 A = magnesium, B = aluminum, C = silver
D.48 3.57×10^3 g
D.49 0.965 g/mL
D.50 (a) 2.72 g/mL (b) 2.72
D.51 49.8 mL
D.52 Ethyl alcohol; 63 mL vs. 50 mL
D.53 18.3 g/cm^3
D.54 4.83×10^3 mL
D.55 Density of liquid = 0.842 g/mL; density of slug = 2.7 g/mL
D.56 (a) 1.16 (b) 77.7 mL sulfuric acid

Chapter 3

C.20 The following statements are correct: c, f, g, j, l, n, o, q, t.
D.1 (a) 391 K (b) 244°F
D.2 30.3% iron
D.3 2.25 g copper(II) oxide
D.4 21.9 g copper
D.5 60.2 g mercury
D.6 8.50% fat
D.7 (a) 6.9 g oxygen (b) 60.3% magnesium
D.8 21.7 g chlorine
D.9 (a) 5.1×10^{14} J; 1.2×10^{14} cal
(b) 4.0×10^8 gal

Chapter 4

C.30 The following statements are correct: c, e, g, j, k, m, n, p, r, u, v, w, y.
D.1 13.8 g Na
D.2 36% Zn, 64% Cu
D.3 16.8 g CaO
D.4 2.66 g/mL
D.5 54.5 mL Br_2
D.6 (a) 44.4% S (b) Ca (c) 24.0 g S
D.7 18 carat
D.8 75% C
D.9 7.6×10^3 g Au
D.10 1.6×10^3 kg alloy

Chapter 5

C.8 The following statements are correct: a, b, c, f, g, i.
C.28 The following statements are correct: a, c, d, e, f, h, i, j, k, l, n, p, q, r.
C.39 The following statements are correct: b, c, e, f, g, h, k, m.
C.43 The following statements are correct: b, d.
D.1 (a) 5.1×10^5 (b) 2.74×10^{-4}
(c) 1×10^{-6} (d) 5.1×10^2
D.2 1st = 2; 2nd = 8; 3rd = 18; 4th = 32; 5th = 50; 6th = 72
D.3

(a) 8	16	O	8	8
(b) 28	58	Ni	28	30
(c) 80	199	Hg	80	119

D.4 131 amu
D.5 107.87 amu
D.6 24.31 amu

Chapter 6

C.39 The following statements are correct: a, d, e, g, h, i, k, n, o, p, t, u, v.

Chapter 7

C.31 The following statements are correct: a, b, e, h, i, j, l, o, q, r, s, t, u, v, w, x, z, bb, cc, gg, hh, jj.

Chapter 8

30. The following statements are correct: a, d, f, g, i, j, l, m, o, p, r, u, w, x, y, aa.

Chapter 9

B.15 The following statements are correct: a, b, e, g, h, k, l, n.
C.1 (a) 119.0 (b) 142.1 (c) 331.2
(d) 46.0 (e) 60.0 (f) 231.4
(g) 342.0 (h) 342.3 (i) 132.0
C.2 (a) 40.0 g (b) 275.8 g (c) 152.0 g
(d) 96.0 g (e) 146.3 g (f) 122.0 g
(g) 180.0 g (h) 368.2 g (i) 244.3 g
C.3 (a) 0.344 mol Zn (b) 2.83×10^{-2} mol Mg
(c) 7.5×10^{-2} mol Cu (d) 6.49 mol Co
(e) 4.6×10^{-4} mol Sn (f) 28 mol N atoms

C.4 (a) 0.625 mol NaOH (b) 0.275 mol Br_2
(c) 7.18×10^{-3} mol $MgCl_2$
(d) 0.463 mol CH_3OH
(e) 2.03×10^{-2} mol Na_2SO_4 (f) 5.97 mol ZnI_2
C.5 (a) 108 g Au (b) 284 g H_2O
(c) 888 g Cl_2 (d) 252 g NH_4NO_3
(e) 4.17×10^{-2} g H_2SO_4 (f) 11.5 g CCl_4
(g) 0.122 g Ti (h) 8.0×10^{-7} g S
C.6 (a) 7.59×10^{23} molecules O_2
(b) 3.4×10^{23} molecules C_6H_6
(c) 6.02×10^{23} molecules CH_4
(d) 1.65×10^{25} molecules HCl
C.7 (a) 3.441×10^{-22} g Pb
(b) 1.791×10^{-22} Ag
(c) 2.99×10^{-23} g H_2O
(d) 3.77×10^{-22} g $C_3H_5(NO_3)_3$
C.8 (a) 550 g Cu (b) 24.6 kg Au
(c) 1.7×10^{-23} mol C
(d) 8.3×10^{-21} mol CO_2
(e) 0.885 mol S
(f) 42.7 mol NaCl
(g) 1.05×10^{24} atoms Mg
(h) 9.47 mol Br_2
C.9 (a) 6.022×10^{23} CS_2 molecules
(b) 6.022×10^{23} C atoms
(c) 1.204×10^{24} S atoms
(d) 1.806×10^{24} total atoms
C.10 8.43×10^{23} atoms P
C.11 5.88 g Na
C.12 10.8 g/atomic wt.
C.13 5.54×10^{19} m
C.14 1.4×10^{14} dollars per person
C.15 (a) 8.3×10^{16} drops per mile^3
(b) 7.3×10^{6} miles^3
C.16 (a) 10.3 cm^3 Ag (b) 2.18 cm
C.17 (a) 6.02×10^{23} atoms O
(b) 3.75×10^{23} atoms O
(c) 3.60×10^{23} atoms O
(d) 6.0×10^{24} atoms O
(e) 5.36×10^{24} atoms O
(f) 5.0×10^{18} atoms O
C.18 (a) 14.4 g Ag (b) 1.28 g Cl (c) 266 g N
(d) 6.74 g O (e) 6.00 g H
C.19 10.3 mol H_2SO_4
C.20 1.62 mol HNO_3
C.21 (a) H_2O (b) CH_3OH
C.22 41.58 g Fe_2S_3
C.23 (a) 22.4% Na; 77.6% Br
(b) 39.1% K; 1.0% H; 12.0% C; 48.0% O
(c) 34.4% Fe; 65.6% Cl
(d) 16.5% Si; 83.5% Cl
(e) 15.8% Al; 28.1% S; 56.1% O
(f) 63.5% Ag; 8.2% N; 28.3% O
C.24 (a) 47.9% Zn; 52.1% Cl
(b) 18.2% N; 9.1% H; 31.2% C; 41.6% O
(c) 12.3% Mg; 31.3% P; 56.5% O
(d) 21.2% N; 6.1% H; 24.3% S; 48.4% O
(e) 23.1% Fe; 17.4% N; 59.6% O
(f) 54.4% I; 45.6% Cl
C.25 (a) 77.7% Fe (b) 69.92% Fe
(c) 72.3% Fe (d) 15.2% Fe
C.26 Highest percent Cl, LiCl; lowest, $BaCl_2$
C.27 43.7% P; 56.3% O
C.28 24.2% C; 4.0% H; 71.7% Cl
C.29 8.60 g Li
C.30 (a) 76.98% Hg (b) 46.4% O
(c) 17.3% N (d) 2.72% Mg
C.31 (a) H_2O (b) N_2O_3 (c) Equal
(d) $KClO_3$ (e) $KHSO_4$ (f) Na_2CrO_4
C.32 (a) N_2O (b) NO (c) N_2O_5 (d) Na_2CO_3
(e) $NaClO_4$ (f) Mn_3O_4
C.33 (a) CuCl (b) $CuCl_2$ (c) Cr_2S_3
(d) K_3PO_4 (e) $BaCr_2O_7$ (f) PBr_8Cl_3
C.34 SnO_2
C.35 V_2O_5
C.36 No. 19.5 g Zn react with 9.57 g S
(only 9.40 g S is available).
C.37 $C_6H_6O_2$
C.38 $C_6H_{12}O_6$
C.39 $C_9H_8O_4$
C.40 4.77 g O_2
C.41 GaAs
C.42 (a) CCl_4; CCl_4 (b) CCl_3; C_2Cl_6
(c) CCl; C_6Cl_6 (d) C_3Cl_8; C_3Cl_8

Chapter 10

C.20 The following statements are correct:
a, d, e, f, h, i, j, l, n, o, p, q.

Chapter 11

B.1 (a) 0.247 mol KNO_3
(b) 0.0658 mol $Ca(NO_3)_2$
(c) 4.4 mol $(NH_4)C_2O_4$
(d) 25.0 mol $NaHCO_3$
(e) 3.85×10^{-3} mol $ZnCl_2$
(f) 0.056 mol NaOH
(g) 16 mol CO_2
(h) 4.29 mol C_2H_5OH
(i) 0.237 mol H_2SO_4
B.2 (a) 272 g $Fe(OH)_3$ (b) 1.31 g $NiSO_4$
(c) 1.25×10^{5} g $CaCO_3$ (d) 3.60 g $HC_2H_3O_2$

(c) 179 g NH_3 (f) 373 g Bi_2S_3 (g) 2.6 g HCl
(h) 1.35 g $C_6H_{12}O_6$ (i) 1.56×10^3 g Br_2
(j) 18 g K_2CrO_4

B.3 (a) 10.0 g H_2O (b) 25.0 g HCl

B.4 (a) $\frac{6 \text{ mol } CO_2}{2 \text{ mol } C_3H_7OH}$ (b) $\frac{2 \text{ mol } C_3H_7OH}{9 \text{ mol } O_2}$
(c) $\frac{9 \text{ mol } O_2}{6 \text{ mol } CO_2}$ (d) $\frac{8 \text{ mol } H_2O}{2 \text{ mol } C_3H_7OH}$
(e) $\frac{6 \text{ mol } CO_2}{8 \text{ mol } H_2O}$ (f) $\frac{8 \text{ mol } H_2O}{9 \text{ mol } O_2}$

B.5 (a) $\frac{3 \text{ mol } CaCl_2}{1 \text{ mol } Ca_3(PO_4)_2}$ (b) $\frac{6 \text{ mol HCl}}{2 \text{ mol } H_3PO_4}$
(c) $\frac{3 \text{ mol } CaCl_2}{2 \text{ mol } H_3PO_4}$ (d) $\frac{1 \text{ mol } Ca_3(PO_4)_2}{2 \text{ mol } H_3PO_4}$
(e) $\frac{6 \text{ mol HCl}}{1 \text{ mol } Ca_3(PO_4)_2}$ (f) $\frac{2 \text{ mol } H_3PO_4}{6 \text{ mol HCl}}$

B.6 2.80 mol Cl_2
B.7 540 g NaOH
B.8 (a) 8.33 mol H_2O (b) 0.800 mol $Al(OH)_3$
B.9 19.7 g $Zn_3(PO_4)_2$
B.10 (a) 0.500 mol Fe_2O_3 (b) 12.4 mol O_2
(c) 6.20 mol SO_2 (d) 65.6 g SO_2
(e) 0.871 mol O_2 (f) 332 g FeS_2
B.11 4.20 mol HCl
B.12 (a) 2.22 mol H_2 (b) 162 g HCl
B.13 87.4 kg Fe
B.14 117 g steam (H_2O); 271 g Fe
B.15 (a) 52.5 mol O_2 (b) 13.0 g CO_2
(c) 220 g CO_2
B.16 (a) HNO_3 is limiting; KOH in excess
(b) H_2SO_4 is limiting; NaOH in excess
(c) H_2S is limiting; $Bi(NO_3)_3$ in excess
(d) H_2O is limiting; Fe in excess
B.17 (a) 2.0 mol Cu; 2.0 mol $FeSO_4$;
1.0 mol $CuSO_4$ (unreacted)
(b) 15.9 g Cu; 38.1 g $FeSO_4$;
5.97 g Fe (unreacted)
B.18 (a) 3.0 mol CO_2
(b) 9.0 mol CO_2
(c) 1.8 mol CO_2
(d) 6.0 mol CO_2; 8.0 mol H_2O; 4.0 mol O_2
(e) 16.5 g CO_2
(f) 60.0 g CO_2
(g) 60.0 g CO_2
B.19 57.8%
B.20 45.7 g CH_3OH (alcohol); 4.29 g H_2 unreacted
B.21 95.0% yield of Cu
B.22 (a) 324 g C_2H_5OH
(b) 1.10×10^3 g $C_6H_{12}O_6$
B.23 522 g C
B.24 77.8% CaC_2
B.25 $MgCl_2$ produces more AgCl than $CaCl_2$
B.26 3.73×10^2 kg Li_2O
B.27 14 kg of concentrated H_2SO_4
B.28 13 tablets
B.29 67.2% $KClO_3$
B.30 The following statements are correct:
a, c, d, f.
B.31 The following statements are correct:
a, c, e.

Chapter 12

C.13 The following statements are correct:
b, d, f(1, 4), h, i, n, o, p, q, t.
D.1 (a) 0.941 atm (b) 28.1 in. Hg (c) 13.8 lb/in.^{2}
(d) 715 torr (e) 953 mbar (f) 95.3 kPa
D.2 (a) 0.037 atm (b) 79 atm
(c) 1.05 atm (d) 4.92×10^{-2} atm
D.3 (a) 263 mL (b) 800 mL (c) 132 mL
D.4 (a) 374 mm Hg (b) 711 mm Hg
D.5 1.0×10^2 atm
D.6 (a) 6.60 L (b) 6.17 L (c) 2.42 L (d) 8.35 L
D.7 600 K (or 327°C)
D.8 1.48×10^3 torr
D.9 −62°C
D.10 65 atm
D.11 320°F
D.12 (a) 363 mL (b) 778 mL
D.13 6.10 L
D.14 3.55 atm
D.15 2.4×10^5 L
D.16 33.4 L
D.17 7.39×10^{21} molecules; 2.22×10^{22} atoms;
275 mL at STP
D.18 703 torr
D.19 1100 torr
D.20 2.42 L
D.21 370 torr
D.22 56 L
D.23 (a) 34 mol H_2 (b) 1.2×10^2 g H_2
D.24 4.91 g CO_2
D.25 43.2 g/mol (mol. wt)
D.26 (a) 22.4 L (b) 8.30 L (c) 44.6 L
D.27 2.69×10^{22} molecules NH_3
D.28 44.6 mol Cl_2
D.29 39.9 g/mol (mol. wt)
D.30 (a) 3.74 g/L (b) 0.18 g/L
(c) 3.58 g/L (d) 2.50 g/L
D.31 (a) 1.70 g/L (b) 1.54 g/L

D.32 $-78°C$
D.33 (a) 8.4 L H_2 (b) 8.20 g CH_4 (c) 8.47 g/L
(d) 63.5 g/mol (mol. wt)
D.34 164 g/mol (mol. wt)
D.35 57 L
D.36 123 L
D.37 2.81 K
D.38 0.127 mol
D.39 279 L H_2
D.40 (a) 5.5 mol NH_3 (b) 5.6 mol NH_3
(c) 215 L NO (d) 0.640 L NO (e) 2.4 L NO
(f) 1.1×10^2 g O_2 (g) 9.7 g NH_3
D.41 (a) 308 L O_2 (b) 224 L SO_2
D.42 9.0 atm
D.43 $\frac{\text{He}}{N_2} = \frac{2.6}{1.0}$ (rate of effusion)
D.44 (a) $He/CH_4 = 2{:}1$
(b) 67 cm from the He end
D.45 C_4H_8 (molecular formula)
D.46 (a) 10 mol CO_2; 3 mol O_2; no CO
(b) 29 atm
D.47 76.0% $KClO_3$

Chapter 13

C.47 The following statements are correct:
a, b, c, f, h, l, m, o, p, s, t, u, w, y.
D.1 (a) 0.420 mol $CoCl_2 \cdot 6H_2O$
(b) 0.262 mol $FeI_2 \cdot 4H_2O$
D.2 (a) 2.52 mol H_2O (b) 1.05 mol H_2O
D.3 51.1% H_2O
D.4 48.6% H_2O
D.5 $Pb(C_2H_3O_2)_2 \cdot 3H_2O$
D.6 $FePO_4 \cdot 4H_2O$
D.7 3.11×10^5 J
D.8 5.5×10^4 J
D.9 1.6×10^5 cal (6.7×10^5 J)
D.10 4.07×10^4 J
D.11 Yes, sufficient ice
D.12 Not sufficient steam
D.13 2.29×10^6 J
D.14 43.9 kJ
D.15 40.2 g H_2O
D.16 18.0 mL vs. 22.4 L
D.17 (a) 6.00 mol O_2 (b) 162 L O_2
D.18 6.2 L O_2
D.19 (a) 18.0 g H_2O (b) 36.0 g H_2O
(c) 0.783 g H_2O (d) 18.0 g H_2O
(e) 0.447 g H_2O (f) 0.167 g H_2O
D.20 6.97×10^{18} molecules/second
D.21 54 g H_2SO_4
D.22 (a) Yes, O_2 remains (b) 20.0 mL O_2

Chapter 14

C.27 The following statements are correct:
a, b, f, h, j, k, l, n, p, r, s, t, v.
D.1 (a) 20.0% NaBr (b) 10.7% K_2SO_4
(c) 7.41% $Mg(NO_3)_2$
D.2 (a) 240 g solution (b) 544 g solution
D.3 (a) 23.1% NaCl (b) 22% $HC_2H_3O_2$
(c) 15.3% $C_6H_{12}O_6$; 84.7% H_2O
D.4 (a) 3.3 g KCl (b) 37.5 g K_2CrO_4
(c) 6.4 g $NaHCO_3$
D.5 455 g solution
D.6 (a) 4.5 g NaCl (b) 449 g H_2O
D.7 19 g H_2O needs to evaporate
D.8 (a) 22.0% CH_3OH (b) 33.6% NaCl
D.9 (a) 25.0% CH_3OH (b) 22% CCl_4
D.10 214 mL
D.11 (a) 424 g HNO_3 (b) 1.18 L
D.12 (a) 0.40 *M* (b) 3.8 *M* NaCl (c) 2.5 *M* HCl
(d) 0.59 *M* $BaCl_2 \cdot 2H_2O$
D.13 (a) 0.327 *M* Na_2CrO_4
(b) 1.81 *M* $C_6H_{12}O_6$
(c) 2.19×10^{-3} *M* $Al_2(SO_4)_3$
(d) 0.172 *M* $Ca(NO_3)_2$
D.14 (a) 40 mol LiCl (b) 0.075 mol H_2SO_4
(c) 3.5×10^{-4} mol NaOH (d) 16 mol $CoCl_2$
D.15 (a) 8.8×10^3 g NaCl (b) 13 g HCl
(c) 4.6×10^2 g H_2SO_4 (d) 8.58 g $Na_2C_2O_4$
D.16 (a) 1.68×10^3 mL (b) 3.91×10^4 mL
(c) 1.05×10^3 mL (d) 7.81×10^3 mL
D.17 6.72 *M* HNO_3
D.18 (a) 6.0 *M* HCl (b) 0.064 *M* $ZnSO_4$
(c) 1.6 *M* HCl
D.19 (a) 200 mL of 12 *M* HCl
(b) 20 mL of 15 *M* NH_3
(c) 16 mL of 16 *M* HNO_3
(d) 69 mL of 18 *M* H_2SO_4
D.20 8.2×10^3 mL (8.2 L)
D.21 (a) 0.48 *M* H_2SO_4 (b) 0.74 *M* H_2SO_4
(c) 1.83 *M* H_2SO_4
D.22 540 mL (water to be added)
D.23 0.32 *M* HNO_3
D.24 Take 31.3 mL of 5.00 *M* KOH and dilute with water to a volume of 250 mL
D.25 (a) 7.60 g $BaCrO_4$
(b) 15 mL of 1.0 *M* $BaCl_2$
D.26 (a) 33.3 mL of 0.250 *M* H_3PO_4
(b) 1.10 g $Mg_3(PO_4)_2$

D.27 (a) 0.300 mol H_2 (b) 7.80 L H_2
D.28 2.08 *M* HCl
D.29 (a) 0.67 mol KCl
(b) 0.33 mol $CrCl_3$
(c) 0.30 mol $FeCl_2$
(d) 69 mL of 0.060 *M* $K_2Cr_2O_7$
(e) 35 mL of 6.0 *M* HCl
D.30 (a) 1.3 mol Cl_2 (b) 16 mol HCl
(c) 1.3×10^2 mL of 6.0 *M* HCl (d) 3.2 L Cl_2
D.31 (a) HCl: 36.5 g/eq. wt; NaOH: 40.0 g/eq. wt
(b) HCl: 36.5 g/eq. wt; $Ba(OH)_2$: 85.7 g/eq. wt
(c) H_2SO_4: 49.1 g/eq. wt;
$Ca(OH)_2$: 37.0 g/eq. wt
(d) H_2SO_4: 98.1 g/eq. wt; KOH: 56.1 g/eq. wt
(e) H_3PO_4: 49.0 g/eq. wt; LiOH: 23.9 g/eq. wt
D.32 (a) 4.0 *N* HCl (b) 0.243 *N* HNO_3
(c) 6.0 *N* H_2SO_4 (d) 5.55 *N* H_3PO_4
(e) 0.250 *N* $HC_2H_3O_2$
D.33 1.86 *N* H_2SO_4
D.34 $Mg(OH)_2$ tablet
D.35 (a) 9.943 mL NaOH (b) 7.261 mL NaOH
(c) 20.19 mL NaOH
D.36 (a) 4.38 *m* CH_3OH (b) 10.0 *m* C_6H_6
(c) 5.6 *m* $C_6H_{12}O_6$
D.37 (a) CH_3OH (b) Both the same
D.38 6.2 *m* H_2SO_4; 5.0 *M* H_2SO_4
D.39 Solution will freeze in 20.0°F temperature
D.40 (a) 101.5°C (b) 2.9 *m* $C_{12}H_{22}O_{11}$
D.41 (a) 10.8 *m* $C_2H_6O_2$ (b) 105.5°C
(c) −20.1°C
D.42 (a) 0.545 *m* $C_{10}H_8$ (b) 2.7°C (c) 81.5°C
D.43 153 g/mol
D.44 163 g/mol
D.45 500 g H_2O
D.46 (a) 8.04×10^3 g $C_2H_6O_2$
(b) 7.24×10^3 mL $C_2H_6O_2$
(c) −4.0°F
D.47 60 g NaOH solution
D.48 96.5 g NaOH solution
D.49 (a) 1.6×10^2 g sugar (b) 0.47 *M* (c) 0.516 *m*
D.50 Molecular formula is $C_8N_4H_2$

Chapter 15

C.30 The following statements are correct:
a, c, f, i, j, k, l, n, p, r, t.
D.1 (a) 0.015 *M* NaCl = 0.015 *M* Na^+,
0.015 *M* Cl^-
(b) 4.25 *M* $NaKSO_4$ = 4.25 *M* Na^+,
4.25 *M* K^+, 4.25 *M* SO_4^{2-}
(c) 0.75 *M* $ZnBr_2$ = 0.75 *M* Zn^{2+},
1.5 *M* Br
(d) 1.65 *M* $Al_2(SO_4)_3$ = 3.30 *M* Al^{3+},
4.95 *M* SO_4^{2-}
(e) 0.20 *M* $CaCl_2$ = 0.20 *M* Ca^{2+},
0.40 *M* Cl^-
(f) 0.265 *M* K^+, 0.265 *M* I^-
(g) 0.682 *M* NH_4^+, 0.341 *M* SO_4^{2-}
(h) 0.0627 *M* Mg^{2+}, 0.125 *M* ClO_3
D.2 (a) 0.035 g Na^+, 0.053 g Cl^-
(b) 9.78 g Na^+, 16.6 g K^+, 40.8 g SO_4^{2-}
(c) 4.9 g Zn^{2+}, 12 g Br^-
(d) 8.91 g Al^{3+}, 47.6 g SO_4^{2-}
(e) 0.80 g Ca^{2+}, 1.4 g Cl^-
(f) 1.04 g K^+, 3.36 g I^-
(g) 1.23 g NH_4^+, 3.28 g SO_4^{2-}
(h) 0.152 g Mg^{2+}, 1.04 g ClO_3^-
D.3 0.260 *M* Ca^{2+}
D.4 (a) 1.0 *M* Na^+, 1.0 *M* Cl^-
(b) 0.50 *M* Na^+, 0.50 *M* Cl^-
(c) 1.0 *M* K^+, 0.50 *M* Ca^{2+}, 2.0 *M* Cl^-
(d) 0.20 *M* K^+, 0.20 *M* H^+, 0.40 *M* Cl^-
(e) No ions left in solution
(f) 0.67 *M* Na^+, 0.67 *M* NO_3^-
D.5 3.0×10^3 mL
D.6 (a) 0.684 *M* HCl (b) 0.542 *M* HCl
(c) 0.367 *M* HCl (d) 0.147 *M* NaOH
(e) 0.964 *M* NaOH (f) 0.4750 *M* NaOH
D.7 0.309 *M* $Ba(OH)_2$
D.8 (a) 40.8 mL of 0.245 *M* HCl
(b) 1.57×10^3 mL of 0.245 *M* HCl
D.9 0.201 *M* HCl
D.10 437 mL HCl
D.11 0.673 g KOH
D.12 88.0% NaOH
D.13 7.0% NaCl
D.14 (a) 0.468 L H_2 (b) 0.936 L H_2
D.15 (a) 2 (b) 0.0 (c) 8.19 (d) 7 (e) 0.30 (f) 4.0
D.16 (a) 3.4 (b) 2.6 (c) 4.3 (d) 10.5
D.17 4
D.18 13.9 L of 18.0 *M* H_2SO_4
D.19 The solution will be basic.
D.20 0.3586 *N* NaOH
D.21 22.7 mL of 0.325 *N* HNO_3
D.22 22.7 mL of 0.325 *N* H_2SO_4
D.23 0.4536 *N* H_3PO_4; 0.1512 *M* H_3PO_4

Chapter 16

C.24 The following statements are correct:
c, d, g, i, j, k, l, n, q, s, t, u, v, z.

D.1 4.20 mol HI
D.2 (a) 3.16 mol HI
(b) 3.4 mol HI; 0.30 mol H_2; 0.57 mol I_2
(c) $K_{eq} = 57$
D.3 2.75 mol H_2; 0.538 mol I_2; 0.500 mol HI
D.4 $K_{eq} = 29$
D.5 128 times as fast
D.6 HClO, 3.5×10^{-8}; $HC_3H_5O_2$, 1.3×10^{-5}; HCN, 4.0×10^{-10}
D.7 (a) $[H^+] = 2.1 \times 10^{-3}\ M$ (b) pH = 2.7
(c) 0.84% ionized
D.8 $K_a = 2.7 \times 10^{-5}$
D.9 $K_a = 7 \times 10^{-10}$
D.10 (a) 0.42% ionized; pH = 2.4
(b) 1.3% ionized; pH = 2.9
(c) 4.2% ionized; pH = 3.4
D.11 $K_a = 1.1 \times 10^{-7}$
D.12 $K_a = 7.3 \times 10^{-6}$
D.13 $[H^+] = 6.0\ M$; $[OH^-] = 1.7 \times 10^{-15}\ M$;
pH = −0.78; pOH = 14.78
D.14 (a) pH = 4.0; pOH = 10.0
(b) pOH = 2.0; pH = 12.0
(c) pOH = 2.6; pH = 11.4
(d) pH = 4.2; pOH = 9.8
(e) pOH = 4.9; pH = 9.1
D.15 (a) $[OH^-] = 1.0 \times 10^{-10}$
(b) $[OH^-] = 3.6 \times 10^{-9}$
(c) $[OH^-] = 2.5 \times 10^{-6}$
D.16 (a) $[H^+] = 1.7 \times 10^{-8}$ (b) $[H^+] = 1 \times 10^{-6}$
(c) $[H^+] = 2.2 \times 10^{-9}$
D.17 (a) 1.5×10^{-9} (b) 1.9×10^{-12}
(c) 1.2×10^{-23} (d) 2.6×10^{-13}
(e) 3.1×10^{-70} (f) 1.7×10^{-10}
(g) 2.4×10^{-5} (h) 5.13×10^{-17}
(i) 1.81×10^{-18}
D.18 (a) 4.5×10^{-5} mol/L
(b) 7.6×10^{-10} mol/L
(c) 1.6×10^{-2} mol/L
(d) 1.2×10^{-4} mol/L
D.19 (a) 8.9×10^{-4} g $BaCO_3$
(b) 9.3×10^{-9} g $AlPO_4$
(c) 0.50 g Ag_2SO_4
(d) 7.0×10^{-4} g $Mg(OH)_2$
D.20 (a) 2.1×10^{-4} mol Ca^{2+}/L;
4.2×10^{-4} mol F^-/L
(b) 8.2×10^{-3} g CaF_2
D.21 (a) Precipitate formed (b) Precipitate formed
(c) No precipitate formed
D.22 (a) $3.0 \times 10^{-8}\ M\ SO_4^{2-}$
(b) 7.0×10^{-7} g $BaSO_4$
D.23 $BaSO_4$ will precipitate first
D.24 (a) 5.0×10^{-12} mol AgBr
(b) 2.5×10^{-12} mol AgBr
D.25 No precipitate of $PbCl_2$ will form.
D.26 (a) $[H^+] = 3.6 \times 10^{-5}$; pH = 4.4
(b) $[H^+] = 1.8 \times 10^{-5}$; pH = 4.7
D.27 (a) Change in pH = 5.3 units
(b) Change in pH = 0.02 units

Chapter 17

C.17 The following statements are correct: a, c, e, g, j, k, m, p, q, r, and s.
D.1 0.0772 mol NO
D.2 20.2 L Cl_2
D.3 17 g $KMnO_4$
D.4 66.2 mL $K_2Cr_2O_7$ solution
D.5 10.0 mL $K_2Cr_2O_7$ solution
D.6 91.3% KI
D.7 149 g Cu
D.8 4.22 L NO
D.9 5.56 mol H_2

APPENDIX VII

Periodic Table of the Elements with Group Numbering Recommended by the IUPAC and the American Chemical Society

Key: Atomic number — 11; Name — Sodium; Symbol — **Na**; Electron structure — 2 8 1; Atomic weight — 22.98977

[a]Mass number of most stable or best-known isotope

[b]Mass of the isotope of longest half-life

Period	Group 1		Transition elements → 3	4	5	6	7	8	9
1	1 Hydrogen **H** 1.0079 (1)								
2	3 Lithium **Li** 6.941 (2 1)	4 Beryllium **Be** 9.01218 (2 2)							
3	11 Sodium **Na** 22.98977 (2 8 1)	12 Magnesium **Mg** 24.305 (2 8 2)							
4	19 Potassium **K** 39.098 (2 8 8 1)	20 Calcium **Ca** 40.08 (2 8 8 2)	21 Scandium **Sc** 44.9559 (2 8 9 2)	22 Titanium **Ti** 47.90 (2 8 10 2)	23 Vanadium **V** 50.9414 (2 8 11 2)	24 Chromium **Cr** 51.996 (2 8 13 1)	25 Manganese **Mn** 54.9380 (2 8 13 2)	26 Iron **Fe** 55.847 (2 8 14 2)	27 Cobalt **Co** 58.9332 (2 8 15 2)
5	37 Rubidium **Rb** 85.4678 (2 8 18 8 1)	38 Strontium **Sr** 87.62 (2 8 18 8 2)	39 Yttrium **Y** 88.9059 (2 8 18 9 2)	40 Zirconium **Zr** 91.22 (2 8 18 10 2)	41 Niobium **Nb** 92.9064 (2 8 18 12 1)	42 Molybdenum **Mo** 95.94 (2 8 18 13 1)	43 Technetium **Tc** 98.9062[b] (2 8 18 14 1)	44 Ruthenium **Ru** 101.07 (2 8 18 15 1)	45 Rhodium **Rh** 102.9055 (2 8 18 16 1)
6	55 Cesium **Cs** 132.9054 (2 8 18 18 8 1)	56 Barium **Ba** 137.34 (2 8 18 18 8 2)	57 Lanthanum **La*** 138.9055 (2 8 18 18 9 2)	72 Hafnium **Hf** 178.49 (2 8 18 32 10 2)	73 Tantalum **Ta** 180.9479 (2 8 18 32 11 2)	74 Wolfram (Tungsten) **W** 183.85 (2 8 18 32 12 2)	75 Rhenium **Re** 186.2 (2 8 18 32 13 2)	76 Osmium **Os** 190.2 (2 8 18 32 14 2)	77 Iridium **Ir** 192.22 (2 8 18 32 17 0)
7	87 Francium **Fr** (223)[a] (2 8 18 32 18 8 1)	88 Radium **Ra** 226.0254[b] (2 8 18 32 18 8 2)	89 Actinium **Ac**** (227)[a] (2 8 18 32 18 9 2)	104 Unnilquadium **Unq** (261)[a] (2 8 18 32 32 10 2)	105 Unnilpentium **Unp** (262)[a] (2 8 18 32 32 11 2)	106 Unnilhexium **Unh** (263)[a] (2 8 18 32 32 12 2)	107 —		109 —

* Lanthanide series 6	58 Cerium **Ce** 140.12 (2 8 18 20 8 2)	59 Praseodymium **Pr** 140.9077 (2 8 18 21 8 2)	60 Neodymium **Nd** 144.24 (2 8 18 22 8 2)	61 Promethium **Pm** (145)[a] (2 8 18 23 8 2)	62 Samarium **Sm** 150.4 (2 8 18 24 8 2)
** Actinide series 7	90 Thorium **Th** 232.0381[b] (2 8 18 32 18 10 2)	91 Protactinium **Pa** 231.0359[b] (2 8 18 32 20 9 2)	92 Uranium **U** 238.029 (2 8 18 32 21 9 2)	93 Neptunium **Np** 237.0482 (2 8 18 32 22 9 2)	94 Plutonium **Pu** (242)[a] (2 8 18 32 23 9 2)

Atomic weights are based on carbon-12. Atomic weights in parentheses indicate the most stable or best-known isotope. Slight disagreement exists as to the exact electronic configuration of several of the high-atomic-number elements. Names and symbols for elements 104, 105, and 106 are unofficial.

10	11	12	13	14	15	16	17	18
								2 Helium **He** 4.00260 2
			5 Boron **B** 10.81 2 3	6 Carbon **C** 12.011 2 4	7 Nitrogen **N** 14.0067 2 5	8 Oxygen **O** 15.9994 2 6	9 Fluorine **F** 18.99840 2 7	10 Neon **Ne** 20.179 2 8
			13 Aluminum **Al** 26.98154 2 8 3	14 Silicon **Si** 28.086 2 8 4	15 Phosphorus **P** 30.97376 2 8 5	16 Sulfur **S** 32.06 2 8 6	17 Chlorine **Cl** 35.453 2 8 7	18 Argon **Ar** 39.948 2 8 8
28 Nickel **Ni** 58.71 2 8 16 2	29 Copper **Cu** 63.546 2 8 18 1	30 Zinc **Zn** 65.38 2 8 18 2	31 Gallium **Ga** 69.72 2 8 18 3	32 Germanium **Ge** 72.59 2 8 18 4	33 Arsenic **As** 74.9216 2 8 18 5	34 Selenium **Se** 78.96 2 8 18 6	35 Bromine **Br** 79.904 2 8 18 7	36 Krypton **Kr** 83.80 2 8 18 8
46 Palladium **Pd** 106.4 2 8 18 18 0	47 Silver **Ag** 107.868 2 8 18 18 1	48 Cadmium **Cd** 112.40 2 8 18 18 2	49 Indium **In** 114.82 2 8 18 18 3	50 Tin **Sn** 118.69 2 8 18 18 4	51 Antimony **Sb** 121.75 2 8 18 18 5	52 Tellurium **Te** 127.60 2 8 18 18 6	53 Iodine **I** 126.9045 2 8 18 18 7	54 Xenon **Xe** 131.30 2 8 18 18 8
78 Platinum **Pt** 195.09 2 8 18 32 17 1	79 Gold **Au** 196.9665 2 8 18 32 18 1	80 Mercury **Hg** 200.59 2 8 18 32 18 2	81 Thallium **Tl** 204.37 2 8 18 32 18 3	82 Lead **Pb** 207.2 2 8 18 32 18 4	83 Bismuth **Bi** 208.9804 2 8 18 32 18 5	84 Polonium **Po** (210)[a] 2 8 18 32 18 6	85 Astatine **At** (210)[a] 2 8 18 32 18 7	86 Radon **Rn** (222)[a] 2 8 18 32 18 8

Inner transition elements

63 Europium **Eu** 151.96 2 8 18 25 8 2	64 Gadolinium **Gd** 157.25 2 8 18 25 9 2	65 Terbium **Tb** 158.9254 2 8 18 27 8 2	66 Dysprosium **Dy** 162.50 2 8 18 28 8 2	67 Holmium **Ho** 164.9304 2 8 18 29 8 2	68 Erbium **Er** 167.26 2 8 18 30 8 2	69 Thulium **Tm** 168.9342 2 8 18 31 8 2	70 Ytterbium **Yb** 173.04 2 8 18 32 8 2	71 Lutetium **Lu** 174.97 2 8 18 32 9 2
95 Americium **Am** (243)[a] 2 8 18 32 25 8 2	96 Curium **Cm** (247)[a] 2 8 18 32 25 9 2	97 Berkelium **Bk** (249)[a] 2 8 18 32 26 9 2	98 Californium **Cf** (251)[a] 2 8 18 32 27 9 2	99 Einsteinium **Es** (254)[a] 2 8 18 32 28 9 2	100 Fermium **Fm** (253)[a] 2 8 18 32 29 9 2	101 Mendelevium **Md** (256)[a] 2 8 18 32 30 9 2	102 Nobelium **No** (254)[a] 2 8 18 32 31 9 2	103 Lawrencium **Lr** (257)[a] 2 8 18 32 32 9 2

Index

Names and Formulas of Common Ions

Positive Ions	
Ammonium	NH_4^+
Copper(I) (Cuprous)	Cu^+
Hydrogen	H^+
Potassium	K^+
Silver	Ag^+
Sodium	Na^+
Barium	Ba^{2+}
Cadmium	Cd^{2+}
Calcium	Ca^{2+}
Cobalt(II)	Co^{2+}
Copper(II) (Cupric)	Cu^{2+}
Iron(II) (Ferrous)	Fe^{2+}
Lead(II)	Pb^{2+}
Magnesium	Mg^{2+}
Manganese(II)	Mn^{2+}
Mercury(II) (Mercuric)	Hg^{2+}
Nickel(II)	Ni^{2+}
Tin(II) (Stannous)	Sn^{2+}
Zinc	Zn^{2+}
Aluminum	Al^{3+}
Antimony(III)	Sb^{3+}
Arsenic(III)	As^{3+}
Bismuth(III)	Bi^{3+}
Chromium(III)	Cr^{3+}
Iron(III) (Ferric)	Fe^{3+}
Titanium(III) (Titanous)	Ti^{3+}
Manganese(IV)	Mn^{4+}
Tin(IV) (Stannic)	Sn^{4+}
Titanium(IV) (Titanic)	Ti^{4+}
Antimony(V)	Sb^{5+}
Arsenic(V)	As^{5+}

Negative Ions	
Acetate	$C_2H_3O_2^-$
Bromate	BrO_3^-
Bromide	Br^-
Chlorate	ClO_3^-
Chloride	Cl^-
Chlorite	ClO_2^-
Cyanide	CN^-
Fluoride	F^-
Hydride	H^-
Bicarbonate (Hydrogen carbonate)	HCO_3^-
Bisulfate (Hydrogen sulfate)	HSO_4^-
Bisulfite (Hydrogen sulfite)	HSO_3^-
Hydroxide	OH^-
Hypochlorite	ClO^-
Iodate	IO_3^-
Iodide	I^-
Nitrate	NO_3^-
Nitrite	NO_2^-
Perchlorate	ClO_4^-
Permanganate	MnO_4^-
Thiocyanate	SCN^-
Carbonate	CO_3^{2-}
Chromate	CrO_4^{2-}
Dichromate	$Cr_2O_7^{2-}$
Oxalate	$C_2O_4^{2-}$
Oxide	O^{2-}
Peroxide	O_2^{2-}
Silicate	SiO_3^{2-}
Sulfate	SO_4^{2-}
Sulfide	S^{2-}
Sulfite	SO_3^{2-}
Arsenate	AsO_4^{3-}
Borate	BO_3^{3-}
Phosphate	PO_4^{3-}
Phosphite	PO_3^{3-}